411 Bloomfield St #3
Hoboken NJ

# APPLIED NUMERICAL METHODS WITH PERSONAL COMPUTERS

## McGraw-Hill Chemical Engineering Series



### BUILDING THE LITERATURE OF A PROFESSION

Fifteen prominent chemical engineers first met in New York more than 60 years ago to plan a continuing literature for their rapidly growing profession. From industry came such pioneer practitioners as Leo H. Baekeland, Arthur D. Little, Charles L. Reese, John V. N. Dorr, M. C. Whitaker, and R. S. McBride. From the universities came such eminent educators as William H. Walker, Alfred H. White, D. D. Jackson, J. H. James, Warren K. Lewis, and Harry A. Curtis. H. C. Parmelee, then editor of *Chemical and Metallurgical Engineering*, served as chairman and was joined subsequently by S. D. Kirkpatrick as consulting editor.

After several meetings, this committee submitted its report to the McGraw-Hill Book Company in September 1925. In the report were detailed specifications for a correlated series of more than a dozen texts and reference books which have since become the McGraw-Hill Series in Chemical Engineering and which became the cornerstone of the chemical engineering curriculum.

From this beginning there has evolved a series of texts surpassing by far the scope and longevity envisioned by the founding Editorial Board. The McGraw-Hill Series in Chemical Engineering stands as a unique historical record of the development of chemical engineering education and practice. In the series one finds the milestones of the subject's evolution: industrial chemistry, stoichiometry, unit operations and processes, thermodynamics, kinetics, and transfer operations.

Chemical engineering is a dynamic profession, and its literature continues to evolve. McGraw-Hill and its consulting editors remain committed to a publishing policy that will serve, and indeed lead, the needs of the chemical engineering profession during the years to come.

THE SERIES

**Bailey and Ollis:** *Biochemical Engineering Fundamentals*
**Bennett and Myers:** *Momentum, Heat, and Mass Transfer*
**Beveridge and Schechter:** *Optimization: Theory and Practice*
**Brodkey and Hershey:** *Transport Phenomena: A Unified Approach*
**Carberry:** *Chemical and Catalytic Reaction Engineering*
**Constantinides:** *Applied Numerical Methods with Personal Computers*
**Coughanowr and Koppel:** *Process Systems Analysis and Control*
**Douglas:** *Conceptual Design of Chemical Processes*
**Edgar and Himmelblau:** *Optimization of Chemical Processes*
**Fahien:** *Fundamentals of Transport Phenomena*
**Finlayson:** *Nonlinear Analysis in Chemical Engineering*
**Gates, Katzer, and Schuit:** *Chemistry of Catalytic Processes*
**Holland:** *Fundamentals of Multicomponent Distillation*
**Holland and Liapis:** *Computer Methods for Solving Dynamic Separation Problems*
**Katz, Cornell, Kobayashi, Poettmann, Vary, Elenbaas, and Weinaug:** *Handbook of Natural Gas Engineering*
**King:** *Separation Processes*
**Luyben:** *Process Modeling, Simulation, and Control for Chemical Engineers*
**McCabe, Smith, J. C., and Harriott:** *Unit Operations of Chemical Engineering*
**Mickley, Sherwood, and Reed:** *Applied Mathematics in Chemical Engineering*
**Nelson:** *Petroleum Refinery Engineering*
**Perry and Chilton (Editors):** *Chemical Engineers' Handbook*
**Peters:** *Elementary Chemical Engineering*
**Peters and Timmerhaus:** *Plant Design and Economics for Chemical Engineers*
**Probstein and Hicks:** *Synthetic Fuels*
**Reid, Prausnitz, and Sherwood:** *The Properties of Gases and Liquids*
**Resnick:** *Process Analysis and Design for Chemical Engineers*
**Satterfield:** *Heterogeneous Catalysis in Practice*
**Sherwood, Pigford, and Wilke:** *Mass Transfer*
**Smith, B. D.:** *Design of Equilibrium Stage Processes*
**Smith, J. M.:** *Chemical Engineering Kinetics*
**Smith, J. M. and Van Ness:** *Introduction to Chemical Engineering Thermodynamics*
**Treybal:** *Mass Transfer Operations*
**Valle-Riestra:** *Project Evolution in the Chemical Process Industries*
**Van Ness and Abbott:** *Classical Thermodynamics of Nonelectrolyte Solutions: With Applications to Phase Equilibria*
**Van Winkle:** *Distillation*
**Volk:** *Applied Statistics for Engineers*
**Walas:** *Reaction Kinetics for Chemical Engineers*
**Wei, Russell, and Swartzlander:** *The Structure of the Chemical Processing Industries*
**Whitwell and Toner:** *Conservation of Mass and Energy*

# APPLIED NUMERICAL METHODS WITH PERSONAL COMPUTERS

**Alkis Constantinides**

*Professor of Chemical and Biochemical Engineering*
*Rutgers*
*The State University of New Jersey*

**McGraw-Hill Book Company**

New York St. Louis San Francisco Auckland Bogotá Hamburg
London Madrid Mexico Milan Montreal
New Delhi Panama Paris São Paulo Singapore Sydney Tokyo Toronto

This book was set in Times Roman by Intercontinental Photocomposition Limited.
The editor was B. J. Clark.
The cover was designed by Nadja Furlan-Lorbek.
The production supervisor was Leroy A. Young.
Project supervision was done by The Total Book.
R. R. Donnelley & Sons Company was printer and binder.

**APPLIED NUMERICAL METHODS WITH PERSONAL COMPUTERS**



34567890 DOCDOC 892109

ISBN 0-07-012463-9

**Library of Congress Cataloging-in-Publication Data**

Constantinides, A.
Applied numerical methods with personal computers.

Includes bibliographies and index.
1. Numerical analysis--Data processing. 2. BASIC
(Computer program language) 3. Engineering mathe-
matics--Data processing. I. Title.
QA297.C649 1987 519.4'028'5 86-10295
ISBN 0-07-012463-9

# ABOUT THE AUTHOR

**Dr. Alkis Constantinides** is Professor of Chemical and Biochemical Engineering at Rutgers University. He received the D.E.Sc. degree in chemical engineering from Columbia University in 1970, and the M.S. and B.Ch.E. degrees from Ohio State University in 1964. He joined the faculty of Rutgers University in 1969.

During his tenure at Rutgers University he has taught graduate and undergraduate courses in chemical and biochemical engineering, including kinetics, thermodynamics, transport phenomena, process design, applied numerical methods, and fermentation biotechnology. In his teaching and research he has made extensive use of computers of all kinds, from macros to micros. Since the advent of personal computers, he has enthusiastically adopted the personal computer as an invaluable educational and research tool for engineers and scientists. From 1976 to 1985, Professor Constantinides was the Director of the Graduate Program in Chemical and Biochemical Engineering. He has been a member of the University Senate, the University Research Council, the Executive Council of the Graduate School, and the Planning Committee of the College of Engineering.

His research interests are in the fields of biochemical engineering and biotechnology, with emphasis on the application of numerical methods in modeling, optimization, and control of fermentations and enzyme bioreactors. For his work in these areas, he has received extensive support in the form of research grants from the National Science Foundation, the State of New Jersey, and the biochemical industry. Dr. Constantinides is the author of numerous papers published in leading biochemical engineering journals, and he is the coeditor of three volumes on "Biochemical Engineering" published by the New York Academy of Sciences.

Dr. Constantinides has industrial experience in process development and design with Exxon Research and Engineering, Procter and Gamble, and Borg-Warner Corporation. He also consults for industry in the areas of biochemical engineering, technology assessment, and applied numerical methods.

He is a member of the American Institute of Chemical Engineers, the American Chemical Society, the Society for Industrial Microbiology, and the New York Academy of Sciences.

To Paul

# CONTENTS

# PREFACE

The purpose of this textbook is to present the theory and application of numerical methods for the solution of engineering problems using personal computers. The first chapter provides the background on the fundamental characteristics of personal computers: the hardware, the operating systems, and the software. A detailed section on BASIC programming language is presented to familiarize the reader with the programming language used in this book. The subsequent chapters are devoted to the derivation of a variety of numerical methods and their application for the solution of engineering problems. These algorithms encompass linear and nonlinear algebraic equations, finite difference methods, ordinary and partial differential equations, and linear and nonlinear regression analysis.

Several worked examples are given in each chapter to demonstrate the numerical techniques. Most of these examples require computer programs for their solution. These programs are written in the advanced BASIC language (BASICA interpreter of the PC-DOS and MS-DOS, version 2.00 or higher). All the programs that appear in the text are included on the diskette which accompanies this book. The programs are interactive and user-friendly. They have been written in a general manner so that they may be applied to the solution of other problems that fall in the same category of application as the worked examples. The programs are given in the source code so that they can be modified easily to suit the needs of the user. The programs are described in detail in order to provide the reader with thorough background and understanding of the BASIC programming language.

The material in this book has been used in graduate and undergraduate courses in the Department of Chemical and Biochemical Engineering at Rutgers University. The numerical methods under each topic are arranged in order of difficulty, with the more advanced techniques covered in the last sections of each chapter. A one-semester graduate level course in applied numerical methods would cover all the material in this book. An undergraduate course (junior or senior level) would cover the first five or six sections in each chapter, at the discretion of the professor.

I gratefully acknowledge the many helpful suggestions provided by Professors R. Vichnevetsky and B. Davidson during the writing of this book. I owe special gratitude to Professors M. G. Salvadori and J. L. Spencer of Columbia University, who first introduced me to the field of applied numerical methods.

McGraw-Hill and the author would like to thank the following reviewers who evaluated the project: Henry R. Bungay, Rensselaer Polytechnic Institute; J. W. Cadman, University of Maryland; C. L. Caenepeel, California State Polytechnic University; James R. Fair, University of Texas at Austin; and James Wei, Massachusetts Institute of Technology.

The manuscript of this book was typed by Donna Foster whose excellent typing skills made the preparation of this book so much easier. The contribution of the graduate students at Rutgers University who tested and proofread the manuscript is also acknowledged.

I dedicate this book to my son Paul, who is the inspiration in all of my work.

*Alkis Constantinides*

CHAPTER

# ONE

# INTRODUCTION TO PERSONAL COMPUTERS AND BASIC PROGRAMMING

## 1-1 MAJOR USES OF PERSONAL COMPUTERS

The advent of the personal computer has revolutionized the application of computers in science, engineering, and business. In fact, there is no aspect of our daily lives that is not directly, or indirectly, affected by the ubiquitous personal computer. Educational institutions, from kindergarten to universities, use personal computers to teach writing, reading, arithmetic, art, science, mathematics, music, history, geography, economics, business, architecture, and engineering. The business establishment makes extensive use of personal computers in word processing, desk-top publishing, accounting, financial planning, stock analysis, project management, data-base management, scheduling, business graphics, commercial art, banking, and communications. The engineering and science professions use personal computers for computations, computer-aided design, robotics, data acquisition and control of experiments, process monitoring and control, data management and statistical analysis, energy management, scientific graphics, image processing, satellite tracking, weather prediction, and, finally, as all-inclusive engineering work stations.

Other professions are also finding many useful applications for personal computers. In medicine, computers are used for teaching, diagnosis, analysis, patient monitoring, patient report generation, tabulating and graphing of analytical laboratory data, and, last but not least, billing. In agriculture, the personal computer is used for farm management, feed-lot scheduling, crop management, and livestock diet planning and breeding. In the home, the personal computer is used for a variety of tasks, from balancing the checkbook to playing games.

New and rapidly developing applications of personal computers are in voice recognition and speech synthesis, language translation, handwriting recognition and transcription, televideo communications, music composition, and artificial intelligence.

It has been predicted that in the next decade there will be "a computer on every desk." In 1985 there were approximately 11 million personal computers in the United States, and the projections show that the number may grow to 36 million by 1990.

## 1-2 DEVELOPMENT OF PERSONAL COMPUTERS—A BRIEF HISTORY†

The evolution of semiconductor technology, which occurred in the 1960s, laid the foundation for the development of personal computers. The push to make more powerful and versatile computers in less space and at a lower cost resulted in the successful manufacture of the integrated circuit. Many transistors and diodes were manufactured on the same tiny chip of silicon. As component density became greater, new technology was developed to pack hundreds of components into extremely small spaces, and in 1967 the first *large-scale-integration* (LSI) assembly was produced [2, 3].

It was this technology that enabled engineers at Intel to build the first programmable *microprocessor chip* for desk calculators in 1971. At that time calculators were built from specialized circuit chips that could perform only a single function. However, the first microprocessor could be programmed to perform multiple specialized calculator functions. This microprocessor (the Intel 4004) was very limited in the number of instructions it could execute and it could manipulate only 4 binary digits (bits) at a time. This was followed by the development of more powerful microprocessors that could operate on 8 bits (1 byte).‡ The first personal computer, the Altair 8800, which used the Intel 8080 microprocessor (an 8-bit chip) was introduced in 1975. This machine was intended for the hobbyist and was originally sold in kit form.

The real personal computer revolution was initiated in 1977 by the introduction of the Apple and the TRS-80 personal computers. These were easy-to-use "plug-in" machines which did not require knowledge of electronics or, for that matter, previous computer training. By 1978 three companies dominated the personal computer market: Apple Computer, Tandy (Radio Shack), and Commodore. The Apple II series of computers became the most popular of these machines with thousands of applications in schools, business, and the home. The success of the Apple Computer Company was quickly followed by other companies that introduced lower-cost units for the home

† Adapted from Ref. 1, by permission of the publisher.

‡ 4 bits = 1 nibble
8 bits = 1 byte
16 bits = 1 word

(Atari, Coleco, Commodore, etc.). The systems sold in the late 1970s and early 1980s were 8-bit machines.

A new generation of personal computers was introduced in the early 1980s. These systems are built around microprocessor chips that can operate on 16 bits at a time. By far the most popular model is the IBM PC personal computer which uses the Intel 8088 microprocessor (Fig. 1-1). Other suppliers also produce desktop and portable 16-bit machines. Thousands of application programs are now available for the IBM PC and its family of models. In order to be able to use this wealth of software, designers of many competitive brands have elected to use the same microprocessor and operating instructions used by IBM. The result has been the swift creation of an IBM PC–compatible industry. Hundreds of firms in this sector are generating programs and products that work with, look like, or plug into IBM PC's and the many similar systems of other suppliers.

In 1984, Apple Computer Company introduced the Macintosh personal computer, with enhanced graphics capabilities, using the Motorola 68000 microprocessor, a 32-bit chip (Fig. 1-2).

In 1984–1985, a new generation of personal computers with multiuser/multitasking capabilities was introduced. The leader among these is the IBM AT, which is based on the more powerful Intel 80286 microprocessor (Fig. 1-3). Simultaneously, other multiuser/multitasking machines, such as the AT&T Unix PC with its Motorola 68010 microprocessor (a 32-bit chip), entered the market. In 1986, IBM introduced the PC RT, a reduced instruction-set computer (RISC) using a 32-bit RISC processor. In September, 1986, Compaq Computer Corporation introduced the Compaq Deskpro 386 which uses the most advanced microprocessor, the Intel 80386.

The top five personal computer manufacturers in 1986 were IBM, Apple, Compaq, Tandy, and Hewlett-Packard.

**Figure 1-1** The IBM PC. *(Courtesy IBM Corporation.)*

**Figure 1-2** The Macintosh. *(Courtesy Apple Computer, Inc.)*

**Figure 1-3** The IBM AT. *(Courtesy IBM Corporation.)*

## 1-3 OPERATING CHARACTERISTICS OF PERSONAL COMPUTERS–THE HARDWARE

A personal computer is the smallest general-purpose data processing system that can execute programmed instructions to perform a wide variety of tasks. The personal computer is centered around the *central processing unit* (CPU) with a number of other devices which perform input, storage, and output functions (Fig. 1-4).

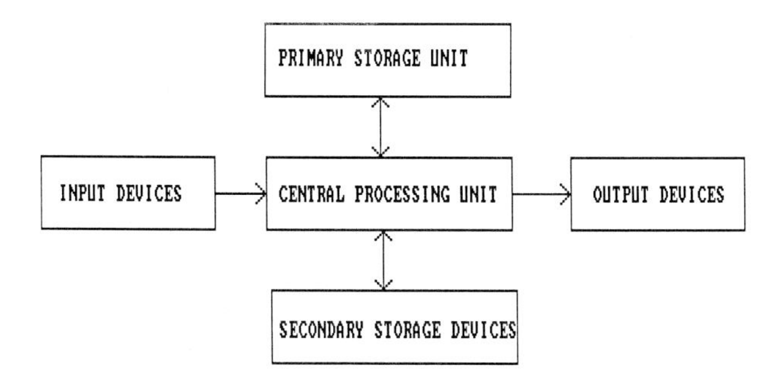


**Figure 1-4** Basic organization of personal computers.

The CPU of a personal computer is the microprocessor, which performs the arithmetic, logic, and control functions of the computer. Several support circuit chips perform timing and control the flow of information. Additional *read-only memory* (ROM) chips have permanently stored preprogrammed instructions used in the start-up and operation of the computer. The primary storage function of the computer is performed by the *random access memory* (RAM) chips. All the above parts are mounted on a heavy plastic circuit board, the system board (Fig. 1-5). A printed pattern of conductors interconnects the chips and supplies them with power from a power source.

Most personal computers are self-contained units which are light enough to be moved easily and many are truly portable. Personal computers were originally designed as single-user systems, but recent advances in microprocessor design and software development have made possible the introduction of multiuser/multitasking applications.

A typical configuration of a personal computer, showing the internal hardware and several peripheral devices, is given in Fig. 1-6. The main information link between internal components of the personal computer is the bus. The bus consists of a set of wires which carry address information, data, control signals, and power to all the interconnected components of a system. Most personal computers have an "open-bus" architecture, which means that several devices can be connected directly into the bus, via available expansion slots.

When the computer is turned on, the start-up ROM sends its preprogrammed set of instructions to the microprocessor which in turn instructs the *disk controller* to look for information in the specified *disk drive*. The *diskette* in the disk drive should contain the *disk operating system* (DOS), which is loaded into the primary memory of the computer. Once in RAM, the disk operating system keeps track of everything in memory and manages the flow of information to and from peripheral devices, such as keyboard, display, disk drives, and printers.

The characteristics and functions of the major internal hardware compo-

**Figure 1-5** The IBM AT system board. *(From* The Peter Norton Programmer's Guide to the IBM PC, *Ref. 4. By permission of the publisher. © by the author and the publisher.)*

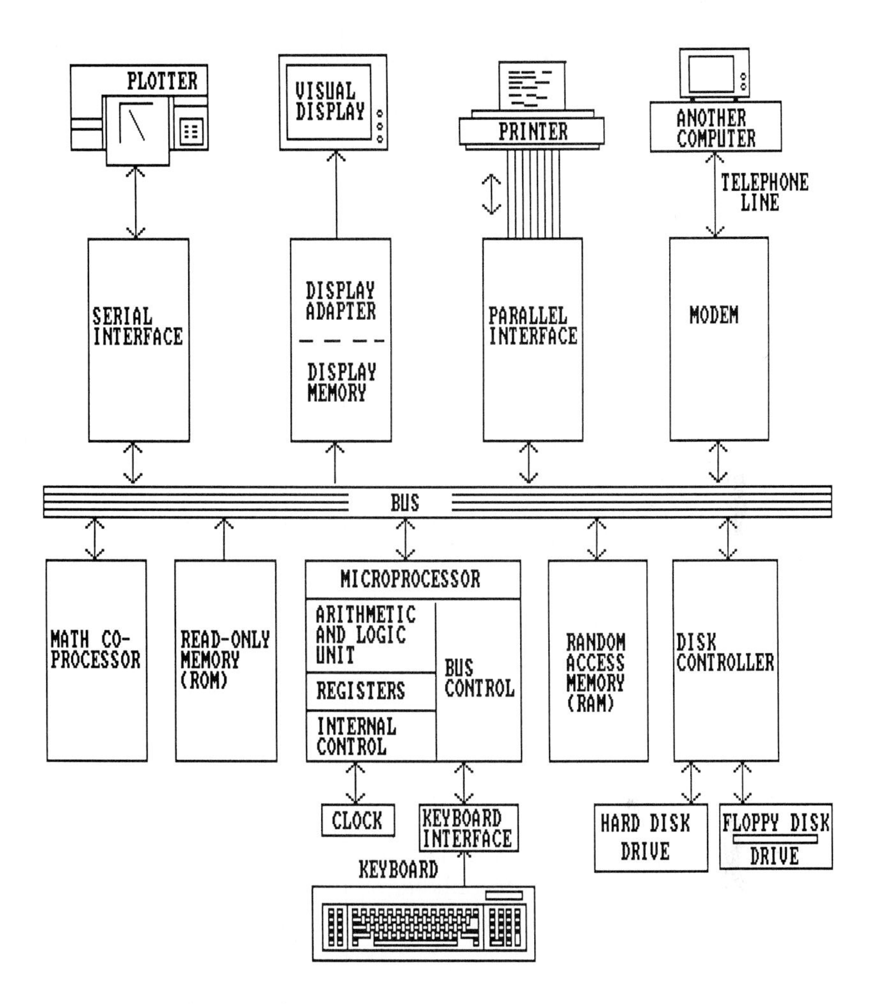


**Figure 1-6** Hardware configuration of a personal computer with selected peripheral devices. The math coprocessor, hard disk drive, modem, and plotter are optional equipment.

nents and peripherals of the personal computer are discussed in more detail in the following subsections. The operating characteristics of the software—the languages and operating systems—which enable the computer to perform its applications, are discussed in Secs. 1-4 and 1-5.

## 1-3.1 Microprocessors

The microprocessor includes the arithmetic and logic unit, the internal registers, and the control and timing block. It is the "central brain" of the computing system and controls all other functions performed by the system. The microprocessor calls for instructions from the memory, decodes them, and executes them. It references the memory and the various input-output ports as required for the execution of various instructions [5].

The "first-generation" personal computers were built around 8-bit microprocessors. The most commonly used chips were the Zilog Z80, the MOS Technology 6502, the Intel 8080, and the Motorola 6809. All these have built-in 8-line *data buses*. This means that they can retrieve from storage, manipulate, and process a single 8-bit *byte* of data at a time. A 16-line *address bus* is also built into these chips to determine the primary storage location of the instructions and data that are needed. With 16 address lines, the microprocessor can identify a maximum of $2^{16}$ or 65,536 separate storage locations (commonly abbreviated as 64 kilobytes, or 64K).

All 16-bit personal computers are also built around a few popular microprocessors. The Intel 8088 is used in the IBM PC, the Compaq, the HP 150, the DEC Rainbow 100, and many other personal computers. The data path between the 8088 and primary storage is only 8 bits wide and so all operations to and from storage are done 8 bits at a time. Once retrieved, however, data is processed 16 bits at a time internally in the 8088. The 8088 is thus referred to as an 8/16-bit chip; it functions like a fast 8-bit microprocessor with an extended set of instructions. A "stablemate" of the 8088 is Intel's 8086. Unlike the 8088, the internal and external data paths of the 8086 are all 16 bits wide. The 8086 is thus a true 16/16-bit chip. An expanded built-in address bus with 20 lines permits both the 8088 and the 8086 to identify about a million (1 megabyte, or 1M) separate primary storage locations [1].

Another popular microprocessor is the Motorola 68000, which is used in the Macintosh, the TRS-80 Model 16, and the Cromenco. The 68000 has a 16-bit external bus and 32-bit internal registers, so it is characterized as a 16/32-bit chip. Its address bus has 24 lines, giving it the ability to address up to 16 million locations of primary storage (16M).

Enhanced versions of the Intel and Motorola 16-bit chips discussed above are also used. The Intel 80186, which has the processing power of the 8086 plus additional support circuitry, is used in Radio Shack's Model 2000 computer. The Intel 80286, which essentially combines two 8086s with special features necessary for *multitasking* and *virtual memory addressing*, is the CPU of the IBM AT, the Compaq Deskpro 286, the Intel 310/380 series, and other advanced-technology microcomputers. Multitasking is the ability of the microprocessor to perform several tasks at a time, thus speeding up considerably the execution of instructions. Virtual memory addressing allows the computer to use more memory than it has in its primary memory banks, by using an elaborate hardware-software design which stores some parts of the program in primary memory and other parts on the disk. The operating system keeps track of the location of all segments of the program and switches these in and out of the microprocessor as needed for smooth and efficient execution of the program.

Microprocessors usually work with integers. Calculations involving floating-point numbers are handled by software subroutines, which are more time-consuming than hardware operations. However, *math coprocessors* are available which perform hardware-based floating-point calculations including trigonometric and logarithmic functions. These coprocessors can speed up the execution of computation-intensive programs considerably (by a factor of 2 or more). The Intel 8087 and 80287 are the corresponding math coprocessors for the 8088 and 80286 microprocessors, respectively. The math coprocessors are optional hardware on most second- or third-generation personal computers and require additional software for their support. Increasing use of these coprocessors is being made in the more recently developed software for engineering applications.

Concentrated development effort by the semiconductor companies is being focused on the 32/32-bit microprocessors, which will power the next generation of personal computers, the supermicrocomputers. The Western Electric 32000 chip, a full 32-bit microprocessor, is now used in AT&T's 3B2/300 series computers. Digital Equipment Corporation's 32-bit microprocessor is used in the MicroVAX II supermicro. Intel's 80386 and Motorola's 68020 have recently become available. Compaq, Intel, and AT&T are developing personal computers based on the 80386 microprocessor. Compaq led the group by its introduction of the Deskpro 386 in late 1986. More of these advanced machines will likely be introduced in early 1987.

### 1-3.2 Clocks

The CPU operates in a cyclical manner, i.e., it fetches an instruction from the program memory, decodes it, and executes the particular operations specified by the instructions. The next instruction is then fetched, and the process is repeated until the entire program is executed. This entire sequence is synchronized by a timing clock which is usually built into the microprocessor. The clock frequency is provided by an external source, an accurately controlled crystal [5]. Thus the speed with which an instruction is executed is directly related to the clock speed. Different versions of a particular microprocessor may be produced with different clock speeds. Most of today's personal computers function in the 2- to 8-MHz range, but the newest chips can operate at speeds up to 25 MHz [1].

### 1-3.3 Read-Only Memory (ROM)

The read-only memory of the computer contains permanently recorded programs that are needed to start and operate the computer and its peripheral devices. Personal computers come with different amounts of ROM. The IBM PC has 40K of ROM while the IBM AT, with its more complex hardware, uses 64K of ROM. There are four elements to the ROM in the IBM PC family: the start-up programs which do the work of getting the computer started; the ROM-BIOS (an acronym for Basic Input/Output Systems), which is a collection of machine-language routines that provide support services for continuing operation of the computer; the ROM-BASIC, which provides the core of the

BASIC programming language; and the ROM extensions which are programs that are added to the main ROM when certain optional equipment is added to the computer [4].

### 1-3.4 Random Access Memory (RAM)

The semiconductor chips which constitute the primary storage elements of personal computers are usually referred to as random access memory (RAM) chips. The program or programmer can randomly select and use any of the storage locations on a chip to directly store (write) and retrieve (read) data and instructions. Information stored in RAM is volatile, i.e., it is lost when the power in the computer is turned off.

There are two types of RAM chips that can be used for computer primary memory: dynamic RAM and static RAM. In dynamic RAM the charge in the memory capacitors leaks off and must be recharged periodically [at least once every 2 milliseconds (ms)]. Static RAM chips do not require this periodic recharge because they maintain their charge as long as the power in the computer stays on. Dynamic RAM chips are less expensive and more compact than the static RAM ones, therefore they are used in most personal computers.

Each cell circuit in dynamic RAM chips contains one transistor and one capacitor which can store 1 bit of information. Each chip contains thousands of such cells. The most widely used RAM chips today can store either 64 kilobits or 256 kilobits of information. Chips with 1 megabit capacity are commercially available and will soon be used in personal computers. Since most personal computers handle information in 8-bit bytes, the RAM chips are usually arranged in modules of 9 (one for each bit and one for error checking). Therefore, a 64K memory will consist of nine 64-kilobit chips. A 256K memory will require thirty-six 64-kilobit chips, or nine 256-kilobit chips. Most personal computers have 128K to 256K of memory installed on the system board (see Fig. 1-5) and additional memory that can be added using memory expansion boards which plug into the expansion slots on the system board.

The limit in the amount of memory that can be added to a personal computer is determined by the ability of the microprocessor to address this memory. The Intel 8088 and 8086 microprocessors (found in the IBM PC and other compatible personal computers) can address 1024K of memory. Of this memory, 640K is allocated to RAM, and the rest is reserved for video display and ROM. Table 1-1 shows the memory allocation for the IBM PC.

The Intel 80286 microprocessor, used in the IBM AT and other advanced technology personal computers, has a 24-line address bus which enables it to address up to 16 megabytes of memory. Actually, the 80286 has two modes of operation: the "real address mode," and the "protected address mode."† In the first mode it can address 1 megabyte of memory, while in the second mode it can reach 16 megabytes of memory. In addition, virtual memory support is

† For complete explanation of the operations of the Intel 80286 microprocessor, see Ref. 6.

**Table 1-1 System memory map for the IBM PC**

| Memory start/end | Function |
|---|---|
| 0/64K | Working RAM on system board; generally used by system software |
| 64K/256K | Working RAM on system board |
| 256K/640K | Working RAM on expansion boards |
| 640K/704K | Display memory expansion for enhanced graphics adapter |
| 704K/768K | Conventional display memory |
| 768K/960K | ROM memory expansion and control |
| 960K/1024K | ROM BIOS, ROM BASIC, diagnostics |

integrated within the 80286, making it able to use up to 1 gigabyte of virtual address space. Effective use of such large memories is slowly being made as software development catches up with the advances in hardware technology.

Extra RAM memory in personal computers can be used to emulate a disk storage device. The "virtual disks" (or "RAM disks") can be used for superfast storage and retrieval of files during execution of a program that frequently needs to write or read information to and from a disk, or for temporary storage of utility programs that are accessed often, such as the external DOS commands (see Sec. 1-4.3).

### 1-3.5 Mass Storage Devices

All computers must have a facility to store large amounts of information in ways that make it readily accessible to the CPU, easily transportable between compatible computers, and permanently storable outside the computer for use at a later time. This is accomplished by *mass storage devices* which provide the *secondary memory* of the computer. These devices are mainly *floppy disk drives*, *hard disk drives*, and *magnetic tape drives*.

Floppy disk drives record data in serial bit streams on the coated surfaces of thin plastic diskettes. The coating is the familiar brown layer of metal oxide particles used on magnetic tapes for tape recorders. Data is recorded on the diskette surface in a set of concentric circles called tracks, and the data on each track is further divided into equal-size pieces called sectors (Fig. 1-7). Magnetic read-write heads, like those used in tape recorders, retrieve and record the information on the surface of the diskette as the latter spins at 300 rpm underneath the heads. To move from one track to another, the heads are moved closer to or farther away from the center of the spinning diskette by a stepper motor [7].

The standard-size diskette is the $5\frac{1}{4}$-in floppy, which is used by the majority of personal computers. However, the $3\frac{1}{2}$-in minidiskette has recently gained popularity through its use in the Macintosh, Hewlett-Packard, and IBM

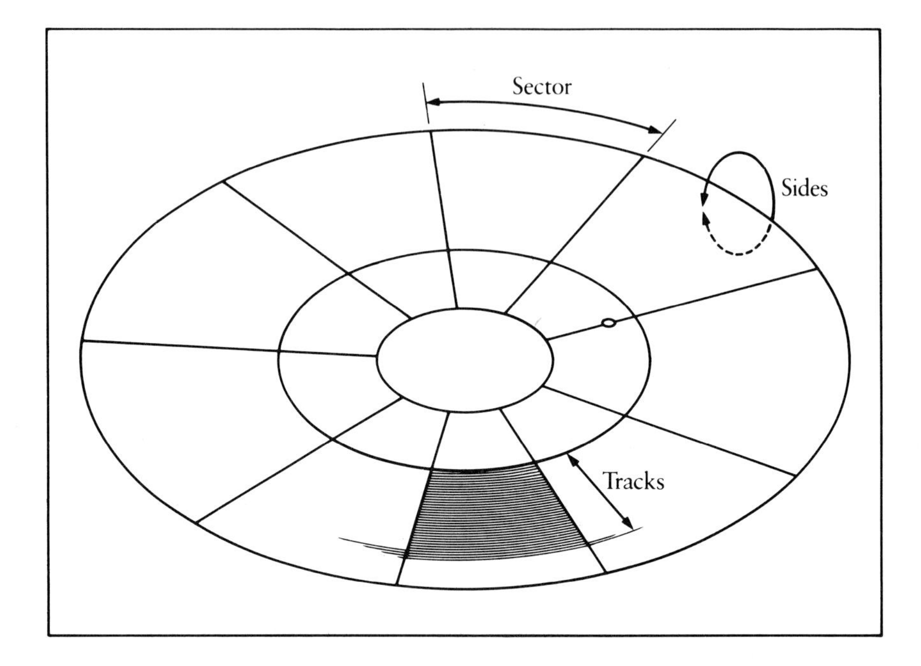


**Figure 1-7** Arrangement of floppy disk sectors and tracks. *(From* The Peter Norton Programmer's Guide to the IBM PC, *Ref. 4, Bellevue, WA, 1985. By permission of the publishers. © by the author and the publisher.)*

convertible personal computers. The amount of storage capacity on these diskettes varies, depending on the number of tracks, number of sectors, and density of recording, and whether both sides of the diskette are used for storing information. The latest version of disk drives used in the IBM PC can record in *double density* on two sides with nine sectors and 40 tracks, giving the diskette a 360K capacity. (The same disk drives can also read-write on single sides of the diskette and/or using only eight sectors, to conform with earlier versions of the operating system.) The IBM AT has a higher-capacity floppy disk drive. This is a *quad-density*, double-sided, 15-sector, 80-track diskette that can store 1.2 M of information.

Larger storage capacities are provided by hard disks. These range from 5 megabytes to 1 gigabyte. Hard disk drives use rigid metal platters coated with a metal oxide layer and spin at typically 3600 rpm. The read-write heads float above the spinning magnetic surface and require high precision and a complete freedom from dust particles. For this reason, most hard disks are hermetically sealed to prevent dust particles from entering and causing a head crash onto the disk. The hard-disk technology was originally developed by IBM and was given the code name Winchester, by which it is commonly known today [1, 7].

Since most hard disks are not removable from their drives, it may be necessary to have backup storage systems to ensure against loss of the information stored on the disks and to provide for permanent archiving of this information. This is commonly done by using magnetic tape drives to transfer the information from the hard disk to magnetic tapes which can be stored outside the computer.

*Optical disk* storage devices, similar to videodisks, which use lasers to store and retrieve digital information, are becoming available for use with personal computers. These devices have very large storage capacities (1 gigabyte or more) and fast access times. However, at the present time, recording on the optical disks must be done at the factory using powerful lasers. The videodisk players connected to the computer have read-only capability. This technology, however, is changing rapidly; prototypes of erasable optical disks are becoming available.

### 1-3.6 Input-Output Devices

A large variety of *input-output* (I/O) devices may be connected to a personal computer. The most widely used device for entering information into the computer is the keyboard. This contains the standard set of alphanumeric characters, punctuation keys, special control keys, function keys, and a numeric keypad. The keyboard interfaces with the CPU through a special keyboard controller.

Other input devices are mice, light pens, touchscreens, digitizing pads, bar code readers, optical character readers, scanners, and voice recognition devices. A *mouse* is an input device that moves the cursor on the display screen as the mouse is moved across a flat surface, such as a desk. The mouse is equipped with buttons that are pressed to make selections. *Light pens* and *touchscreens* allow computer users to point at images on the display screen to make choices that control the computer, using either a photocell pen or simply their own fingers. A *digitizing pad* is a flat plate equipped with an electronic stylus which can be used to enter graphical information with great precision and performs a variety of graphics and drafting functions, such as that done in computer-aided design. *Bar code readers* are photosensitive devices that can read a series of specially drawn parallel lines of varying thickness. *Optical character readers* scan pages of printed text, convert each character to its corresponding digital signal, and send this information to the personal computer for manipulation and/or storage. *Scanners* are similar to optical character readers with the exception that instead of recognizing individual characters, they read the images dot-by-dot, thus enabling them to copy pictures as well as text. *Voice recognition devices* use a microphone (or telephone) and special interfaces which identify voice patterns and convert the spoken words into the corresponding preprogrammed commands.

The personal computer can manifest its output in a variety of forms: visual, printed, graphical, and electronic. The *visual display screen* is usually a *cathode-ray tube* (CRT) monitor which displays the text and graphics generated by the computer. Television sets, equipped with a radio-frequency modulator, can be used as visual displays. However, dedicated monitors with better text resolutions are preferred for use with personal computers. Two major categories are the monochrome monitor and the color monitor. Monochrome monitors come in two varieties: for text only or for text and graphics. Color monitors can also be divided into two groups: those that operate on a *composite video signal* and those that use a *red-green-blue* (RGB) signal. Both have text and graphics capabilities. The RGB monitors have better definition of color and text, but are more expensive. Alternatives to the cathode-ray-tube monitor are *liquid crystal displays* (LCD) which are used in lap-size portable personal computers.

Hard-copy output of computer-generated information can be obtained using a printer or a plotter. Several types of printers are available; they are suited for different kinds of work, such as printing letters of typewritten quality, quickly printing rough drafts, intermixing graphics and text, producing documents of typeset quality, or printing color graphics. The two most widely used types of printers are letter quality and dot matrix printers. However, laser printers, thermal transfer printers, and inkjet printers are quickly gaining in popularity.

*Letter quality* printers use a daisy wheel or a thimble letterhead to produce fully formed characters like those of a typewriter. They can print at speeds ranging between 18 and 80 characters per second. *Dot matrix* printers form the characters by printing a pattern of dots corresponding to the shape of the desired character. Dot matrix printers are faster (up to 400 characters per second) and more flexible than letter quality printers. They can generate compressed, expanded or bold characters, and graphics images. Their print is not as sharply defined as those of the letter quality printers, but some have special techniques, such as passing over the same line twice with a slight shift in placement, to produce better-defined characters.

*Laser printers* use a beam of laser light to electrostatically etch characters on a photoconductive drum, which is identical to the print drum used in photocopiers. The laser beam scans across the drum discharging rows of dots, at a density of 300 dots per inch, to form the characters and images of an entire page. Black powdered toner adheres to areas on the drum that the laser beam has charged. When a sheet of paper passes over the print drum, the toner-defined dots adhere to the paper by pressure and heat, thus producing a printed page. Laser printers combine the advantages of excellent quality of print (almost to typeset standards), graphics printing capability, speed (8 to 12 pages per minute), and quiet operation. Naturally, they are more expensive than the other types of printers described above.

*Thermal transfer* printers form the characters by using heat to melt tiny dots of resin onto the paper. The print quality is similar to that of dot matrix printers. *Inkjet* printers squirt tiny droplets of ink onto the paper to form the characters. They use a print head with 6 to 24 nozzles mounted on a carriage which moves across the page to create the images. A new class of inkjet printers with a large stationary array of 2000 to 3000 nozzles are being developed. Inkjet printers are quieter than letter quality and dot matrix ones, and some can print up to 270 characters per second. Inkjet printers are best suited for printing color graphics and other multicolor images.

*Plotters* are output devices which permit the production of high-quality drawings and text using pens. Unlike graphics printers, which are raster devices that print horizontal lines of between 70 to 150 dots per inch, plotters are vector devices that draw up to 1000 line segments per inch. Plotters draw smooth, continuous lines, while graphics printers print broken lines. Since plotters use pens, they can fill with solid color the shapes that a printer would fill with a pattern of dots. Plotters can draw on paper as well as on transparency film.

Personal computers share their output with humans in the visual form (printed or graphical) as described above. In addition, these computers can share their information with other computers in an electronic form. This communication can take place via several pathways. The simplest such pathway is to connect two personal computers to each other through their serial interfaces. This requires no specialized devices and can be easily accomplished with the appropriate communications software. A direct link between a personal computer and a mainframe computer, via a coaxial cable, can be established through a specialized circuit board, with special software which matches the communications protocols of the interconnected computers.

A more widely used means of communication between computers is via the telephone lines with *modems* connected to each computer (Fig. 1-6). Modems are electronic devices which convert the computer's digital signals to the telephone's analog signals, and vice versa. This capacity to modulate and demodulate audio signals gives the modem its name. Modems can be connected directly to the computer bus through a slot in the system board or can interface with the computer via the serial interface. Modems also require software for establishing the communication protocols between the interconnected computers.

Another means of communication between computers is the *local area network* (LAN). Local area networks enable several computers to be interconnected to each other and to share common peripheral devices and common data bases of information. The components on the local area system are connected through adapter boards that plug into the slots on the system board. The data is transmitted through coaxial cables, or fiber optics cables for superfast communication (Fig. 1-8).

### 1-3.7 Interfaces

All input-output devices interface with the central processing unit of the computer via special *interfaces*, such as the keyboard interface, the disk controller, the display adapter, the serial interface, the parallel interface, the network interface, etc. The computer and the peripheral input-output devices seldom operate at the same speed. The peripherals which could be

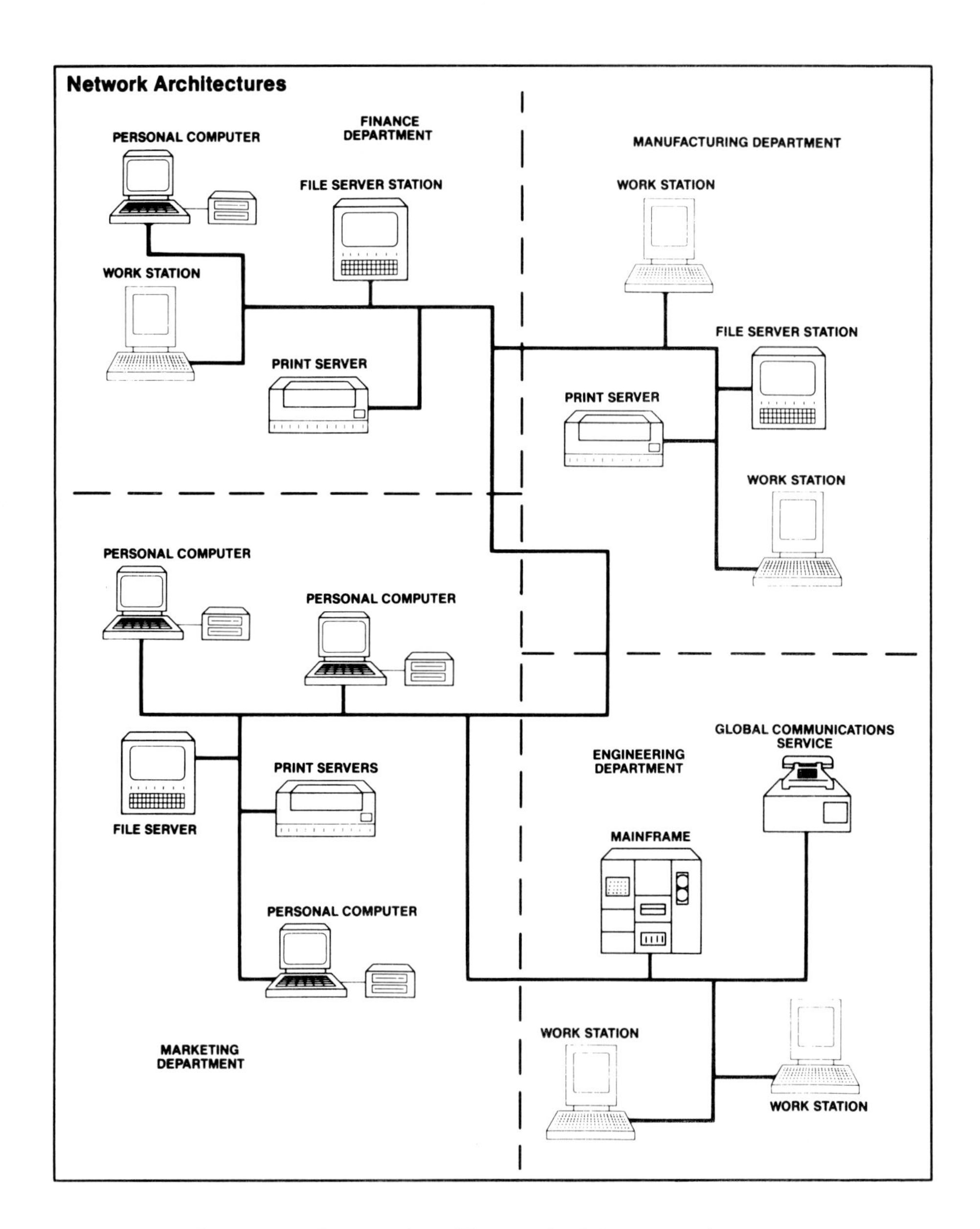


**Figure 1-8** Local area network connection. *(Courtesy Intel Corporation.)*

electromechanical devices, e.g., printers and disk drives, usually operate at much slower speeds than the all-electronic CPU. The speed and/or timing differences between CPU and I/O devices must be reconciled. Furthermore, the data formats of the peripherals may be different from the format used by the computers. The interface hardware, together with the appropriate software, performs the vital task of establishing and controlling the communication between the CPU and the peripheral devices. It is outside the scope of this book to describe the electronic operation of interfaces. We simply outline their basic function and describe several classes of interfaces available for the personal computer. A block diagram of a typical interface is shown in Fig. 1-9. The CPU side performs the following functions [8]:

- Decodes and responds to the unique address from the CPU that identifies the associated peripheral device.
- Decodes and responds to CPU control signals indicating the direction of the data transfer to be performed.
- Waits for the CPU to request a data transfer and
  1. Transfers commands or output data from the data bus to an output register.
  2. Transfers device status or input data from an input register to the data bus.
- May generate an interrupt request signal when the peripheral is ready to perform a data transfer in either direction.

The device side of the interface, shown in Fig. 1-9, performs the following functions:

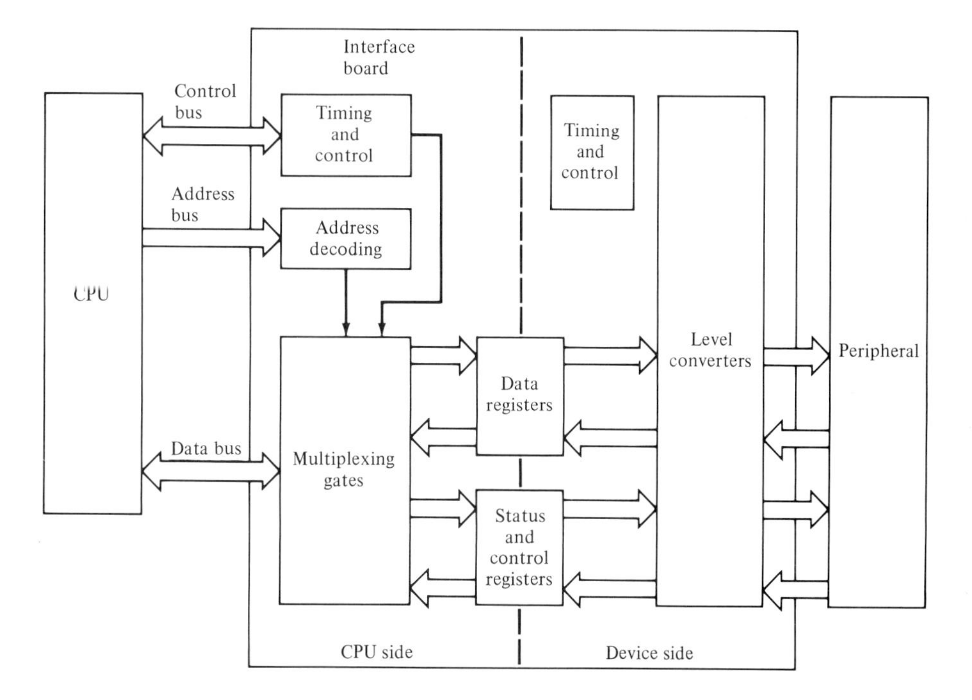


**Figure 1-9** Typical CPU/peripheral interface; block diagram. *(From I. Flores and C. Terry, Ref. 8. By permission of the publisher.)*

- Converts all signals from the voltage levels used in the computer to those used by the peripheral, and vice versa.
- Interrogates the device to determine the busy-ready status of various functions, and sets or resets corresponding bits in its status register, which is accessible to the CPU.
- Notifies an output device that data is ready in the output register; may wait for acknowledgement before sending a data strobe to perform the transfer, but the data ready signal is sometimes also used as the strobe.
- Performs input and error detection (parity), and sets or resets corresponding bits in the status register.

We classify interfaces into two categories: specialized and general interfaces. The interfaces in the first category can handle only one type of device, such as a keyboard, a display, or a disk drive. Those in the second category are more flexible in that they can be connected to more than one type of peripheral device. The serial and parallel interfaces belong to this category.

The *keyboard interface* manages the electronic signals generated by the keyboard device. Its main duty is to watch the keys and to report to ROM-BIOS whenever a key has been activated. It has limited diagnostic and error-checking capability and a buffer which temporarily stores the key strokes. The *disk controller* is also a specialized interface which handles the flow of data to and from the disk drives. This controller performs several functions such as selecting the drive to be used in multidrive systems, positioning the read-write head of the drive over a specified track, selecting the particular sector in the track to be read or written to, specifying the starting address of the memory buffer to be used in the subsequent read or write command, and finally reading or writing the designated sector [8].

The *display adapter* (commonly known as the *graphics board*) converts digital signals into video signals. Display adapters usually have their own RAM chips (typically 64K) for temporary storage of the display data. Display adapters vary widely in their features and capabilities. Most adapters operate in text-and-graphics mode, but some are designed for text-only mode. Color graphics boards can drive color monitors as well as monochrome monitors. Most IBM-compatible boards generate both a composite video signal and a red-green-blue signal (RGB). These boards usually operate in three modes: text, medium-resolution graphics, and high-resolution graphics. The number of colors available in each mode and the number of picture elements (pixels) per screen varies considerably among different computer models and display adapters.

The IBM Color/Graphics Adapter, and other similar adapters, generate a $40 \times 25$ or an $80 \times 25$ character display in text mode; a $320 \times 200$ pixel graphics display, with $40 \times 25$ characters, in medium-resolution graphics mode; and a $640 \times 200$ graphics display, with $80 \times 25$ characters, in high-resolution mode. The text mode can have 16 foreground and 8 background colors; the medium-resolution mode uses 4 colors; and the high-resolution mode has only 2 colors. Enhanced-resolution boards, with up to $1024 \times 1024$ pixels and a large number of colors, are available for use in high-precision applications, such as computer-aided design graphics.

*Serial* and *parallel interfaces* are two that have attained the highest degree of standardization; therefore, they have been adopted by a wide variety of peripheral devices for communication with personal computers. This standardization enables a serial interface, for example, to be connected to a printer, a modem, and a plotter, as well as many laboratory instruments. A serial interface controls the bidirectional flow of data between the CPU and the peripheral device. The microprocessor produces an 8-bit byte in parallel and directs it to the serial board over eight parallel bus lines; in turn, the serial interface lines up the byte into a single stream of bits for serial transmission to the peripheral device over a single line. The most popular serial interface is the RS-232C, a standard originally established by the Electronic Industries Association to facilitate the transmission of data over telephone lines. Most of the serial ports installed on personal computers, printers, and several other devices used with personal computers follow this RS-232C standard. This standard guarantees hardware compatibility between the interconnected units. The communications protocol between computer and peripheral must be set by software to accomplish full operational compatibility.

A *parallel interface* also controls the bidirectional flow of data between the CPU and the peripheral devices. Parallel interfaces are designed to transfer the data 8 or more bits at a time over the same number of parallel wires. There are several types of parallel interfaces: the Centronics, the IEEE-696, and the IEEE-488 are the three best known. Most of today's personal computers use the Centronics printer interface which was developed by the Centronics Data Computer Corporation for its popular series of printers and has been adopted by many other printer manufacturers. If both the computer and the printer have compatible Centronics parallel connections or ports, then a hardware link-up can be made. But problems can still be encountered if the program instructions (or *driver program*) used by the computer to communicate with a printer cannot be recognized by the printer.

Finally, more specialized interfaces exist which can be used for transmission of data between analog-type devices and personal computers. In process control applications, many control devices and instruments operate on analog signals. These are continuously variable signals that do not have discrete levels or states. The analog-to-digital (A/D) interface converts continuous voltage, or current, signals to digital signals for input to the computer. The reverse operation is performed by the digital-to-analog (D/A) interface which decodes the digital signal produced by the computer to an analog signal that can be received by the analog device.

The interface boards used for the direct link between personal computers and mainframes are specialized circuit boards which emulate the characteristics of a specific mainframe terminal. Most of these are designed to let a single personal computer communicate with a mainframe. The software supplied with these interfaces enables the user to set the communication protocols between the terminal (personal computer) and the mainframe computer.

## 1-4 OPERATING CHARACTERISTICS OF PERSONAL COMPUTERS—THE SOFTWARE

A *computer program* is a set of instructions which enables the computer hardware to perform a designated task. All the programs that are needed to run a particular computer are collectively known as the *software*. Computer software is essential to the operation of the computer in that the hardware is incapable of doing anything unless there are programs to direct it to perform its intended functions. Software programs fall into two categories:

1. *Operating systems software*—those programs designed to control the execution of other programs and utilize the hardware efficiently
2. *Applications software*—those programs designed to use the capability of the computer to solve specific user-oriented problems

Computer programs must be written in a *programming language* that the computer understands or has a facility to translate.

### 1-4.1 Programming Languages

The electronic circuitry of the digital computer functions on a set of binary logic operations, which may be represented by the binary numbers 0 and 1. Therefore, the native language of the computer—the *machine language*—is based on the binary number system. The earliest forms of computer programs were written using binary numbers, but machine language programs written today use mainly hexadecimal arithmetic.

A typical programming instruction has two parts. The first part is the *operation code* that tells the computer what functions to perform. The second part consists of the *operands* that give the computer the data to use to perform this operation, or the address in storage where a data or an instruction is located [9]. A complete program consists of an ordered set of instructions that is sequentially executed by the computer.

A programmer writing a machine language program must remember all the operation codes in binary, or hexadecimal, arithmetic and must keep track of all the storage locations of the data and instructions. Machine language programs make the most efficient use of the computer's hardware capabilities in speed and memory allocation; however, such programs are very tedious to write.

To make this task easier, *assembly languages* which use *mnemonic*† operation codes and *symbolic* addresses have been developed. Mnemonic codes consist of letters or words which are easier to remember than the corresponding binary or hexadecimal numbers. Symbolic addressing allows the programmer to give a symbol rather than an actual numeric address. A starting

† The word mnemonic is derived from the Greek word μνήμη, which means memory.

address for the program is specified, and the *assembler* automatically assigns storage locations for the data and instructions. Since the computer operates only in machine language, an assembler is needed to convert a *source program*, consisting of assembly language commands, into an *object program* in machine language. In addition, a *linker* is used to link several object programs together, and produce an *executable program*.

Figure 1-10 illustrates the procedure involved in the assembly, linking, and execution of an assembly language program. The individual steps are

1. The assembler, which resides on the disk storage device, is loaded into the computer.
2. The assembler loads the assembly language source program from the disk. This program has been written by the programmer earlier, using a text editor, and has been saved on the disk.
3. The assembler converts the assembly language instructions into machine language. If during this step the assembler finds any commands that cannot be translated into machine language, it generates error messages. In that case, the programmer has to exit from the assembler, use the editor to correct the program, and return to step 1. When the program is successfully assembled it is saved as an object program on the disk.
4. The linker program is loaded into the computer.
5. The linker loads the object program and converts it into an executable program. This is a *relocatable* machine language program that the operating system of the computer can store at any convenient memory location during the execution phase. In addition, the linker can be used to load and link together several object programs that have been assembled separately.
6. The linker stores the executable program on the disk.
7. In the last phase, the executable program is loaded from the disk and executed in the computer.
8. The executing program may obtain input data from the keyboard, from a previously prepared file on the disk, or from other input devices.
9. The program directs its results to output devices, such as the display monitor, the printer, and/or the plotter. The program may also save the results in output files on the disk.

Assembly languages are easier to use than machine languages; however, they have the fundamental limitation that they are machine oriented, i.e., each type of microprocessor has its own assembly language commands. For this reason, *high-level languages* have been developed which are applications oriented. These languages use commands which are closer to spoken language and standard mathematical notation, and are more universally transportable between different types of computers. Languages such as BASIC, FORTRAN, Pascal, and C are used in microcomputers, minicomputers, and mainframes.

High-level languages also must be translated into machine language before they can be understood by the computer. There are two types of

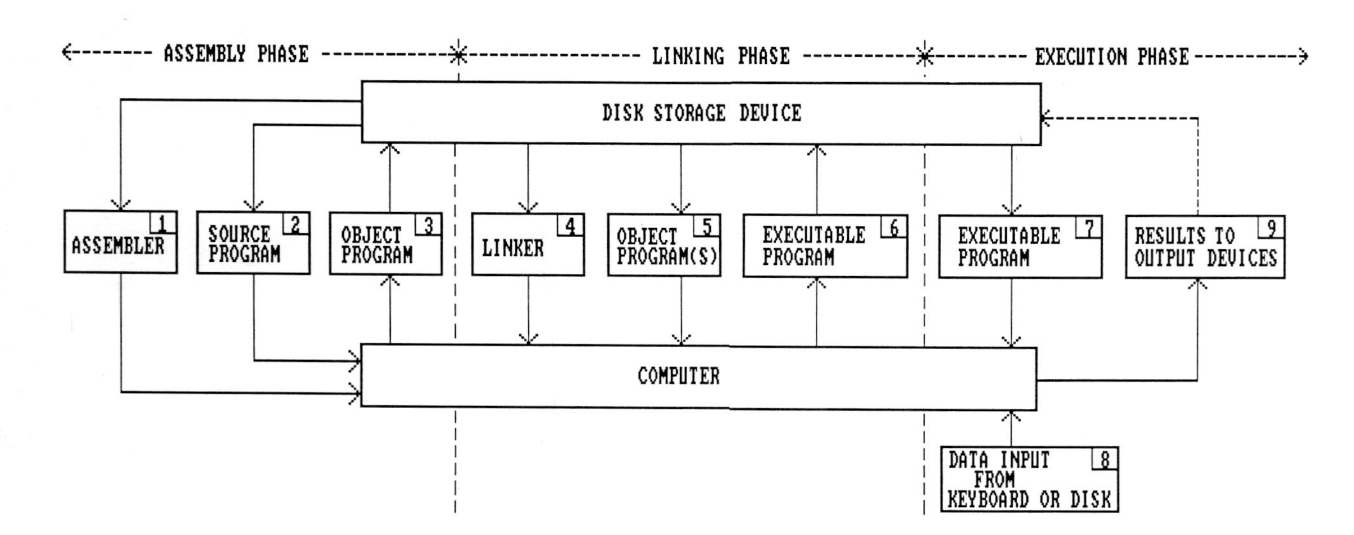


**Figure 1-10** Assembly, linking, and execution phases for source programs written in assembly language.

translators available: *compilers* and *interpreters*. The main difference between compilers and interpreters is the way in which they deal with the source statements. The compiler translates the entire program into a machine language object program in much the same way an assembler handles assembly language programs. Figure 1-11 illustrates the procedure involved in compiling, linking, and executing programs written in high-level languages. The only differences between this and the assembly procedure are in step 1, where a compiler is used instead of an assembler, and in step 5, where the linker links together the object program with library routines which perform such intrinsic functions as SIN, COS, EXP, LOG, etc.

Interpreters, on the other hand, work with one source statement at a time, translating each statement whenever it is encountered while executing the program. Interpretive languages are much easier to use than compiler ones, in that they are more interactive and they involve fewer steps in the preparation of the program. However, execution times are lengthier since the interpreter translates each statement every time it is encountered. The programming and execution of an interpretive language program can be viewed as a single phase. Figure 1-12 outlines the steps involved in this procedure:

1. The interpreter is loaded into the computer from the disk storage device.
2. The source program is entered from the keyboard using the interpreter itself as the editor for the preparation of the program. At any time during this procedure, the programmer can execute this program (or sections of it) to observe its performance and to detect any errors. The interpreter locates syntax errors and produces error messages giving the line number and type of error that was detected. Corrections of the errors can be done immediately and interactively without leaving the interpreter.
3. During execution, the program may obtain input data from the keyboard, the disk, or other input devices.
4. The program directs its results to the output devices.
5. Finally, the completed program may be saved on a disk for future use. This step is shown with dashed lines in Fig. 1-12 to indicate that it is an optional but highly recommended step.

Of all the high-level languages available, *interpretive BASIC* is the language of choice used in the majority of personal computers. The word BASIC is an acronym for Beginners All-purpose Symbolic Instruction Code. It was originally developed at Dartmouth College in the 1960s as a general-purpose computer language, particularly in time-sharing systems. Its interactive style, its simplicity, and the compactness of its interpreter made it well-suited for use in microcomputers. BASIC has changed considerably from its original version. Its most notable extension, a direct consequence of its use with personal computers, was the incorporation of graphics commands, which enable the programmer to draw pictures and construct graphs. There are several "dialects" of BASIC, all very similar in their commands for mathematical

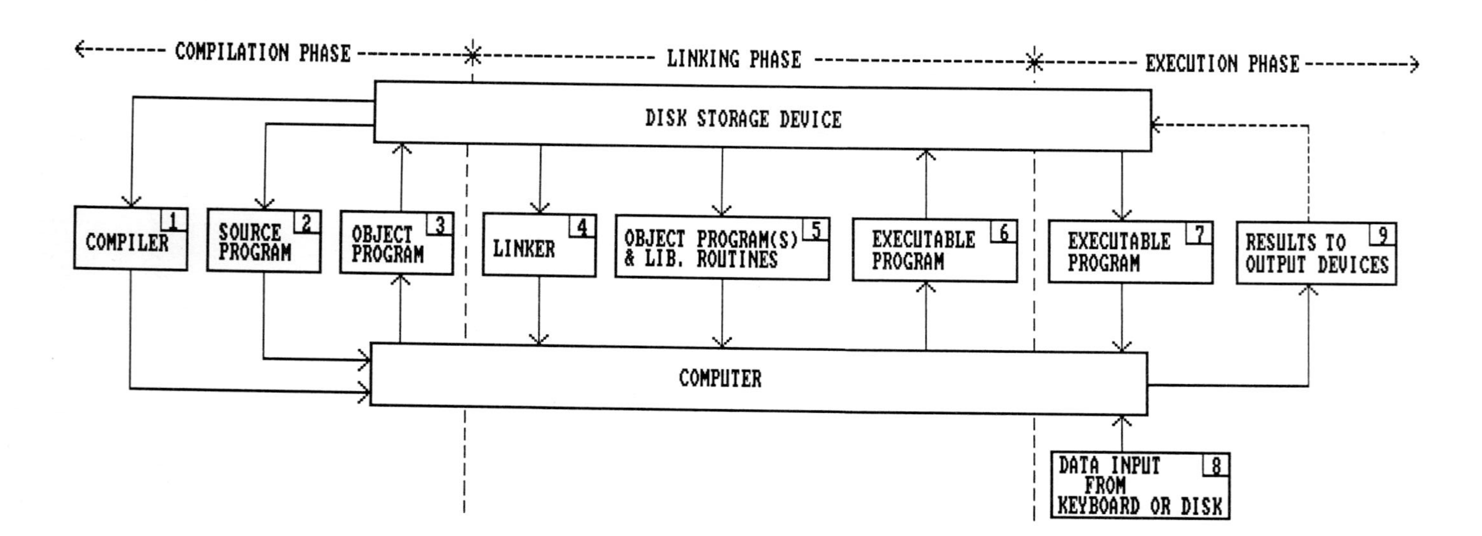


**Figure 1-11** Compilation, linking, and execution phases for source programs written in high-level languages.

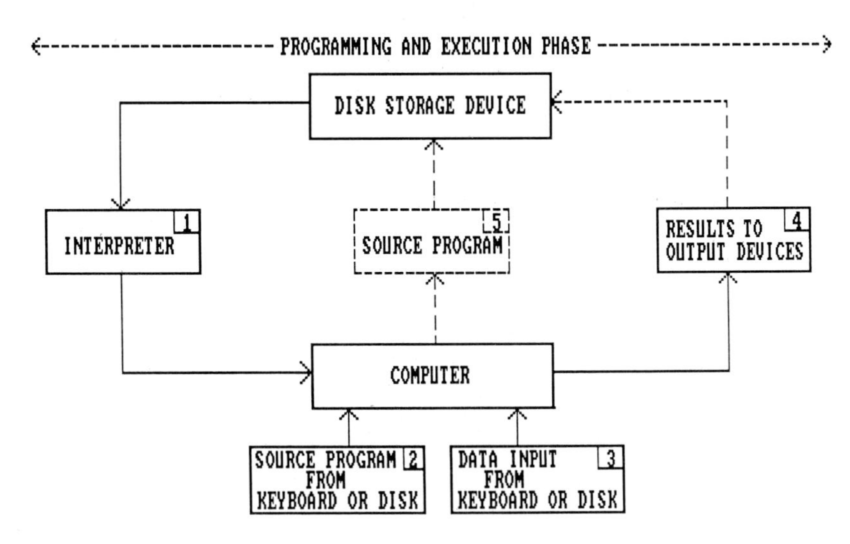


**Fig. 1-12** Programming and execution phase for source programs written in an interpretive language, such as BASIC.

operations and functions, but quite different from each other in the way they handle graphics, file structure commands, and logical operations. At the present time, BASIC is the only high-level language readily available in both interpreter and compiler versions. The most widespread version of this language is the interpreter BASIC written by Microsoft. It is used in the IBM PC, IBM AT, and all compatible computers. A slightly different version, also written by Microsoft, is used in the Apple II and TRS-80 series of personal computers. A detailed discussion of the IBM PC version of BASIC is given in Sec. 1-5.

The second most popular language is FORTRAN (from FORmula TRANslator) which was originally developed by IBM in the 1950s. This language has also undergone considerable change. The last two versions, FORTRAN IV and FORTRAN 77, have been used extensively for scientific and engineering applications on mainframe computers. Compilers for FORTRAN 77 are now available for use with microcomputers. FORTRAN has a more powerful structure than BASIC and has been strictly standardized, which makes it quite portable between different computers. However, it lacks the facility to draw graphics—a distinct disadvantage in its use with microcomputers which owe their popularity partly to their graphics capabilities.

Pascal is gaining popularity as a microcomputer language. It was developed in Switzerland in the early 1970s as a teaching tool. Pascal was designed to be completely transportable between computers, but most implementations of Pascal for microcomputers are subsets of the full language with variations from the original. The University of California at San Diego developed the UCSD Pascal version (P-system) which has been implemented by many microcomputers.

A better high-level, structured programming language for many purposes is C, which is used by professional programmers for writing systems programs. The machine language produced by C compilers is compact and efficient. C code is very portable and gives programmers extraordinary flexibility, allowing them to do things which are not possible in other high-level languages. In many ways it can be regarded as a language whose level falls somewhere in between assembly language and other high-level languages. C is intimately related to the UNIX operating system (UNIX is written in C) and is indispensable on computers that run UNIX [7].

Several other high-level languages, such as COBOL (for business applications), ADA (for military applications), and FORTH (for computer game programs), are available for use with microcomputers.

### 1-4.2 Operating Systems†

An *operating system* (OS) is the overall set of program instructions that are needed to manage and coordinate the various parts of a computer system. The operating system acts as an interface between the user—or the user's programs—and the system hardware and peripherals. Operating systems for microcomputers are usually called *disk operating systems* (DOS). In order to carry out their functions, disk operating systems have the following components:

1. A *command processor* which receives, identifies, and executes commands given to the system by the user working at the keyboard, or by the user's programs.
2. A file-management system which controls access to data and programs stored on the system disk. It maintains the disk space by keeping track of all files written to the disk and reuses disk space released by the deletion of files. This is the largest component of the operating system and it is usually referred to as the *basic disk operating system* (BDOS).
3. A logical input-output system consisting of routines, called *monitor calls*, which standardize the way applications programs talk to the input-output hardware, such as the keyboard, display monitor, printer, etc. This segment of the operating system is called *basic input-output system* (BIOS).
4. The hardware input-output routines, called *drivers*, which exert control over the physical input-output components of the system according to instructions from the operating system or from applications programs. These drivers are installation-dependent, e.g., a particular printer may require one driver to operate with an IBM PC and a different driver to operate with an Apple personal computer.

† Adapted from Ref. 10, by permission of the author.

In short, the operating system is responsible for translating user commands that affect the computer's hardware into the actual hardware commands needed to make the user's requests happen. The operating system also contains error recovery procedures, in case the hardware or user does not behave as expected.

Beyond this basic description, the structure, behavior, and appearance of operating systems vary widely. The typical operating system consists of two main levels, the *command language*, sometimes called the *shell*, that permits the user to enter commands from the keyboard, and the *hardware interface*, or *kernel*, that takes care of the actual system manipulation. The command languages of some simple operating systems, like Apple's DOS, are merely extensions to the BASIC language that allow the user to access the hardware. *Menu-based* operating systems, like the UCSD P-system, allow the user to select functions from a menu of options usually displayed on the system screen. *Command-based* operating systems, such as PC-DOS (MS-DOS), CP/M, and Unix, allow the user to type commands to a system monitor.

An operating system's hardware interface is, as might be expected, built around the system hardware. Although some operating systems are written for a particular microcomputer and dedicated to its hardware configurations, these are the exception. Most operating systems are written to be used on a variety of hardware, performing their interfacing function in such a way that they present a standard appearance to user programs. This feature, best represented in operating systems like CP/M, Unix, and the UCSD P-system, allows programs to be written without concern about the ultimate hardware configuration on which they will be run. This program portability between systems allows greater freedom in programming and gives users the option of upgrading to more powerful hardware without the need for extensive program modification.

Some newer features of operating systems, available only in the latest versions of the operating software, are *multiuser* and *multitasking* capabilities. A multiuser operating system is capable of sharing the central processing unit and its peripherals among two or more terminals. To do this the system must have the ability to determine which terminal originated each command and the proper destination of output; the ability to control access to shared devices such as disk drives and printers; and the ability to do multitasking, i.e., to run two or more application programs at a time. Multitasking, which is vital to multiuser systems, can also be quite useful in single user systems. The latter take advantage of multitasking by running tasks requiring no input from or output to the console in the "background," while the job that requires human interaction is running in the "foreground."

Currently, the most widely used operating system for microcomputers is Microsoft's DOS, known as PC-DOS for the IBM PC family of personal computers, and MS-DOS for all other PC-compatible machines. Detailed discussion of this system is deferred to Sec. 1-4.3, while in the remaining paragraphs of the present section we give a brief description of other operating systems.

**Apple DOS** Written specifically for the Apple II microcomputer, Apple DOS is an exception to many of the rules concerning characteristics of operating systems. Since DOS was written for a specific hardware configuration, it cannot be used as the basis for programs that are portable, except among Apple II's. Also, the command language of DOS is a set of commands, added to Apple BASIC, that allow simple interactions with disks and other peripherals. Manipulation of hardware beyond the limited capabilities of the DOS commands must be done using BASIC's PEEK and POKE operations, which look at and change the contents of control registers and single bytes of memory. The Apple DOS system supports only one language, BASIC, although some assemblers are also available. The Apple is also capable of running the P-system (see below).

**TRSDOS** Like Apple DOS, TRSDOS was designed for use with one type of hardware—the Radio Shack TRS-80. It was written for the original TRS-80, the model 1, and was adapted for subsequent Radio Shack systems. It is a command-based system with some worthwhile features, like a built-in real-time clock. However, its lack of compatibility with other computer operating systems limits its potential. The number of commercial quality applications packages available to run under TRSDOS is small compared with CP/M or Apple DOS.

**CP/M and CP/M-86** The CP/M operating system was developed for 8-bit personal computers, around the same time as Apple DOS, but with a significantly different design philosophy. CP/M is a command-based operating system which is designed specifically to allow programs to run, without modification, on microcomputers with differing hardware configurations. This is accomplished by linking a standard set of monitor calls to a machine-dependent set of drivers which can be modified by the user. This allows a sophisticated user, or more likely a hardware manufacturer, to make the necessary changes to CP/M to allow support of a new peripheral like a Winchester disk. The only firm hardware prerequisite to running CP/M is an 8080 or Z80 processor. The architecture of CP/M has been adapted for the more powerful 16-bit microprocessors (the Intel 8086 and 8088) in operating systems such as the CP/M-86, the MS-DOS, and PC-DOS (for the IBM PC).

**P-system** Originally written as a classroom project at the University of California at San Diego, the P-system is a complete programming environment for the Pascal language. It is a menu-driven operating system, in which single-letter commands send the user into separate modules for editing program or text files, manipulating the file directory, compiling a source program, or running a precompiled application program. Within each of these modules, another menu presents the options available.

The strength of the P-system is portability. In order to implement it, the authors invented an imaginary computer, called the P-machine, with its own instruction set and a very rigidly defined hardware architecture. The rest of the P-system—compiler, editor, filer—was then written in the language of the imaginary machine. In order to move the P-system to a new computer, all that is needed is to write a program called an emulator that makes the target computer behave like a P-machine. The rest of the system then runs without change, as do applications programs written for the system. As a result of this flexibility, the P-system is available for a wide range of machines, from Apples to high-performance 16-bit microcomputers. A FORTRAN has been written to run under the P-system, and other languages are being adapted.

The P-system's main weaknesses are size and speed. Since the P-machine adds an extra layer of overhead to the system, programs written in P-system Pascal or FORTRAN run more slowly than those written using compilers that generate code directly for the real system. This is less of a problem on new-generation high-performance microcomputers, which support memory sizes in the megabyte range and run significantly faster than 8-bit machines. The P-system and UNIX are becoming the most popular operating systems for the emerging supermicrocomputers.

**UNIX** The UNIX operating system was first developed at Bell Laboratories in 1969 for larger computers. It was later adapted to run on 16-bit microcomputers using the Motorola 68000 processor. Versions of UNIX are now available for most 16-bit and 32-bit machines. The UNIX system is a general-purpose command-based operating system which gives the user the ability to substantially modify the command environment. The only part of the UNIX system that is unalterable is the kernel, which takes care of such low-level functions as file manipulation, input-output, file and memory security, and usage accounting. Beyond the kernel lie the utility and applications programs, and beyond them is the shell whose basic function is command processing. Much of the power of UNIX lies in the fact that the shell can be modified by the user, who can write shell programs to change or extend almost any aspect of the command language's behavior. Along with UNIX, a special programming language called C was developed. C was, for a long time, the only high-level language available under UNIX. Now other languages are appearing, particularly under the Berkeley version of UNIX. The majority of UNIX is written in C. UNIX's major strength is the adaptability of its command language, the small size of the basic system, and the low overhead of operating system functions on the processor.

### 1-4.3 Introduction to the PC-DOS (MS-DOS)

In 1981, IBM introduced its personal computer, the IBM PC, which quickly became the most popular microcomputer on the market. Along with the PC came the operating systems developed by Microsoft, the PC-DOS 1.00 for use with IBM computers and the MS-DOS 1.00 for use with all other 8086/8088-based PC-compatible machines. Because of the popularity of this class of personal computers, this disk operating system has become the dominant operating system and, consequently, a de facto standard for microcomputers. Several versions of the PC-DOS have been released since the introduction of DOS 1.00, the latest being version 3.20. Table 1-2 shows the chronology of these versions and the associated hardware changes in the IBM PC family of personal computers. All versions are *upward-compatible*, i.e., diskettes prepared and most programs written with the earlier versions are compatible with the later versions of the operating system.

Our discussion in this section centers around the PC-DOS 3.10. The differences between versions 2.00 and 3.10 are not substantial; therefore, this discussion applies to both versions, except for a few commands which we point out as we go along. In addition, the differences in the command language between PC-DOS and MS-DOS are inconsequential; therefore, this discussion applies equally to both.

The PC-DOS is a command-based operating system which allows its user to type commands to a system monitor in order to control the operations of the computer, the execution of applications programs, and the flow of their results. This operating system consists of a large number of files on disks, containing programs which perform the various operating functions of the computer. Table 1-3 lists the directories of the two diskettes which contain the PC-DOS 3.10 operating system and some supplementary programs. The three most essential files are the IBMDOS.COM, which contains the basic disk operating system (BDOS); the IBMBIO.COM, which contains the basic input-output system (BIOS); and the COMMAND.COM, which is the command processor for the system. The first two of these files are "hidden files," i.e., they do not appear in the directory of the disk and they cannot be erased, unless the diskette is reformatted. The rest of the files contain programs for external commands, drivers for peripheral devices, a text editor for writing programs (EDLIN.COM), a debugger for monitoring the execution of programs to be debugged (DEBUG.COM), a linker program for linking separately

**Table 1-2 Versions of the PC-DOS**

| Version | Date | Hardware change |
|---|---|---|
| 1.00 | 8/04/81 | Original PC model (single-sided drive) |
| 1.10 | 5/07/82 | Double-sided diskette drive |
| 2.00 | 3/08/83 | XT model (hard disk drive) |
| 2.10 | 10/20/83 | PCjr and portable PC models (half-high drives) |
| 3.00 | 8/14/84 | AT model (high-capacity diskette drive) |
| 3.10 | 3/07/85 | Networking (network disk drive) |
| 3.20 | 3/18/86 | PC convertible ($3\frac{1}{2}$ in drive) |

*Source*: *The Peter Norton Programmer's Guide to the IBM PC*, Ref. 4.

**Table 1-3 Programs on the PC-DOS 3.10 Diskettes**

| Directory of DOS diskette | | | | |
|---|---|---|---|---|
| ANSI | SYS | 1651 | 3-07-85 | 1:43p |
| ASSIGN | COM | 1509 | 3-07-85 | 1:43p |
| ATTRIB | EXE | 15091 | 3-07-85 | 1:43p |
| BACKUP | COM | 5577 | 3-07-85 | 1:43p |
| BASIC | COM | 17792 | 3-07-85 | 1:43p |
| BASICA | COM | 27520 | 3-07-85 | 1:43p |
| CHKDSK | COM | 9435 | 3-07-85 | 1:43p |
| COMMAND | COM | 23210 | 3-07-85 | 1:43p |
| COMP | COM | 3664 | 3-07-85 | 1:43p |
| DISKCOMP | COM | 4073 | 3-07-85 | 1:43p |
| DISKCOPY | COM | 4329 | 3-07-85 | 1:43p |
| EDLIN | COM | 7261 | 3-07-85 | 1:43p |
| FDISK | COM | 8173 | 3-07-85 | 1:43p |
| FIND | EXE | 6403 | 3-07-85 | 1:43p |
| FORMAT | COM | 9398 | 3-07-85 | 1:43p |
| GRAFTABL | COM | 1169 | 3-07-85 | 1:43p |
| GRAPHICS | COM | 3111 | 3-07-85 | 1:43p |
| JOIN | EXE | 15971 | 3-07-85 | 1:43p |
| KEYBFR | COM | 2473 | 4-12-85 | 4:22p |
| KEYBGR | COM | 2418 | 4-12-85 | 4:23p |
| KEYBIT | COM | 2361 | 4-12-85 | 4:25p |
| KEYBSP | COM | 2451 | 4-12-85 | 4:24p |
| KEYBUK | COM | 2348 | 4-12-85 | 4:26p |
| LABEL | COM | 1826 | 3-07-85 | 1:43p |
| MODE | COM | 5295 | 3-07-85 | 1:43p |
| MORE | COM | 282 | 3-07-85 | 1:43p |
| PRINT | COM | 8291 | 3-07-85 | 1:32p |
| RECOVER | COM | 4050 | 3-07-85 | 1:43p |
| RESTORE | COM | 5410 | 3-07-85 | 1:43p |
| SELECT | COM | 2084 | 3-07-85 | 1:43p |
| SHARE | EXE | 8304 | 3-07-85 | 1:43p |
| SORT | EXE | 1664 | 3-07-85 | 1:43p |
| SUBST | EXE | 16611 | 3-07-85 | 1:43p |
| SYS | COM | 3727 | 3-07-85 | 1:43p |
| TREE | COM | 2831 | 3-07-85 | 1:43p |
| VDISK | SYS | 3307 | 3-07-85 | 1:43p |
| | 36 File(s) | | 61440 bytes free | |

**Table 1-3** (***Continued***)

| Directory of DOS supplemental programs diskette | | | | |
|---|---|---|---|---|
| ART | BAS | 1879 | 3-07-85 | 1:43p |
| BALL | BAS | 1966 | 3-07-85 | 1:43p |
| BASIC | PIF | 369 | 3-07-85 | 1:43p |
| BASICA | PIF | 369 | 3-07-85 | 1:43p |
| CIRCLE | BAS | 1647 | 3-07-85 | 1:43p |
| COLORBAR | BAS | 1427 | 3-07-85 | 1:43p |
| COMM | BAS | 4254 | 3-07-85 | 1:43p |
| DEBUG | COM | 15552 | 3-07-85 | 1:43p |
| DONKEY | BAS | 3572 | 3-07-85 | 1:43p |
| EXE2BIN | EXE | 2816 | 3-07-85 | 1:43p |
| LINK | EXE | 38144 | 3-07-85 | 1:43p |
| MORTGAGE | BAS | 6178 | 3-07-85 | 1:43p |
| MUSIC | BAS | 8575 | 3-07-85 | 1:43p |
| MUSICA | BAS | 13431 | 3-07-85 | 1:43p |
| PIECHART | BAS | 2184 | 3-07-85 | 1:43p |
| SAMPLES | BAS | 2363 | 3-07-85 | 1:43p |
| SPACE | BAS | 1851 | 3-07-85 | 1:43p |
| VDISK | LST | 136313 | 3-07-85 | 1:43p |
| | 18 File(s) | | 108544 bytes free | |

produced object programs (LINK.EXE), two BASIC language interpreters (BASIC.COM and BASICA.COM), several demonstration programs written in BASIC (SAMPLE.BAS, ART.BAS, etc.), and a sample telecommunications program (COMM.BAS).

In the remaining paragraphs of this section we describe the sequence of steps which the DOS takes when the PC is turned on, the various important files used in this flow of events, and a number of useful DOS commands for the operation of the PC. We cover only those commands we feel are needed to give the readers a rudimentary knowledge of the operating system, and to enable them to use the BASICA interpreter (discussed in Sec. 1-5) and the BASIC language programs that appear throughout this text. Detailed discussion of all the DOS commands and advanced features are given in the DOS manuals [12] that accompany the personal computer. The readers of this text are strongly advised to study these manuals for complete instructions of how to utilize the disk operating system.

**Starting the system** We assume that our computer system is a PC with two disk drives (A and B), a keyboard, a monitor, and a printer (Fig. 1-13). The

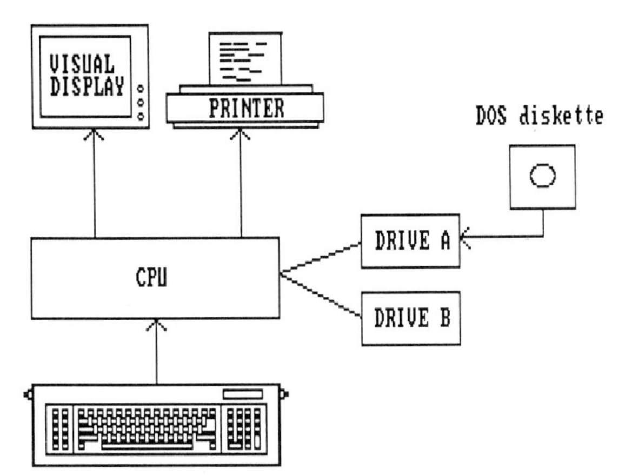


**Figure 1-13** Loading the disk operating system.

diskette containing the DOS is placed in drive A and the power for the computer is turned on. The ROM on the computer's system board contains diagnostic routines which check the hardware (Fig. 1-14). If this check is successful, another part of the ROM, the "boot" ROM, uploads the IBMBIO.COM, the IBMDOS.COM, and the COMMAND.COM system files and executes them. These files are stored in the primary memory (RAM) of the computer and stay there until the power is turned off, or until the computer is reset by the CTRL-ALT-DEL keys, as shown in Fig. 1-14. Next, the DOS looks on the disk in drive A for a file called CONFIG.SYS. This file is optional and is used for changing the standard configuration of systems. If found, the CONFIG.SYS file is executed, otherwise the DOS proceeds to search for the AUTOEXEC.BAT file. This file is also optional; it may contain a set of commands to perform a predetermined sequence of functions, such as setting the clock and calendar, loading device drivers, and executing a series of applications programs. Finally, the operating system issues the prompt

```
A>
```

indicating that it is ready to receive user input, and that drive A is the active (default) drive. The user can switch from one drive to the other by giving the command

```
B:
```

for activating drive B, and

```
A:
```

for returning to drive A.

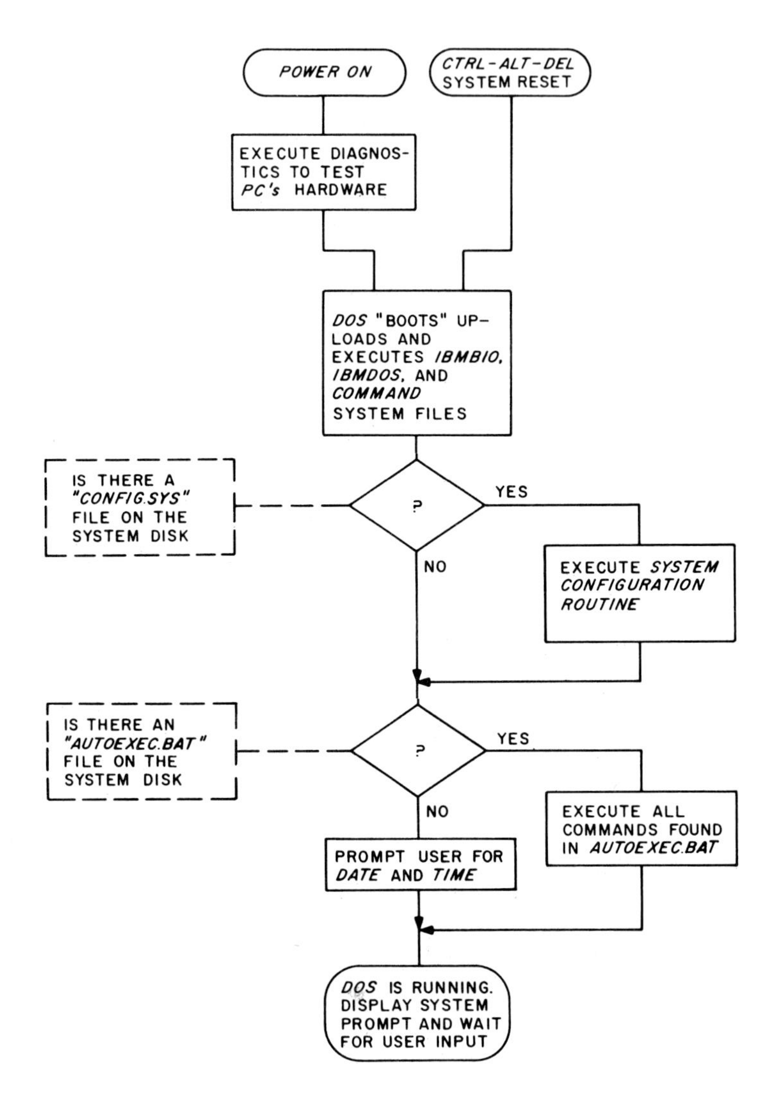


**Figure 1-14** This flowchart outlines the tasks the DOS performs before it issues the system prompt and waits for user input. *(From T. Field, Ref. 11. By permission of the publisher.)*

**File specification** Programs and data are stored on the disk in files whose name must be specified by the user. The complete file specification consists of three parts: the drive specifier, the filename, and extension:

```
d:filename.ext
```

where d: = the drive that contains the file. This specification is optional; if omitted, the DOS looks only in the default drive.

filename = the name used to identify the file. It consists of one to eight characters; certain punctuation characters are invalid (see DOS manual), and certain words have special meaning in DOS and may not be used as filenames (see Device names)

.ext = an optional extension to the filename used to identify various categories of files. It consists of a period followed by one to three characters; certain punctuation characters are invalid.

Certain extensions have special meaning and are reserved for those specifications, e.g.,

| | |
|---|---|
| .COM | for command files |
| .SYS | for system files |
| .EXE | for executable files |
| .BAT | for batch files |
| .BAK | for backup files created by the text editor EDLIN |
| .BAS | for BASIC language file |
| .OBJ | for object files created by compilers or assemblers |
| .MAP | for list file created by the linker |
| .LIB | for libraries of routines used in compilers |
| .FOR | for FORTRAN language source files |

Other extensions, such as .DAT, .PIC, .GAM, etc., may be formulated by the user to help identify specific categories of files.

Files with extensions .COM, .EXE, and .BAT are executable programs. Invoking the filename of one of these programs will cause the DOS to load and execute that program immediately. It is not necessary to enter the extension.

**Device names** PC-DOS reserves the names listed below for specific input-output devices. These names must not be used as filenames or extensions.

| | |
|---|---|
| COM1: or AUX: | Primary asynchronous communications port (serial interface) |
| COM2: | Secondary asynchronous communications port |
| CON: | Keyboard input and screen output |
| LPT1: or PRN: | Primary line printer (output device only) |
| LPT2: and LPT3: | Other line printers (output devices only) |
| NUL: | Nonexistent device for testing applications |

**Description of DOS commands** The PC-DOS version 3.10 has 63 DOS commands. Thirty-five of these are internal and twenty-eight are external commands (see Table 1-4). Internal commands are built in to DOS, i.e., once DOS is loaded, they reside in the primary memory of the computer and can be executed immediately without the DOS diskette. External commands are on the diskette as program files; they require the presence of the diskette when they are being executed. Eight of the internal commands are used for CONFIG.SYS files and seven are batch processing commands.

**Table 1-4 Summary of PC-DOS 3.10 commands**

| Internal commands | | | |
|---|---|---|---|
| BREAK | DATE | PATH | TIME |
| CHDIR | DEL | PROMPT | TYPE |
| CLS | DIR | RENAME | VER |
| COPY | ERASE | RMDIR | VERIFY |
| CTTY | MKDIR | SET | VOL |
| **External commands** | | | |
| ASSIGN | DISKCOPY | JOIN | RESTORE |
| ATTRIB | EXE2BIN | KEYBxx | SELECT |
| BACKUP | FDISK | LABEL | SHARE |
| CHKDSK | FIND | MODE | SORT |
| COMMAND | FORMAT | MORE | SUBST |
| COMP | GRAFTABL | PRINT | SYS |
| DISKCOMP | GRAPHICS | RECOVER | TREE |
| **Internal commands for CONFIG.SYS files** | | | |
| BREAK | COUNTRY | FCBS | LASTDRIVE |
| BUFFERS | DEVICE | FILES | SHELL |
| **Internal commands for batch (.BAT) files** | | | |
| ECHO | GOTO | PAUSE | SHIFT |
| FOR | IF | REM | |

We have chosen 26 of these commands to describe below, with at least one example for each. It should be understood that there are many variations to the formats of these commands, the details of which are fully discussed in the DOS manuals. We list these commands in the order in which the users are most likely to need them or encounter them in their work with the operating system. We assume that the default drive is A, and that it still contains the DOS diskette for accessing the external commands.

| | |
|---|---|
| DIR | Displays the directory of the diskette in the default drive. |
| DIR B: | Displays the directory of the diskette in drive B. |
| CLS | Clears the display screen. |
| FORMAT B: | Formats a new diskette in drive B, creating a 360K, 9-sector, double-sided diskette. |

| Command | Description |
|---|---|
| FORMAT B:/S | Same as above, plus it copies the operating system files (IBMBIO.COM, IBMDOS.COM, and COMMAND .COM) to the new diskette. |
| FORMAT B:/V | The /V parameter allows the user to enter a volume label for identifying the diskette. |
| CHKDSK B: | Analyzes the directories, files, and file allocation table on the diskette in drive B, and produces a disk and primary memory status report. |
| DISKCOPY A: B: | Copies entire contents of the diskette in drive A to the diskette in drive B. If the target diskette in drive B has a different format than the source diskette, it gets reformatted to conform with the format of the source diskette. |
| DISKCOMP A: B: | Compares the contents of the diskette in drive A to those of the diskette in drive B. |
| COPY A:PROGRAM1.BAS B: | Copies a file named PROGRAM1. BAS from drive A to drive B. |
| COPY A:PROGRAM1.BAS B:PROGRAM2.BAS | Copies the file named PROGRAM1.BAS from drive A to drive B and renames it as PROGRAM2.BAS. |
| COMP A:PROGRAM1.BAS B:PROGRAM2.BAS | Compares the contents of PROGRAM1.BAS to those of PROGRAM2.BAS. |
| ATTRIB +R PROGRAM1.BAS | The file PROGRAM1.BAS is given a read-only attribute, i.e., it can be read but not written to. Using the −R parameter removes this restriction. This command is available in DOS 3.00 or higher versions only. |
| TYPE AUTOEXEC.BAT | Types the contents of file AUTOEXEC.BAT. |
| ERASE PROGRAM1.BAS | Erases the specified file from the diskette. |
| DEL PROGRAM1.BAS | Same as ERASE |
| RENAME PROGRAM1.BAS PROGRAM3.BAS | Changes the name of PROGRAM1.BAS to PROGRAM3.BAS |
| DATE 1-29-87 | Sets the date to January 29, 1987. |
| TIME 13:30 | Sets the time to 1:30 P.M. |

| | |
|---|---|
| LABEL A:DISK1 | Labels the diskette in drive A as DISK1. Can be used to change or delete labels. This command is available in DOS 3.00 or higher versions only. |
| VOL A: | Displays the label of the diskette in drive A. |
| RECOVER B:MYPROGR.BAT | Recovers the specified file from a diskette in drive B that has a defective sector. |
| MODE | Sets the way that a printer, a color/graphics monitor adapter, or an asynchronous communications adapter operates (see DOS manual for details). |
| SYS B: | Transfers the operating system files IBMDOS.COM and IBMBIO.COM to the diskette in drive B. |
| MKDIR A:LEVEL1 | Creates a subdirectory called LEVEL1 on the diskette in drive A. |
| RMDIR A:LEVEL1 | Removes the subdirectory called LEVEL1 from the diskette in drive A. |
| TREE A: /F | Displays all the directory paths on drive A, and lists the files in each subdirectory. |
| CHDIR d:path | Changes the directory of specified drive (see DOS manual for details). |
| GRAPHICS | This is a driver that allows the contents of a graphics display screen to be printed on an IBM personal computer printer. Other printers, not compatible with the IBM printers, may require different drivers. |
| GRAFTABL | An extended character set represented by ASCII characters 128–255 is available on the IBM PC in text mode. The command GRAFTABL loads these additional characters for the color/graphics adapter so that they may be used in graphics mode as well. |

Several other commands are available in DOS, and its associated utility programs such as the line editor (EDLIN), the debugger program (DEBUG), the linker program (LINK), and the fixed disk setup program (FDISK). These are fully described in the DOS manuals [12].

**Creating a CONFIG.SYS file** The following sequence of steps creates a CONFIG.SYS file using the line editor EDLIN program:

```
A>EDLIN CONFIG.SYS
New file
*1I
        1:*DEVICE=VDISK.SYS 320 512 64
        2:*FILES=15
        3:*^C
*E

A>
```

This CONFIG.SYS file, which is executed by the DOS during the start-up phase of the computer, performs the following tasks:

1. Uses the VDISK.SYS device driver to simulate a virtual disk drive in the RAM memory of the computer. In this case it allocates 320K of memory to the virtual disk, with 512 bytes per sector, and a maximum of 64 directory entries. The DEVICE command is a new command introduced in PC-DOS version 3.00.
2. FILES=15 specifies the maximum number of files that can be open concurrently. The default value is 8.

Other internal commands that may be used in CONFIG.SYS files are listed in Table 1.4. The EDLIN editing commands used in creating the above CONFIG.SYS file are

1I    Inserts lines of text starting at line 1.
^C    CTRL-C key for stopping the insert command.
E     Ends the editing session and saves the edited file.

Other EDLIN editing commands are described in the DOS manual [12].

**Creating an AUTOEXEC.BAT file** The AUTOEXEC.BAT is an optional batch file which, if present on the disk, is executed by DOS during the start-up procedure. A batch file contains one or more commands that DOS executes one at a time. All batch files must have the filename extension .BAT. The following is an example of the AUTOEXEC.BAT file created using EDLIN:

```
A>EDLIN AUTOEXEC.BAT
New file
*1I
        1:*DATE
        2:*TIME
        3:*MODE COM1:9600,N,8,1,P
        4:*MODE LPT1:=COM1
        5:*GRAFTABL
        6:*LASERJET A
        7:*PENDRV
        8:*COPY A:*.* C:*.*
        9:*BASICA
       10:*^C
*E

A>
```

The above program performs the following tasks:

| | |
|---|---|
| DATE<br>TIME | Gives the user the ability to enter the current data and time. Once entered, the internal clock maintains the correct time and date provided that the computer is not turned off. Battery-operated clocks are standard in the IBM AT and can be installed as optional equipment in the IBM PC. Battery-operated clocks need not be reset every time the computer is started up. |
| MODE COM1:9600,N,8,1,P | Sets the parameters for the asynchronous communication adapter (serial interface) COM1. |
| MODE LPT1:=COM1 | Redirects parallel printer (LPT1) output to the serial interface (COM1) for serial printer output. |
| GRAFTABL | Loads the extended character set for the color/graphics adapter so that it may be used in graphics mode. |
| LASERJET A | Loads a special driver needed for printing a graphics display screen on the HP LaserJet printer used in this text. This is equivalent to the GRAPHICS driver provided with DOS for IBM printers. |
| PENDRV | Loads a special driver for controlling the PENPAD digitizer pad used in drawing the figures in this text. |

| | |
|---|---|
| COPY A:*.* C:*.* | Copies all programs from drive A to the newly created RAM drive C (virtual disk created by the CONFIG.SYS file). The asterisk * is a wild-card character standing for any filename and any extension. The reason for copying all the DOS files to the virtual disk (if you have created one) is so that they remain easily accessible when the DOS diskette is removed from drive A. This step is completely optional and for systems with hard disks it is unnecessary, since the hard disk would most likely contain all such programs. |
| BASICA | Loads the advanced version of the BASIC language interpreter. This interpreter will remain active and give the user the ability to write and/or execute programs in the BASIC language until the user exits the interpreter by the BASIC command SYSTEM which returns control to the AUTOEXEC.BAT file, and subsequently to the DOS, since no other commands are present in this AUTOEXEC.BAT file. |

## 1-5 INTRODUCTION TO BASIC PROGRAMMING LANGUAGE†

In this section we discuss the interpretive BASIC language used with the IBM PC and PC-compatible machines. There are three versions of the BASIC interpreter available:

1. Cassette
2. Disk
3. Advanced

The cassette version is built into the computer in ROM. If the PC is turned on without a diskette in drive A, it defaults to this ROM version of BASIC. The only storage device that can be used to save cassette BASIC information is a cassette tape recorder. Consequently, this version of BASIC is rarely used. The disk and advanced versions of the interpreter are on the PC-DOS diskettes as files BASIC.COM and BASICA.COM, respectively (see Table 1-3). The advanced interpreter employs the most extensive set of commands of all three

† Parts of this section are based on Refs. 13, 14, and 15 by permission of IBM.

versions of the BASIC language, including advanced features of graphics and music, such as VIEW, WINDOW, PAINT, PLAY, etc. All three versions are upward-compatible. In this book we make use of the advanced interpreter, BASICA; our discussion here therefore centers around this version.

The latest release of the BASIC interpreters from IBM is 3.0, which accompanies PC-DOS 3.00 and 3.10. The differences between BASIC 2.0 and BASIC 3.0 are minor (only five new commands were added to BASIC 3.0). Therefore our discussion applies equally well to both 2.0 and 3.0 releases.

### 1-5.1 Starting the Advanced BASIC Interpreter

Assuming that the computer has been started using the DOS diskette (as described in Sec. 1-4.3), and that the diskette containing the file BASICA.COM is still in drive A, load the interpreter by giving the following command in response to the A> prompt of DOS:

```
BASICA
```

The interpreter displays the following message and waits for commands from the user:

```
The IBM Personal Computer Basic
Version A3.10 Copyright IBM Corp. 1981, 1985
60446 Bytes free

Ok







1LIST  2RUN←  3LOAD"  4SAVE"  5CONT←  6,"LPT1  7TRON←  8TROFF←  9KEY  0SCREEN
```

Ok is the prompt used by BASIC to signify that the interpreter is in command mode. Loading BASICA can also be done by an AUTOEXEC.BAT file, as shown in Sec. 1-4.3. To exit from BASICA and return to DOS, simply give the command

```
SYSTEM
```

Several options are available to the user at the time of loading the interpreter. These are not necessary for routine operation of the interpreter but are useful for advanced programming applications. These options, which include the loading of a program to be run immediately, redirection of the standard input-output devices, maximum number of files that can be open at one time, size of buffers, etc., are fully described in the *BASIC Handbook* [13].

The Ok prompt of the BASICA interpreter signifies that the interpreter is ready to receive commands. There are two modes of operation: *direct mode* and *indirect mode*. In the direct mode, the interpreter performs the statements or commands entered by the user immediately. The results of arithmetic and logical operations can be displayed or stored but the instructions are not saved after they are executed. This mode is useful for quick computations or for debugging programs. A simple example of direct mode operation is

```
A=15
Ok
B=37
Ok
C=A*B
Ok
PRINT A,B,C
 15            37            555
Ok
```

In indirect mode (also called *programming mode*), each statement is given a line number. The statements are stored in the memory as a program and are executed later when the RUN command is entered. For example,

```
10 A=15
20 B=37
30 C=A*B
40 PRINT A,B,C
RUN
 15            37            555
Ok
```

The indirect mode is the most useful mode of operation since it allows the user to write and save programs for later use. The program is saved by the SAVE command and by specifying a filename, e.g.,

```
SAVE"TEST
```

If no extension is provided by the user, the interpreter automatically adds the extension .BAS to identify the file as one containing a BASIC language program.

## 1-5.2 The Keyboard: Special Keys and Program Editing

The keyboard of the IBM PC is divided into three areas: the function keys, the typewriter keyboard, and the numeric-editing keypad (see Fig. 1-15). There are 10 function keys, F1 to F10, which have been preprogrammed with the following frequently used BASIC commands:

```
F1    LIST         F2     RUN
F3    LOAD"        F4     SAVE
F5    CONT         F6     ,"LPT1:"
F7    TRON         F8     TROFF
F9    KEY          F10    SCREEN 0,0,0
```

These are "soft keys," i.e., they can be reprogrammed by the users to suit their specific needs (see KEY statement in *BASIC Reference* [14] for details).

Other useful BASIC keywords can be obtained quickly by holding down the ALT key and the letter shown in Table 1-5.

There are 256 characters that the PC can recognize. These have been assigned ASCII values ranging from 0 to 255 (see ASCII Character Codes, in *BASIC Reference* [14]). The first 128 of these can be produced by the keys of the keyboard. The *extended character set* (ASCII 128 to 255) are generated by holding the ALT key and giving the corresponding ASCII number from the numeric keypad. For example

```
ALT 228
```

generates the letter

$$\Sigma$$

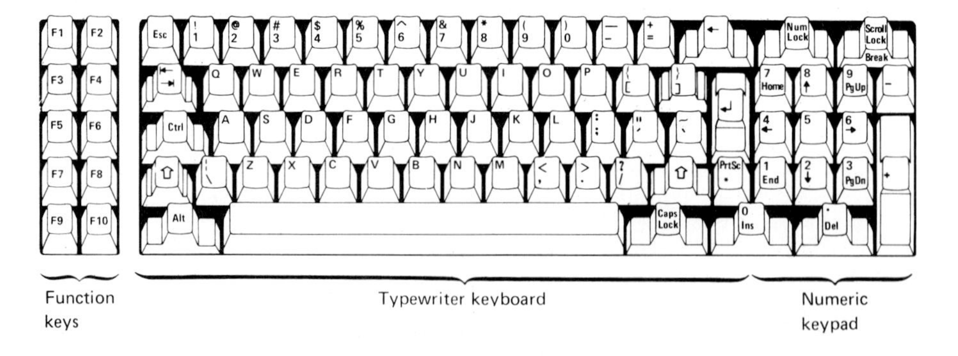


**Figure 1-15** The IBM PC keyboard. *(Courtesy IBM Corporation.)*

**Table 1-5**

| Letter | Word | Letter | Word |
|---|---|---|---|
| A | AUTO | M | MOTOR |
| B | BSAVE | N | NEXT |
| C | COLOR | O | OPEN |
| D | DELETE | P | PRINT |
| E | ELSE | R | RUN |
| F | FOR | S | SCREEN |
| G | GOTO | T | THEN |
| H | HEX$ | U | USING |
| I | INPUT | V | VAL |
| K | KEY | W | WIDTH |
| L | LOCATE | X | XOR |

The same character is produced by

```
PRINT CHR$(228)

Σ
```

Other special key combinations are

| | |
|---|---|
| CRTL-G | Produces a beep. |
| CTRL-Break | Interrupts the execution of a BASIC program and returns to command level. |
| CTRL-Num Lock | Freezes the execution of a program. Any other key restarts it. |
| CTRL-PrtSc | On/off switch which causes any text sent to the screen to be also sent to the printer. |
| Shift-PrtSc | Sends to the printer the contents of the display screen. If graphics are to be printed, a special driver for the printer (such as GRAPHICS.COM for IBM printers) must be loaded from DOS, before the Shift-PrtSc is used. |

The interpreter allows editing of a BASIC program by moving the cursor to any line on the screen and making changes to that line. The changes become effective when the Enter key is pressed. The editing keys are

| | |
|---|---|
| Home | Moves the cursor to the upper left-hand corner of the screen. |

| | |
|---|---|
| CTRL-Home | Clears the screen and moves the cursor to the upper left-hand corner of the screen. |
| $\leftarrow \uparrow \downarrow \rightarrow$ | Moves the cursor in direction of arrow. |
| End | Moves the cursor to the end of the statement line. |
| CTRL-End | Erases to the end of the statement line. |
| Ins | Sets insert mode on or off. |
| Del | Deletes the character at the current cursor position. |
| ← (Backspace) | Deletes the character to the left of the cursor. |
| Esc | Erases the entire line from the screen but not from memory. |
| CTRL-Break | Returns to command level without saving any changes that were made to the line being edited. |
| →\| (Tab) | Moves the cursor to the next tab stop. Tab stops occur every eight characters. |
| CTRL→ | Moves the cursor right to the next word. |
| CTRL← | Moves the cursor left to the beginning of the previous word. |

### 1-5.3 Numeric Constants and String Constants

Numbers can be stored internally by the PC as *integer constants*, *single-precision constants*, or *double-precision constants*. The set of integer constants consists of whole numbers from −32768 to 32767. The set of single-precision constants consists of 0, all numbers from $2.938736 \times 10^{-39}$ to $1.701412 \times 10^{38}$ with at most seven significant figures, and the negatives of these numbers. The set of double-precision constants consists of 0, all numbers from $2.938735877055719 \times 10^{-39}$ to $1.701411834604692 \times 10^{38}$ with at most 17 significant figures, and the negatives of these numbers. All 17 digits are stored but only 16 are displayed for double-precision numbers.

Single- and double-precision constants are sometimes expressed in "scientific," or "floating-point," notation; that is, as a number times a power of 10. For single-precision constants, the letter E followed by the integer $n$ denotes 10 raised to the $n$th power. For double-precision constants, the letter D is used instead of E.

Unless otherwise specified, all constants are assumed to be single-precision. A number is specified as an integer constant by following it with a percent sign (%), and, if given a name, by following the name with a percent sign. A number is specified as a double-precision constant by following it with a number sign (#), writing it with more than seven digits, writing it in floating-point form using the letter D, or, if given a name, by following the name with a number sign. An exclamation sign following a number or a numeric variable denotes a single-precision number. Numbers that result from the arithmetic operations +, −, /, and * inherit as their precision the greatest precision of the numbers involved. Transcendental functions can have calculations performed in double-precision provided that the option /D was selected when BASICA was initially loaded [15].

A *string constant* is a sequence of up to 255 characters enclosed in double quotation marks. The characters can be letters, numbers, or symbols. Any word, phrase, or sentence can be treated as a string constant provided that it is enclosed in quotation marks.

Examples of numeric and string constants are given below.

| | |
|---|---|
| 25% | Integer constant |
| 25.2 | Single-precision constant |
| 5.2D7 | Double-precision constant |
| "PAUL" | String constant |

### 1-5.4 Roundoff Error

The computer uses a finite number of binary digits to handle numbers internally. During floating-point operations the numbers are chopped beyond a certain digit, and the final results are converted to decimal arithmetic, rounded to the nearest significant figure, and presented as output. This chopping and rounding introduces a small *roundoff error* in the numerical results. For example, if the exact value obtained from the addition of two numbers is represented by $(x + y)$ and the floating-point value obtained from this operation by the computer is shown as $\mathrm{fl}(x + y)$, then the *relative roundoff error* is defined as

$$\delta = \left| \frac{(x + y) - \mathrm{fl}(x + y)}{x + y} \right| \tag{1-1}$$

The value of $\delta$ varies depending on the method used by the particular computer and its software to handle floating-point numbers and floating-point operations. One can determine a measure of the maximum achievable accuracy of the computation by calculating the *unit in the last place* ($u$). This is the smallest positive machine number $u$, such as

$$\mathrm{fl}(1 + u) > 1 \tag{1-2}$$

The following BASIC language program determines the value of $u$, for the IBM PC, in single-precision arithmetic, by decreasing the value of $u$ slowly until $\mathrm{fl}(1 + u)$ is no longer recognized by the computer as being greater than 1:

```
10 U=.00001
20 U=.9999*U
30 UP1=U+1
40 IF UP1>1 GOTO 20
50 PRINT "SINGLE PRECISION U= ";U
60 END
RUN
SINGLE PRECISION U= 5.959951E-08
Ok
```

The value of the unit in the last place is directly related to the number of binary digits the computer uses to represent a floating-point number. The PC uses 32 bits (4 bytes) for single-precision numbers. Of these, 24 bits are allocated to the mantissa, including the sign, and 8 bits to the exponent, including its sign. Therefore, the value of $u$ can be determined by

$$u = 2^{-24} = 5.96 \times 10^{-8}$$

which is very close to the value obtained by the above program. In double-precision arithmetic, the PC employs 64 bits (8 bytes) to represent floating-point numbers. Of these, 56 bits are used for the mantissa and 8 bits for the exponent. The value of $u$ in double-precision is

$$u = 2^{-56} = 1.388 \times 10^{-17}$$

which agrees with the value obtained when the above program is run in double-precision.

The value of $u$ is a measure of the maximum achievable accuracy of a single computation done by the computer. In carrying out computer calculations of practical magnitude, a very large number of arithmetic operations are performed (such as in the integration of differential equations), and the roundoff errors can propagate (see Chap. 5).

### 1-5.5 Variables

*Variables* are names used to represent values in a BASIC program. There are two types of variables: numeric and string. A numeric variable always has an arithmetic value. A string variable can only have a character string value. Variable names can be up to 40 characters long. The characters in a name can be letters, numbers, and decimal points. The first character must be a letter. Certain words and letter combinations have a special meaning in BASIC and must not be used as variable names (see *BASIC Handbook* [13]). A variable name determines its type (string or numeric, and if numeric, its precision). String variable names are written with a dollar sign ($) as the last character. Numeric variable names can declare a number as integer (%), single-precision (!), or double-precision (#). Examples of valid variable names are given below.

| | |
|---|---|
| A$ | String variable |
| ITER% | Integer variable |
| EPS! | Single-precision |
| EPS | Defaults to single-precision |
| XP# | Double-precision |

If the type-declaration is not specified, the default is single-precision.

Variable types can also be declared using the DEFtype statements. For example,

| | |
|---|---|
| DEFSNG A-H | The variable names beginning with the letter A through H are declared single-precision variables. |
| DEFINT I-N | The variable names beginning with the letter I through N are declared integers. |
| DEFSTR O-Q | The variable names beginning with the letter O through Q are declared string variables. |
| DEFDBL R-Z | The variable names beginning with the letter R through Z are declared double-precision variables. |

Vectors and matrices may be represented by single variable names provided that they are appropriately dimensioned. For example, a vector containing 100 single-precision numbers may be given the variable name X and dimensioned as

```
DIM X(99)
```

Zero is the minimum position in the vector; therefore X(99) provides space for 100 elements. If desired, the minimum position for arrays can be changed to 1 by the command

```
OPTION BASE 1
```

A matrix of order ($n$rows × $n$columns) may be dimensioned as

```
DIM A(nrows-1, ncolumns-1)
```

Alternatively, it may be dimensioned as

```
DIM A(nrows, ncolumns)
```

if the OPTION BASE 1 is used.

A small vector containing up to 11 elements may be used in a BASIC program without giving the DIMension statement; however, a twelfth, or higher, element in that vector will not be recognized by the interpreter and would cause an error message to be produced and the program execution to be interrupted.

### 1-5.6 Numeric Expressions and Operators

A numeric expression can be a numeric constant, a numeric variable, or a combination of these brought together by numeric operators to yield a single numeric value; e.g.,

```
5.72
A
(B*C)/2
```

There are four categories of numeric operators in the BASIC language:

- Arithmetic operators
- Relational operators
- Logical operator
- Numeric functions

*Numeric operators* perform mathematical or logical operations, most often on numeric values but sometimes on string values. They always produce a numeric value [13].

*Arithmetic operators* perform the usual mathematical operations in the order of preference shown in Table 1-6. In integer division, the operands are rounded to integers before the division is performed, and the quotient is truncated to an integer. Modulo arithmetic gives the integer value that is the remainder in an integer division.

*Relational operators* compare two constants or variables which are both numeric or string. If the result of the comparison is "true," the relational operator has a value of −1; if it is "false," then it has a value of 0. Relational operators are usually used in IF . . . THEN statements to determine the flow of a program. These operators are listed in Table 1-7. In string comparisons, the ASCII value of each character is used to establish the order of the string.

*Logical operators* perform logical, or *boolean*, operations on numeric values. They are also used in IF . . . THEN statements to determine the flow of a program. These operations and the results they return are listed in Table 1-8 (T means "true," or nonzero value; F means "false," or zero value).

**Table 1-6**

| Operator | Operation | Example |
|---|---|---|
| ^ | Exponentiation | $X$^$Y$ |
| − | Negation | $-X$ |
| *, /, \ | Multiplication | $X*Y$ |
| | Floating-point division | $X/Y$ |
| | Integer division | $X\backslash Y$ |
| MOD | Modulo arithmetic | $X$ MOD $Y$ |
| +, − | Addition | $X+Y$ |
| | Subtraction | $X-Y$ |

**Table 1-7**

| Operator | Relation tested | Example |
|---|---|---|
| = | Equality | $X = Y$ |
| <> or >< | Inequality | $X <> Y$ |
| < | Less than | $X < Y$ |
| > | Greater than | $X > Y$ |
| <= or =< | Less than or equal to | $X <= Y$ |
| >= or => | Greater than or equal to | $X >= Y$ |

*Numeric functions* perform predetermined operations on numeric constants, variables, or expressions. BASIC has eleven built-in numeric functions (see Sec. 1-5.8), such as

| | |
|---|---|
| COS(X) | Cosine function |
| LOG(Y) | Natural logarithm |
| SQR(Z) | Square root |

In addition to the built-in functions, other functions can be defined by the user with the DEF FN statement (see Sec. 1-5.8).

**Table 1-8**

NOT (logical complement)

| *X* | NOT *X* |
|---|---|
| T | F |
| F | T |

AND (conjunction)

| *X* | *Y* | *X* AND *Y* |
|---|---|---|
| T | T | T |
| T | F | F |
| F | T | F |
| F | F | F |

OR (disjunction)

| *X* | *Y* | *X* OR *Y* |
|---|---|---|
| T | T | T |
| T | F | T |
| F | T | T |
| F | F | F |

XOR (exclusive OR)

| *X* | *Y* | *X* XOR *Y* |
|---|---|---|
| T | T | F |
| T | F | T |
| F | T | T |
| F | F | F |

EQV (equivalence)

| *X* | *Y* | *X* EQV *Y* |
|---|---|---|
| T | T | T |
| T | F | F |
| F | T | F |
| F | F | T |

IMP (implication)

| *X* | *Y* | *X* IMP *Y* |
|---|---|---|
| T | T | T |
| T | F | F |
| F | T | T |
| F | F | T |

### 1-5.7 String Expressions and Operators

A string expression can be a string constant, a string variable, or a combination of these brought together by string operators which produce a single string value. There are two categories of string operators:

- Concatenation
- String functions

*Concatenation* is the joining together of two strings to produce a longer string, e.g.,

A$ = "JAMES"
B$ = " SMITH"
C$ = A$ + B$

The variable C$ now contains the string "JAMES SMITH".

*String functions* perform predetermined operations on constants, variables, or expressions. BASIC has nine built-in string functions (see Sec. 1-5.8), such

| | |
|---|---|
| LEN(A$) | Gives the length of string A$ |
| LEFT$(A$,2) | Gives the leftmost two characters of string A$ |

In addition to the built-in functions, other string functions can be defined by the user with the DEF FN statement (see Sec. 1-5.8).

### 1-5.8 Summary of Commands, Statements, and Functions

The advanced BASIC interpreter (version 3.0) has over 180 commands, statements, and functions. The distinction between a command and a statement is largely a matter of tradition. Commands generally operate on programs (e.g., LOAD, SAVE, etc.) and are usually entered in direct mode. Statements generally direct program flow from within a program, so they usually appear in indirect mode as part of a program line. Actually, most BASIC commands and statements can be entered in either direct or indirect mode.

Most of the BASIC commands and statements have several extensions and variations. It is not possible to describe all forms of these statements in this chapter. Complete discussions of these, with examples of their usage, are given in Refs. 13 to 19. In this section we list one form of each BASIC command, statement, and function with examples, as given in the IBM *BASIC Quick Reference* manual†. These are subdivided into the following categories:

- General
- Communications
- Conversions
- Development
- Files
- Graphics
- Math functions
- String functions

† By permission of IBM.

## General

**ASC**

```
10 X$="TEST"
20 PRINT ASC(X$)
```

The ASCII code of the first character of X$ is printed. For the capital letter T, the value is 84.

**BEEP**

```
10 IF X<20 THEN BEEP
```

If X is less than 20 the speaker emits a single beep.

**BLOAD**

```
10 DEF SEG=&HB800
20 BLOAD "PICTURE",0
```

Loads a file called PICTURE that has been previously BSAVEd into memory at offset=0 of the segment beginning at HB8000.

**BSAVE**

```
10 DEF SEG=&HB800
20 BSAVE "PICTURE",0,&H4000
```

Saves an image of the memory area in segment HB8000 starting at offset=0 and ending at offset=4000. The image is written to a disk file called PICTURE.

**CALL**

```
10 DEF SEG=&H8000
20 OFS=0
30 CALL OFS(A,B$,C)
```

Calls a machine language subroutine that begins at offset=0 of segment H8000 and passes variables A, B$, and C to the called routine.

**CHAIN**

```
10 CHAIN "A:PROG1",10,ALL
```

Transfers control to PROG1, which begins execution at line 10, and passes ALL variables to PROG1.

**CLEAR**

```
10 CLEAR 32768,2000
```

CLEARs all memory used for data without erasing current program; sets maximum workspace=32768 bytes; sets stack size=2000 bytes.

**CLS**

```
10 CLS
```

Clears the screen; homes the cursor.

**COLOR** (Text Mode)

```
10 COLOR 1,0
```

IBM monochrome display produces green underlined characters on a black background. Color display produces blue foreground characters on a black background.

**COMMON**

```
100 COMMON A, B(), C$
110 CHAIN "A:PROG3"
```

Passes variables A, B(), and C$ to PROG3.

**CSRLIN**

```
10 Y=CSRLIN
```

Stores the value of the current cursor line number.

**DATA**

```
10 DATA 3.05, 5.89, 23.08
```

Stores a list of constant DATA to be accessed by a READ statement.

**DATE$**

```
10 DATE$="10/31/83"
20 PRINT DATE$
```

Can either set date or retrieve date.

**DEF FN**

```
10 PI=3.141593
20 DEF FNAR(R)=PI*RA2
30 INPUT "Radius? ", RAD
40 PRINT "Area=";FNAR(RAD)
```

Defines a function written by you to calculate the area of a circle.

**DEF SEG**

```
10 DEF SEG=&HB800
```

Defines the segment in memory that is currently being accessed. HB8000 begins the area of the color screen buffer.

**DEFtype**

```
10 DEFINT A-D
20 DEFSING F-L,X
30 DEFSTR M-W
```

Declares variable types by the first letter of the variable name.

**DEF USR**

```
10 CLEAR,&H8000
20 DEF SEG
30 DEF USR2=&H8000
40 X=USR2 (Y+2)
```

Defines a machine language subroutine called USR(2) that begins at offset=24000 of the current segment.

**DIM**

```
10 DIM VAR(20)
```

Allocates space in memory for up to 21 subscripts for the variable VAR.

**END**

```
10 END
```

Ends program execution.

**ENVIRON**

```
10 ENVIRON "PATH=C:\"
```

Sets a path parameter in the BASIC environment table to the root directory on the C: drive.

**ENVIRON$**

```
10 PRINT ENVIRON$("PATH")
```

Displays the current contents of the PATH parameter in the BASIC environment table.

**EOF**

```
10 IF EOF THEN END
```

Checks for an end-of-file condition when reading from a sequential file.

**ERASE**

```
10 ERASE TEST1,TEST2
```

Erases arrays TEST1 and TEST2 from memory.

**ERDEV**

```
10 PRINT ERDEV
```

Displays a number containing a device error code.

**ERDEV$**

```
10 PRINT ERDEV$
```

Displays the name of the device causing an error.

**ERR**

```
10 IF ERR=27 THEN GOTO 100
```

If error code=27 (Out of paper) then GOTO the error-handling routine.

**ERL**

```
10 IF ERL=250 THEN RESUME
```

If the error occurred in line 250, ignore the error.

**ERROR**

```
20 ON ERROR GOTO 40
30 IF B>500 THEN ERROR 210
40 IF ERR=210 THEN END
```

Defines a new error code to be handled by your program.

**FOR ... NEXT**

```
10 FOR I=1 TO 500 STEP 2
20 NEXT I
```

Advances a counter two steps at a time each time NEXT is executed until the count reaches 500.

**FRE**

```
PRINT FRE(0)
```

Displays the number of unused bytes of memory.

**GOSUB ... RETURN**

```
20 GOSUB 40
30 PRINT"CHECK":END
40 PRINT"DOUBLE":RETURN
```

Goes to a subroutine at line 40; then returns to the statement following the GOSUB.

**GOTO**

```
10 GOTO 30
20 PRINT"ERROR"
30 PRINT"CONTINUE"
40 PRINT"PROGRAM"
```

Goes to line 30 and continues execution from there.

**IF**

```
10 IF A=20 THEN GOTO 30 ELSE STOP
```

If A=20 is true, then branch to line 30; otherwise, stop execution.

**INKEY$**

```
10 PRINT"PRESS ANY KEY"
20 A$=INKEY$
30 IF A$="" THEN 10
```

Reads a character from the keyboard and assigns it to a variable called A$. If no key is pressed, return to line 10 and loop until a key is pressed.

**INP**

```
10 A=INP(255)
```

Reads input from machine port number 255 and assigns it to variable A.

**INPUT**

```
10 INPUT "WHAT RADIUS";R
```

Waits for keyboard input; displays the quoted phrase; assigns the keyboard input to variable R.

**INPUT$**

```
10 X$=INPUT$(5)
```

Reads a string of 5 characters from the keyboard and assigns them to X$.

**IOCTL**

```
10 OPEN "O", #1,"LPT1:"
20 IOCTL #1, "PL56"
```

Passes control data (PL56) that sets page length to open device #1.

**IOCTL$**

```
10 IF IOCTL$(1)="56" THEN PRINT "OK"
```

Reads the control data string on open device #1.

**KEY**

```
10 KEY 6, "PRINT"
```

Programs function key F6 to display the word PRINT.

**KEY(n)**

```
10 KEY(10) STOP
```

Disables trapping of function key 10.

**LET**

```
10 LET Y=35
```

Assigns the value 35 to the variable Y.

**LINE INPUT**

```
10 LINE INPUT "Address? ";C$
```

Reads an entire line of input, including delimiters, into variable C$.

**LOCATE**

```
10 LOCATE 10,15
```

Positions the cursor at row 10, column 15.

**LPOS**

```
10 PRINT LPOS(0)
```

Displays the position of the print head in the print buffer.

**LPRINT**

```
10 LPRINT "TESTING"
```

Prints the word TESTING on the printer.

**LPRINT USING**

```
10 LPRINT USING "###.##";456.7832
```

Prints the expression 456.7832 on the printer using the quoted format. Printed copy reads 456.78.

**MERGE**

```
10 MERGE "B:PROG2"
```

Merges the lines of PROG2 with the lines of the current program in memory. Duplicate line numbers are replaced by the lines in PROG2.

**MOTOR**

```
10 MOTOR 1
```

Turns on the motor of the cassette player

**ON ERROR**

```
10 ON ERROR GOTO 500
```

When any error occurs, branch to line 500.

**ON...GOSUB**

```
10 ON NUMBER GOSUB 220, 450, 130
```

If NUMBER=1, GOSUB 220; If NUMBER=2, GOSUB 450; If NUMBER=3, GOSUB 130.

**ON... GOTO**

```
10 ON NUMBER GOTO 220, 450, 130
```

Similar to ON GOSUB.

**ON KEY(n)**

```
10 ON KEY(10) GOSUB 1000
```

When function key 10 is pressed, branch to line 1000.

**ON PEN**

```
10 ON PEN GOSUB 2000
```

When light pen is detected, branch to line 2000.

**ON PLAY(n)**

```
10 ON PLAY(5) GOSUB 500
```

When music is playing in background and 5 notes remain in the buffer, branch to line 500.

**ON STRIG(n)**

```
10 ON STRIG(2) GOSUB 1000
```

When joystick trigger button is pressed, branch to line 1000.

**ON TIMER**

```
10 ON TIMER(30) GOSUB 500
```

When 30 seconds have elapsed, branch to line 500.

**OPTION BASE**

```
10 OPTION BASE 1
```

Sets the subscript of the lowest-numbered array element as 1.

**OUT**

```
10 OUT 980,2
```

Sends a value of 2 to machine port 980.

**PEEK**

```
10 PEEK (&H410)
```

Reads the value stored in memory location H410.

**PEN**

```
10 PRINT PEN(6)
```

Displays which row light pen was on when activated.

**PLAY**

```
10 PLAY "XA$;"
```

Plays music as instructed by contents of string XA$.

**PLAY(n)**

```
10 PRINT PLAY(0)
```

Displays the number of notes remaining in the music background buffer.

**POKE**

```
10 SCREEN 1
20 DEF SEG
30 POKE &HFE,2
```

Writes the value 2 into memory location HFE.

**POS**

```
10 IF POS(0)>60
   THEN PRINT CHR$(13)
```

If the cursor column position is beyond 60, then perform a carriage return.

**PRINT**

```
10 PRINT "ANYTHING"
```

Displays the word ANYTHING on the screen.

**PRINT USING**

```
10 PRINT USING "###.###";456.2341
```

Displays the expression 456.2341 on the screen using the quoted format. Screen displays 456.234.

**RANDOMIZE**

```
10 RANDOMIZE (20000)
```

Reseeds the random number generator with the number 20000 to produce a new sequence of numbers.

**READ**

```
10 READ A(I)
```

Reads a value from a DATA statement and assigns it to the variable A(I).

**RESTORE**

```
10 RESTORE
```

Causes the next READ statement to begin reading at the first DATA statement in the program.

**RESUME**

```
10 RESUME 120
```

Following an error recovery, causes execution to resume at line 120.

**RETURN**

```
10 RETURN
```

Causes program to return to the line following the GOSUB that initiated the branch to this subroutine.

**RND**

```
10 Y=INT(RND*7)
```

Yields random integers in the range 0 to 6.

**SCREEN**

```
10 PRINT SCREEN(5,10)
```

Displays the ASCII code for the character on the screen at row 5, column 10.

**SHELL**

```
10 SHELL
```

Loads DOS from BASIC; current program remains in memory.

**SOUND**

```
10 SOUND 220.000, 18
```

Produces a sound of 220 cps for a duration of 18 clock ticks (1 second).

**SPC**

```
10 PRINT SPC(20)
```

Prints 20 spaces.

**STICK**

```
10 P=STICK(0)
```

Returns x-coordinate of joystick A and assigns it to variable P.

**STRIG**

```
10 STRIG ON
```

Enables status of joystick buttons to be read.

**STRIG(n)**

```
10 STRIG(2) ON
```

Enables trapping of joystick button B1 by the ON STRIG(n) statement.

**SWAP**

```
10 SWAP X$,Y$
```

Exchanges the values of variables X$ and Y$.

**SYSTEM**

```
10 SYSTEM
```

Returns to DOS.

**TAB**

```
10 PRINT TAB(25)
```

Moves cursor to position 25.

**TIME$**

```
10 TIME$="10:23:00"
20 PRINT TIME$
```

Either sets or retrieves the current time.

**TIMER**

```
10 PRINT TIMER
```

Displays the number of seconds elapsed since midnight or since system reset.

**USR**

```
10 DEF USR0=&HF000
20 C=USR0(B/2)
```

Calls the machine language subroutine USR0 and supplies the argument (B/2).

**VARPTR**

```
10 PRINT VARPTR(X)
```

Displays the memory location of the variable X.

**VARPTR$**

```
10 PLAY "X"+VARPTR$(A$)
```

Returns a 3-byte string of the memory location of the variable A$. Is equivalent to PLAY "XA$;".

**WAIT**

```
10 WAIT 32,2
```

Causes program to wait until port 32 receives a bit-value of 1 in the second bit position.

**WHILE and WEND**

```
10 X=0
20 WHILE X=0
30 INPUT X
40 S=S+X
50 WEND
60 PRINT "SUM=";S
```

Causes the statements between WHILE and WEND to loop until a value for X is input.

**WIDTH**

```
10 WIDTH 40
```

Sets the screen width to 40 characters per line.

**WRITE**

```
10 A=80:B=90:C$="THE END"
20 WRITE A,B,C$
```

Similar to PRINT, but inserts commas and quotes. Example will display 80,90,THE END

## Communications

**COM(n)**

```
10 COM(1) ON
```

Enables trapping of communications adapter #1.

**ON COM(n)**

```
10 ON COM(1) GOSUB 1000
```

When activity is detected in communications adapter #1 branch to line 1000.

**OPEN "COM...**

```
10 OPEN "COM:" AS 1
```

Opens communications adapter #1 for communications.

## Conversions

**CDBL**

```
10 PRINT CDBL(A)
```

Converts the value of A to a double-precision number.

**CHR$**

```
10 PRINT CHR$(66)
```

Converts the value 66 to the equivalent ASCII code character which is a capital letter B.

**CINT**

```
10 PRINT CINT(45.67)
```

Converts the value 45.67 to the integer value 46.

**CSNG**

```
10 PRINT CSNG(45.34536789)
```

Converts the double-precision value 45.34536789 to the single-precision value 45.3453.

**CVD**

```
10 Y=CVD(N$)
```

Converts the 8-byte string variable N$ to a double-precision numeric variable.

**CVI**

```
10 X=CVI(X$)
```

Converts the 2-byte string variable X$ to an integer variable.

**CVS**

```
10 Y=CVS(Y$)
```

Converts the 4-byte string variable Y$ to a single-precision numeric variable.

**HEX$**

```
10 H$=HEX$(16)
```

Converts the decimal value 16 to the hexadecimal value 10.

**MKD$**

```
10 D$=MKD$(AMT)
```

Converts the value of the single-precision variable AMT to a string variable.

**MKI$**

```
10 R$=MKI$(STEP)
```

Converts the value of the integer variable STEP to a string variable.

**MKS$**

```
10 K$=MKS$(BALANCE)
```

Converts the single-precision variable BALANCE to a string variable.

**OCT$**

```
10 PRINT OCT$(24)
```

Converts the decimal value (24) to the equivalent octal value (30).

## Development

**AUTO**

```
10 AUTO 100,50
```

Automatically numbers each new program line using steps of 50 and beginning with line 100.

**CONT**

Continues program execution after a pause.

**DELETE**

Erases lines 40-100.

**EDIT**

Displays line 35 for editing.

**LIST**

Displays a listing of lines 20-200 of the program in memory.

**LLIST**

Prints a complete program listing.

**LOAD**

Loads the program PROG3 into memory.

**NAME**

Renames a file.

**NEW**

Clears current program and all variables from memory.

**REM**

Inserts a nonexecutable remark.

**RENUM**

Renumber the entire program to start with a line number of 1000 with each line incremented by 10.

**RUN**

```
10 RUN
```

Executes the current program in memory.

**SAVE**

```
10 SAVE "PROG4",A
```

Saves the program in memory using the name PROG4 as an ASCII file.

**STOP**

```
10 STOP
```

Halts program execution.

**TRON and TROFF**

```
10 TRON
20 REM code to be traced
30 TROFF
```

Turns program trace on and off.

## Files

**CHDIR**

```
10 CHDIR "\"
```

Changes directory to the root directory.

**CLOSE**

```
10 CLOSE
```

Closes all open files and devices.

**FIELD**

```
10 FIELD 1, 20 AS N$,
   30 AS A$
```

Allocates space for variables in random file buffer #1.

**FILES**

```
10 FILES "B:*.*"
```

Displays all files on drive B.

**GET**

```
10 OPEN "A:CUST" AS #1
20 GET 1
```

Reads a record from random file #1.

**INPUT#**

```
10 OPEN "O",#1,"DATA"
20 INPUT#1,INFO$
```

Reads the variable INFO$ from the open file.

**KILL**

```
10 KILL"B:DATA.BAS"
```

Erases the file on drive B called DATA.BAS.

**LINE INPUT#**

```
10 LINE INPUT#1, ADDRESS$
```

Reads an entire line from the open file and assigns it to the variable ADDRESS$.

**LOC**

```
10 PRINT LOC(1)
```

Displays the current position in the file opened as #1.

**LOF**

```
10 PRINT LOF(1)
```

Displays length of the file opened as #1.

**LSET and RSET**

```
10 LSET N$=NA$
```

Moves the contents of NA$ into the random file buffer named N$ and left-justifies that field.

**MKDIR**

```
10 MKDIR "SALES"
```

Creates a directory called SALES.

**OPEN**

```
10 OPEN "O",#1,"DATA"
```

Opens a device or file called DATA to receive output as file #1.

**PRINT #**

```
10 PRINT #1,DATE$;TIME$
```

Writes the variables DATE$ and TIME$ to the sequential file open as #1.

**PRINT # USING**

```
10 PRINT #1 USING "###.##";AMOUNT
```

Writes the value of the variable AMOUNT to the sequential file open as #1 using the quoted format.

**PUT**

```
10 PUT #1
```

Writes a record that has been LSET or RSET into random buffer #1, a random file.

**RESET**

```
10 RESET
```

Closes all open disk files.

**RMDIR**

```
10 RMDIR "SALES"
```

Removes the directory called SALES.

**WRITE #**

```
10 WRITE #1,NAME$,AGE$
```

Writes the variables NAME$ and AGE$ to the sequential file opened as #1. Automatically inserts commas between items and quotes around strings.

## Graphics

### CIRCLE

```
10 CIRCLE (160, 100),50
```

Draws a circle with center at X=160, Y=100 and radius=50.

### COLOR

```
10 COLOR 9,0
```

Sets background color to light blue and selects palette 0 in SCREEN 1.

### DRAW

```
10 SCREEN 1
20 DRAW "U20 R20 D20 L20
```

Draws a box 20 units wide and high.

### GET

```
10 GET (10,10)-(20,20), "PICTURE"
```

Saves the contents of the screen within the rectangle whose opposite corners are (10,10) and (20,20) into an array named PICTURE.

### LINE

```
10 LINE (1,1)-(5,5),2,B
```

Draws a box using color attribute 2 whose opposite corners are (1,1) and (5,5).

### PAINT

```
10 PAINT (15,15),2
```

Fills the interior of the box in the Line example with color attribute 2 starting at point (15,15).

### PMAP

```
10 WINDOW (-1,-1)-(1,1)
20 PMAP(1,0):PMAP(0,1)
```

Translates the center point of the WINDOW from world coordinates of (0,0) to physical coordinates of (160,100).

### POINT

```
10 C=POINT(15,15)
```

Reads the color attribute of the point at screen location (15,15) and assigns it to variable C.

### PSET and PRESET

```
10 PSET (15,15)
20 PRESET (15,15)
```

PSET draws a point at coordinates (15,15) in the foreground color. PRESET removes the same point.

### PUT

```
10 PUT (40,40),"PICTURE"
```

Takes the bit image that was saved with GET in the array called PICTURE and puts it on the screen with the upper left corner of the image at location (40,40).

### SCREEN

```
10 SCREEN 1,0
```

Switches screen to medium resolution graphics mode and enables color burst.

### VIEW

```
10 VIEW (5,5)-(100,100)
```

Defines a rectangular section of the screen whose opposite corners are (5,5) and (100,100) as a viewport into which the contents of WINDOW are mapped.

### WINDOW

```
10 WINDOW (-6,-6)-(6,6)
20 CIRCLE (4,4),5,1
```

Produces a circle of radius=5 that fills most of the screen because WINDOW redefined the coordinates of the screen to range from only +6 to −6.

## Math Functions

### ABS

```
10 PRINT ABS(7*(-5))
```

Displays the absolute value (35) of the stated expression.

### ATN

```
10 PRINT ATN(5)
```

Displays the arctangent in radians of the number 5.

### COS

```
10 PRINT COS(3.14)
```

Displays the cosine in radians of an angle equal to 3.14 radians.

### EXP

```
10 PRINT EXP(1)
```

Displays the value of the number e raised to the first power.

### FIX

```
10 PRINT FIX(45.67)
```

Displays the integer digits (45) of the number 45.67.

### INT

```
10 PRINT INT(-2.89)
```

Displays the largest integer (−3) that is less than or equal to (−2.89).

### LOG

```
10 PRINT LOG(45/7)
```

Displays the natural logarithm (1.860752) of the expression (45/7).

**SGN**

```
1Ø PRINT SGN(X)
```

Displays the sign of the variable X.

**SIN**

```
1Ø PRINT SIN(3.14)
```

Displays the sine in radians of an angle equal to 3.14 radians.

**SQR**

```
1Ø PRINT SQR(81)
```

Displays the square root (9) of the value 81.

**TAN**

```
1Ø PRINT TAN(3.14)
```

Print the tangent in radians of an angle equal to 3.14 radians.

## String Functions

**INSTR**

```
1Ø A$="ABCDEFG"
2Ø B$="B"
PRINT INSTR(A$,B$)
```

Searches for the first occurrence of B$ within A$ and displays the position (2) where the match begins.

**LEFT$**

```
1Ø A$="BASIC EXAMPLE"
2Ø PRINT LEFT$(A$,2)
```

Displays the leftmost 2 characters (BA) of the string A$.

**LEN**

```
1Ø X$="IBM PC"
2Ø PRINT LEN(A$)
```

Displays the length (6) of the string X$.

**MID$**

```
1Ø EX$="QRSTUVW"
2Ø PRINT MID$(EX$,2,3)
```

Starting with the second character, displays the next 3 characters (RST).

**RIGHT$**

```
1Ø SAM$="SAMPLE"
2Ø PRINT RIGHT$(SAM$,3)
```

Displays the rightmost 3 characters (PLE) of string SA.

**SPACE$**

```
1Ø SP$=SPACE$(25)
2Ø PRINT SP$;"TEST"
```

Prints a string of 25 spaces; then prints the word TEST.

**STR$**

```
1Ø PRINT LEN(STR$([illegible]))
```

Treats the numeric expression 321 as a string expression and displays the length.

**STRING$**

```
1Ø X$=STRING$(1Ø,45)
2Ø PRINT X$;"TEST";X$
```

Displays a string consisting of ASCII character 45 (−) repeated 1Ø times; then prints the word TEST; then displays 1Ø more dashes.

**VAL**

```
1Ø PRINT VAL("29 EAST AVE.")
```

Displays the numeric value of the given string (29).

# 1-6 THE APPLIED NUMERICAL METHODS SOFTWARE THAT ACCOMPANIES THIS BOOK, AND THE PERSONAL COMPUTER SYSTEM USED FOR DEVELOPING IT

The personal computer system used in developing all the software and most of the figures that appear in this book was the IBM AT, with the Hewlett-Packard LaserJet printer, the Hewlett-Packard 7475A plotter, and the Pencept Penpad digitizer (Fig. 1-16). An IBM PC was used in the early stages of this work.

All the computer programs in this book have been written in advanced BASIC and can be run using the advanced BASIC interpreter (BASICA, version 2.0 or higher). The diskette that accompanies the book contains all

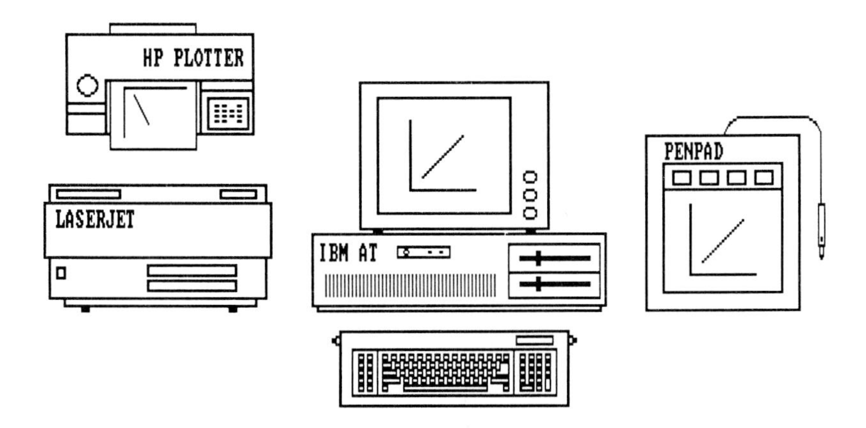


**Figure 1-16** The personal computer system used in writing this book was the IBM AT with the HP Laserjet Printer, the HP 7475A plotter, and the Pencept Penpad digitizer.

these programs. We will refer to this as the "Applied Numerical Methods" diskette. The programs on the diskette are divided into two groups, as shown in Table 1-9 on p. 64. The first group of programs resides in the root directory of the diskette; these are the computer programs for the worked examples of Chaps. 2 through 6. The second group of programs resides in a subdirectory called NLR; these constitute the Nonlinear Regression program of Chap. 7. A brief description of each program and data file is given in Table 1-10 on pp. 65–66. A more detailed discussion of the programs is covered in the appropriate chapter.

The Applied Numerical Methods diskette is not a system diskette (it does not contain the files IBMDOS.COM, IBMBIO.COM, COMMAND.COM, or the BASICA.COM interpreter); therefore, it cannot be used alone to boot the computer. It is recommended that each of the two groups of programs be copied to a separate, newly formatted system diskette. If your computer has a fixed disk, you may copy the two groups of programs to two different subdirectories of the fixed disk. Follow instructions A or B below, whichever is applicable to your case.

## A. Copying the Programs to Two Separate System Diskettes

1. Use DOS 2.00, or higher, to start your computer.
2. Format two new diskettes using the /S option:

```
FORMAT B:/S
```

3. Copy the file BASICA.COM from the DOS diskette to the newly formatted diskette:

```
COPY BASICA.COM B:
```

(Repeat this for each diskette.)

4. Place the Applied Numerical Methods diskette in drive A and one of the newly formatted diskettes in drive B.
5. Copy the files in the root directory of drive A to drive B:

```
COPY A:*.* B:*.*
```

When finished copying, remove the diskette from drive B and label it "Programs for Chaps. 2 to 6."

6. Place the second newly formatted diskette in drive B.
7. Copy the files in the NLR subdirectory of drive A to drive B:

```
COPY A:\NLR\*.* B:*.*
```

When finished copying, remove the diskette from drive B and label it "Nonlinear Regression Program for Chap. 7."

8. Place the original Applied Numerical Methods diskette in a safe place. Work with the newly copied diskettes. Both these diskettes are self-booting through their individual AUTOEXEC.BAT files. The latter contain the following commands:

```
DATE
TIME
BASICA MENU
```

The first two commands are for entering the correct date and time. These steps may be skipped if your computer is equipped with a battery-operated clock/calendar (you may want to modify the AUTOEXEC.BAT file to load any drivers which might be needed for your system). The command BASICA MENU loads the BASICA interpreter and executes the program MENU.BAS. This menu program displays the directory of the diskette and instructs you how to execute the programs.

## B. Copying the Programs to the Fixed Disk

1. Start your computer from the fixed disk, using DOS 2.00 or higher.
2. Create two subdirectories, called CHAPT26 and NLR, on the fixed disk:

```
MD\CHAPT26
MD\NLR
```

3. Copy the file BASICA.COM from the root directory of the fixed disk (or from the DOS diskette) to each of the new subdirectories:

```
COPY C:\BASICA.COM C:\CHAPT26\*.*
COPY C:\BASICA.COM C:\NLR\*.*
```

4. Place the Applied Numerical Methods diskette in drive A.
5. Copy the files in the root directory of drive A to the CHAPT26 subdirectory of the fixed disk:

```
COPY A:*.* C:\CHAPT26\*.*
```

6. Copy the files in the NLR subdirectory of drive A to the NLR subdirectory of the fixed disk:

```
COPY A:\NLR\*.* C:\NLR\*.*
```

7. Place the original Applied Numerical Methods diskette in a safe place. Work from the fixed disk. To run any of the program from Chaps. 2 to 6, first change to the CHAPT26 subdirectory:

```
CD\CHAPT26
```

and then execute the AUTOEXEC.BAT file:

```
AUTOEXEC
```

To run the Nonlinear Regression Program, change to the NLR subdirectory:

```
CD\NLR
```

and then execute the AUTOEXEC.BAT file:

```
AUTOEXEC
```

**Table 1-9 Directory listing of the Applied Numerical Methods diskette**

| Programs for Chaps. 2 to 6<br>Root directory | Nonlinear Regression Program for Chap. 7<br>Path:\NLR |
|---|---|
| AUTOEXEC.BAT | AUTOEXEC.BAT |
| MENU .BAS | READ .ME |
| ROOT .BAS | MENU .BAS |
| POLY .BAS | MAIN .BAS |
| GRAEFFE .BAS | INPUT .BAS |
| NEWTON .BAS | COMPAR .BAS |
| GAUSS .BAS | DATIN .BAS |
| JORDAN .BAS | GJLNB .BAS |
| SEIDEL .BAS | JSRKB .BAS |
| EIGEN .BAS | MUS .BAS |
| QR .BAS | PHLPA .BAS |
| INTEGR .BAS | SA12V .BAS |
| ODE .BAS | SETPAR .BAS |
| BOUNDARY.BAS | SUMS .BAS |
| PEN .EQU | SUMS1 .BAS |
| ELLIPTIC .BAS | OUTPUT .BAS |
| PARABOL .BAS | CONS .DAT |
| | DATA .DAT |
| | DIST .DAT |
| | GUESS .DAT |
| | INITCOND .DAT |
| | OUTPUT .DAT |
| | MODEL .EQU |

**Table 1-10 Description of programs on the Applied Numerical Methods diskette**

| Program name | Example | Description |
|---|---|---|
| | | I. Programs for Chaps. 2 to 6 |
| AUTOEXEC.BAT | — | This program self-boots the diskette, loads the BASICA interpreter, and runs the program MENU.BAS. |
| MENU.BAS | — | Displays the directory of the available BASIC language programs for Chaps. 2 to 6 and gives instructions on how to load and run these programs. |
| ROOT.BAS | 2-1 | Linear interpolation and Newton-Raphson methods applied to the solution of nonlinear functions. |
| POLY.BAS | 2-2 | Newton-Raphson with synthetic division applied to $n$th-degree polynomials and to the Beattie-Bridgeman equation. |
| GRAEFFE.BAS | 2-3 | Graeffe's root-squaring method for real and complex roots of polynomials. |
| NEWTON.BAS | 2-4 | Newton's method for real and complex roots of polynomials and transfer functions. |
| GAUSS.BAS | 3-1 | Gauss elimination method for simultaneous linear algebraic equations. |
| JORDAN.BAS | 3-2 | Gauss-Jordan reduction method for simultaneous linear algebraic equations and matrix inversion. |
| SEIDEL.BAS | 3-3 | Gauss-Seidel substitution method for diagonal systems of linear algebraic equations. |
| EIGEN.BAS | 3-4 | Calculation of eigenvalues and eigenvectors using the Faddeev-Leverrier/Newton-Raphson/Gauss methods. |
| QR.BAS | 3-5 | Calculation of eigenvalues and eigenvectors using elementary similarity transformations to convert the matrix to Hessenberg form, and the $QR$ algorithm with plane rotations. |
| INTEGR.BAS | 5-6 | Integration formulas: Trapezoidal, Simpson's $\frac{1}{3}$, and Simpson's $\frac{3}{8}$ rules. |
| ODE.BAS | 5-7 | Fourth-order Runge-Kutta and Euler Predictor-Corrector methods for integrating simultaneous ordinary differential equations. |
| BOUNDARY.BAS | 5-8 | Boundary-value problems: The Newton method. |
| PEN.EQU | 5-8 | Contains the differential and algebraic equations needed for the solution of Example 5-8. |
| ELLIPTIC.BAS | 6-1 | Elliptic partial differential equations. |
| PARABOL.BAS | 6-2 | Parabolic partial differential equations. |
| | | II. Nonlinear Regression Program for Chap. 7 |
| AUTOEXEC.BAT | — | This program self-boots the diskette, loads the BASICA interpreter, and runs the program MENU.BAS. |
| READ.ME | — | This is a text file which contains a brief description of the Nonlinear Regression program. It can be read from DOS using the command TYPE READ.ME. |

(*continued on next page*)

**Table 1-10** (*Continued*)

| Program name | Example | Description |
|---|---|---|
| MENU.BAS | — | Displays the directory of the Nonlinear Regression program for Chap. 7, and executes that program by running MAIN.BAS. |
| MAIN.BAS | 7-1 | The main program calls all the other programs and performs the Marquardt method. |
| INPUT.BAS | 7-1 | This program enters all the data, constants, and equations for the nonlinear regression. |
| COMPAR.BAS | 7-1 | Compares the variance due to lack-of-fit with the variance due to experimental error, and performs the F-tests. |
| DATIN.BAS | 7-1 | Determines the variance of the experimental data and weighting factors. |
| GJLNB.BAS | 7-1 | Uses the Gauss-Jordan reduction method to obtain the parameter increment vector. |
| JSRKB.BAS | 7-1 | Integration routine which utilizes the fourth-order Runge-Kutta method. This routine contains the state and variational equations which are specified by the user through the INPUT.BAS program. |
| MUS.BAS | 7-1 | Calculates the residuals between the group means of the experimental points and the predicted curve. |
| PHLPA.BAS | 7-1 | A matrix inversion routine to determine the inverse of the $A$-transpose-$A$ matrix. |
| SA12V.BAS | 7-1 | A statistical analysis routine that performs a series of tests on the predicted parameters. |
| SETPAR.BAS | 7-1 | Searches the sum of squares space at the end of the axes of the hyperellipsoid. |
| SUMS.BAS | 7-1 | This routine calculates the residuals between the experimental and the predicted points, and the sum of squared residuals. |
| SUMS1.BAS | 7-1 | Similar to SUMS. |
| OUTPUT.BAS | 7-1 | Prints tables and draws graphs of final values of variables and data. |
| CONS.DAT | 7-1 | Random access file for all nonlinear regression constants. |
| DATA.DAT | 7-1 | Random access file that contains all the experimental data and weights. |
| DIST.DAT | 7-1 | Read-only file that contains the $F$-distribution table. |
| GUESS.DAT | 7-1 | Sequential file that contains the initial parameter guesses. |
| INITCOND.DAT | 7-1 | Sequential file for all the initial conditions of the differential equations. |
| OUTPUT.DAT | 7-1 | Sequential output file for the matrix which contains the profiles of the state and variational variables over the entire integration interval. |
| MODEL.EQU | 7-1 | File containing the state and variational equations. |

# REFERENCES


1. Sanders, D. H.: *Computers Today*, 2d ed., McGraw-Hill Book Company, New York, 1985.
2. Boyce, J. C.: *Microprocessor and Microcomputer Basics*, Prentice-Hall, Inc., Englewood Cliffs, N.J., 1979.
3. Toong, H. D., and Gupta, A.: "Personal Computers," *Scientific American*, December 1982, p. 87.
4. *The Peter Norton Programmer's Guide to the IBM PC*, Microsoft Press, Bellevue, Wash., 1985.
5. Khambata, A. J.: *Microprocessors/Microcomputers, Architecture, Software and Systems*, John Wiley & Sons, Inc., New York, 1982.
6. Wells, P.: "The 80286 Microprocessor," *BYTE*, November 1984, p. 231.
7. Sargent III, M., and Shoemaker, R. L.: *The IBM Personal Computer from the Inside Out*, Addison-Wesley Publishing Company, Reading, Mass., 1984.
8. Flores, I., and Terry, C.: *Microcomputer Systems*, Van Nostrand Reinhold Company, New York, 1982.
9. Frates, J., and Moldrup, W.: *Computers and Life—An Interactive Approach*, Prentice-Hall, Inc., Englewood Cliffs, N.J., 1983.
10. Stillman, R.: "Lecture on Operating Systems," given in the course *Personal Computers*, A. Constantinides (course director), The Center for Professional Advancement, East Brunswick, N.J., May 1984.
11. Field, T.: "Installable Device Drivers for PC-DOS 2.00," *BYTE*, November 1983, p. 188.
12. "Disk Operating System Version 3.10," 1st ed., IBM Corp., Boca Raton, Fla., February, 1985.
13. *BASIC Handbook, General Programming Information*, 3d ed., IBM Corp., Boca Raton, Fla., May 1984.
14. *BASIC Reference*, 3d ed., IBM Corp., Boca Raton, Fla., May 1984.
15. *BASIC Quick Reference*, 1st ed., IBM Corp., Boca Raton, Fla., May 1984.
16. Schneider, D. I.: *Handbook of BASIC for the IBM PC*, Brady Communications Company, Inc., Bowie, Md., 1985.
17. Goldstein, L. J., and Goldstein, M.: *IBM PC, An Introduction to the Operating System, Basic Programming, and Applications*, Brady Communications Company, Inc., Bowie, Md., 1984.
18. Goldstein, L.J.: *Advanced BASIC and Beyond for the IBM PC*, Brady Communications Company, Inc., Bowie, Md., 1984.
19. Waite, M., and Morgan, C. L.: *Graphics Primer for the IBM PC*, Osborne/McGraw-Hill, Berkeley, Calif., 1983.

CHAPTER

# TWO

# NUMERICAL SOLUTION OF NONLINEAR ALGEBRAIC EQUATIONS

## 2-1 INTRODUCTION

Many problems in engineering and science require the solution of nonlinear algebraic equations. Several examples of such problems drawn from the field of chemical engineering and from other application areas are discussed in this section. The methods of solution are developed in the remaining sections of the chapter and specific examples of the solutions are demonstrated using the personal computer.

In thermodynamics, the pressure-volume-temperature relationships of real gases are described by the equations of state. There are several semitheoretical or empirical equations, such as the Redlich-Kwong, the Beattie-Bridgeman, and the Benedict-Webb-Rubin equations, which have been used extensively in chemical engineering [1]. For example, the Beattie-Bridgeman equation of state has the form

$$PV = RT + \frac{\beta}{V} + \frac{\gamma}{V^2} + \frac{\delta}{V^3} \tag{2-1}$$

where $P$, $V$, and $T$ are the pressure, specific volume, and temperature, respectively. $R$ is the gas constant, and $\beta$, $\gamma$, $\delta$ are empirical functions of temperature, specific for each gas. Equation (2-1) is a fourth-degree polynomial in $V$ and can be easily rearranged into the canonical form for a polynomial, which is

$$PV^4 - RTV^3 - \beta V^2 - \gamma V - \delta = 0 \tag{2-2}$$

Therefore, the problem of finding the specific volume of a gas at a given temperature and pressure reduces to the problem of finding the appropriate root of a polynomial equation.

In the calculations for multicomponent separations, it is often necessary to estimate the minimum reflux ratio of a multistage distillation column. A method developed for this purpose by Underwood [2], and described in detail by Treybal [3], requires the solution of the equation

$$\sum_{j=1}^{n} \frac{\alpha_j z_{jF} F}{\alpha_j - \phi} - F(1 - q) = 0 \tag{2-3}$$

where $F$ is the molar feed flow rate, $n$ is the number of components in the feed, $z_{jF}$ is the mole fraction of each component in the feed, $q$ is the quality of the feed, $\alpha_j$ is the relative volatility of each component at average column conditions, and $\phi$ is the root of the equation. The feed flow rate, composition, and quality are usually known, and the average column conditions can be approximated. Therefore, $\phi$ is the only unknown in Eq. (2-3). Since this equation is a polynomial in $\phi$ of degree $n$, there are $n$ possible values of $\phi$ (roots) that satisfy the equation.

The Fanning friction factor $f$ for turbulent flow of an incompressible fluid in a smooth pipe is given by the nonlinear equation

$$\sqrt{\frac{2}{f}} = \frac{1}{k} \ln\left(N_{\text{Re}} \sqrt{\frac{f}{8}}\right) + B - A \tag{2-4}$$

where $A$, $B$, and $k$ are constants and $N_{\text{Re}}$ is the Reynolds number [4]. This equation does not readily rearrange itself into a polynomial form; however, it can be arranged so that all the nonzero terms are on the left side of the equation as follows:

$$\sqrt{\frac{2}{f}} - \frac{1}{k} \ln\left(N_{\text{Re}} \sqrt{\frac{f}{8}}\right) - B + A = 0 \tag{2-5}$$

The method of differential operators is applied in finding analytical solutions of $n$th-order linear homogeneous differential equations. The general form of an $n$th-order linear homogeneous differential equation is

$$a_n \frac{d^n y}{dx^n} + a_{n-1} \frac{d^{n-1} y}{dx^{n-1}} + \cdots + a_1 \frac{dy}{dx} + a_0 y = 0 \tag{2-6}$$

By defining $D$ as the differentiation with respect to $x$

$$D = \frac{d}{dx} \tag{2-7}$$

Eq. (2-6) can be written as

$$[a_n D^n + a_{n-1} D^{n-1} + \cdots + a_1 D + a_0] y = 0 \tag{2-8}$$

where the bracketed term is called the differential operator. In order for Eq. (2-8) to have a nontrivial solution, the differential operator must be equal to zero:

$$a_n D^n + a_{n-1} D^{n-1} + \cdots + a_1 D + a_0 = 0 \tag{2-9}$$

This, of course, is a polynomial equation in $D$ whose roots must be evaluated in order to construct the complementary solution of the differential equation.

The field of process dynamics and control often requires the location of the roots of transfer functions which usually have the form of polynomial equations. In kinetics and reactor design, the simultaneous solution of rate equations and energy balances results in mathematical models of simultaneous nonlinear and transcendental equations. Methods of solution for these and other such problems are developed in this chapter.

## 2-2 TYPES OF ROOTS AND THEIR APPROXIMATION

All the nonlinear equations presented in Sec. 2-1 are of the general form

$$f(x) = 0 \tag{2-10}$$

where $x$ is a single variable which can have multiple values (roots) that satisfy this equation. The function $f(x)$ may assume a variety of nonlinear functionalities ranging from that of a polynomial equation whose canonical form is

$$f(x) = a_n x^n + a_{n-1} x^{n-1} + \cdots + a_1 x + a_0 = 0 \tag{2-11}$$

to the transcendental equations, which involve trigonometric, exponential, and logarithmic terms. The roots of these functions could be

1. Real and distinct
2. Real and repeated
3. Complex conjugates
4. A mixture of all the above

The real parts of the roots may be positive or negative.

Figure 2-1 graphically demonstrates all the above cases using fourth-degree polynomials. Figure 2-1*a* is a plot of the polynomial

$$x^4 + 6x^3 + 7x^2 - 6x - 8 = 0 \tag{2-12}$$

which has four real and distinct roots at $-4$, $-2$, $-1$, and 1, as indicated by the intersections of the function with the $x$ axis. Figure 2-1*b* is a graph of the polynomial

$$x^4 + 7x^3 + 12x^2 - 4x - 16 = 0 \tag{2-13}$$

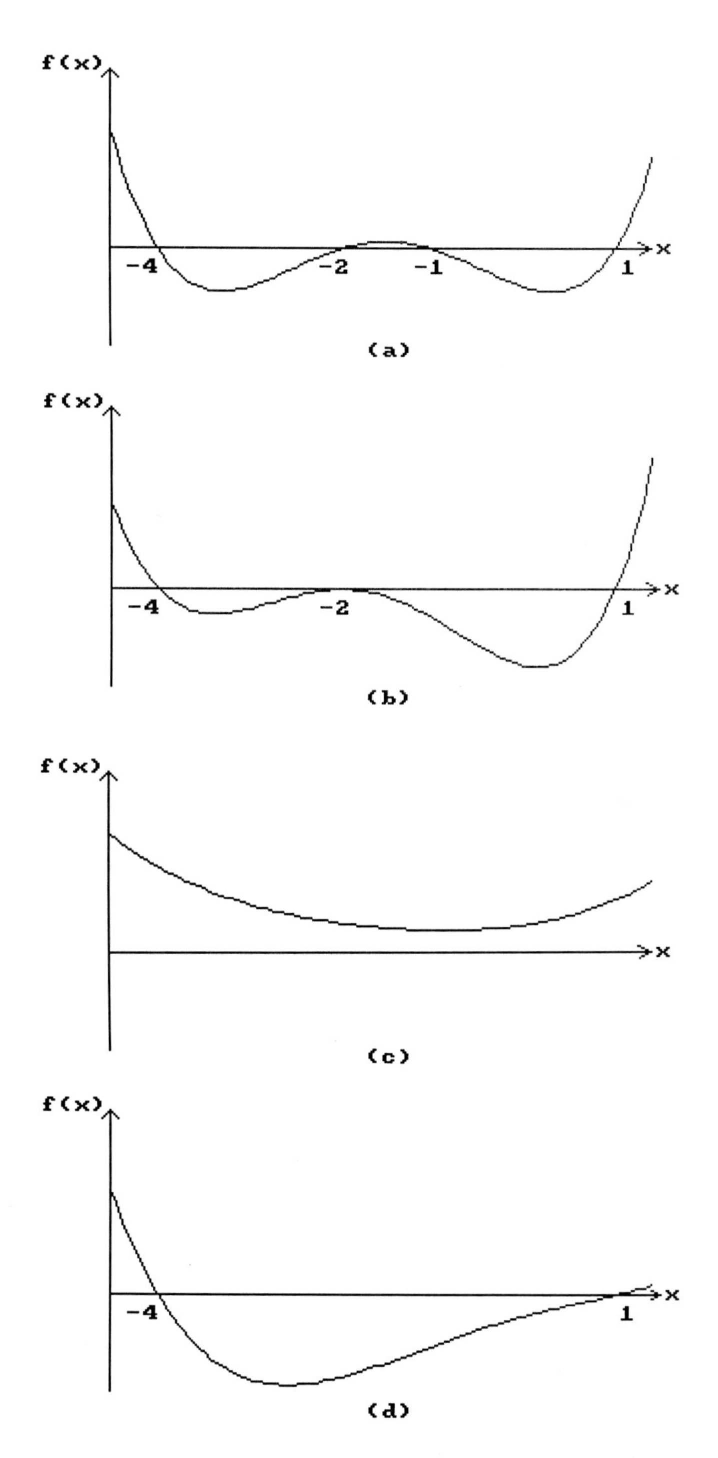


**Figure 2-1** Roots of fourth-degree polynomial equations: (a) four real distinct, (b) two real and two repeated, (c) four complex, (d) two real and two complex.

which has two real and distinct roots at −4 and 1, and two real and repeated roots at −2. The point of tangency with the $x$ axis indicates the presence of the repeated roots. At this point $f(x) = 0$ and $f'(x) = 0$. Figure 2-1$c$ is a plot of the polynomial

$$x^4 - 6x^3 + 18x^2 - 30x + 25 = 0 \tag{2-14}$$

which has only complex roots at $1 \pm 2i$ and $2 \pm i$. In this case, no intersection with the $x$ axis of the cartesian coordinate system occurs since all the roots are located in the complex plane. Finally, Figure 2-1$d$ demonstrates the presence of two real and two complex roots with the polynomial

$$x^4 + x^3 - 5x^2 + 23x - 20 = 0 \tag{2-15}$$

whose roots are −4, 1, and $1 \pm 2i$. As expected, the function crosses the $x$ axis only at two positions: −4 and 1.

The roots of an $n$th-degree polynomial such as Eq. (2-11) may be verified using Newton's relations [5] which are

Newton's 1st relation:

$$\sum_{i=1}^{n} x_i = -\frac{a_{n-1}}{a_n} \tag{2-16}$$

where $x_i$ are the roots of the polynomial.

Newton's 2nd relation:

$$\sum_{i,\, j=1}^{n} x_i x_j = \frac{a_{n-2}}{a_n} \tag{2-17}$$

Newton's 3rd relation:

$$\sum_{i,\, j,\, k=1}^{n} x_i x_j x_k = -\frac{a_{n-3}}{a_n} \tag{2-18}$$

$$\vdots$$

Newton's $n$th relation:

$$x_1 x_2 x_3 \ldots x_n = (-1)^n \frac{a_0}{a_n} \tag{2-19}$$

where $i \neq j \neq k \neq \cdots$ for all the above equations which contain products of roots.

In certain problems it may be necessary to locate all the roots of the equation, including the complex roots. This is the case in finding the zeros and poles of transfer functions in process control applications, and in formulating the analytical solution of linear $n$th-order differential equations. On the other hand, different problems may require the location of only one of the roots. For example, in the solution of the equation of state, the positive real root is the one of interest. In any case, the physical constraints of the problem may dictate the feasible region of search where only a subset of the total number of roots may be located. In addition, the physical characteristics of the problem may provide an approximate value of the desired root.

The most effective way of finding the roots of nonlinear equations is to devise iterative algorithms which start at an initial estimate of a root and converge to the exact value of the desired root in a finite number of steps. Once a root is located, it may be removed by synthetic division if the equation is of the polynomial form. Otherwise, convergence on the same root may be avoided by initiating the search for subsequent roots in different regions of the feasible space.

For equations of the polynomial form, Descartes' rule of sign may be used to determine the number of positive and negative roots. This rule states: The number of positive roots is equal to the number of sign changes in the coefficients of the equation (or less than that by an even integer); the number of negative roots is equal to the number of sign repetitions in the coefficients (or less than that by an even integer). Zero coefficients are counted as positive [5]. The purpose of the qualifier "less than that by an even integer" is to allow for the existence of conjugate pairs of complex roots. The reader is encouraged to apply Descartes' rule to Eqs. (2-12) to (2-15) to verify the results already shown.

If the problem to be solved is a purely mathematical one, i.e., the model whose roots are being sought has no physical origin, then brute-force methods would have to be used to establish approximate starting values of the roots for the iterative techniques. Two categories of such methods will be mentioned here. The first one is a truncation method applicable to equations of the polynomial form. For example, the following polynomial

$$a_4x^4 + a_3x^3 + a_2x^2 + a_1x + a_0 = 0 \tag{2-20}$$

may have its lower-powered terms truncated

$$a_4x^4 + a_3x^3 \cong 0 \tag{2-21}$$

to yield an approximation of one of the roots

$$x \cong -\frac{a_3}{a_4} \tag{2-22}$$

Alternatively, if the higher-powered terms are truncated

$$a_1x + a_0 \cong 0 \tag{2-23}$$

the approximate root is

$$x \cong -\frac{a_0}{a_1} \tag{2-24}$$

This technique applied to the Beattie-Bridgeman equation [Eq. (2-2)] results in

$$V \cong \frac{RT}{P} \tag{2-25}$$

This, of course, is the well-known *ideal gas law*, which is an excellent approximation of the pressure-volume-temperature relationship of real gases at low pressures. On the other end of the polynomial, truncation of the higher-powered terms results in

$$V \cong -\frac{\delta}{\gamma} \tag{2-26}$$

giving a value of $V$ very close to zero. In this case, the physical considerations of the problem dictate that Eq. (2-25) should be used to initiate the iterative search technique for the real root.

Another method of locating initial estimates of the roots is to scan the entire region of search by small increments and to observe the steps in which a change of sign in the function $f(x)$ occurs. This signals that the function $f(x)$ crosses the $x$ axis within the particular step. The scan method may be a rather time-consuming procedure for polynomials whose roots lie in a large region of search. A variation of this search is the method of bisection which divides the interval of search by 2 and always retains that half of the search interval in which the change of sign has occurred. When the range of search has been narrowed down sufficiently, a more accurate search technique would then be applied within that step in order to refine the value of the root.

More efficient methods based on linear interpolation of the function (*method of false position*) and the tangential descent of the function (Newton-Raphson method) will be described in the next two sections of this chapter.

## 2-3 THE METHOD OF LINEAR INTERPOLATION (METHOD OF FALSE POSITION)

The simplest iterative root-finding technique is one based on linear interpolation between two points on the function which have been found by a scan to lie on either side of a root. For example, $x_1$ and $x_2$ in Fig. 2-2*a* are positions on opposite sides of the root $x^*$ of the nonlinear function $f(x)$. The points $(x_1, f(x_1))$ and $(x_2, f(x_2))$ are connected by a straight line, which we will call a chord, whose equation is

$$y(x) = ax + b \tag{2-27}$$

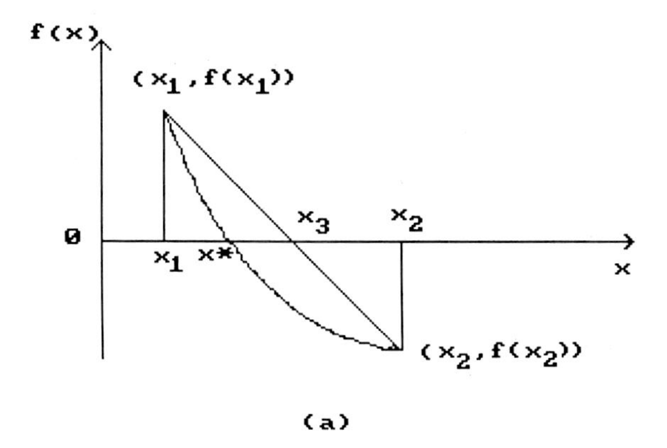


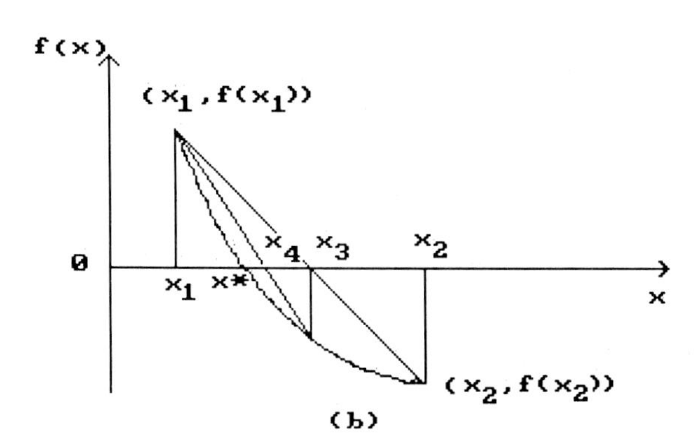


**Figure 2-2** Method of linear interpolation.

Since this chord passes through the two points $(x_1, f(x_1))$ and $(x_2, f(x_2))$, its slope is

$$a = \frac{f(x_2) - f(x_1)}{x_2 - x_1} \tag{2-28}$$

and its $y$ intercept is

$$b = f(x_1) - ax_1 \tag{2-29}$$

Equation (2-27) then becomes

$$y(x) = \left[\frac{f(x_2) - f(x_1)}{x_2 - x_1}\right]x + \left\{f(x_1) - \left[\frac{f(x_2) - f(x_1)}{x_2 - x_1}\right]x_1\right\} \tag{2-30}$$

Locating $x_3$ using Eq. (2-30), where $y(x_3) = 0$,

$$x_3 = x_1 - \frac{f(x_1)}{[f(x_2) - f(x_1)]/(x_2 - x_1)} \tag{2-31}$$

Note that for the shape of curve chosen on Fig. 2-2, $x_3$ is nearer to the root $x^*$ than either $x_1$ or $x_2$. This, of course, will not always be the case with all functions. Discussion of the criteria for convergence will be given in the next section.

Now repeating the above operation and connecting the points $(x_1, f(x_1))$ and $(x_3, f(x_3))$ with a new chord, as shown in Fig. 2-2*b*, we obtain the value of $x_4$:

$$x_4 = x_1 - \frac{f(x_1)}{[f(x_3) - f(x_1)]/(x_3 - x_1)} \tag{2-32}$$

which is nearer to the root than $x_3$. Successive application of the general formula

$$x_{n+1} = x_1 - \frac{f(x_1)}{[f(x_n) - f(x_1)]/(x_n - x_1)} \tag{2-33}$$

may result in improved approximations of the root of the function. This method is known by several names: method of chords, linear interpolation, false position (*regula falsi*). Its simplicity of calculation (no need for evaluating derivatives of the function) gave it its popularity in the early days of numerical computations. However, its accuracy and speed of convergence are hampered by the choice of $x_1$ which forms the pivot point for all subsequent iterations.

This method will be demonstrated in Example 2.1 and its efficiency will be compared to that of the Newton-Raphson method which is described in the next section.

## 2-4 THE NEWTON-RAPHSON METHOD

The best known, and possibly the most widely used, technique for locating roots of nonlinear algebraic equations is the Newton-Raphson method. This method is based on a Taylor series expansion of the nonlinear function $f(x)$ around an initial estimate $(x_1)$ of the root:

$$f(x) = f(x_1) + f'(x_1)(x - x_1) + \frac{f''(x_1)(x - x_1)^2}{2!} + \frac{f'''(x_1)(x - x_1)^3}{3!} + \cdots \tag{2-34}$$

Since what is being sought is the value of $x$ which forces the function $f(x)$ to assume a zero value, the left side of Eq. (2-34) is set to zero and the resulting equation is solved for $x$. However, the right-hand side is an infinite series. Therefore a finite number of terms must be retained and the remaining terms must be truncated. Retaining only the first two terms on the right-hand side of the Taylor series is equivalent to linearizing the function $f(x)$. This operation results in

$$x = x_1 - \frac{f(x_1)}{f'(x_1)} \tag{2-35}$$

i.e., the value of $x$ is calculated from $x_1$ by correcting this initial guess by $f(x_1)/f'(x_1)$. The geometrical significance of this correction is shown in Fig. 2-3*a*. The value of $x$ is obtained by moving from $x_1$ to $x$ in the direction of the tangent $f'(x_1)$ of the function $f(x_1)$.

Since the Taylor series was truncated, retaining only two terms, the new value $x$ will not satisfy Eq. (2-10). We will designate this value as $x_2$ and reapply the Taylor series linearization at $x_2$ (shown in Fig. 2-3*b*) to obtain $x_3$. Repetitive application of this step converts Eq. (2-35) to an iterative formula:

$$x_{n+1} = x_n - \frac{f(x_n)}{f'(x_n)} \tag{2-36}$$

In contrast to the method of linear interpolation discussed in Sec. 2-3, the Newton-Raphson method uses the newly found position as the starting point for each subsequent iteration.

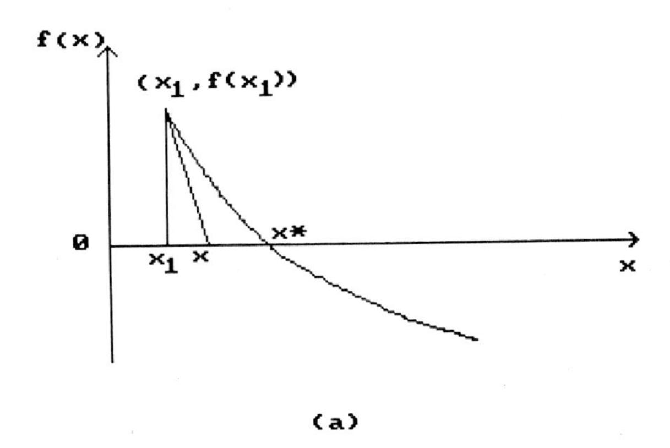


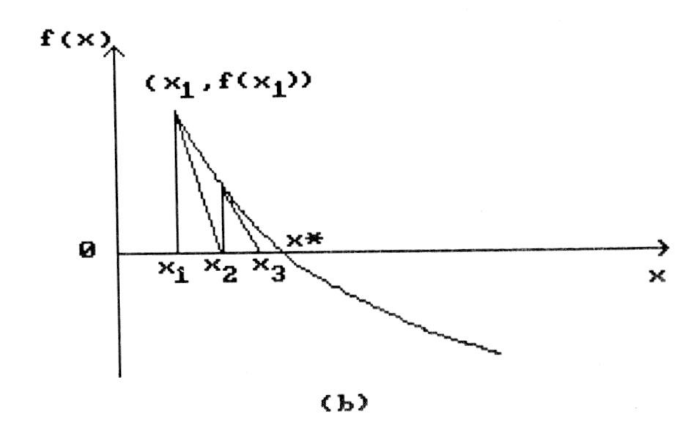


**Figure 2-3** The Newton-Raphson method.

In the discussion for both methods (linear interpolation and Newton-Raphson) a certain shape of the function was used to demonstrate how these techniques converge toward a root in the space of search. However, the shapes of nonlinear functions may vary drastically and convergence is not always guaranteed. As a matter of fact, divergence is more likely to occur, as shown in Figure 2-4, unless extreme care is taken in the choice of the initial starting points.

To investigate the convergence behavior of the Newton-Raphson method one has to examine the term $[-f(x_n)/f'(x_n)]$ in Eq. (2-36). This is the error term or correction term applied to the previous estimate of the root at each iteration. A function with a strong vertical trajectory near the root will cause the denominator of the error term to be large, therefore the convergence will be quite fast [6]. If, however, $f(x)$ is nearly horizontal near the root, the convergence will be slow. If at any point during the search, $f'(x_n) = 0$, the method would fail due to division by zero. Inflection points on the curve, within the region of search, are also troublesome and may cause the search to diverge.

A sufficient, but not necessary, condition for convergence of the Newton-Raphson method was stated by Lapidus [6] as follows: "If $f'(x)$ and $f''(x)$ do

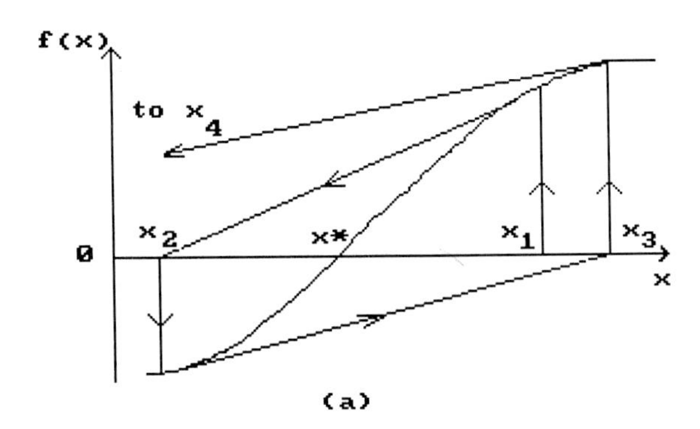


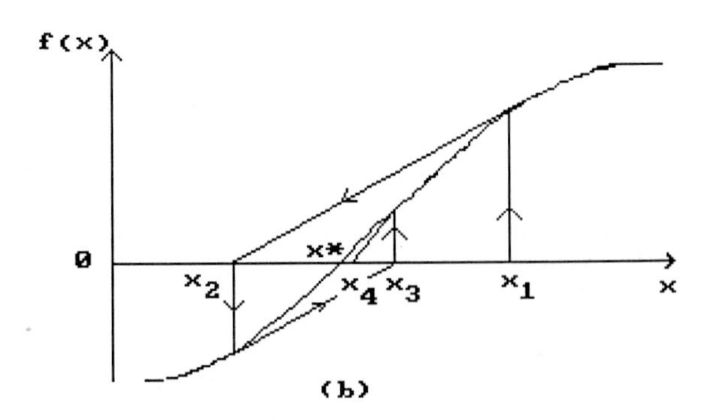


**Figure 2-4** Choice of initial guess affects convergence of Newton-Raphson method: (a) divergence, (b) convergence.

not change sign in the interval $(x_1, x^*)$ and if $f(x_1)$ and $f''(x_1)$ have the same sign, the iteration will always converge to $x^*$." These convergence criteria may be easily programmed as part of the computer program which performs the Newton-Raphson search, and a warning may be issued or other appropriate action may be taken by the computer if the conditions are violated.

A more accurate extension of the Newton-Raphson method is Newton's 2nd-order method which truncates the right-hand side of the Taylor series [Eq. (2-34)] after the third term to yield the equation

$$\frac{f''(x_1)}{2!}(\Delta x_1)^2 + f'(x_1)\,\Delta x_1 + f(x_1) = 0 \tag{2-37}$$

where $\Delta x_1 = x - x_1$. This is a quadratic equation in $\Delta x_1$ whose solution is given by

$$\Delta x_1 = \frac{-f'(x_1) \pm \sqrt{[f'(x_1)]^2 - 2f''(x_1)f(x_1)}}{f''(x_1)} \tag{2-38}$$

The general iterative formula for this method would be

$$\overset{+}{x}_{n+1} = x_n - \frac{f'(x_n)}{f''(x_n)} + \frac{\sqrt{[f'(x_n)]^2 - 2f''(x_n)f(x_n)}}{f''(x_n)} \tag{2-39a}$$

or

$$\overset{-}{x}_{n+1} = x_n - \frac{f'(x_n)}{f''(x_n)} - \frac{\sqrt{[f'(x_n)]^2 - 2f''(x_n)f(x_n)}}{f''(x_n)} \tag{2-39b}$$

The choice between (2-39*a*) and (2-39*b*) will be determined by exploring both values of $\overset{+}{x}_{n+1}$ and $\overset{-}{x}_{n+1}$ and determining which one results in the function $f(\overset{+}{x}_{n+1})$ or $f(\overset{-}{x}_{n+1})$ being closer to zero.

An alternative to the above exploration will be to treat Eq. (2-37) as another nonlinear equation in $\Delta x$ and to apply the Newton-Raphson method for its solution:

$$F(\Delta x) = \frac{f''(x_1)}{2!}(\Delta x_1)^2 + f'(x_1)\,\Delta x_1 + f(x_1) = 0 \tag{2-40}$$

where

$$\Delta x_{n+1} = \Delta x_n - \frac{F(\Delta x_n)}{F'(\Delta x_n)} \tag{2-41}$$

Two nested Newton-Raphson algorithms would have to be programmed together as follows:

1. Assume a value of $x_1$.
2. Calculate $\Delta x_1$ from Eq. (2-35).
3. Calculate $\Delta x_2$ from Eq. (2-41).
4. Calculate $x_2$ from $x_2 = x_1 + \Delta x_2$.
5. Repeat steps 2 to 4 until convergence is achieved.

**Example 2-1 The linear interpolation and Newton-Raphson methods applied to the solution of nonlinear functions** This example illustrates the use of the linear interpolation and Newton-Raphson methods in locating roots of nonlinear functions. For comparison purposes a well-known function was chosen to demonstrate the advantages and disadvantages of each method. This is the sinusoidal wave function

$$f(x) = \sin x = 0$$

whose roots exist at integer multiples of $\pi$. The program is written in a general fashion so that the user can easily enter other functions whose roots are to be determined.

METHOD OF SOLUTION Equations (2-33) and (2-36) are used for linear interpolation and Newton-Raphson, respectively. The iterative procedure stops when the absolute value of the function is less than the convergence criterion, or when the number of iterations exceeds 21. The program prints out the convergence results and draws a graph which demonstrates geometrically how each method arrives at the answer.

PROGRAM DESCRIPTION This computer program, called ROOT.BAS, is on the Applied Numerical Methods diskette which accompanies this book. The program consists of the sections listed in Table E2-1.

Since this is the first program in this book, a detailed description of each section is given for the benefit of the reader.

*Title* Line 0 enters into text mode (SCREEN 0), sets the width of the screen to 80 columns (WIDTH 80), clears the screen (CLS), moves the paper on printer (if available) to top of new page [PRINT CHR$(12)], and erases the function key display (KEY OFF). The rest of the lines in this section print the title of the program.

*Main program* Line 110 dimensions the matrix A which contains all the values of the function to be plotted by the plotting routine.

**Table E2-1**

| Program section | Line numbers |
|---|---|
| Title | 0–85 |
| Main program | 100–720 |
| Subroutines | |
| 1. Linear interpolation method | 1000–1150 |
| 2. Newton-Raphson method | 2000–2150 |
| 3. Plotting | 10000–10530 |

Lines 120–180 define the function, its first derivative, and a name to be used for labeling the $y$ axis (not to exceed 16 characters). The sine function, shown in this example, may be replaced by any function whose roots are to be determined, provided that the function and its derivatives are entered correctly, i.e., the left side of statements 160–180 must remain as they are and the right side of statements 160–170 must be functions of $X$ only.

Lines 230–305 give the user the choice of either the linear interpolation or the Newton-Raphson method.

Lines 320–410 enable the user to enter all the search variables.

Lines 430–520 calculate the function in the range of search and store the values of $x$ and $f(x)$ in the first two columns of matrix A.

Line 540 branches program execution to either the linear interpolation or the Newton-Raphson subroutine.

Line 570 calls the plotting subroutine.

Lines 660–710 give the user the option to run another case or to end the program.

*Subroutine 1—linear interpolation method* Lines 1020, 1030, and 1050 print the values of $x$ and $f(x)$ as the convergence proceeds from the initial value (X1) to the newly calculated values (X2).

Lines 1040 and 1120 mark the beginning and end of the iteration loop which has an upper limit of 21 iterations.

Lines 1060–1070 evaluate the function at X1 and X2 and store these values in matrix A. The third column of matrix A stores the values of the $x$ variable and the fourth column contains the values of the function.

Line 1080 checks the function for convergence within the convergence criterion, EPS, entered as input to the program by the user.

Lines 1090–1110 calculate a new value of X2 using the linear interpolation equation [Eq. (2-33)].

Line 1130 prints a message when no convergence is attained and stops the execution of the program. The user may choose to continue execution of the program by pressing the CONT key (F5).

*Subroutine 2—Newton-Raphson method* This subroutine differs from the previous one in only two respects: (1) in the way it stores the consecutive values of $x$ and $f(x)$ in matrix A (lines 2050–2060) and (2) in the use of Eq. (2-36) to recalculate the new values of $x$ and $f(x)$ (lines 2090–2110).

*Subroutine 3—plotting* This is a general plotting subroutine which can be used with other programs with little or no modification. It plots the numbers contained in the columns of matrix A. Column 2, which contains $f(x)$, is plotted versus column 1, which contains $x$ (all rows from 0 to 100), and column 4 is plotted versus column 3 (rows 0 to ITER). This subroutine requires the following input from the main program: matrix A, the name of the function (NAM$), the title of the graph (TIT$), and the number of rows to be plotted from columns 3 and 4 (ITER).

Line 10020 clears the screen and changes to medium resolution graphics (SCREEN 1).

Lines 10030–10100 locate the maxima and minima of the *x* and *y* axes for scaling the graph.

Lines 10110–10120 expands the *y* axis to eliminate crowding.

Line 10130 uses the VIEW statement of advanced basic (BASICA) to set aside part of the screen as a viewport.

Line 10140 uses the WINDOW statement to redefine the coordinates of the viewport. After these two statements are executed, everything drawn by the LINE and PSET statements appear within the viewport, provided that their coordinates are within those defined by WINDOW. Any point or line outside these coordinates will be invisible.

Lines 10150–10390 draw and label the axes and print the title of the graph.

Lines 10400–10460 draw the function (column 2 versus column 1).

Lines 10470–10510 draw the path to the root (column 4 versus column 3).

Line 10520 returns execution to the main program.

PROGRAM

```
0 SCREEN 0:WIDTH 80:PRINT CHR$(12):CLS:KEY OFF
5 PRINT "*********************************************************************"
10 PRINT"*                                                                   *"
15 PRINT"*                            EXAMPLE 2-1                            *"
20 PRINT"*                                                                   *"
25 PRINT"*       THE LINEAR INTERPOLATION AND NEWTON-RAPHSON METHODS         *"
30 PRINT"*                                                                   *"
35 PRINT"*         APPLIED TO THE SOLUTION OF NONLINEAR FUNCTIONS            *"
40 PRINT"*                                                                   *"
45 PRINT"*                            (ROOT.BAS)                             *"
50 PRINT"*                                                                   *"
85 PRINT"*********************************************************************"
100 '**************************** Main Program ****************************
110 DIM A(100,4)
120 'The following are the function,
130 'its derivative, and its name,
140 'which can be defined by the user
150 '
160 DEF FNF(X)=SIN(X)
170 DEF FNFP(X)=COS(X)
180 NAM$="SINE"
190 '
200 '
210 '
220 '
230 'Choose method to use
240 '
250 PRINT"METHOD TO USE:"
260 PRINT:PRINT"     1. LINEAR INTERPOLATION"
270 PRINT:PRINT"     2. NEWTON-RAPHSON"
280 PRINT:PRINT"YOUR CHOICE";:INPUT METH
290 IF METH<1 OR METH>2 GOTO 250
300 IF METH=1 THEN TIT$="LINEAR INTERPOLATION METHOD"
305 IF METH=2 THEN TIT$="NEWTON-RAPHSON METHOD"
310 '
320 'Input all program variables
330 '
340 PRINT: PRINT "DEFINE THE RANGE OF SEARCH:"
350 PRINT "LOWER LIMIT";: INPUT MINX
360 PRINT "HIGHER LIMIT";: INPUT MAXX
370 IF METH=2 GOTO 400
380 PRINT:PRINT "GIVE TWO STARTING VALUES FOR THE SEARCH.":
     PRINT"ONE SHOULD BE LOWER AND THE OTHER HIGHER THAN THE ROOT:":
     INPUT X1:INPUT X2
390 GOTO 410
400 PRINT:PRINT"GIVE YOUR BEST GUESS OF THE ROOT";:INPUT X1
410 PRINT:PRINT "GIVE THE CONVERGENCE VALUE OF F";: INPUT EPS
420 '
430 'Calculate the function in the range of search
440 '
470 DX = (MAXX -MINX) / 100
480 FOR K = 0 TO 100
490 X = MINX + K * DX
500 A(K ,1) = X
510 A(K ,2) = FNF(X)
520 NEXT K
```

```
530 '
540 ON METH GOSUB 1000,2000 'Root-finding routine
550 '
560 PRINT:PRINT"PRESS ANY KEY TO CONTINUE":HK$=INPUT$(1)
570 GOSUB 10000 'Plotting routine
650 '
660 'Option to reset conditions
670 '
680 PRINT: PRINT "DO YOU WANT TO RESET CONDITIONS(Y/N)";: INPUT V$
690 IF V$ = "N" OR V$="n" THEN  GOTO 710
700 PRINT CHR$(12):SCREEN 0:WIDTH 80:GOTO 250
710 END
720 '
1000 '************** Subroutine 1: Linear Interpolation Method **************
1010 '
1020 PRINT:PRINT"CONVERGENCE TO ROOT:":PRINT
1030 PRINT TAB(3) "x=";X1 TAB(20) "f(x)=";FNF(X1)
1040 FOR ITER=0 TO 40 STEP 2
1050 PRINT TAB(3) "x=";X2 TAB(20) "f(x)=";FNF(X2)
1060 A(ITER,3)=X1:A(ITER,4)=FNF(X1)
1070 A(ITER+1,3)=X2:A(ITER+1,4)=FNF(X2)
1080 IF  ABS (FNF(X2))< EPS GOTO 1140
1090 F1=FNF(X1)
1100 F2=FNF(X2)
1110 X2=X1-(F1/((F2-F1)/(X2-X1)))
1120 NEXT ITER
1130 PRINT"EXCEEDS ITERATION LIMIT:NO CONVERGENCE":
      PRINT"ITER=";ITER: PRINT"EPS=";EPS:PRINT"F=";F2:
      PRINT"FUNCTION KEY F5 WILL CONTINUE PROGRAM":STOP
1140 RETURN
1150 '
2000 '************** Subroutine 2: Newton-Raphson Method *******************
2010 '
2020 PRINT:PRINT"CONVERGENCE TO ROOT:":PRINT
2030 FOR ITER=0 TO 40 STEP 2
2040 PRINT TAB(3) "x=";X1 TAB(20) "f(x)=";FNF(X1)
2050 A(ITER,3)=X1:A(ITER,4)=0
2060 A(ITER+1,3)=X1:A(ITER+1,4)=FNF(X1)
2070 IF  ABS (FNF(X1))< EPS GOTO 2140
2090 F1=FNF(X1)
2100 FP=FNFP(X1)
2110 X1=X1-F1/FP
2120 NEXT ITER
2130 PRINT"EXCEEDS ITERATION LIMIT:NO CONVERGENCE":
      PRINT"ITER=";ITER:PRINT"EPS=";EPS:PRINT"F=";F1:
      PRINT"FUNCTION KEY F5 WILL CONTINUE PROGRAM":STOP
2140 RETURN
2150 '
10000 '****************** Subroutine 3: Plotting ****************************
10010 '
10020 CLS:SCREEN 1
10030 'Locate maxima and minima for scaling axes
10040 MINX=A(0,1):MAXX=A(0,1):MINY =A(0,2):MAXY =A(0,2)
10050 FOR K = 0 TO 100
10060 IF A(K ,1) < MINX THEN MINX = A(K ,1)
10070 IF A(K ,1) > MAXX THEN MAXX = A(K ,1)
10080 IF A(K ,2) < MINY THEN MINY = A(K ,2)
10090 IF A(K ,2) > MAXY THEN MAXY = A(K ,2)
10100 NEXT K
```

```
10110 'Expand the y-axis
10115 IF MINY>0 THEN MAXY=1.1*MAXY: MINY=.9*MINY
10120 IF MAXY<0 THEN MAXY=.9*MAXY: MINY=1.1*MINY
10130 VIEW (40,5)-(315,145)
10140 WINDOW (MINX,MINY)-(MAXX,MAXY)
10150 'Draw the axes
10160 LINE(MINX,0)-(MAXX,0)
10170 LINE(0,MINY)-(0,MAXY)
10180 'Label the axes
10190 VP=1+(5+140*(MAXY/((MAXY-MINY))))/8
10200 HP=-1+(40+275*(ABS(MINX)/((MAXX-MINX))))/8
10210 IF (MINX/MAXX)>0 THEN LINE(MINX,MINY)-(MINX,MAXY):HP=4
10215 IF (MINY/MAXY)>0 THEN LINE(MINX,MINY)-(MAXX,MINY)
10220 LOCATE VP,40:PRINT"x"
10230 LOCATE 1,HP:PRINT"f(x)"
10240 'Center vertical axis title
10250 IF LEN(NAM$)>16 GOTO 10310
10260 V1=3+(14-LEN(NAM$))/2
10270 FOR I=1 TO LEN(NAM$)
10280 LOCATE V1+I,2:PRINT MID$(NAM$,I,1)
10290 NEXT I
10300 GOTO 10340
10310 FOR I=1 TO 16
10320 LOCATE 3+I,2:PRINT MID$(NAM$,I,1)
10330 NEXT I
10340 'Center horizontal title
10350 IF LEN(TIT$)>35 GOTO 10390
10360 LOCATE 22,5+(36-LEN(TIT$))/2
10370 PRINT TIT$
10380 GOTO 10400
10390 LOCATE 22,5:PRINT MID$(TIT$,1,36)
10400 'Draw the function
10410 X=A(0,1):Y=A(0,2)
10420 PSET(X,Y)
10430 FOR I=1 TO 100
10440 X=A(I,1) :Y=A(I,2)
10450 LINE -(X,Y)
10460 NEXT I
10470 'Locate root
10480 FOR I=0 TO ITER
10490 LINE(A(I,3),A(I,4))-(A(I+1,3),A(I+1,4))
10500 FOR PAUSE=1 TO 500:NEXT PAUSE
10510 NEXT I
10520 RETURN
10530 '
```

## Results

```
*******************************************************************
*                                                                 *
*                          EXAMPLE 2-1                            *
*                                                                 *
*     THE LINEAR INTERPOLATION AND NEWTON-RAPHSON METHODS         *
*                                                                 *
*        APPLIED TO THE SOLUTION OF NONLINEAR FUNCTIONS           *
*                                                                 *
*                           (ROOT.BAS)                            *
*                                                                 *
*******************************************************************
METHOD TO USE:


     1. LINEAR INTERPOLATION


     2. NEWTON-RAPHSON


YOUR CHOICE? 1



DEFINE THE RANGE OF SEARCH:
LOWER LIMIT? 0
HIGHER LIMIT? 8



GIVE TWO STARTING VALUES FOR THE SEARCH.
ONE SHOULD BE LOWER AND THE OTHER HIGHER THAN THE ROOT:
? 2
? 6.2



GIVE THE CONVERGENCE VALUE OF F? 0.00001



CONVERGENCE TO ROOT:

  x= 2                f(x)= .9092975
  x= 6.2              f(x)=-8.308959E-02
  x= 5.848347         f(x)=-.421264
  x= 4.629936         f(x)=-.9966028
  x= 3.254732         f(x)=-.1128983
  x= 3.116151         f(x)= 2.543906E-02
  x= 3.148276         f(x)=-6.683149E-03
  x= 3.139898         f(x)= 1.694829E-03
  x= 3.142027         f(x)=-4.337708E-04
  x= 3.141482         f(x)= 1.107772E-04
  x= 3.141621         f(x)=-2.845924E-05
  x= 3.141585         f(x)= 7.303554E-06

PRESS ANY KEY TO CONTINUE
```

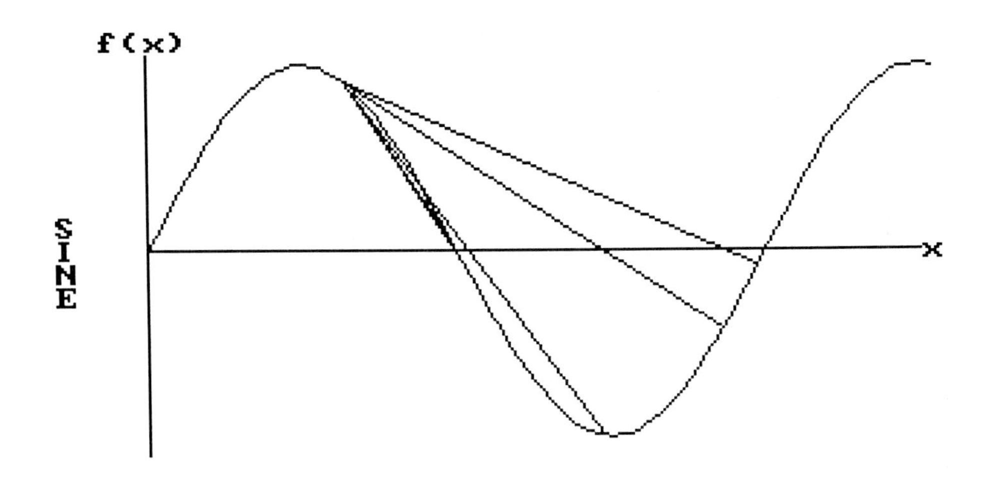


**LINEAR INTERPOLATION METHOD**

**DO YOU WANT TO RESET CONDITIONS(Y/N)? Y**

```
METHOD TO USE:

     1. LINEAR INTERPOLATION

     2. NEWTON-RAPHSON

YOUR CHOICE? 2

DEFINE THE RANGE OF SEARCH:
LOWER LIMIT? 0
HIGHER LIMIT? 8

GIVE YOUR BEST GUESS OF THE ROOT? 2

GIVE THE CONVERGENCE VALUE OF F? 0.00001

CONVERGENCE TO ROOT:

  x= 2                f(x)= .9092975
  x= 4.18504          f(x)=-.8641442
  x= 2.467893         f(x)= .6238813
  x= 3.266187         f(x)=-.1242717
  x= 3.140944         f(x)= 6.486496E-04
  x= 3.141593         f(x)=-8.742278E-08

PRESS ANY KEY TO CONTINUE
```

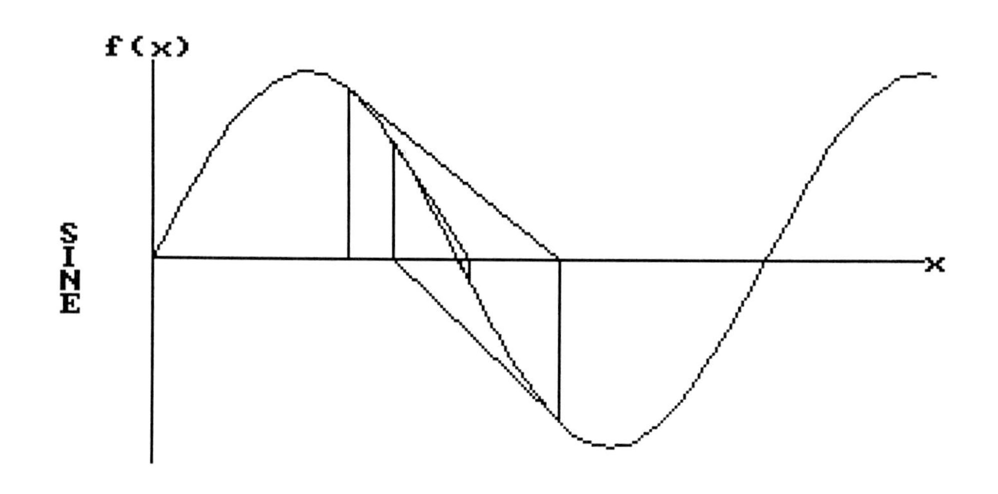


```
                NEWTON-RAPHSON METHOD
DO YOU WANT TO RESET CONDITIONS(Y/N)? Y

METHOD TO USE:

     1. LINEAR INTERPOLATION

     2. NEWTON-RAPHSON

YOUR CHOICE? 2

DEFINE THE RANGE OF SEARCH:
LOWER LIMIT? 0
HIGHER LIMIT? 8

GIVE YOUR BEST GUESS OF THE ROOT? 1.85

GIVE THE CONVERGENCE VALUE OF F? 0.00001

CONVERGENCE TO ROOT:

   x= 1.85           f(x)= .9612752
   x= 5.33806        f(x)=-.8105707
   x= 6.722134       f(x)= .4249882
   x= 6.252637       f(x)=-3.054365E-02
   x= 6.283195       f(x)= 9.711589E-06

PRESS ANY KEY TO CONTINUE
```

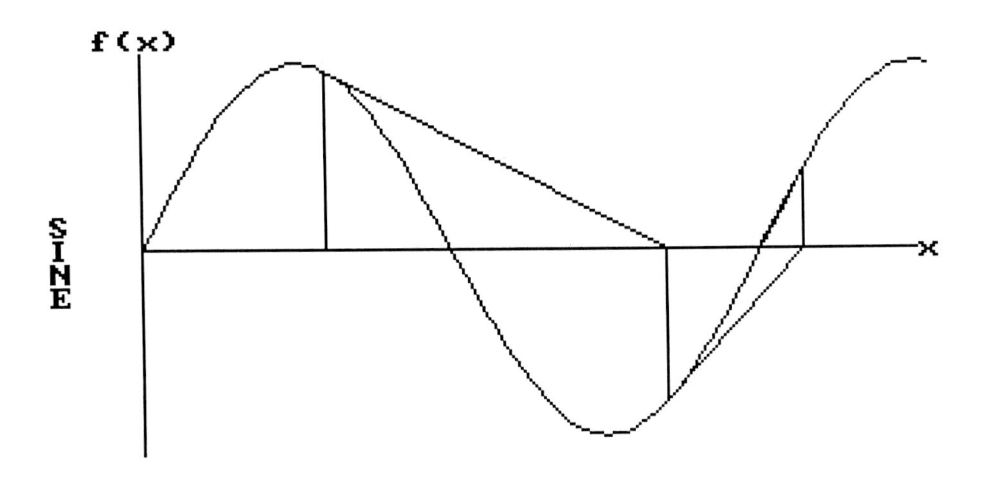


NEWTON-RAPHSON METHOD

DO YOU WANT TO RESET CONDITIONS(Y/N)? N■

Discussion of results The first case to be executed uses the linear interpolation method. The range of search is defined to be 0 to 8. The two starting values required by the linear interpolation method are set at 2 and 6.2. They bracket the root at $\pi$. These values were selected so that the resulting graph would give a good visual demonstration of how the method converges to the root. Each successive new value of $x$ is closer to the correct root until the convergence criterion.

$$|f(x)| < 0.00001$$

is met. This is accomplished in ten steps, with the final value of $x =$ 3.141585.

The second case to be executed uses the Newton-Raphson method. The same range of search is used and the starting value is set at 2. The graphical solution demonstrates that each successive value of $x$ is obtained from the $x$-axis intersection of the tangent of $f(x_i)$. The same convergence criterion is met within five iterations with the final value of $x = 3.141593$. This is within roundoff error of the value of $\pi$ which is 3.14159265.

The final case demonstrates the importance of choosing the starting value for the search. A value of 1.85 causes the Newton-Raphson method to converge to a different root ($x = 2\pi$) of this transcendental function.

The user of this program is encouraged to examine the convergence from different starting values and to experiment with finding the roots of other functions by replacing lines 160–180 in the main program.

## 2-5 SYNTHETIC DIVISION ALGORITHMS

If the nonlinear equation being solved is of the polynomial form, each real root (located by one of the methods already discussed) can be removed from the polynomial by synthetic division, thus reducing the degree of the polynomial to $n-1$. Each successive application of the synthetic division algorithm will reduce the degree of the polynomial further, until all real roots have been located.

A simple computational algorithm for synthetic division has been given by Lapidus [6]. Consider the fourth-degree polynomial

$$f(x) = a_4x^4 + a_3x^3 + a_2x^2 + a_1x + a_0 = 0 \tag{2-42}$$

whose first real root has been determined to be $x^*$. This root can be factored out as follows:

$$f(x) = (x - x^*)(b_3x^3 + b_2x^2 + b_1x + b_0) = 0 \tag{2-43}$$

In order to determine the coefficients $(b_i)$ of the third-degree polynomial first multiply out Eq. (2-43) and rearrange in descending powers of $x$:

$$f(x) = b_3x^4 + (b_2 - b_3x^*)x^3 + (b_1 - b_2x^*)x^2 + (b_0 - b_1x^*)x - b_0x^* \tag{2-44}$$

Equating Eqs. (2-42) and (2-44), the coefficients of like powers of $x$ must be equal to each other, i.e.,

$$\begin{aligned} a_4 &= b_3 \\ a_3 &= b_2 - b_3x^* \\ a_2 &= b_1 - b_2x^* \\ a_1 &= b_0 - b_1x^* \\ a_0 &= -b_0x^* \end{aligned} \tag{2-45}$$

Solving Eqs. (2-45) for $b_i$,

$$\begin{aligned} b_3 &= a_4 \\ b_2 &= a_3 + b_3x^* \\ b_1 &= a_2 + b_2x^* \\ b_0 &= a_1 + b_1x^* \end{aligned} \tag{2-46}$$

In general notation, for a polynomial of $n$th degree the new coefficients after application of synthetic divisions are given by

$$\left.\begin{aligned} b_{n-1} &= a_n \\ b_{n-1-r} &= a_{n-r} + b_{n-r}x^* \end{aligned}\right\} \qquad r = 1, 2, \ldots, (n-1) \tag{2-47}$$

The polynomial is then reduced by one degree

$$n = n - 1 \tag{2-48}$$

and the newly calculated coefficients are renamed

$$a_i = b_i \qquad i = 1, 2, \ldots, n \tag{2-49}$$

This procedure is repeated until all real roots are extracted. When this is accomplished, the remainder polynomial will contain the complex roots. The presence of a pair of complex roots will give a quadratic equation which can be easily solved by the quadratic formula. However, two or more pairs of complex roots require the application of more elaborate techniques such as Graeffe's and Newton's methods, which are developed in the next two sections.

**Example 2-2 The Newton-Raphson method with synthetic division applied to *n*th-degree polynomials and to the Beattie-Bridgeman equation** Develop a computer program which uses the Newton-Raphson method with synthetic division to determine all the roots of a polynomial equation. This polynomial may have one pair of complex roots. Apply this program to the solution of the following problems:

*a*. Find all the roots of the fourth-degree polynomial

$$x^4 + 6x^3 + 7x^2 - 6x - 8 = 0$$

The roots of this polynomial are known to be 1, −1, −2, and −4 (see Sec. 2-2). This exercise will serve to verify the accuracy of the program.

*b*. Calculate the specific volume of a pure gas, at a given temperature and pressure, by using the Beattie-Bridgeman equation of state

$$PV = RT + \frac{\beta}{V} + \frac{\gamma}{V^2} + \frac{\delta}{V^3}$$

The variables $\beta$, $\gamma$, and $\delta$ are empirical functions of temperature:

$$\beta = RTB_0 - A_0 - \frac{Rc}{T^2}$$

$$\gamma = -RTB_0 b + \alpha A_0 - \frac{RB_0 c}{T^2}$$

$$\delta = \frac{RB_0 bc}{T^2}$$

The values of the constants $A_0$, $B_0$, $\alpha$, $b$, and $c$ are empirically determined for each gas.

Calculate the specific volume of $n$-butane at 425 K and at pressures of 1, 10, 20, 30, and 40 atm. Compare these results with the ones obtained from using the ideal gas law

$$PV = RT$$

What conclusion do you draw from this comparison?

The constants for $n$-butane are

$$A_0 = 17.7940 \qquad B_0 = 0.24620$$

$$\alpha = 0.12161 \qquad b = 0.09423 \qquad c = 350.0 \times 10^4$$

The units for each variable are

$$P = \text{atmospheres}$$

$$V = \text{liters (L)/mol}$$

$$T = \text{kelvins}$$

The gas constant is

$$R = 0.08206\ (\text{L} \cdot \text{atm})/(\text{mol} \cdot \text{K})$$

METHOD OF SOLUTION Equation (2-36) is used for the Newton-Raphson evaluation of each root. In addition, Eqs. (2-47) through (2-49) are used to perform synthetic division in order to extract each root from the polynomial and to reduce the latter by one degree. The determination of each root is shown in tabular and in graphical form. When the $n$th-degree polynomial has been reduced to a quadratic

$$a_2x^2 + a_1x + a_0 = 0$$

the program uses the quadratic solution formula

$$x_{1,2} = \frac{-a_1 \pm \sqrt{a_1^2 - 4a_2a_0}}{2a_2}$$

to check for the existence of a pair of complex roots.

PROGRAM DESCRIPTION This computer program, which we call POLY.BAS, consists of the sections in Table E2-2$a$.

**Table E2-2*a***

| Program section | Line numbers |
|---|---|
| Title | 0–95 |
| Main program | 100–1100 |
| Subroutines | |
| 1. Newton-Raphson method | 2000–2150 |
| 2. Define the polynomial | 3000–3110 |
| 3. Define the Beattie-Bridgeman equation | 4000–4220 |
| 4. Plotting | 10000–10530 |

The plotting subroutine is identical to that used in Example 2-1. The title section and the Newton-Raphson subroutine have been changed slightly to conform with the new problem. The main program, which incorporates the general form of the polynomial and the synthetic division algorithm, is discussed in more detail below. Two input subroutines have been added. The first defines the $n$th-degree polynomial and the second defines the variables of the Beattie-Bridgeman equation.

*Main program* Line 110 dimensions the matrix A and the vectors AA, B, and XW. The matrix contains all the values of the function to be plotted by the plotting subroutine. The vector AA contains the coefficients $a_i$ of the polynomial equation. The vector B contains the values of the new coefficients after the application of synthetic division. The vector XW stores the values of all the roots as they are determined. All the vectors are given the dimension of 20, i.e., up to a twentieth-degree polynomial can be handled. This dimension may be changed by the user to accommodate higher-degree polynomials if needed.

Lines 140–220 constitute a small subroutine which evaluates the $n$th-degree polynomial F and its derivative FP.

Lines 230–290 enable the user to choose between finding the solution of a general $n$th-degree polynomial or of the Beattie-Bridgeman equation of state, which is a fourth-degree polynomial.

Lines 295, 305, and 415 set and check the flag variable RST$ which controls whether the search variables will be redefined after each root is evaluated.

Line 300 defines the title of the graph.

Lines 320–415 enable the user to enter all the search variables.

Lines 430–520 evaluate the function in the range of search and store the values of $x$ and $f(x)$ in the first two columns of matrix A.

Line 540 branches program execution to the Newton-Raphson subroutine.

Line 570 diverts program execution to bypass plotting if the no-graph option has been chosen.

Line 580 branches execution to the plotting subroutine.

Line 590 diverts program execution to bypass synthetic division if only the first root is desired.

Line 600 stores the root just calculated into the vector XW.

Lines 620–720 perform the synthetic division algorithm, thus reducing the polynomial by one degree and evaluating the coefficients of the reduced polynomial.

Lines 740–900 first verify whether the polynomial has been reduced to a quadratic and then determine, using the quadratic solution formula, whether there is a pair of complex roots. If such a pair is detected, then all the roots are printed out. Otherwise, the program proceeds with the Newton-Raphson determination.

Line 910 introduces a momentary pause between calculation of each root so that the program user can study the graphical solution before it is cleared.

Line 920 checks whether all the roots have been determined. If not, then execution is returned to line 300 for calculation of the remaining roots.

Lines 940–990 print out all the roots.

Lines 1010–1100 give the user the options to reset the conditions, to enter another polynomial for solution, or to end the program.

*Subroutine 1—Newton-Raphson method* This has been changed from Example 2-1 to reflect the fact that the polynomial function is designated by F, and its derivative by FP, and that they are calculated by the subroutine on line 140.

*Subroutine 2—define the polynomial* This subroutine enables the program user to define the degree of the polynomial and to enter the values of the coefficients. The user is also given the choice of whether to evaluate only the first root or all the roots, and whether to have a graphical solution.

*Subroutine 3—define the Beattie-Bridgeman equation* In this subroutine, the user enters the values of the constants $A_0$, $B_0$, $\alpha$, and $b$ and $c$, the temperature and pressure. The functions $b$, $\gamma$, and $\delta$ and the ideal-gas-law volume are calculated and printed. The choice of roots and graphical solution is also given.

*Subroutine 4—plotting* This subroutine is identical to the one used in Example 2-1.

PROGRAM

```
0 SCREEN 0:WIDTH 80:PRINT CHR$(12):CLS:KEY OFF
5 PRINT "********************************************************************"
10 PRINT"*                                                                  *"
15 PRINT"*                          EXAMPLE 2-2                             *"
20 PRINT"*                                                                  *"
25 PRINT"*        THE NEWTON-RAPHSON METHOD WITH SYNTHETIC DIVISION         *"
30 PRINT"*                                                                  *"
35 PRINT"*                           APPLIED TO                             *"
40 PRINT"*                                                                  *"
45 PRINT"*    nth DEGREE POLYNOMIALS AND TO THE BEATTIE-BRIDGEMAN EQUATION  *"
50 PRINT"*                                                                  *"
55 PRINT"*                           (POLY.BAS)                             *"
60 PRINT"*                                                                  *"
95 PRINT"********************************************************************"
100 '**************************** Main Program ***************************
110 DIM A(100,4),AA(20),B(20),XW(20)
120 GOTO 250
130 '
140 'Evaluation of the polynomial and its derivative
150 '
160 F = 0:FP=0
170 FOR KK = N TO 0 STEP  - 1
180 F = F + AA(KK) * X ^ KK
190 IF KK=0 GOTO 220
200 FP= FP+ KK*AA(KK) * X ^ (KK-1)
210 NEXT KK
220 RETURN
225 '
230 'Choose problem to solve
240 '
250 PRINT"PROBLEM TO SOLVE:"
260 PRINT:PRINT"     1. nth DEGREE POLYNOMIAL"
270 PRINT:PRINT"     2. BEATTIE-BRIDGEMAN EQUATION"
280 PRINT:PRINT"YOUR CHOICE";:INPUT METH
285 IF METH<1 OR METH>2 GOTO 250
290 ON METH GOSUB 3000,4000
295 RST$="Y"
300 TIT$="ROOT"+STR$(NW-N+1)
305 IF RST$="N" OR RST$="n"  GOTO 430
310 '
320 'Input the search variables
330 '
340 PRINT :PRINT"DEFINE THE RANGE OF SEARCH:"
350 PRINT "LOWER LIMIT";: INPUT MINX
360 PRINT "HIGHER LIMIT";: INPUT MAXX
400 PRINT:PRINT"GIVE YOUR BEST GUESS OF THE ROOT";:INPUT X1
410 PRINT:PRINT "GIVE THE CONVERGENCE VALUE OF F";: INPUT EPS
414 IF ALLROOT$="N" OR ALLROOT$="n" GOTO 430
415 PRINT:PRINT"DO YOU WANT TO BE ABLE TO RESET THE":
     PRINT"ABOVE VALUES AFTER EACH ROOT(Y/N)";:INPUT RST$
420 '
430 'Calculate the function in the range of search
440 '
470 DX = (MAXX -MINX) / 100
480 FOR K = 0 TO 100
490 X = MINX + K * DX
```

```
500 A(K,1)=X
510 GOSUB 140:A(K,2)=F
520 NEXT K
530 '
540 GOSUB 2000 'Root-finding routine
550 '
560 PRINT:PRINT"PRESS ANY KEY TO CONTINUE":HK$=INPUT$(1)
570 IF GRAPH$="N" OR GRAPH$="n" GOTO 590
580 GOSUB 10000 'Plotting routine
590 IF ALLROOT$="N" OR ALLROOT$="n" GOTO 1010
600 XW(N) = X1
610 '
620 'Synthetic division
630 '
640 IF N = 1 GOTO 960
650 B(N-1)=AA(N)
660 FOR R = 1 TO N - 1
670 B(N - 1 - R) = AA(N - R) + B(N - R) * X1
680 NEXT
690 N = N - 1
700 FOR JJ = N TO 0 STEP  - 1
710 AA(JJ) = B(JJ)
720 NEXT
730 '
740 'Check for complex roots
750 '
760 IF N<>2 GOTO 910
770 WW=AA(1)^2-4*AA(2)*AA(0)
780 IF WW>=0 GOTO 910
790 XW(2)=-AA(1)/(2*AA(2))
800 XW(1)=SQR(-WW)/(2*AA(2))
810 PRINT "THE ";NW;" ROOTS ARE:"
820 FOR JK = NW TO 3 STEP  - 1
830 PRINT XW(JK)
840 NEXT
850 '
860 'Print complex roots
870 '
880 PRINT XW(2);"+";XW(1);"i"
890 PRINT XW(2);"-";XW(1);"i"
900 GOTO 1030
910 FOR PAUSE=1 TO 4000:NEXT PAUSE
920 IF N > 0 THEN SCREEN 0:WIDTH 80:GOTO 300
930 '
940 'Print the roots
950 '
960 PRINT "THE ";NW;" ROOTS ARE:"
970 FOR JK = NW TO 1 STEP  - 1
980 PRINT XW(JK);"  ";
990 NEXT
1000 '
1010 'Option to reset conditions
1020 '
1030 IF METH=1 THEN GOTO 1070
1040 PRINT "DO YOU WANT TO RESET CONDITIONS(Y/N)";: INPUT V$
1050 IF V$ = "N" OR V$="n" THEN  GOTO 1070
1060 SCREEN 0:WIDTH 80:PRINT CHR$(12):GOTO 290
1070 PRINT "DO YOU WANT TO RERUN THE PROGRAM(Y/N)";: INPUT V$
1080 IF V$ = "N" OR V$="n" THEN  GOTO 1100
```

```
1090 SCREEN 0:WIDTH 80:PRINT CHR$(12):GOTO 250
1100 END
2000 '************** Subroutine 1: Newton-Raphson Method ******************
2010 '
2020 PRINT CHR$(12):PRINT"CONVERGENCE TO ROOT";(NW-N+1):PRINT
2030 FOR ITER=0 TO 40 STEP 2
2035 X=X1:GOSUB 140
2040 PRINT TAB(3) "x=";X1 TAB(20) "f(x)=";F
2050 A(ITER,3)=X1:A(ITER,4)=0
2060 A(ITER+1,3)=X1:A(ITER+1,4)=F
2070 IF  ABS (F)< EPS GOTO 2140
2110 X1=X1-F/FP
2120 NEXT ITER
2130 PRINT"EXCEEDS ITERATION LIMIT:NO CONVERGENCE":
      PRINT"ITER=";ITER:PRINT"EPS=";EPS:PRINT"F=";F:
      PRINT"FUNCTION KEY F5 WILL CONTINUE PROGRAM":STOP
2140 RETURN
2150 '
3000 '************** Subroutine 2: Define the polynomial ******************
3010 '
3020 PRINT: PRINT "GIVE THE DEGREE OF THE POLYNOMIAL";: INPUT N:NW = N
3030 PRINT:PRINT "GIVE THE VALUES OF THE COEFFICIENTS"
3040 FOR K = N TO 0 STEP  - 1
3050 PRINT "A";K;"=";: INPUT AA(K)
3060 NEXT
3070 '
3080 PRINT:PRINT"DO YOU WANT TO SEE ALL ROOTS(Y/N)";:INPUT ALLROOT$
3090 PRINT:PRINT"DO YOU WANT GRAPHICAL SOLUTION(Y/N)";:INPUT GRAPH$
3100 RETURN
3110 '
4000 '******** Subroutine 3: Define the Beattie-Bridgeman equation ********
4010 '
4020 N=4:NW=N
4030 PRINT:PRINT"HAVE YOU ENTERED THE BEATTIE-BRIDGEMAN":PRINT"CONSTANTS(Y/N)";
4040 INPUT VK$
4050 IF VK$="Y" OR VK$="y" GOTO 4090
4060 PRINT:PRINT"SPECIFY BEATTIE-BRIDGEMAN CONSTANTS:"
4070 INPUT "AO=";AO:INPUT "BO=";BO:INPUT "ALPHA=";ALPHA
4080 INPUT "B=";B:INPUT "C=";C
4090 PRINT:PRINT"SPECIFY TEMPERATURE(KELVIN)=";:INPUT T
4100 PRINT "SPECIFY PRESSURE(ATM)";: INPUT P
4110 R = .08206
4120 BETA = R * T * BO - AO - R * C / (T ^ 2)
4130 V=R*T/P
4140 GAMMA =  - R * T * BO * B + ALPHA * AO - R * BO * C / (T ^ 2)
4150 DLTA = R * BO * B * C / (T ^ 2)
4160 AA(4)=P:AA(3)=-R*T:AA(2)=-BETA:AA(1)=-GAMMA:AA(0)=-DLTA
4170 PRINT : PRINT "IDEAL GAS VOLUME=";V
4180 PRINT: PRINT"BETA=";BETA:PRINT"GAMMA=";GAMMA:PRINT"DLTA=";DLTA:PRINT
4190 PRINT:PRINT"DO YOU WANT TO SEE ALL ROOTS(Y/N)";:INPUT ALLROOT$
4200 PRINT:PRINT"DO YOU WANT GRAPHICAL SOLUTION(Y/N)";:INPUT GRAPH$
4210 RETURN
4220 '
10000 '******************* Subroutine 4: Plotting ***********************
10010 '
10020 CLS:SCREEN 1
10030 'Locate maxima and minima for scaling axes
10040 MINX=A(0,1):MAXX=A(0,1):MINY =A(0,2):MAXY =A(0,2)
10050 FOR K = 0 TO 100
```

```
10060 IF A(K ,1) < MINX THEN MINX = A(K ,1)
10070 IF A(K ,1) > MAXX THEN MAXX = A(K ,1)
10080 IF A(K ,2) < MINY THEN MINY = A(K ,2)
10090 IF A(K ,2) > MAXY THEN MAXY = A(K ,2)
10100 NEXT K
10110 'Expand the y-axis
10115 IF MINY>0 THEN MAXY=1.1*MAXY: MINY=.9*MINY
10120 IF MAXY<0 THEN MAXY=.9*MAXY: MINY=1.1*MINY
10130 VIEW (40,5)-(315,145)
10140 WINDOW (MINX,MINY)-(MAXX,MAXY)
10150 'Draw the axes
10160 LINE(MINX,0)-(MAXX,0)
10170 LINE(0,MINY)-(0,MAXY)
10180 'Label the axes
10190 VP=1+(5+140*(MAXY/((MAXY-MINY))))/8
10200 HP=-1+(40+275*(ABS(MINX)/((MAXX-MINX))))/8
10210 IF (MINX/MAXX)>0 THEN LINE(MINX,MINY)-(MINX,MAXY):HP=4
10215 IF (MINY/MAXY)>0 THEN LINE(MINX,MINY)-(MAXX,MINY)
10220 LOCATE VP,40:PRINT"x"
10230 LOCATE 1,HP:PRINT"f(x)"
10240 'Center vertical axis title
10250 IF LEN(NAM$)>16 GOTO 10310
10260 V1=3+(14-LEN(NAM$))/2
10270 FOR I=1 TO LEN(NAM$)
10280 LOCATE V1+I,2:PRINT MID$(NAM$,I,1)
10290 NEXT I
10300 GOTO 10340
10310 FOR I=1 TO 16
10320 LOCATE 3+I,2:PRINT MID$(NAM$,I,1)
10330 NEXT I
10340 'Center horizontal title
10350 IF LEN(TIT$)>35 GOTO 10390
10360 LOCATE 22,5+(36-LEN(TIT$))/2
10370 PRINT TIT$
10380 GOTO 10400
10390 LOCATE 22,5:PRINT MID$(TIT$,1,36)
10400 'Draw the function
10410 X=A(0,1):Y=A(0,2)
10420 PSET(X,Y)
10430 FOR I=1 TO 100
10440 X=A(I,1) :Y=A(I,2)
10450 LINE -(X,Y)
10460 NEXT I
10470 'Locate root
10480 FOR I=0 TO ITER
10490 LINE(A(I,3),A(I,4))-(A(I+1,3),A(I+1,4))
10500 FOR PAUSE=1 TO 500:NEXT PAUSE
10510 NEXT I
10520 RETURN
10530 '
```

## RESULTS

```
*******************************************************************
*                                                                 *
*                           EXAMPLE 2-2                           *
*                                                                 *
*        THE NEWTON-RAPHSON METHOD WITH SYNTHETIC DIVISION        *
*                                                                 *
*                            APPLIED TO                           *
*                                                                 *
*   nth DEGREE POLYNOMIALS AND TO THE BEATTIE-BRIDGEMAN EQUATION  *
*                                                                 *
*                            (POLY.BAS)                           *
*                                                                 *
*******************************************************************

PROBLEM TO SOLVE:

     1. nth DEGREE POLYNOMIAL

     2. BEATTIE-BRIDGEMAN EQUATION

YOUR CHOICE? 1


GIVE THE DEGREE OF THE POLYNOMIAL? 4
GIVE THE VALUES OF THE COEFFICIENTS
A 4 =? 1
A 3 =? 6
A 2 =? 7
A 1 =? -6
A 0 =? -8
DO YOU WANT TO SEE ALL ROOTS(Y/N)? Y
DO YOU WANT GRAPHICAL SOLUTION(Y/N)? Y
DEFINE THE RANGE OF SEARCH:
LOWER LIMIT? -5
HIGHER LIMIT? 2
GIVE YOUR BEST GUESS OF THE ROOT? 1.8
GIVE THE CONVERGENCE VALUE OF F? 0.00001
DO YOU WANT TO BE ABLE TO RESET THE
ABOVE VALUES AFTER EACH ROOT(Y/N)? N


CONVERGENCE TO ROOT 1

  x= 1.8             f(x)= 49.3696
  x= 1.310455        f(x)= 12.61003
  x= 1.069158        f(x)= 2.226354
  x= 1.0045          f(x)= .1356335
  x= 1.000021        f(x)= 6.217957E-04
  x= 1               f(x)= 0

PRESS ANY KEY TO CONTINUE
```

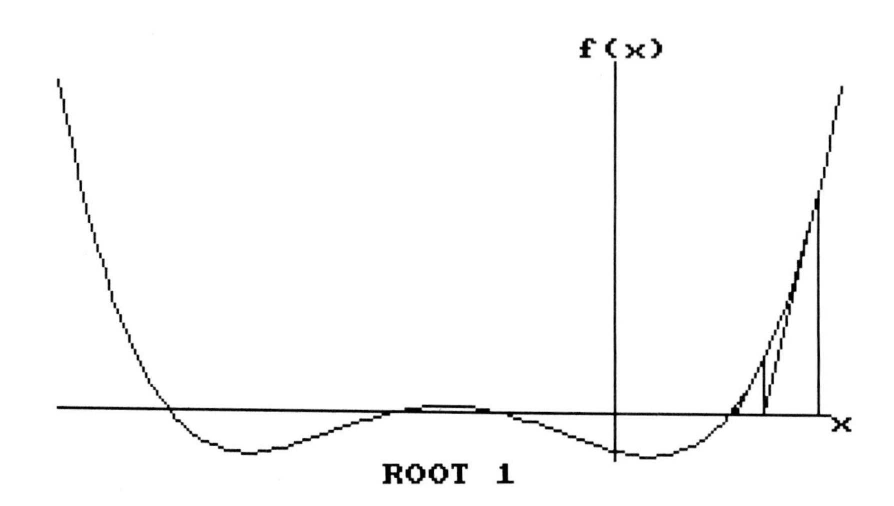


```
CONVERGENCE TO ROOT 2
  x= 1              f(x)= 30
  x= 3.225809E-02   f(x)= 8.458931
  x=-.5529433       f(x)= 2.229958
  x=-.8636941       f(x)= .4857679
  x=-.9808542       f(x)= .0589099
  x=-.9995306       f(x)= 1.409531E-03
  x=-.9999998       f(x)= 9.536743E-07
PRESS ANY KEY TO CONTINUE
```

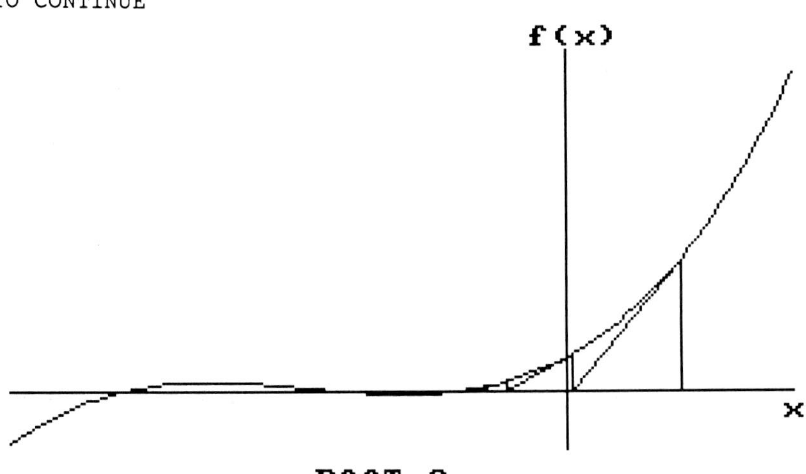


```
CONVERGENCE TO ROOT 3
  x=-.9999998       f(x)= 3.000002
  x=-1.75           f(x)= .562501
  x=-1.975          f(x)= 5.062485E-02
  x=-1.999695       f(x)= 6.098748E-04
  x=-2              f(x)= 0
PRESS ANY KEY TO CONTINUE
```

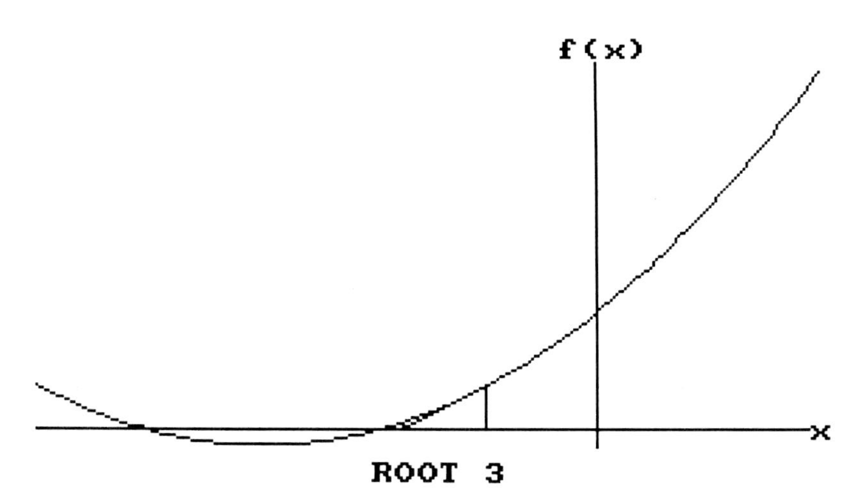


```
CONVERGENCE TO ROOT 4
   x=-2                 f(x)= 2
   x=-4                 f(x)= 0
```

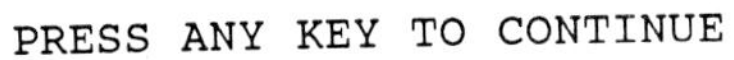

```
PRESS ANY KEY TO CONTINUE
```

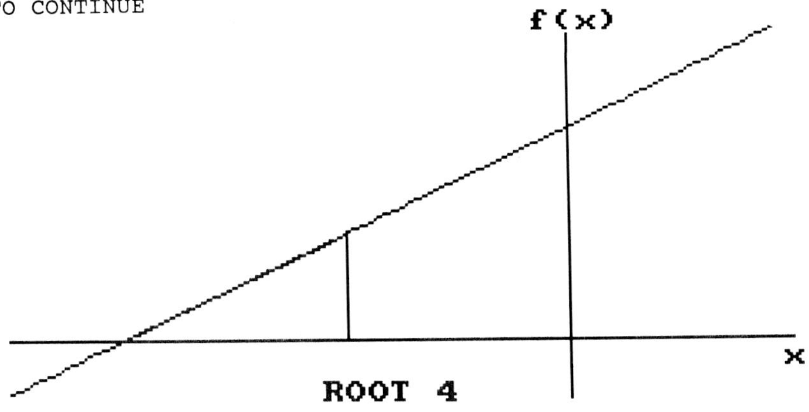


```
THE  4  ROOTS ARE:
 1   -.9999998   -2   -4
DO YOU WANT TO RERUN THE PROGRAM(Y/N)? N
Ok

*******************************************************************
*                                                                 *
*                          EXAMPLE 2-2                            *
*                                                                 *
*       THE NEWTON-RAPHSON METHOD WITH SYNTHETIC DIVISION         *
*                                                                 *
*                          APPLIED TO                             *
*                                                                 *
*   nth DEGREE POLYNOMIALS AND TO THE BEATTIE-BRIDGEMAN EQUATION  *
*                                                                 *
*                          (POLY.BAS)                             *
*                                                                 *
*******************************************************************
```

```
PROBLEM TO SOLVE:
     1. nth DEGREE POLYNOMIAL
     2. BEATTIE-BRIDGEMAN EQUATION
YOUR CHOICE? 2
HAVE YOU ENTERED THE BEATTIE-BRIDGEMAN
CONSTANTS(Y/N)? N
SPECIFY BEATTIE-BRIDGEMAN CONSTANTS:
AO=? 17.794
BO=? 0.2462
ALPHA=? 0.12161
B=? 0.09423
C=? 350.0E4
SPECIFY TEMPERATURE(KELVIN)=? 425
SPECIFY PRESSURE(ATM)? 1
IDEAL GAS VOLUME= 34.8755
BETA=-10.79774
GAMMA= .9633568
DLTA= 3.688917E-02
DO YOU WANT TO SEE ALL ROOTS(Y/N)? Y
DO YOU WANT GRAPHICAL SOLUTION(Y/N)? Y
DEFINE THE RANGE OF SEARCH:
LOWER LIMIT? 0
HIGHER LIMIT? 40
GIVE YOUR BEST GUESS OF THE ROOT? 30
GIVE THE CONVERGENCE VALUE OF F? 0.05
DO YOU WANT TO BE ABLE TO RESET THE
ABOVE VALUES AFTER EACH ROOT(Y/N)? N
CONVERGENCE TO ROOT 1
  x= 30              f(x)=-121949.5
  x= 38.42015        f(x)= 216926.5
  x= 35.4582         f(x)= 39519.13
  x= 34.62698        f(x)= 2595.15
  x= 34.56426        f(x)= 14.22107
  x= 34.56391        f(x)=-3.836546E-02
PRESS ANY KEY TO CONTINUE
```

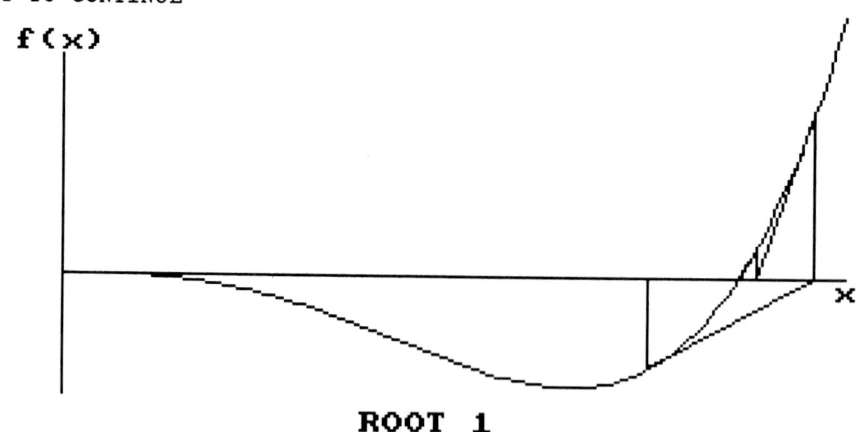

```
CONVERGENCE TO ROOT 2

  x= 34.56391        f(x)= 40920.97
  x= 23.07726        f(x)= 12124.72
  x= 15.4195         f(x)= 3592.502
  x= 10.31435        f(x)= 1064.441
  x= 6.910947        f(x)= 315.3869
  x= 4.642053        f(x)= 93.44619
  x= 3.129514        f(x)= 27.6867
  x= 2.121229        f(x)= 8.202915
  x= 1.449124        f(x)= 2.430273
  x= 1.001124        f(x)= .7200827
  x= .7024292        f(x)= .2135075
  x= .5029603        f(x)= 6.351036E-02
  x= .3687948        f(x)= 1.913665E-02

PRESS ANY KEY TO CONTINUE
```

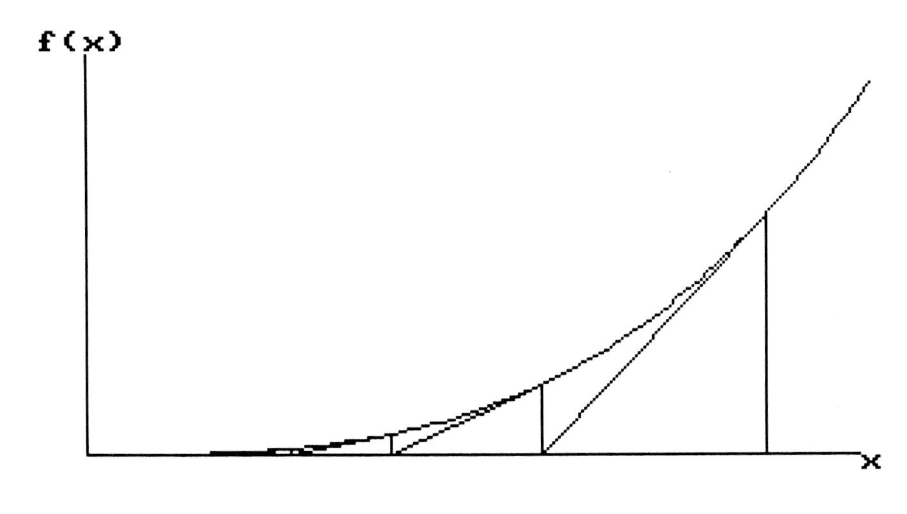


**ROOT 2**

```
THE  4  ROOTS ARE:
 34.56391
 .3687948
-2.860132E-02 + .2195008 i
-2.860132E-02 - .2195008 i
DO YOU WANT TO RESET CONDITIONS(Y/N)? Y

HAVE YOU ENTERED THE BEATTIE-BRIDGEMAN
CONSTANTS(Y/N)? Y

SPECIFY TEMPERATURE(KELVIN)=? 425
SPECIFY PRESSURE(ATM)? 10

IDEAL GAS VOLUME= 3.48755

BETA=-10.79774
GAMMA= .9633568
DLTA= 3.688917E-02
```

```
DO YOU WANT TO SEE ALL ROOTS(Y/N)? N

DO YOU WANT GRAPHICAL SOLUTION(Y/N)? N

DEFINE THE RANGE OF SEARCH:
LOWER LIMIT? 0
HIGHER LIMIT? 4

GIVE YOUR BEST GUESS OF THE ROOT? 3.5

GIVE THE CONVERGENCE VALUE OF F? 0.001

CONVERGENCE TO ROOT 1

  x= 3.5            f(x)= 134.2016
  x= 3.235796       f(x)= 24.6081
  x= 3.160916       f(x)= 1.645277
  x= 3.155148       f(x)= 9.118699E-03
  x= 3.155115       f(x)=-2.823025E-05

PRESS ANY KEY TO CONTINUE
DO YOU WANT TO RESET CONDITIONS(Y/N)? Y


HAVE YOU ENTERED THE BEATTIE-BRIDGEMAN
CONSTANTS(Y/N)? Y

SPECIFY TEMPERATURE(KELVIN)=? 425
SPECIFY PRESSURE(ATM)? 20

IDEAL GAS VOLUME= 1.743775

BETA=-10.79774
GAMMA= .9633568
DLTA= 3.688917E-02


DO YOU WANT TO SEE ALL ROOTS(Y/N)? N

DO YOU WANT GRAPHICAL SOLUTION(Y/N)? N

DEFINE THE RANGE OF SEARCH:
LOWER LIMIT? 0
HIGHER LIMIT? 2

GIVE YOUR BEST GUESS OF THE ROOT? 1.7

GIVE THE CONVERGENCE VALUE OF F? 0.001

CONVERGENCE TO ROOT 1

  x= 1.7            f(x)= 25.22955
  x= 1.500428       f(x)= 6.386478
  x= 1.403848       f(x)= 1.081348
  x= 1.379544       f(x)= 5.821123E-02
  x= 1.378079       f(x)= 2.120957E-04

PRESS ANY KEY TO CONTINUE
DO YOU WANT TO RESET CONDITIONS(Y/N)? Y
```

```
HAVE YOU ENTERED THE BEATTIE-BRIDGEMAN
CONSTANTS(Y/N)? Y

SPECIFY TEMPERATURE(KELVIN)=? 425
SPECIFY PRESSURE(ATM)? 30

IDEAL GAS VOLUME= 1.162517

BETA=-10.79774
GAMMA= .9633568
DLTA= 3.688917E-02


DO YOU WANT TO SEE ALL ROOTS(Y/N)? N

DO YOU WANT GRAPHICAL SOLUTION(Y/N)? N

DEFINE THE RANGE OF SEARCH:
LOWER LIMIT? 0
HIGHER LIMIT? 1.5

GIVE YOUR BEST GUESS OF THE ROOT? 1.1

GIVE THE CONVERGENCE VALUE OF F? 0.001

CONVERGENCE TO ROOT 1

  x= 1.1              f(x)= 9.4724
  x= .9305888         f(x)= 2.810195
  x= .8192205         f(x)= .7582178
  x= .7584979         f(x)= .1554028
  x= .7380193         f(x)= 1.420083E-02
  x= .735738          f(x)= 1.632199E-04

PRESS ANY KEY TO CONTINUE
DO YOU WANT TO RESET CONDITIONS(Y/N)? Y


HAVE YOU ENTERED THE BEATTIE-BRIDGEMAN
CONSTANTS(Y/N)? Y

SPECIFY TEMPERATURE(KELVIN)=? 425
SPECIFY PRESSURE(ATM)? 40

IDEAL GAS VOLUME= .8718875

BETA=-10.79774
GAMMA= .9633568
DLTA= 3.688917E-02


DO YOU WANT TO SEE ALL ROOTS(Y/N)? N

DO YOU WANT GRAPHICAL SOLUTION(Y/N)? N

DEFINE THE RANGE OF SEARCH:
LOWER LIMIT? 0
HIGHER LIMIT? 1
```

```
GIVE YOUR BEST GUESS OF THE ROOT? .9

GIVE THE CONVERGENCE VALUE OF F? 0.001

CONVERGENCE TO ROOT 1

   x= .9                  f(x)= 8.662021
   x= .7280155            f(x)= 2.764109
   x= .5966554            f(x)= .8938306
   x= .4934439            f(x)= .2981002
   x= .4068161            f(x)= .1057316
   x= .3241401            f(x)= .0391612
   x= .2446819            f(x)= 6.332182E-03
   x= .2288747            f(x)=-1.228526E-04

PRESS ANY KEY TO CONTINUE
DO YOU WANT TO RESET CONDITIONS(Y/N)? N
DO YOU WANT TO RERUN THE PROGRAM(Y/N)? N
Ok
```

### Discussion of results

*Part a* The first polynomial to be evaluated by this program is

$$x^4 + 6x^3 + 7x^2 - 6x - 8 = 0$$

It was mentioned in Sec. 2-2 that this polynomial has roots at 1, −1, −2, and −4. All four of these roots are determined by the program, and the graphical sequence of root finding, alternating with synthetic division, is demonstrated. The range of search is defined to be −5 to 2 in order to encompass all four roots without the need to redefine the range of search. The first guess of the root is chosen at 1.8. The Newton-Raphson method converges to the root at 1 in five steps. The first graphical solution (ROOT 1) shows the location of all four roots and draws the Newton-Raphson trajectory to the first root.

Synthetic division reduces the polynomial to a third-degree equation as can be observed in the second graphical solution (ROOT 2) which shows the function crossing the $x$ axis at three positions. The search for the second root begins with the value of the first root as the starting position and arrives at the value of −1.0 (the computer gives the root as −0.9999998 due to roundoff error). After synthetic division, the third root is located at −2 and subsequently the fourth root is found at −4, as demonstrated by the graphical solutions titled ROOT 3 and ROOT 4, respectively.

This exercise establishes the ability of the program to locate the real roots of a polynomial.

*Part b* In this example we use the program to calculate the specific volume of a gas using the Beattie-Bridgeman equation of state. Since this equation can be arranged in the canonical form of a fourth-degree polynomial, the same method and program as in part *a* can be used. Additional information such as the temperature, pressure, and constants are entered through the subroutine.

The Beattie-Bridgeman equation has one real root which is of interest, the one located near the value of $V$ given by the ideal gas law; therefore, the latter is used to estimate the initial guess of the root. A pair of complex roots are also present in the Beattie-Bridgeman equation.

At 425 K and 1 atm pressure the ideal gas law calculates the specific volume to be 34.8755 L/mol. Using a starting value of 30 and a range of 0 to 40, the Newton-Raphson method finds the root of the Beattie-Bridgeman equation at 34.5639 in five iterations. Following synthetic division, the program locates the rest of the roots as

$$0.36879$$

$$-0.02860 + 0.21950i$$

$$-0.02860 - 0.21950i$$

This verifies the ability of the program to determine one pair of complex roots.

The specific volumes for the remaining pressures are determined without showing graphical solutions and by locating only one root for each pressure—the one closest to the ideal gas volume.

Direct comparison between the Beattie-Bridgeman and the ideal gas volumes is made in Table E2-2*b*. As expected, the deviation from ideality increases as the pressure of the gas increases.

**Table E2-2*b***

| | Specific volume, L/mol | |
|---|---|---|
| Pressure, atm | Beattie-Bridgeman | Ideal gas |
| 1 | 34.5639 | 34.8755 |
| 10 | 3.1551 | 3.4875 |
| 20 | 1.3781 | 1.7438 |
| 30 | 0.7357 | 1.1625 |
| 40 | 0.2289 | 0.8719 |

## 2-6 GRAEFFE'S METHOD FOR REAL AND COMPLEX ROOTS

Up to this point our discussion of root-finding methods has concentrated on equations which have mainly real roots. A method which is universal in its ability to locate all roots of polynomial equations, whether they may be real, complex, or repeated, is known as Graeffe's method. This method is based on the transformation of the original polynomial into one whose roots can be directly evaluated from the coefficients of the transformed polynomial.

Consider the polynomial equation

$$f(x) = a_n x^n + a_{n-1}x^{n-1} + \cdots + a_1 x + a_0 = 0 \tag{2-50}$$

whose roots can be arranged in descending order according to their absolute vaues:

$$|x_1| \geq |x_2| \geq |x_3| \geq \cdots \geq |x_n| \tag{2-51}$$

If Eq. (2-50) can be converted to an equation which has roots that are much larger than each other in absolute value, i.e.,

$$|x_1^m| \gg |x_2^m| \gg |x_3^m| \gg \cdots \gg |x_n^m| \tag{2-52}$$

then Newton's relations (see Sec. 2-2) can be used to calculate the roots directly.

Graeffe's method first substitutes $-x$ for $x$ in the polynomial to form

$$f(-x) = (-1)^n a_n x^n + (-1)^{n-1} a_{n-1} x^{n-1} + \cdots (-1) a_1 x + a_0 = 0 \tag{2-53}$$

and then multiplies the two polynomials together to obtain a new transformed polynomial

$$f_2(-x^2) = f(x)f(-x) = 0 \tag{2-54}$$

This new function, $f_2(-x^2)$, has only even powers of $x$, and its roots are the negative squares $(-x^2)$ of the roots of the original polynomial $f(x)$. Therefore, the ordering of roots shown by Eq. (2-51) becomes even more accentuated for the roots $|-x_i^2|$.

If the substitution-and-squaring operation is repeated again on each succeeding new function, a series of polynomials is developed each having roots which are the negatives of the squares of the roots of the previous polynomial:

$$f_4(-x^4) = f_2(x^2)f_2(-x^2) = 0 \tag{2-55}$$

$$f_8(-x^8) = f_4(x^4)f_4(-x^4) = 0 \tag{2-56}$$

$$f_{16}(-x^{16}) = f_8(x^8)f_8(-x^8) = 0 \tag{2-57}$$

$$\vdots$$

$$f_m(-x^m) = f_{m/2}(x^{m/2})f_{m/2}(-x^{m/2}) = 0 \tag{2-58}$$

where $m = 2^r$ and $r$ = number of times the transformation operation has been applied.

The roots of the final polynomial [Eq. (2-58)] would then have an ordering as depicted by

$$|x_1^m| \gg |x_2^m| \gg |x_3^m| \gg \cdots \gg |x_n^m| \tag{2-52}$$

The entire Graeffe transformation is summarized by the following equation, which is a modification of an equation given by Carnahan et al. [7]:

$$_rA_k = {}_{r-1}A_k^2 + \sum_{l=1}^{k} 2(-1)^l \,{}_{r-1}A_{k+l}\,{}_{r-1}A_{k-l} \qquad 0 \le k \le n,\ k+l \le n,\ k-l \ge 0 \tag{2-59}$$

where $_0A_k = a_k$ (the coefficients of the original polynomial) and $_rA_k$ = the coefficients obtained from the $r$th application of the transformation.

The steps of this transformation are demonstrated in Table 2-1. The coefficients $a_k$ of the original polynomial are first squared. To these squared terms are added the product $(-2) \times$ (the coefficient to the left of $a_k$) $\times$ (the coefficient to the right of $a_k$) and the product $(2) \times$ (the coefficient two positions over to the left of $a_k$) $\times$ (the coefficient two positions over to the right of $a_k$). This is continued until all the possible products have been added.

As was mentioned earlier in this chapter, the polynomial may have real, complex, and repeated roots. The following four cases will be discussed and their solution will be shown using Graeffe's method: (1) real and distinct roots, (2) one pair of complex roots or repeated real roots, (3) multiple pairs of complex roots or repeated real roots, with different moduli, and (4) multiple pairs of complex roots or repeated real roots, with identical moduli.

Repetition of the Graeffe transformation a sufficiently large number of times (5 to 10) will cause the coefficients to grow very rapidly in magnitude. Depending on the types of roots present in the polynomial, it is possible for the squared term, $_{r-1}A_k^2$, in Eq. (2-59) to grow much more rapidly than the summation of the product terms in that equation. In such a case, the new coefficients $_rA_k$ are pure squares of $_{r-1}A_k$. However, the presence of complex or repeated roots prevents these coefficients from becoming pure squares. It is this behavior of the coefficients that gives the Graeffe method its power to identify each of the four categories of roots mentioned in the previous paragraph, and to determine the values of the real and complex roots.

**Real and distinct roots** In the case of real and distinct roots, application of Graeffe's root-squaring transformation yields a polynomial whose coefficients $_rA_k$ are *all pure squares*. This can be detected easily during the computational algorithm and the program can automatically exit from Graeffe's procedure and apply Newton's relations.

Application of Newton's 1st relation [Eq. (2-16)] to the transformed polynomial yields

**Table 2-1 Application of Graeffe's root-squaring transformation**

| $r$ | $n$ | $n-1$ | $n-2$ | $n-3$ | $\cdots$ | 3 | 2 | 1 | 0 |
|---|---|---|---|---|---|---|---|---|---|
| 0 | $a_n$ | $a_{n-1}$ | $a_{n-2}$ | $a_{n-3}$ | $\cdots$ | $a_3$ | $a_2$ | $a_1$ | $a_0$ |
| | $a_n^2$ | $a_{n-1}^2$ | $a_{n-2}^2$ | $a_{n-3}^2$ | $\cdots$ | $a_3^2$ | $a_2^2$ | $a_1^2$ | $a_0^2$ |
| | | $-2a_na_{n-2}$ | $-2a_{n-1}a_{n-3}$ | $-2a_{n-2}a_{n-4}$ | $\cdots$ | $-2a_4a_2$ | $-2a_3a_1$ | $-2a_2a_0$ | |
| | | | $+2a_na_{n-4}$ | $+2a_{n-1}a_{n-5}$ | $\cdots$ | $+2a_5a_1$ | $+2a_4a_0$ | | |
| | | | | $-2a_na_{n-6}$ | $\cdots$ | $-2a_6a_0$ | | | |
| 1 | ${}_1A_n$ | ${}_1A_{n-1}$ | ${}_1A_{n-2}$ | ${}_1A_{n-3}$ | $\cdots$ | ${}_1A_3$ | ${}_1A_2$ | ${}_1A_1$ | ${}_1A_0$ |
| | ${}_1A_n^2$ | ${}_1A_{n-1}^2$ | ${}_1A_{n-2}^2$ | ${}_1A_{n-3}^2$ | $\cdots$ | ${}_1A_3^2$ | ${}_1A_2^2$ | ${}_1A_1^2$ | ${}_1A_0^2$ |
| | | $-2A_nA_{n-2}$ | $-2A_{n-1}A_{n-3}$ | $-2A_{n-2}A_{n-4}$ | $\cdots$ | $-2A_4A_2$ | $-2A_3A_1$ | $-2A_2A_0$ | |
| | | | $+2A_nA_{n-4}$ | $+2A_{n-1}A_{n-5}$ | $\cdots$ | $+2A_5A_1$ | $+2A_4A_0$ | | |
| | | | | $-2A_nA_{n-6}$ | $\cdots$ | $-2A_6A_0$ | | | |
| 2 | ${}_2A_n$ | ${}_2A_{n-1}$ | ${}_2A_{n-2}$ | ${}_2A_{n-3}$ | $\cdots$ | ${}_2A_3$ | ${}_2A_2$ | ${}_2A_1$ | ${}_2A_0$ |
| | $\vdots$ | $\vdots$ | $\vdots$ | $\vdots$ | | $\vdots$ | $\vdots$ | $\vdots$ | $\vdots$ |
| $r$ | ${}_rA_n$ | ${}_rA_{n-1}$ | ${}_rA_{n-2}$ | ${}_rA_{n-3}$ | $\cdots$ | ${}_rA_3$ | ${}_rA_2$ | ${}_rA_1$ | ${}_rA_0$ |

*Note*: The presubscripts on ${}_rA_k$ have been omitted in the products to eliminate congestion.

$$\sum_{i=1}^{n} -x_i^m = -\frac{{}_rA_{n-1}}{{}_rA_n} \tag{2-60}$$

But since inequality (2-52) is true, the first root is predominant and the summation term may be replaced by $-x_1^m$:

$$-x_1^m = -\frac{{}_rA_{n-1}}{{}_rA_n} \tag{2-61}$$

Application of Newton's 2nd relation [Eq. (2-17)] results in

$$\sum_{\substack{i,\, j=1 \\ i \neq j}}^{n} (-x_i^m)(-x_j^m) = \frac{{}_rA_{n-2}}{{}_rA_n} \tag{2-62}$$

Invoking inequality (2-52) simplifies the above equation to

$$(-x_1^m)(-x_2^m) = \frac{{}_rA_{n-2}}{{}_rA_n} \tag{2-63}$$

Similarly, application of Newton's 3rd relation [Eq. (2-18)] gives

$$(-x_1^m)(-x_2^m)(-x_3^m) = -\frac{{}_rA_{n-3}}{{}_rA_n} \tag{2-64}$$

Finally, application of Newton's $n$th relation [Eq. (2-19)] yields

$$(-x_1^m)(-x_2^m)\cdots(-x_n^m) = (-1)^n \frac{{}_rA_0}{{}_rA_n} \tag{2-65}$$

Each of the roots of the initial polynomial can be solved explicitly by combining Eqs. (2-61), (2-63), (2-64), and (2-65):

$$\begin{aligned} x_1 &= \pm\sqrt[m]{\frac{{}_rA_{n-1}}{{}_rA_n}} \\ x_2 &= \pm\sqrt[m]{\frac{{}_rA_{n-2}}{{}_rA_{n-1}}} \\ x_3 &= \pm\sqrt[m]{\frac{{}_rA_{n-3}}{{}_rA_{n-2}}} \\ &\vdots \\ x_n &= \pm\sqrt[m]{\frac{{}_rA_0}{{}_rA_1}} \end{aligned} \tag{2-66}$$

The sign of each of the above roots must be determined by substitution into the original polynomial, $f(x) = 0$.

**One pair of complex roots or repeated real roots** When one of the coefficients ${}_rA_k$ of the new polynomial, say the ${}_rA_s$ coefficient, does not become a pure square but is flanked by pure squares, e.g.,

| ${}_rA_{s+1}$ | ${}_rA_s$ | ${}_rA_{s-1}$ |
|---|---|---|
| Pure square | Nonsquare | Pure square |

then this signifies the presence of a pair of complex roots (or a pair of repeated real roots) [5].

The modulus of a pair of complex roots is defined as

$$R^2 = (\alpha + \beta i)(\alpha - \beta i) = \alpha^2 + \beta^2 \tag{2-67}$$

The ratio of the two pure squares which flank the nonsquare gives the $m$th power of the modulus:

$$(R^2)^m = \frac{{}_rA_{s-1}}{{}_rA_{s+1}} \tag{2-68}$$

Combining Eqs. (2-67) and (2-68):

$$\alpha^2 + \beta^2 = \sqrt[m]{\frac{{}_rA_{s-1}}{{}_rA_{s+1}}} \tag{2-69}$$

There are two unknown quantities $(\alpha, \beta)$ in Eq. (2-69). One more equation is needed to determine both $\alpha$ and $\beta$. This is provided by Newton's 1st relation applied to the original polynomial:

$$\sum_{i=1}^{n} x_i = -\frac{a_{n-1}}{a_n} \tag{2-70}$$

It is assumed that all the real and distinct roots have been determined as described earlier, so the only unknowns in Eq. (2-70) are the pair of complex roots $(\alpha + \beta i)$ and $(\alpha - \beta i)$. Therefore, Eqs. (2-69) and (2-70) can be solved simultaneously for $\alpha$ and $\beta$.

**Multiple pairs of complex roots or repeated real roots with different moduli** It is possible to have more than one nonsquare coefficient flanked by pure squares. For example, if there are two nonsquares at positions $s_1$ and $s_2$, respectively, the coefficients of the transformed polynomial will be

| ${}_rA_{s_1+1}$ | ${}_rA_{s_1}$ | ${}_rA_{s_1-1}$ | $\cdots$ | ${}_rA_{s_2+1}$ | ${}_rA_{s_2}$ | ${}_rA_{s_2-1}$ | $\cdots$ |
|---|---|---|---|---|---|---|---|
| Pure square | Non-square | Pure square | | Pure square | Non-square | Pure square | |

This implies that there are two pairs of complex (or repeated real) roots. Each pair is distinct from the other, i.e., it has a different modulus.

The two moduli are given by

$$R_{s_1}^2 = (\alpha_{s_1} + \beta_{s_1} i)(\alpha_{s_1} - \beta_{s_1} i) = \alpha_{s_1}^2 + \beta_{s_1}^2 \tag{2-71}$$

$$(R_{s_1}^2)^m = \frac{{}_rA_{s_1-1}}{{}_rA_{s_1+1}} \tag{2-72}$$

$$R_{s_2}^2 = (\alpha_{s_2} + \beta_{s_2} i)(\alpha_{s_2} - \beta_{s_2} i) = \alpha_{s_2}^2 + \beta_{s_2}^2 \tag{2-73}$$

$$(R_{s_2}^2)^m = \frac{{}_rA_{s_2-1}}{{}_rA_{s_2+1}} \tag{2-74}$$

Since four unknowns are involved ($\alpha_{s_1}, \alpha_{s_2}, \beta_{s_1}, \beta_{s_2}$), four equations are needed for complete solution of this problem. Combine Eq. (2-71) with Eq. (2-72) to obtain

$$\alpha_{s_1}^2 + \beta_{s_1}^2 = \sqrt[m]{\frac{{}_rA_{s_1-1}}{{}_rA_{s_1+1}}} \tag{2-75}$$

and Eq. (2-73) with Eq. (2-74) to obtain

$$\alpha_{s_2}^2 + \beta_{s_2}^2 = \sqrt[m]{\frac{{}_rA_{s_2-1}}{{}_rA_{s_2+1}}} \tag{2-76}$$

These two equations, in conjunction with Newton's 1st and 2nd relations,

$$\sum_{i=1}^{n} x_i = -\frac{a_{n-1}}{a_n} \tag{2-77}$$

$$\sum_{\substack{i,\, j=1 \\ i \neq j}}^{n} x_i x_j = \frac{a_{n-2}}{a_n} \tag{2-78}$$

provide a set of four equations in four unknowns which may be solved simultaneously. The product term in Newton's 2nd relation makes this set of equations nonlinear. For the above case of two pairs of complex or repeated roots, in addition to real distinct roots, the simultaneous solution of Eqs. (2-75) to (2-78) yields a quadratic equation for one of the $\alpha$'s:

$$\alpha_{s_1}^2 + \frac{1}{2}\left(\frac{a_{n-1}}{a_n} + \sum_i^{\text{real}} x_i\right)\alpha_{s_1}$$

$$+ \frac{1}{4}\left[\frac{a_{n-2}}{a_n} + \left(\frac{a_{n-1}}{a_n} + \sum_i^{\text{real}} x_i\right)\left(\sum_i^{\text{real}} x_i\right) - \sum_{\substack{i,\, j \\ i \neq j}}^{\text{real}} x_i x_j - R_{s_1}^2 - R_{s_2}^2\right] = 0 \tag{2-79}$$

and a first-degree equation for the other $\alpha$:

$$2\alpha_{s_2} = -\frac{a_{n-1}}{a_n} - \sum_{i}^{\text{real}} x_i - 2\alpha_{s_1} \tag{2-80}$$

The values of $\alpha$ in Eqs. (2-79) and (2-80) were labeled $\alpha_{s_1}$ and $\alpha_{s_2}$, respectively. However, the equations cannot distinguish between $\alpha_{s_1}$ and $\alpha_{s_2}$. These values could be the reverse, i.e., $\alpha_{s_2}$ obtained from Eq. (2-79) and $\alpha_{s_1}$ from Eq. (2-80). The only way to distinguish between them is by substitution of the complex roots into the polynomial.

The summation terms in Eqs. (2-79) and (2-80) involve only the real distinct roots of the polynomial. If no real distinct roots exist, then these terms vanish. The values of the two moduli $R_{s_1}^2$ and $R_{s_2}^2$, in Eq. (2-79), are calculated from Eqs. (2-72) and (2-74), respectively. The solution of Eq. (2-79) for the value of $\alpha_{s_1}$ is obtained by the quadratic equation formula. The solution of Eq. (2-80) for $\alpha_{s_2}$ is obtained directly after substitution of $\alpha_{s_1}$. The values of $\beta_{s_1}$ and $\beta_{s_2}$ are calculated from Eqs. (2-75) and (2-76). At this point, the substitution $\alpha_{s_1} + \beta_{s_1} i$ into the polynomial will verify whether $\alpha_{s_1}$ and $\alpha_{s_2}$ were labeled correctly. If this root does not satisfy the polynomial equation, the values of $\alpha_{s_1}$ and $\alpha_{s_2}$ are swapped, the values of $\beta_{s_1}$ and $\beta_{s_2}$ are recalculated from (2-75) and (2-76), and the substitution into the polynomial is repeated to verify the roots.

If Graeffe's root-squaring process yields three (or more) nonsquares, all separated by pure squares, this would imply the presence of three (or more) pairs of complex roots or repeated real roots of different moduli. The required number of equations for the solution of this problem can be formulated by involving the higher-order Newton's relations. The resulting set of simultaneous equations would be nonlinear of degree higher than 2 and would require the application of a root-finding method for its complete solution. In essence, Graeffe's root-squaring method is circular on itself, i.e., it reduces the $n$th-degree polynomial to an $(n - m)$th-degree problem after evaluating the real roots and one or two pairs of complex (or repeated) roots. This is probably the main disadvantage in Graeffe's method, which has prevented it from becoming more widely used.

**Multiple pairs of complex roots or repeated real roots with identical moduli** When $(2p - 1)$ nonsquares are flanked by two pure squares

$$\underset{\text{Pure square}}{{}_rA_{s+1}} \qquad \underbrace{{}_rA_s \qquad {}_rA_{s-1} \qquad \cdots}_{\text{Nonsquares}} \qquad \underset{\text{Pure squares}}{{}_rA_{s-(2p-1)}}$$

then there are $p$ pairs of complex (or repeated real) roots with identical moduli. The common modulus of these roots is given by

$$(R^2)^{pm} = \frac{{}_rA_{s-(2p-1)}}{{}_rA_{s+1}} \tag{2-81}$$

where
$$R^2 = \alpha_i^2 + \beta_i^2 \qquad i = 1 \text{ to } p \tag{2-82}$$

Note that Eq. (2-82) can be written $p$ times. The $p$ equations from (2-82) and the first $p$ Newton's relations can then be combined to solve for all the roots. As mentioned in the previous case, if two pairs of roots exist, solution is tractable. For three or more pairs of roots the solution becomes increasingly difficult.

**Example 2-3 Graeffe's root-squaring method for real and complex roots of polynomials** Develop a computer program which uses Graeffe's root-squaring method to determine real and complex roots of polynomials. The program should be able to calculate $n$ number of real roots and up to two pairs of complex and/or repeated roots. Apply this program to the solution of the following polynomial equations:

*a.* $x^5 - 5x^4 - 15x^3 + 85x^2 - 26x - 120 = 0$
*b.* $x^4 + 7x^3 + 12x^2 - 4x - 16 = 0$
*c.* $x^4 - 15x^3 + 138x^2 - 324x + 200 = 0$
*d.* $x^4 - 8x^3 + 42x^2 - 80x + 125 = 0$
*e.* $x^5 - 10x^4 + 42x^3 - 102x^2 + 145x - 100 = 0$

METHOD OF SOLUTION The application of Graeffe's root-squaring method causes the coefficients of the transformed polynomials to grow very rapidly in magnitude, often exceeding the limits of floating-point numbers that can be handled by digital computers. The IBM PC can work with floating-point numbers in the range of 2.9E − 39 to 1.7E + 38 (positive or negative). Any number whose absolute value is larger than 1.7E + 38 is reset to 1.7E + 38 and an "Overflow" error message is given by the computer. Also, any number whose absolute value is smaller than 2.9E − 39 is automatically treated as zero, without any warning message given. These serious limitations have been alleviated in this computer program by the following techniques:

1. The coefficients of the original polynomial are separated into a mantissa part and an exponent part (see lines 220–300 of the program). The mantissa part, which ranges between ±10, is stored in matrix A and the exponent part in matrix FACT.
2. After each step of the Graeffe method the newly calculated mantissa parts of the coefficients are examined. If their absolute values are greater than 10, these values are divided by 10 and the corresponding exponent factors are incremented by 1. If the absolute values of the mantissas are less than 1, they are multiplied by 10 and the corresponding exponent factors are decreased by 1 (see lines 610–620).

3. The matrix FACT, containing the exponent factors, is handled as a floating-point matrix. Therefore, the exponent factors can have maximum values of $\pm 1.7E38$, i.e., the coefficients can reach the phenomenally large values of $10 \times 10^{1.7 \times 10^{38}}$. Such large numbers are not necessary, even for Graeffe's method. Therefore we have chosen an upper limit of 999 for the exponent factor. If one of the exponent factors exceeds this limit (see line 640), the iteration is terminated and the roots are evaluated.
4. The matrices A and FACT are handled together, throughout this program, so that the exponent part of each coefficient is always accounted for.

During execution, the program keeps account of the number and location of pure square and nonsquare coefficients. In the case of all pure squares, the roots are all real and distinct and the program uses Eqs. (2-66) to evaluate these roots. In the case of only one nonsquare, one pair of complex or repeated roots exists and the program uses Eqs. (2-69) and (2-70) to evaluate the roots. When two nonsquares are present, the polynomial has two pairs of complex or repeated roots with different moduli; Eqs. (2-72), (2-74), (2-75), (2-76), (2-79), and (2-80) are used. If three neighboring nonsquares are detected, the polynomial has two pairs of complex or repeated roots with identical moduli; Eqs. (2-79) to (2-82) are used.

A step-by-step description of the program is given in the next section.

PROGRAM DESCRIPTION This program, which we call GRAEFFE.BAS, consists of the sections listed in Table E2-3.

*Main program* Lines 120–190 are used to input the order of the polyno-

**Table E2-3**

| Program section | Line numbers |
|---|---|
| Title | 0–90 |
| Main program | 100–1360 |
| Subroutines | |
| 1. Checking sign of real roots | 1500–1630 |
| 2. Calculation of the sum of the product of the real roots | 2000–2090 |
| 3. Checking complex roots | 3000–3270 |
| 4. One pair of complex or repeated roots | 4000–4220 |
| 5. Two pairs of complex or repeated roots with different moduli | 5000–5410 |
| 6. Two pairs of complex or repeated roots with identical moduli | 6000–6530 |

mial and its coefficients. Line 200 inputs the convergence criterion for the polynomial function $f(x)$. A value of 0.0001 is recommended. If convergence fails, this value may be raised by the user.

Lines 220–300 perform the factorization of the coefficients into the mantissa part and the exponent part as described in the previous section.

Lines 330–400 print the values of the original coefficients—mantissa and exponent are printed together.

Line 450 is the beginning of the Graeffe root-squaring iteration. A maximum of 20 iterations is specified. This may be changed by the user.

Lines 500–650 perform the root-squaring calculations and generate the new transformed coefficients—both the mantissa and the exponent of each coefficient are evaluated. If the exponent value is higher than 999, the root-squaring iteration is interrupted and execution continues at line 880.

Lines 670–750 determine whether the coefficients are pure squares. If the value of the square of the coefficient is 1000 times larger than the sum of the product terms added to it to obtain the transformed coefficient, then the latter is considered to be a pure square.

Lines 770–820 count the number of pure squares and nonsquares. If all coefficients become pure squares, the root-squaring iteration is interrupted and execution of the program continues at line 900.

Lines 860–890 print the appropriate message to indicate why the iteration was terminated and reduce the iteration counter by one.

Lines 900–910 print the status (pure square or nonsquare) of the coefficients.

Lines 930–970 locate the position of the nonsquares.

Lines 990–1080 evaluate the real roots by using Eq. (2-66).

Line 1090 branches the execution of the program to subroutine 1 which determines the sign of the real root by direct substitution into the original polynomial equation.

Line 1100 calculates the sum of the real roots. This sum is utilized by the program in Newton's 1st relation, when pairs of repeated or complex roots are present in addition to the real distinct roots.

Line 1140 branches execution of the program to subroutine 2 which calculates the sum of the product of the real roots. This sum is utilized in Newton's 2nd relation, when pairs of repeated or complex roots are present in addition to the real distinct roots.

Line 1160 branches execution of the program to subroutines 4, 5, or 6 depending on the number of nonsquares. If NSQUARES = 1, execution goes to subroutine 4, which calculates one pair of complex or repeated roots. If NSQUARES = 2, execution branches to subroutine 5, which calculates two pairs of complex or repeated roots with different moduli. Finally, if NSQUARES = 3, subroutine 6 is called to calculate two pairs of complex or repeated roots with identical moduli.

Lines 1180–1340 print out all the roots, real or complex. The values of the roots are rounded in lines 1250 and 1300 to eliminate the possibility of the computer printing values, such as 1.999998, due to truncation error of the computer.

*Subroutine 1* This subroutine evaluates the polynomials using the positive and negative values of each root. It chooses the one that satisfies the convergence criterion and gives the lowest value of the function. A warning is printed if neither positive nor negative values satisfy the convergence criterion.

*Subroutine 2* This subroutine calculates the sum of the product of the real roots.

*Subroutine 3* Checks the validity of the complex roots by direct substitution of the complex roots in the original polynomial. A warning is printed if the complex root does not satisfy the convergence criterion.

*Subroutine 4* This subroutine is used in the case where one pair of complex or repeated roots exists, possibly in addition to real distinct roots.

Line 4030 is the solution of Eq. (2-70) for $\alpha$, the real part of the roots.

Line 4070 evaluates the modulus of the roots using Eq. (2-68).

Line 4080 calculates $\beta$, the complex part of the root, using Eq. (2-68).

Lines 4110–4150 determine whether the pair of roots are complex or real repeated. If the roots are repeated, the complex part is zero and the square root of the modulus would give the absolute value of the repeated root. In line 4110, the square root of the modulus is substituted in the original polynomial by calling subroutine 1. If neither the positive nor the negative values satisfy the convergence criterion, the pair must be complex. If both positive and negative values satisfy the convergence criterion, the roots are repeated but of opposite sign. If only the positive or the negative value satisfies the convergence criterion, the roots are repeated and of the same sign.

Lines 4180–4200 check the validity of the complex pair of roots by direct substitution in the original polynomial (subroutine 4).

Line 4210 returns execution to the main program.

*Subroutine 5* This subroutine calculates two pairs of complex or repeated roots with different moduli. This case is recognized by the presence of two nonsquare coefficients separated by pure squares. The subroutine locates each nonsquare sequentially and uses the same method employed in subroutine 4 to determine whether the pair is complex or repeated. If this subroutine locates a pair of repeated roots, it adds these roots to the sum of the real roots, it cancels the nonsquare corresponding to the repeated pair, and calls subroutine 4 which in turn evaluates the remaining pair of complex or repeated roots. However, if neither pair consists of repeated roots, subroutine 5 continues execution as follows.

Lines 5240–5260, together with 5050, are the solution of Eqs. (2-79) and (2-80). They yield the values of $\alpha_{s_1}$ and $\alpha_{s_2}$.

Lines 5270–5320 calculate the value of $\beta_{s_1}$ from Eq. (2-75).

Lines 5330–5360 verify the pairing of $\alpha_{s_1}$ and $\beta_{s_1}$ by calling subroutine 3. If the pair does not satisfy the convergence criterion, $\alpha_{s_1}$ and $\alpha_{s_2}$ are swapped, the value of $\beta_{s_1}$ is recalculated, and subroutine 3 is called again to verify the new pair.

Lines 5370–5380 calculate the value of $\beta_{s_2}$ from Eq. (2-76).

Line 5400 returns execution to the main program.

*Subroutine 6* This subroutine calculates two pairs of complex or repeated roots with identical moduli. This case is recognized by the presence of three neighboring nonsquare coefficients. It is possible for the Graeffe method to develop three nonneighboring nonsquares; however, such a case would signify the presence of three pairs of complex or repeated roots with different moduli. This latter case is not solved by this program.

Lines 6020–6070 verify the presence of neighboring nonsquares. If the nonsquares are not neighboring, a message is printed (line 6520) and execution is returned to the main program (line 6530).

Lines 6080–6230 locate the first nonsquare and ascertain whether this corresponds to a set of repeated or complex roots. The method is similar to that used in subroutine 5, with two differences: (1) since the moduli in this case are identical, Eq. (2-81) is used to calculate the modulus (line 6160), and (2) since the moduli are identical, only one test for repeated roots can be performed. If this test discovers a pair of repeated roots with the same sign or a pair of complex roots, execution continues at line 6240. If the test concludes that the pair of roots are repeated but of opposite sign, execution transfers to line 6380. The latter case also includes the situation in which the pair of roots is complex with zero real component.

Lines 6250–6270, together with 6100, are the solution of Eqs. (2-79) and (2-80). They yield the values of the two $\alpha$'s.

Lines 6280–6360 calculate the values of the two $\beta$'s from Eq. (2-82). No verification of the pairing of the $\alpha$'s and $\beta$'s is needed because the moduli are identical.

Line 6360 returns execution to the main program.

Lines 6380–6500 evaluate the cases of repeated roots with opposite sign and complex roots with zero real part. This situation is treated separately because the opposite sign in the repeated pair causes the two values of $\alpha$ to cancel each other in Eq. (2-80) (note the difference between lines 6400 and 6270). The value of $\beta$ is calculated using Eq. (2-82) in line 6440. In order to differentiate between repeated roots of opposite sign and complex roots with zero real part, a substitution into the original polynomial is performed by subroutine 3 (lines 6450–6460).

Line 6510 rcturns cxecution to the main program.

PROGRAM

```
0 SCREEN 0:WIDTH 80:CLS:KEY OFF
10 PRINT"**********************************************************************"
20 PRINT"*                                EXAMPLE 2-3                         *"
30 PRINT"*                                                                    *"
40 PRINT"*                      GRAEFFE'S ROOT-SQUARING METHOD                *"
50 PRINT"*                                                                    *"
60 PRINT"*                               (GRAEFFE.BAS)                        *"
90 PRINT"**********************************************************************"
100 '**************************** Main Program ****************************
110 '
120 'Define the polynomial
130 '
140 PRINT"DEGREE OF POLYNOMIAL";:INPUT N
150 DIM A(20,N),C$(N),SUM(N),FACT(20,N),LC(N),R(N),X(N),XI(N),ROOT$(N),SUMI(N),S
UMC(N)
160 FOR K=N TO 0 STEP -1
170 PRINT"  COEFFICIENT ";K;
180 INPUT A(0,K)
190 NEXT K
200 PRINT "GIVE THE CONVERGENCE VALUE OF F";: INPUT EPS
210 '
220 'Extract factor of ten and reduce coefficients by this factor,
230 'in order to be able to handle very large numbers.
240 'Keep account of factors, and increase or decrease accordingly.
250 '
260 FOR K=N TO 0 STEP -1
270 IF A(0,K)=0 GOTO 300
280 FACT(0,K)=INT(LOG(ABS(A(0,K)))/LOG(10))
290 A(0,K)=A(0,K)/(10^FACT(0,K))
300 NEXT K
310 PRINT:PRINT"ROOT-SQUARING PROCESS:"
320 '
330 'Print the original coefficients with the factors
340 '
350 PRINT:PRINT" r";:FOR K=N TO 0 STEP -1:PRINT"       A       ";:NEXT K:PRINT
360 FOR K=N TO 0 STEP -1:PRINT"          r ";K ;:NEXT K:PRINT
370 PRINT USING "## ";R;
380 FOR K=N TO 0 STEP -1
390 PRINT USING "###.###";A(0,K);:PRINT"E";:PRINT USING "### ";FACT(0,K);
400 NEXT K
410 PRINT
420 '
430 'Beginning of major iteration
440 '
450 FOR R=1 TO 20
460 PRINT USING "## ";R;
470 '
480 'Calculate new coefficients
490 '
500 FOR I=N TO 0 STEP -1
510 FACT(R,I)=2*FACT(R-1,I)
520 SUM(I)=0
530 FOR L=1 TO I
540 IF (I-L)<0 OR (I+L)>N THEN GOTO 590
550 FA=FACT(R-1,I+L)+FACT(R-1,I-L)
560 FB=FACT(R,I)-FA
570 IF FB>20 GOTO 590
580 SUM(I) =SUM(I)+2*(-1)^L*A(R-1,I+L)*A(R-1,I-L)/(10^FB)
```

```
590 NEXT L
600 A(R,I)=A(R-1,I)^2+SUM(I)
610 IF ABS(A(R,I))>10 THEN A(R,I)=A(R,I)/10:FACT(R,I)=FACT(R,I)+1:GOTO 610
620 IF ABS(A(R,I))<1  THEN A(R,I)=A(R,I)*10:FACT(R,I)=FACT(R,I)-1
630 PRINT USING "###.###";A(R,I);:PRINT"E";:PRINT USING "### ";FACT(R,I);
640 IF ABS(FACT(R,I))>999 THEN GOTO 880
650 NEXT I
660 '
670 'Check whether coefficients are pure squares
680 '
690 PS$="PURE SQUARE":NS$="non-square"
700 C$(N)=PS$:C$(0)=PS$
710 FOR I=N-1 TO 1 STEP -1
720 IF SUM(I)=0 GOTO 750
730 W=(A(R-1,I)^2)/SUM(I)
740 IF ABS(W)>1000 THEN C$(I)=PS$ ELSE C$(I)=NS$
750 NEXT I
760 PRINT
770 SQUARES=0
780 FOR I=N TO 0 STEP -1
790 IF C$(I)=PS$ THEN SQUARES=SQUARES+1
800 NEXT I
810 NSQUARES=N+1-SQUARES
820 IF SQUARES=N+1 THEN GOTO 900
830 NEXT R
840 'End of major iteration
850 '
860 PRINT:PRINT"ITERATIONS EXCEEDED. POSSIBILITY OF COMPLEX OR REPEATED ROOTS."
870 GOTO 890
880 PRINT: PRINT"FACTOR EXCEEDS 999. POSSIBILITY OF COMPLEX OR REPEATED ROOTS."
890 R=R-1
900 PRINT:PRINT"THE COEFFICIENTS ARE:":PRINT
910 PRINT"  ";:FOR I=N TO 0 STEP -1:PRINT C$(I);", ";:NEXT:PRINT
920 '
930 'Find location of nonsquares
940 '
950 FOR K=N TO 0 STEP -1
960 IF C$(K)<>PS$ THEN LC(K)=K
970 NEXT K
980 '
990 'Evaluate the real roots and the sum of these roots
1000 '
1010 M=2^R
1020 PRINT:PRINT "THE NUMBER OF SQUARING: r =";R,"THE POWER: m =";M
1030 PRINT:PRINT"CALCULATION OF ROOTS:"
1040 SUMRT=0
1050 FOR K=N TO 1 STEP -1
1060 NR=N+1-K
1070 IF C$(K-1)<>PS$ THEN K=K-1:GOTO 1110
1080 X(NR)=(A(R,K-1)/A(R,K))^(1/M)*10^((FACT(R,K-1)-FACT(R,K))/M)
1090 GOSUB 1500 'Check the sign of the root
1100 SUMRT=SUMRT+X(NR)
1110 NEXT K
1120 '
1130 'Calculating the sum of the product of the real roots
1140 GOSUB 2000
1150 'Branch to subroutine for complex or repeated roots
1160 ON NSQUARES GOSUB 4000,5000,6000
1170 IF NSQUARES>3 THEN PRINT"**** PROGRAM CANNOT DETERMINE MORE THAN TWO PAIRS
OF COMPLEX ROOTS ****"
```

```
1180 'Print all the roots
1190 PRINT:PRINT"THE";N;"ROOTS ARE:"
1200 FOR K=1 TO N
1210 YY=1000
1220 IF ABS(X(K)*YY)>32000 THEN YY=YY/10:GOTO 1220
1230 IF X(K)=0 THEN GOTO 1260
1240 IF ABS(X(K)*YY)<1000 THEN YY=YY*10:GOTO 1240
1250 X(K)=CINT(X(K)*YY)/YY
1260 YY=1000
1270 IF ABS(XI(K)*YY)>32000 THEN YY=YY/10:GOTO 1270
1280 IF XI(K)=0 THEN GOTO 1310
1290 IF ABS(XI(K)*YY)<1000 THEN YY=YY*10:GOTO 1290
1300 XI(K)=CINT(XI(K)*YY)/YY
1310 PRINT TAB(20);X(K);
1320 IF XI(K)>0 THEN PRINT"+";XI(K);"i"
1330 IF XI(K)<0 THEN PRINT"-";-XI(K);"i"
1340 NEXT K
1350 END
1360 '
1500 '************* Subroutine 1: Checking sign of real roots ****************
1510 '
1520 FUP=A(0,0)*(10^FACT(0,0)):FUN=A(0,0)*(10^FACT(0,0))
1530 FOR I=N TO 1 STEP -1
1540 FUP=FUP+A(0,I)*(10^FACT(0,I))*X(NR)^I
1550 FUN=FUN+A(0,I)*(10^FACT(0,I))*(-X(NR))^I
1560 NEXT I
1570 PRINT:PRINT"  FUNCTION WITH POSITIVE VALUE OF (";X(NR);")="; FUP
1580 PRINT"  FUNCTION WITH NEGATIVE VALUE OF (";X(NR);")="; FUN
1590 IF ABS(FUP)> EPS AND ABS(FUN) > EPS THEN PRINT "* WARNING: CONVERGENCE NOT
SATISFIED BY REAL ROOT *"
1600 'Choose root which gives lowest value of function
1610 IF ABS(FUN)<ABS(FUP) THEN X(NR)=-X(NR)
1620 RETURN
1630 '
2000 '* Subroutine 2: Calculation of the sum of the product of the real roots *
2010 '
2020 PRODRT=0
2030 FOR K=1 TO N-1
2040 FOR KK=K+1 TO N
2050 PRODRT=PRODRT+X(K)*X(KK)
2060 NEXT KK
2070 NEXT K
2080 RETURN
2090 '
3000 '************* Subroutine 3: Checking complex roots *********************
3010 '
3020 FOR KK=0 TO N: SUMI(KK)=0: NEXT KK
3030 FOR KK=0 TO N: SUMC(KK)=0: NEXT KK
3040 SUMC(0)=1
3050 FOR I=1 TO N+1
3060 SUMC(1)=I
3070 FOR KK=2 TO N
3080 SUMC(KK)=SUMC(KK)+SUMC(KK-1)
3090 NEXT KK
3100 FOR J=0 TO N+1-I
3110 SUMI(J)=SUMI(J)+A(0,I-1+J)*(10^FACT(0,I-1+J))*(ALPHA^(I-1))*SUMC(J)
3120 NEXT J
3130 NEXT I
3140 SUMII=0
3150 FOR I=1 TO N STEP 2
```

```
3160 SUMII=SUMII+(BETA^I)*(SUMI(I))*(-1)^((I-1)/2)
3170 NEXT I
3180 SUMII2=0
3190 FOR I=2 TO N STEP 2
3200 SUMII2=SUMII2+(BETA^I)*(SUMI(I))*(-1)^(I/2)
3210 NEXT I
3220 SUMALL=SUMI(0)+SUMII+SUMII2
3230 IF ABS(SUMALL)<EPS THEN RETURN
3240 PRINT"* WARNING: CONVERGENCE NOT SATISFIED BY COMPLEX ROOT *"
3250 PRINT"* VALUE OF FUNCTION=";SUMALL;" >";EPS; "*"
3260 RETURN
3270 '
4000 '********* Subroutine 4: One pair of complex or repeated roots ***********
4010 '
4020 PRINT:PRINT"CALCULATION OF ONE PAIR OF COMPLEX OR REPEATED ROOTS:"
4030 ALPHA=((-A(0,N-1)/A(0,N))*10^(FACT(0,N-1)-FACT(0,N)) - SUMRT)/2
4040 FOR K=N TO 1 STEP -1
4050 NR=N+1-K
4060 IF LC(K)=0 THEN GOTO 4170
4070 R(K)=(A(R,K-1)/A(R,K+1))^(1/M)*10^((FACT(R,K-1)-FACT(R,K+1))/M)
4080 BETA=SQR(ABS(R(K)-ALPHA^2))
4090 'Check for repeated roots
4100 PRINT"  CHECK FOR REPEATED ROOTS:"
4110 X(NR)=SQR(R(K)):GOSUB 1500
4120 IF ABS(FUP) > EPS AND ABS(FUN) > EPS THEN PRINT"* ROOTS ARE COMPLEX *":GOTO
 4190
4130 IF ABS(FUP)<EPS AND ABS(FUN)<EPS THEN PRINT"** ROOTS ARE REAL AND REPEATED
BUT OF OPPOSITE SIGN **":X(NR-1)=-X(NR)
4140 IF ABS(FUP)<EPS AND ABS(FUN)>EPS THEN PRINT"** ROOTS ARE REAL AND REPEATED
AND OF THE SAME SIGN **":X(NR-1)=X(NR)
4150 IF ABS(FUP)>EPS AND ABS(FUN)<EPS THEN PRINT"** ROOTS ARE REAL AND REPEATED
AND OF THE SAME SIGN **":X(NR-1)=X(NR)
4160 GOTO 4210
4170 NEXT K
4180 'Check convergence with complex roots
4190 GOSUB 3000
4200 X(NR)=ALPHA:XI(NR)=-BETA:X(NR-1)=ALPHA:XI(NR-1)=BETA
4210 RETURN
4220 '
5000 'Subroutine 5:Two pairs of complex or repeated roots with different moduli
5010 '
5020 RSUM=0
5030 PRINT:PRINT"TWO PAIRS OF COMPLEX OR REPEATED ROOTS WITH DIFFERENT MODULI:"
5040 B=((A(0,N-1)/A(0,N))*10^(FACT(0,N-1)-FACT(0,N))+SUMRT)/2
5050 PRINT"  CHECK FOR REPEATED ROOTS:"
5060 FOR K=N TO 1 STEP -1
5070 NR=N+1-K
5080 IF LC(K)=0 THEN GOTO 5230
5090 R(K)=(A(R,K-1)/A(R,K+1))^(1/M)*10^((FACT(R,K-1)-FACT(R,K+1))/M)
5100 RSUM=RSUM + R(K)
5110 'Check for repeated roots
5120 X(NR)=SQR(R(K)):GOSUB 1500
5130 IF ABS(FUP) > EPS AND ABS(FUN) > EPS THEN PRINT"* ROOTS ARE COMPLEX *":GOTO
 5230
5140 IF ABS(FUP)<EPS AND ABS(FUN)<EPS THEN PRINT"** ROOTS ARE REAL AND REPEATED
BUT OF OPPOSITE SIGN **":X(NR-1)=-X(NR)
5150 IF ABS(FUP)<EPS AND ABS(FUN)>EPS THEN PRINT"** ROOTS ARE REAL AND REPEATED
AND OF THE SAME SIGN **":X(NR-1)=X(NR)
5160 IF ABS(FUP)>EPS AND ABS(FUN)<EPS THEN PRINT"** ROOTS ARE REAL AND REPEATED
AND OF THE SAME SIGN **":X(NR-1)=X(NR)
```

```
5170 'Cancel the nonsquare corresponding to the repeated pair
5180 'and go to the subroutine for one pair of roots
5190 LC(K)=0
5200 SUMRT=SUMRT+X(NR)+X(NR-1)
5210 GOSUB 4000
5220 GOTO 5400
5230 NEXT K
5240 C=((A(0,N-2)/A(0,N))*10^(FACT(0,N-2)-FACT(0,N))+2*B*SUMRT-PRODRT-RSUM)/4
5250 ALPHA1=(-B+SQR(ABS(B^2-4*C)))/2
5260 ALPHA2=-B-ALPHA1
5270 COUNT=0
5280 FOR K=N TO 1 STEP -1
5290 NR=N+1-K
5300 IF LC(K)=0 THEN GOTO 5390
5310 COUNT=COUNT+1
5320 IF COUNT=1 THEN BETA1=SQR(ABS(R(K)-ALPHA1^2))
5330 'Check convergence with complex roots
5340 ALPHA=ALPHA1: BETA=BETA1: GOSUB 3000
5350 IF ABS(SUMALL)>EPS  THEN PRINT "* SWAP THE VALUES OF ALPHA1 AND ALPHA2 AND
TRY AGAIN *":SWAP ALPHA1,ALPHA2: GOTO 5270
5360 IF COUNT=1 THEN X(NR)=ALPHA1:XI(NR)=-BETA1:X(NR-1)=ALPHA1:XI(NR-1)=BETA1
5370 IF COUNT=2 THEN BETA2=SQR(ABS(R(K)-ALPHA2^2))
5380 IF COUNT=2 THEN X(NR)=ALPHA2:XI(NR)=-BETA2:X(NR-1)=ALPHA2:XI(NR-1)=BETA2
5390 NEXT K
5400 RETURN
5410 '
6000 'Subroutine 6:Two pairs of complex or repeated roots with identical moduli
6010 '
6020 'Check for neighboring nonsquares
6030 W=0
6040 FOR K=N TO 1 STEP -1
6050 IF LC(K)<>0 AND LC(K-1)<>0 THEN W=1
6060 NEXT K
6070 IF W=0 THEN GOTO 6520
6080 PRINT:PRINT"TWO PAIRS OF COMPLEX OR REPEATED ROOTS WITH IDENTICAL MODULI:"
6090 P=(NSQUARES+1)/2
6100 B=((A(0,N-1)/A(0,N))*10^(FACT(0,N-1)-FACT(0,N))+SUMRT)/2
6110 RSUM=0
6120 PRINT"  CHECK FOR REPEATED ROOTS:"
6130 FOR K=N TO 1 STEP -1
6140 NR=N+1-K
6150 IF LC(K)=0 THEN GOTO 6230
6160 R(K)=(A(R,K+1-2*P)/A(R,K+1))^(1/(P*M))*10^((FACT(R,K+1-2*P)-FACT(R,K+1))/(P
*M))
6170 RSUM=RSUM + P*R(K)
6180 X(NR)=SQR(R(K)):GOSUB 1500
6190 IF ABS(FUP) > EPS AND ABS(FUN) > EPS THEN PRINT"* ROOTS ARE COMPLEX *":GOTO
 6240
6200 IF ABS(FUP)<EPS AND ABS(FUN)<EPS THEN PRINT"** ROOTS ARE REAL AND REPEATED
BUT OF OPPOSITE SIGN **":X(NR-1)=-X(NR):GOTO 6380
6210 IF ABS(FUP)<EPS AND ABS(FUN)>EPS THEN PRINT"** ROOTS ARE REAL AND REPEATED
AND OF THE SAME SIGN **":GOTO 6240
6220 IF ABS(FUP)>EPS AND ABS(FUN)<EPS THEN PRINT"** ROOTS ARE REAL AND REPEATED
AND OF THE SAME SIGN **":GOTO 6240
6230 NEXT K
6240 'Repeated roots of same sign or complex roots
6250 C=((A(0,N-2)/A(0,N))*10^(FACT(0,N-2)-FACT(0,N))+2*B*SUMRT-PRODRT-RSUM)/4
6260 ALPHA1=(-B+SQR(ABS(B^2-4*C)))/2
6270 ALPHA2=-B-ALPHA1
6280 FOR K=N TO 1 STEP -1
```

```
6290 NR=N+1-K
6300 IF LC(K)=0 THEN GOTO 6360
6310 BETA1=SQR(ABS(R(K)-ALPHA1^2))
6320 X(NR)=ALPHA1:XI(NR)=-BETA1:X(NR-1)=ALPHA1:XI(NR-1)=BETA1
6330 BETA2=SQR(ABS(R(K)-ALPHA2^2))
6340 X(NR+2)=ALPHA2:XI(NR+2)=-BETA2:X(NR+1)=ALPHA2:XI(NR+1)=BETA2
6350 GOTO 6370
6360 NEXT K
6370 RETURN
6380 'Repeated roots of opposite sign or complex roots with zero real part
6390 COUNT=0
6400 ALPHA2=-B
6410 FOR K=N TO 1 STEP -1
6420 NR=N+1-K
6430 IF LC(K)=0 THEN GOTO 6500
6440 BETA2=SQR(ABS(R(K)-ALPHA2^2))
6450 ALPHA=ALPHA2: BETA=BETA2: GOSUB 3000
6460 IF ABS(SUMALL)>EPS THEN PRINT"* SWAP THE VALUES OF ALPHA2 AND BETA2 AND TRY
 AGAIN *":SWAP ALPHA2,BETA2: COUNT=COUNT+1: GOTO 6450
6470 X(NR+2)=ALPHA2:XI(NR+2)=-BETA2:X(NR+1)=ALPHA2:XI(NR+1)=BETA2
6480 IF BETA2=0 AND COUNT>0 THEN X(NR+2)=ALPHA2:X(NR+1)=-ALPHA2
6490 GOTO 6510
6500 NEXT K
6510 RETURN
6520 PRINT"** PROGRAM CANNOT DETERMINE MORE THAN TWO PAIRS OF COMPLEX ROOTS **"
6530 RETURN
```

RESULTS

```
*****************************************************************
*                          EXAMPLE 2-3                          *
*                                                               *
*               GRAEFFE'S ROOT-SQUARING METHOD                  *
*                                                               *
*                         (GRAEFFE.BAS)                         *
*****************************************************************
DEGREE OF POLYNOMIAL? 5
  COEFFICIENT  5 ? 1
  COEFFICIENT  4 ? -5
  COEFFICIENT  3 ? -15
  COEFFICIENT  2 ? 85
  COEFFICIENT  1 ? -26
  COEFFICIENT  0 ? -120
GIVE THE CONVERGENCE VALUE OF F? 0.002
ROOT-SQUARING PROCESS:
 r      A            A            A            A            A            A
         r 5          r 4          r 3          r 2          r 1          r 0
 0   1.000E  0  -5.000E  0  -1.500E  1   8.500E  1  -2.600E  1  -1.200E  2
 1   1.000E  0   5.500E  1   1.023E  3   7.645E  3   2.108E  4   1.440E  4
 2   1.000E  0   9.790E  2   2.477E  5   1.691E  7   2.240E  8   2.074E  8
 3   1.000E  0   4.630E  5   2.871E 10   1.753E 14   4.317E 16   4.300E 16
 4   1.000E  0   1.569E 11   6.621E 20   2.825E 28   1.849E 33   1.849E 33
 5   1.000E  0   2.330E 22   4.295E 41   7.959E 56   3.418E 66   3.418E 66
 6   1.000E  0   5.421E 44   1.845E 83   6.334E113   1.168E133   1.168E133
 7   1.000E  0   2.939E 89   3.403E166   4.012E227   1.365E266   1.365E266
THE COEFFICIENTS ARE:

  PURE SQUARE, PURE SQUARE, PURE SQUARE, PURE SQUARE, PURE SQUARE, PURE SQUARE,

THE NUMBER OF SQUARING: r = 7             THE POWER: m = 128

CALCULATION OF ROOTS:

  FUNCTION WITH POSITIVE VALUE OF ( 4.999999 )=-2.136231E-04
  FUNCTION WITH NEGATIVE VALUE OF ( 4.999999 )=-2239.996

  FUNCTION WITH POSITIVE VALUE OF ( 4.000002 )=-80.00004
  FUNCTION WITH NEGATIVE VALUE OF ( 4.000002 )=-1.670837E-03

  FUNCTION WITH POSITIVE VALUE OF ( 2.999999 )= 3.814697E-05
  FUNCTION WITH NEGATIVE VALUE OF ( 2.999999 )= 480

  FUNCTION WITH POSITIVE VALUE OF ( 2 )= 2.288818E-05
  FUNCTION WITH NEGATIVE VALUE OF ( 2 )= 280.0001

  FUNCTION WITH POSITIVE VALUE OF ( 1 )=-80
  FUNCTION WITH NEGATIVE VALUE OF ( 1 )=-7.629395E-06

THE 5 ROOTS ARE:
                     5
                    -4
                     3
                     2
                    -1
Ok
```

```
*******************************************************************
*                            EXAMPLE 2-3                          *
*                                                                 *
*                 GRAEFFE'S ROOT-SQUARING METHOD                  *
*                                                                 *
*                          (GRAEFFE.BAS)                          *
*******************************************************************
DEGREE OF POLYNOMIAL? 4
  COEFFICIENT  4 ? 1
  COEFFICIENT  3 ? 7
  COEFFICIENT  2 ? 12
  COEFFICIENT  1 ? -4
  COEFFICIENT  0 ? -16
GIVE THE CONVERGENCE VALUE OF F? 0.0001

ROOT-SQUARING PROCESS:

 r      A            A            A            A            A
        r 4          r 3          r 2          r 1          r 0
 0   1.000E  0   7.000E  0   1.200E  1  -4.000E  0  -1.600E  1
 1   1.000E  0   2.500E  1   1.680E  2   4.000E  2   2.560E  2
 2   1.000E  0   2.890E  2   8.736E  3   7.398E  4   6.554E  4
 3   1.000E  0   6.605E  4   3.369E  7   4.329E  9   4.295E  9
 4   1.000E  0   4.295E  9   5.630E 14   1.845E 19   1.845E 19
 5   1.000E  0   1.845E 19   1.585E 29   3.403E 38   3.403E 38
 6   1.000E  0   3.403E 38   1.255E 58   1.158E 77   1.158E 77
 7   1.000E  0   1.158E 77   7.877E115   1.341E154   1.341E154
 8   1.000E  0   1.341E154   3.100E231   1.798E308   1.798E308
 9   1.000E  0   1.798E308   4.788E462   3.232E616   3.232E616
10   1.000E  0   3.232E616   1.131E925   1.045E%1233
 FACTOR EXCEEDS 999. POSSIBILITY OF COMPLEX OR REPEATED ROOTS.


THE COEFFICIENTS ARE:

  PURE SQUARE, PURE SQUARE, non-square, PURE SQUARE, PURE SQUARE,

THE NUMBER OF SQUARING: r = 9             THE POWER: m = 512

CALCULATION OF ROOTS:

  FUNCTION WITH POSITIVE VALUE OF ( 4 )= 864
  FUNCTION WITH NEGATIVE VALUE OF ( 4 )= 0

  FUNCTION WITH POSITIVE VALUE OF ( .9999998 )=-1.049042E-05
  FUNCTION WITH NEGATIVE VALUE OF ( .9999998 )=-6.000003

CALCULATION OF ONE PAIR OF COMPLEX OR REPEATED ROOTS:
  CHECK FOR REPEATED ROOTS:

  FUNCTION WITH POSITIVE VALUE OF ( 2 )= 96
  FUNCTION WITH NEGATIVE VALUE OF ( 2 )= 0
** ROOTS ARE REAL AND REPEATED AND OF THE SAME SIGN **

THE 4 ROOTS ARE:
                     -4
                     -2
                     -2
                      1
Ok
```

```
*******************************************************************
*                        EXAMPLE 2-3                              *
*                                                                 *
*              GRAEFFE'S ROOT-SQUARING METHOD                     *
*                                                                 *
*                        (GRAEFFE.BAS)                            *
*******************************************************************
DEGREE OF POLYNOMIAL? 4
  COEFFICIENT  4 ? 1
  COEFFICIENT  3 ? -15
  COEFFICIENT  2 ? 138
  COEFFICIENT  1 ? -324
  COEFFICIENT  0 ? 200
GIVE THE CONVERGENCE VALUE OF F? 0.0001

ROOT-SQUARING PROCESS:

 r      A            A            A            A            A
         r 4          r 3          r 2          r 1          r 0
 0   1.000E  0  -1.500E  1   1.380E  2  -3.240E  2   2.000E  2
 1   1.000E  0  -5.100E  1   9.724E  3   4.978E  4   4.000E  4
 2   1.000E  0  -1.685E  4   9.971E  7   1.700E  9   1.600E  9
 3   1.000E  0   8.439E  7   1.000E 16   2.570E 18   2.560E 18
 4   1.000E  0  -1.288E 16  10.000E 31   6.554E 36   6.554E 36
 5   1.000E  0  -3.417E 31  10.000E 63   4.295E 73   4.295E 73
 6   1.000E  0  -1.883E 64  10.000E127   1.845E147   1.845E147
 7   1.000E  0   1.547E128  10.000E255   3.403E294   3.403E294
 8   1.000E  0   3.921E255  10.000E511   1.158E589   1.158E589
 9   1.000E  0  -1.846E512  10.000E%1023
FACTOR EXCEEDS 999. POSSIBILITY OF COMPLEX OR REPEATED ROOTS.

THE COEFFICIENTS ARE:

  PURE SQUARE, non-square, PURE SQUARE, PURE SQUARE, PURE SQUARE,

THE NUMBER OF SQUARING: r = 8             THE POWER: m = 256

CALCULATION OF ROOTS:

  FUNCTION WITH POSITIVE VALUE OF ( 2 )= 0
  FUNCTION WITH NEGATIVE VALUE OF ( 2 )= 1536

  FUNCTION WITH POSITIVE VALUE OF ( 1 )= 0
  FUNCTION WITH NEGATIVE VALUE OF ( 1 )= 678.0001

CALCULATION OF ONE PAIR OF COMPLEX OR REPEATED ROOTS:
  CHECK FOR REPEATED ROOTS:

  FUNCTION WITH POSITIVE VALUE OF ( 10 )= 5760
  FUNCTION WITH NEGATIVE VALUE OF ( 10 )= 42240
* WARNING: CONVERGENCE NOT SATISFIED BY REAL ROOT *
* ROOTS ARE COMPLEX *

THE 4 ROOTS ARE:
                    6 + 8 i
                    6 - 8 i
                    2
                    1
Ok
```

```
******************************************************************
*                         EXAMPLE 2-3                            *
*                                                                *
*               GRAEFFE'S ROOT-SQUARING METHOD                   *
*                                                                *
*                        (GRAEFFE.BAS)                           *
******************************************************************
DEGREE OF POLYNOMIAL? 4
  COEFFICIENT  4 ? 1
  COEFFICIENT  3 ? -8
  COEFFICIENT  2 ? 42
  COEFFICIENT  1 ? -80
  COEFFICIENT  0 ? 125
GIVE THE CONVERGENCE VALUE OF F? 0.0001

ROOT-SQUARING PROCESS:

 r      A           A            A            A            A
         r 4         r 3          r 2          r 1          r 0
 0   1.000E  0  -8.000E  0   4.200E  1  -8.000E  1   1.250E  2
 1   1.000E  0  -2.000E  1   7.340E  2  -4.100E  3   1.563E  4
 2   1.000E  0  -1.068E  3   4.060E  5  -6.127E  6   2.441E  8
 3   1.000E  0   3.286E  5   1.522E 11  -1.607E 14   5.960E 16
 4   1.000E  0  -1.965E 11   2.328E 22   7.676E 27   3.553E 33
 5   1.000E  0  -7.955E 21   5.421E 44  -1.065E 56   1.262E 67
 6   1.000E  0  -1.021E 45   2.939E 89  -2.338E111   1.593E134
 7   1.000E  0   4.545E 89   8.636E178  -8.817E223   2.538E268
 8   1.000E  0   3.386E178   7.458E357   3.390E447   6.441E536
 9   1.000E  0  -1.377E358   5.562E715   1.883E894   4.149E%1073
FACTOR EXCEEDS 999. POSSIBILITY OF COMPLEX OR REPEATED ROOTS.

THE COEFFICIENTS ARE:

  PURE SQUARE, non-square, PURE SQUARE, non-square, PURE SQUARE,

THE NUMBER OF SQUARING: r = 8             THE POWER: m = 256

CALCULATION OF ROOTS:

TWO PAIRS OF COMPLEX OR REPEATED ROOTS WITH DIFFERENT MODULI:
  CHECK FOR REPEATED ROOTS:

  FUNCTION WITH POSITIVE VALUE OF ( 4.999999 )= 399.9998
  FUNCTION WITH NEGATIVE VALUE OF ( 4.999999 )= 3199.999
* WARNING: CONVERGENCE NOT SATISFIED BY REAL ROOT *
* ROOTS ARE COMPLEX *

  FUNCTION WITH POSITIVE VALUE OF ( 2.236068 )= 91.67184
  FUNCTION WITH NEGATIVE VALUE OF ( 2.236068 )= 628.3283
* WARNING: CONVERGENCE NOT SATISFIED BY REAL ROOT *
* ROOTS ARE COMPLEX *

THE 4 ROOTS ARE:
                    3 + 4 i
                    3 - 4 i
                    1 + 2 i
                    1 - 2 i
Ok
```

```
*******************************************************************
*                          EXAMPLE 2-3                            *
*                                                                 *
*               GRAEFFE'S ROOT-SQUARING METHOD                    *
*                                                                 *
*                         (GRAEFFE.BAS)                           *
*******************************************************************
DEGREE OF POLYNOMIAL? 5
  COEFFICIENT  5 ? 1
  COEFFICIENT  4 ? -10
  COEFFICIENT  3 ? 42
  COEFFICIENT  2 ? -102
  COEFFICIENT  1 ? 145
  COEFFICIENT  0 ? -100
GIVE THE CONVERGENCE VALUE OF F? 0.0001

ROOT-SQUARING PROCESS:

 r      A            A            A            A            A            A
         r 5          r 4          r 3          r 2          r 1          r 0
 0    1.000E  0  -1.000E  1   4.200E  1  -1.020E  2   1.450E  2  -1.000E  2
 1    1.000E  0   1.600E  1   1.400E  1   0.224E  3   0.625E  3   1.000E  4
 2    1.000E  0   2.280E  2  -5.722E  3   3.527E  5  -4.089E  6   1.000E  8
 3    1.000E  0   6.343E  4  -1.363E  8   1.232E 11  -5.381E 13   1.000E 16
 4    1.000E  0   4.296E  9   2.832E 15   1.778E 21   4.321E 26   1.000E 32
 5    1.000E  0   1.845E 19  -7.249E 30   1.571E 42  -1.688E 53   1.000E 64
 6    1.000E  0   3.403E 38  -5.415E 60   3.905E 83  -2.935E105   1.000E128
 7    1.000E  0   1.158E 77  -2.364E122   1.887E167  -6.948E211   1.000E256
 8    1.000E  0   1.341E154   1.219E244   5.084E333   1.052E423   1.000E512
 9    1.000E  0   1.798E308   1.213E487   2.882E666   0.909E845   1.000E%1024
FACTOR EXCEEDS 999. POSSIBILITY OF COMPLEX OR REPEATED ROOTS.


THE COEFFICIENTS ARE:

  PURE SQUARE, PURE SQUARE, non-square, non-square, non-square, PURE SQUARE,

THE NUMBER OF SQUARING: r = 8             THE POWER: m = 256

CALCULATION OF ROOTS:

  FUNCTION WITH POSITIVE VALUE OF ( 4.000002 )= 6.103516E-05
  FUNCTION WITH NEGATIVE VALUE OF ( 4.000002 )=-8584.01


TWO PAIRS OF COMPLEX OR REPEATED ROOTS WITH IDENTICAL MODULI:
  CHECK FOR REPEATED ROOTS:

  FUNCTION WITH POSITIVE VALUE OF ( 2.236068 )=-10.29416
  FUNCTION WITH NEGATIVE VALUE OF ( 2.236068 )=-1709.706
* WARNING: CONVERGENCE NOT SATISFIED BY REAL ROOT *
* ROOTS ARE COMPLEX *


THE 5 ROOTS ARE:
                    4
                    2 + 1 i
                    2 - 1 i
                    1 + 2 i
                    1 - 2 i
Ok
```

### Discussion of results

*Part a* In this case all the transformed coefficients calculated by Graeffe's root-squaring algorithm become pure squares after seven steps. This is an indication that all the roots are real and distinct. The values of the roots calculated by the program are

$$5, \quad -4, \quad 3, \quad 2, \quad -1$$

Application of Descartes' rule to this polynomial finds three changes in sign, i.e., three positive roots; and two sign repetitions, i.e., two negative roots.

*Part b* In this case, the program detects one nonsquare coefficient surrounded by pure squares. The nonsquare coefficient does not become pure square even after ten iterations. The exponential factor exceeds 999 and the Graeffe procedure is terminated. There is one pair of repeated roots which causes the appearance of the single nonsquare coefficient. The values of the roots calculated by the program are

$$-4, \quad -2, \quad -2, \quad 1$$

Application of Descartes' rule verifies that there are three negative and one positive roots.

*Part c* This polynomial also yields one nonsquare coefficient surrounded by pure squares. However, the nonsquare in this case is due to the presence of a pair of complex conjugate roots. The roots are

$$6 \pm 8i, \quad 2, \quad 1$$

Descartes' rule detects four changes in sign, i.e., four positive roots (or less than that by an even integer, to account for the complex pair).

*Part d* This polynomial yields two nonsquare coefficients, each surrounded by pure squares. There are two pairs of complex roots with different moduli, as shown below:

$$3 \pm 4i, \quad 1 \pm 2i$$

*Part e* In this case, Graeffe's method detects three neighboring nonsquares, an indication that there are two pairs of complex (or repeated) roots with identical moduli. Since this is a fifth-degree polynomial, there is an additional real root. The five roots are

$$4, \quad 2 \pm 1i, \quad 1 \pm 2i$$

It is evident, from the above cases, that the program can successfully determine real, complex, and repeated roots of polynomial equations.

## 2-7 NEWTON'S METHOD FOR SIMULTANEOUS NONLINEAR ALGEBRAIC EQUATIONS

If the mathematical model involves two (or more) simultaneous nonlinear algebraic equations in two (or more) unknowns, the Newton-Raphson method can be extended to solve these equations simultaneously. In what follows we will first develop the Newton-Raphson method for two equations and then expand the algorithm to a system of $k$ equations.

The model for two unknowns will have the general form

$$\begin{aligned} f_1(x_1, x_2) &= 0 \\ f_2(x_1, x_2) &= 0 \end{aligned} \tag{2-83}$$

where $f_1$ and $f_2$ are nonlinear function of variables $x_1$ and $x_2$. Both these functions may be expanded in two-dimensional Taylor series around an initial estimate of $x_1^{(1)}$ and $x_2^{(1)}$:

$$f_1(x_1, x_2) = f_1(x_1^{(1)}, x_2^{(1)}) + \left.\frac{\partial f_1}{\partial x_1}\right|_{x^{(1)}}(x_1 - x_1^{(1)}) + \left.\frac{\partial f_1}{\partial x_2}\right|_{x^{(1)}}(x_2 - x_2^{(1)}) + \cdots \tag{2-84a}$$

$$f_2(x_1, x_2) = f_2(x_1^{(1)}, x_2^{(1)}) + \left.\frac{\partial f_2}{\partial x_1}\right|_{x^{(1)}}(x_1 - x_1^{(1)}) + \left.\frac{\partial f_2}{\partial x_2}\right|_{x^{(1)}}(x_2 - x_2^{(1)}) + \cdots \tag{2-84b}$$

The superscript (1) will be used to designate the iteration number of the estimate.

Setting the left sides of Eqs. (2-84) to zero and truncating the second-order and higher derivatives of the Taylor series, we obtain the following equations:

$$\left.\frac{\partial f_1}{\partial x_1}\right|_{x^{(1)}}(x_1 - x_1^{(1)}) + \left.\frac{\partial f_1}{\partial x_2}\right|_{x^{(1)}}(x_2 - x_2^{(1)}) = -f_1(x_1^{(1)}, x_2^{(1)}) \tag{2-85a}$$

$$\left.\frac{\partial f_2}{\partial x_1}\right|_{x^{(1)}}(x_1 - x_1^{(1)}) + \left.\frac{\partial f_2}{\partial x_2}\right|_{x^{(1)}}(x_2 - x_2^{(1)}) = -f_2(x_1^{(1)}, x_2^{(1)}) \tag{2-85b}$$

If we define the correction variable $\delta$ as

$$\delta_1^{(1)} = x_1 - x_1^{(1)} \tag{2-86a}$$

$$\delta_2^{(1)} = x_2 - x_2^{(1)} \tag{2-86b}$$

then Eqs. (2-85) simplify to

$$\left.\frac{\partial f_1}{\partial x_1}\right|_{x^{(1)}} \delta_1^{(1)} + \left.\frac{\partial f_1}{\partial x_2}\right|_{x^{(1)}} \delta_2^{(1)} = -f_1(x_1^{(1)}, x_2^{(1)}) \tag{2-87a}$$

$$\left.\frac{\partial f_2}{\partial x_1}\right|_{x^{(1)}} \delta_1^{(1)} + \left.\frac{\partial f_2}{\partial x_2}\right|_{x^{(1)}} \delta_2^{(1)} = -f_2(x_1^{(1)}, x_2^{(1)}) \tag{2-87b}$$

Equations (2-87) are a set of simultaneous linear algebraic equations, where the unknowns are $\delta_1^{(1)}$ and $\delta_2^{(1)}$. These equations can be written in matrix format as follows:

$$\begin{bmatrix} \left.\frac{\partial f_1}{\partial x_1}\right|_{x^{(1)}} & \left.\frac{\partial f_1}{\partial x_2}\right|_{x^{(1)}} \\ \left.\frac{\partial f_2}{\partial x_1}\right|_{x^{(1)}} & \left.\frac{\partial f_2}{\partial x_2}\right|_{x^{(1)}} \end{bmatrix} \begin{bmatrix} \delta_1^{(1)} \\ \delta_2^{(1)} \end{bmatrix} = - \begin{bmatrix} f_1^{(1)} \\ f_2^{(1)} \end{bmatrix} \tag{2-88}$$

Since this set contains only two equations in two unknowns, it can be readily solved by the application of Cramer's rule (see Chap. 3) to give the first set of values for the correction vector

$$\delta_1^{(1)} = -\frac{\left[f_1 \frac{\partial f_2}{\partial x_2} - f_2 \frac{\partial f_1}{\partial x_2}\right]}{\left[\frac{\partial f_1}{\partial x_1}\frac{\partial f_2}{\partial x_2} - \frac{\partial f_1}{\partial x_2}\frac{\partial f_2}{\partial x_1}\right]} \tag{2-89a}$$

$$\delta_2^{(1)} = -\frac{\left[f_2 \frac{\partial f_1}{\partial x_1} - f_1 \frac{\partial f_2}{\partial x_1}\right]}{\left[\frac{\partial f_1}{\partial x_1}\frac{\partial f_2}{\partial x_2} - \frac{\partial f_1}{\partial x_2}\frac{\partial f_2}{\partial x_1}\right]} \tag{2-89b}$$

The superscripts, indicating the iteration number of the estimate, have been omitted from the right-hand side of Eqs. (2-89) in order to avoid over-crowding.

The new estimate of the solution may now be obtained from the previous estimate by adding to it the correction vector

$$x_i^{(n+1)} = x_i^{(n)} + \delta_i^{(n)} \tag{2-90}$$

This equation is merely a rearrangement and generalization to the $(n+1)$st iteration of Eqs. (2-86).

The method just described for two nonlinear equations is readily expandable to the case of $k$ simultaneous nonlinear equations in $k$ unknowns

$$\begin{aligned} f_1(x_1, \ldots, x_k) &= 0 \\ &\vdots \\ f_k(x_1, \ldots, x_k) &= 0 \end{aligned} \tag{2-91}$$

The linearization of this set by the application of the Taylor series expansion produces Eq. (2-92).

$$\begin{bmatrix} \dfrac{\partial f_1}{\partial x_1} & \cdots & \dfrac{\partial f_1}{\partial x_k} \\ \cdots & \cdots & \cdots \\ \dfrac{\partial f_k}{\partial x_1} & \cdots & \dfrac{\partial f_k}{\partial x_k} \end{bmatrix} \begin{bmatrix} \delta_1 \\ \vdots \\ \delta_k \end{bmatrix} = - \begin{bmatrix} f_1 \\ \vdots \\ f_k \end{bmatrix} \tag{2-92}$$

In matrix/vector notation this condenses to

$$\mathbf{J}\boldsymbol{\delta} = -\mathbf{f} \tag{2-93}$$

where $\mathbf{J}$ is the *Jacobian* matrix containing the partial derivatives, $\boldsymbol{\delta}$ is the correction vector, and $\mathbf{f}$ is the vector of functions. Equation (2-93) represents a set of linear algebraic equations whose solution will be discussed in Chap. 3.

## 2-8 NEWTON'S METHOD FOR REAL AND COMPLEX ROOTS

The problem of finding the complex roots of polynomial equations can be handled by Newton's method for simultaneous equations. The real and imaginary parts of the root are treated separately and a search is conducted on both the real and imaginary parts simultaneously using Newton's method.

Consider the polynomial

$$a_N x^N + a_{N-1} x^{N-1} \cdots a_1 x + a_0 = 0 \tag{2-94}$$

which can be written as

$$f(x) = \sum_{n=0}^{N} a_n x^n = 0 \tag{2-95}$$

The complex root which satisfies Eq. (2-95) is of the form

$$x = \alpha + \beta i \tag{2-96}$$

Substitution of (2-96) into (2-95) results in an equation which contains both real and imaginary terms

$$f(x) = u(\alpha, \beta) + v(\alpha, \beta) i \tag{2-97}$$

The real terms, $u(\alpha, \beta)$, are given by the equation

$$u(\alpha, \beta) = \sum_{n=0}^{N} a_n \alpha_n = 0 \tag{2-98}$$

and the imaginary terms by

$$v(\alpha, \beta) = \sum_{n=0}^{N} a_n \beta_n = 0 \tag{2-99}$$

The $\alpha_n$ and $\beta_n$ expressions are evaluated as follows:

$$n = 0 \qquad \begin{cases} \alpha_0 = 1.0 \\ \beta_0 = 0.0 \end{cases} \tag{2-100}$$

$$n > 0 \qquad \begin{cases} \alpha_n = \alpha\alpha_{n-1} - \beta\beta_{n-1} \\ \beta_n = \alpha\beta_{n-1} + \beta\alpha_{n-1} \end{cases} \tag{2-101}$$

Equations (2-98) and (2-99) constitute two simultaneous nonlinear equations in two unknowns, $\alpha$ and $\beta$. This problem is identical to the one described in Sec. 2-7; therefore Newton's method can be applied. The correction vector for $\alpha$ and $\beta$ can be obtained by application of Eqs. (2-89*a*) and (2-89*b*):

$$\delta\alpha = -\frac{\left[u \dfrac{\partial v}{\partial \beta} - v \dfrac{\partial u}{\partial \beta}\right]}{\left[\dfrac{\partial u}{\partial \alpha}\dfrac{\partial v}{\partial \beta} - \dfrac{\partial u}{\partial \beta}\dfrac{\partial v}{\partial \alpha}\right]} \tag{2-102a}$$

$$\delta\beta = -\frac{\left[v \dfrac{\partial u}{\partial \alpha} - u \dfrac{\partial v}{\partial \alpha}\right]}{\left[\dfrac{\partial u}{\partial \alpha}\dfrac{\partial v}{\partial \beta} - \dfrac{\partial u}{\partial \beta}\dfrac{\partial v}{\partial \alpha}\right]} \tag{2-102b}$$

Equations (2-102*a*) and (2-102*b*) are simplified by the substitution of the Cauchy-Riemann equations [8]

$$\frac{\partial u}{\partial \alpha} = \frac{\partial v}{\partial \beta} \qquad \text{and} \qquad \frac{\partial u}{\partial \beta} = -\frac{\partial v}{\partial \alpha} \tag{2-103}$$

to yield

$$\delta\alpha = \frac{\left[v \dfrac{\partial u}{\partial \beta} - u \dfrac{\partial u}{\partial \alpha}\right]}{\left[\left(\dfrac{\partial u}{\partial \alpha}\right)^2 + \left(\dfrac{\partial u}{\partial \beta}\right)^2\right]} \tag{2-104a}$$

$$\delta\beta = -\frac{\left[u \dfrac{\partial u}{\partial \beta} + v \dfrac{\partial u}{\partial \alpha}\right]}{\left[\left(\dfrac{\partial u}{\partial \alpha}\right)^2 + \left(\dfrac{\partial u}{\partial \beta}\right)^2\right]} \tag{2-104b}$$

The partial derivatives are evaluated by differentiating Eqs. (2-98) and (2-99) to obtain

$$\frac{\partial u}{\partial \alpha} = \sum_{n=1}^{N} n a_n \alpha_{n-1} \tag{2-105a}$$

$$\frac{\partial u}{\partial \beta} = -\sum_{n=1}^{N} n a_n \beta_{n-1} \tag{2-105b}$$

Finally, the corrected values of $\alpha$ and $\beta$ are calculated from

$$\alpha^{(i+1)} = \alpha^{(i)} + \delta\alpha^{(i)} \tag{2-106a}$$

$$\beta^{(i+1)} = \beta^{(i)} + \delta\beta^{(i)} \tag{2-106b}$$

where the superscript $(i)$ designates the iteration number. For this method, an initial guess of $\alpha$ and $\beta$ are necessary to start the iterative process.

The above method is based on subroutine POLRT, originally written in FORTRAN and described in *Scientific Subroutine Package* [9]. This subroutine was translated by the author into BASIC language for the IBM PC. Several modifications were made to increase its efficiency. The BASIC language version is used as a subroutine in Example 2-4.

**Example 2-4 Newton's method for real and complex roots of polynomials and transfer functions** Consider the isothermal continuous stirred tank reactor (CSTR) shown in Fig. E2-4. Components A and R are fed to the reactor at rates of $Q$ and $(q - Q)$, respectively. The following complex reaction scheme develops in the reactor:

$$A + R \rightarrow B$$

$$B + R \rightarrow C$$

$$C + R \rightleftharpoons D$$

$$D + R \rightarrow E$$

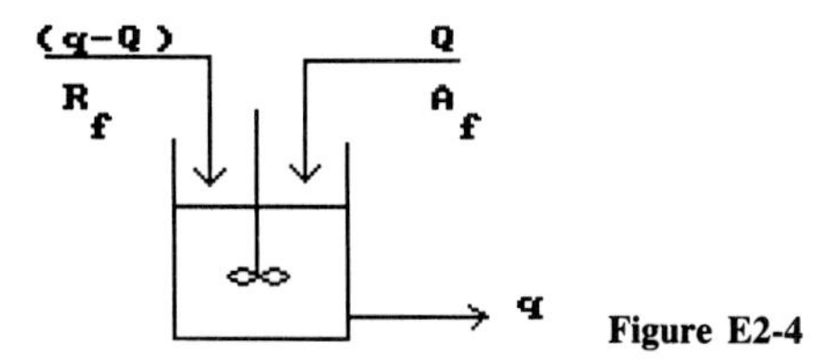


Figure E2-4

This problem was analyzed by Douglas [10] in order to illustrate the various techniques for designing simple feedback control systems. In his analysis of this system, Douglas made the following assumptions:

1. Component R is present in the reactor in sufficiently large excess so that the reaction rates can be approximated by first-order expressions.
2. The feed compositions of components B, C, D, and E are zero.
3. A particular set of values is chosen for feed concentrations, feed rates, kinetic rate constants, and reactor volume.
4. Disturbances are due to changes in the composition of component R in the vessel.

The control objective is to maintain the composition of component C in the reactor as close as possible to the steady-state design value, despite the fact that disturbances enter the system. This objective is accomplished by measuring the actual composition of C and using the difference between the desired and measured values to manipulate the inlet flow rate $Q$ of component A.

Douglas developed the following transfer function for the reactor with a proportional control system:

$$K_c \frac{2.98(s + 2.25)}{(s + 1.45)(s + 2.85)^2(s + 4.35)} = -1$$

where $K_c$ is the gain of the proportional controller. This control system is stable for values of $K_c$, which yield roots of the transfer function having negative real parts.

Using Newton's method, determine the roots of the transfer function for a range of values of the proportional gain $K_c$ and calculate the critical value of $K_c$ above which the system becomes unstable. Write the program so that it can be used to solve either $n$th-degree polynomials or transfer functions of the type shown in the above equation.

METHOD OF SOLUTION The numerator and the denominator of the transfer function are multiplied out to yield

$$\frac{(2.98s + 6.705)K_c}{s^4 + 11.50s^3 + 47.49s^2 + 83.06326s + 51.23268} = -1$$

A first-degree polynomial is present in the numerator and a fourth-degree polynomial in the denominator. To convert this to the canonical form of a polynomial, we multiply through by the denominator and rearrange to obtain

$$[s^4 + 11.50s^3 + 47.49s^2 + 83.06326s + 51.23268] + [2.98s + 6.705]K_c = 0$$

It is obvious that once a value of $K_c$ is chosen, the two bracketed terms of this equation can be added to form a single fourth-degree polynomial whose roots can be evaluated.

When $K_c = 0$, the transfer function has the following four negative real roots which can be found by inspection of the original transfer function:

$$s_1 = -1.45 \qquad s_2 = -2.85 \qquad s_3 = -2.85 \qquad s_4 = -4.35$$

These are called the poles of the open-loop transfer function.

The value of $K_c$ which causes one or more of the roots of the transfer function to become positive (or have positive real parts) is called the critical value of the proportional gain. This critical value is calculated as follows:

1. A range of search for $K_c$ is established.
2. The bisection method is used to search this range.
3. Newton's method is used to evaluate all the roots of the transfer function at each step of the bisection search.
4. The roots are checked for positive real parts. The range of $K_c$, over which the change from negative to positive roots occurs, is retained.
5. Steps 2 to 4 are repeated until successive values of $K_c$ change by less than a convergence criterion, $\epsilon$.

PROGRAM DESCRIPTION This program, which we call NEWTON.BAS, consists of the sections listed in Table E2-4. A general description of each section of this program is given below.

*Main program* The program gives the user a choice of solving either an $n$th-degree polynomial or a transfer function. If the first choice is taken, the user is asked to input the degree of the polynomial and the values of the coefficients. In turn, the program calls subroutine 2 (Newton's method), which calculates the roots and then calls subroutine 1, which prints the roots.

If a transfer function is to be solved, the user is asked to enter information for two polynomials: the one residing in the denominator of the transfer function and the one in the numerator of the function. In addition the user enters a range of search and a convergence criterion for the value of the proportional gain ($K_c$). It is good practice to choose zero for the minimum value of the range; thus the poles of the open-loop

**Table E2-4**

| Program section | Line numbers |
|---|---|
| Title | 0–90 |
| Main program | 100–920 |
| Subroutines | |
| 1. Print the roots | 1000–1240 |
| 2. Newton's method | 2000–3070 |

transfer function are evaluated in the first step of the search. The maximum value must be higher than the critical value, otherwise the search will not arrive at the critical value.

The two polynomials are combined, as shown in the Method of Solution, using $K_c$ as the multiplier of the polynomial from the numerator of the transfer function. Subroutine 2 is called to calculate the roots of the overall polynomial function and the sign of all the roots is checked for positive real parts. This procedure is performed at the minimum, maximum, and midpoints of the range of search of $K_c$. That half of the interval in which the change from negative to positive roots (stable to unstable system) occurs is retained by the bisection algorithm. This new interval is bisected again and the evaluation of the roots is repeated, until the convergence criterion is met.

The values of the roots and the stability/instability status is printed out at each step of the search.

*Subroutine 1* This subroutine prints out all the roots after they have been rounded to eliminate the computer truncation error.

*Subroutine 2* This subroutine uses Newton's method, as described in Sec. 2-8, to evaluate the real and complex roots of the polynomial. The program starts out by setting the initial guesses of the real and imaginary part of the root to low values near zero. It applies Eqs. (2-100) and (2-101) to calculate the $\alpha_n$ and $\beta_n$ expressions. It uses Eqs. (2-98) and (2-99) to calculate the functions $u$ and $v$, and Eqs. (2-105$a$) and (2-105$b$) to evaluate the partial derivatives of $u$ and $v$. Equations (2-106$a$) and (2-106$b$) are used to correct the values of $\alpha$ and $\beta$. The program has a maximum limit of 500 iterations and a maximum of five different starting values for $\alpha$ and $\beta$. The polynomial is reduced as each root is located. The final iterations on each root are performed using the original polynomial rather than the reduced polynomial to avoid accumulated error in the polynomial.

PROGRAM

```
0 SCREEN 0:WIDTH 80:CLS:KEY OFF
5 PRINT "*********************************************************************"
10 PRINT"*                                                                   *"
15 PRINT"*                            EXAMPLE 2-4                            *"
20 PRINT"*                                                                   *"
25 PRINT"*             NEWTON'S METHOD FOR REAL AND COMPLEX ROOTS            *"
30 PRINT"*                                                                   *"
35 PRINT"*               OF POLYNOMIALS AND TRANSFER FUNCTIONS               *"
40 PRINT"*                                                                   *"
45 PRINT"*                           (NEWTON.BAS)                            *"
50 PRINT"*                                                                   *"
90 PRINT"*********************************************************************"
100 '************************* Main Program ******************************
110 '
120 DIM XCOF(37),XCOFN(37),XCOFD(37),COF(37),ROOTR(36),ROOTI(36)
130 PRINT:PRINT"PROBLEM TO SOLVE:"
140 PRINT:PRINT"     1. nth DEGREE POLYNOMIAL"
150 PRINT:PRINT"     2. TRANSFER FUNCTION"
160 PRINT:PRINT"YOUR CHOICE";:INPUT PROBLEM
170 IF PROBLEM < 1 OR PROBLEM > 2 THEN 160
180 ON PROBLEM GOTO 200, 300
190 '
200 'Define the nth degree polynomial
210 '
220 PRINT:INPUT "DEGREE OF POLYNOMIAL";M
230 IF M < = 0  THEN 870: IF M > 36 THEN 890
240 N=M
250 FOR I = M + 1  TO 1 STEP -1
260 PRINT:PRINT "    COEFFICIENT OF X TO THE POWER ";I-1;:INPUT XCOF(I)
270 NEXT I
280 GOTO 590
290 '
300 'Define the polynomial in the denominator of the function
310 '
320 PRINT:INPUT "DEGREE OF POLYNOMIAL IN THE DENOMINATOR OF THE FUNCTION"; M
330 IF M < = 0  THEN 870: IF M > 36 THEN 890
340 N = M
350 FOR I = M + 1  TO 1 STEP -1
360 PRINT:PRINT "    COEFFICIENT OF X TO THE POWER ";I-1;:INPUT XCOFD(I)
370 NEXT I
380 '
390 'Define the polynomial in the numerator of the function
400 '
410 PRINT:INPUT "DEGREE OF POLYNOMIAL IN THE NUMERATOR OF THE FUNCTION"; MM
420 IF MM < = 0  THEN 870: IF MM > 36  THEN 890
430 FOR I = MM + 1  TO 1 STEP -1
440 PRINT:PRINT "    COEFFICIENT OF X TO THE POWER ";I-1;:INPUT XCOFN(I)
450 NEXT I
460 '
470 'Bisection search
480 '
490 PRINT:PRINT"THE RANGE OF SEARCH FOR THE GAIN (Kc):"
500 PRINT:INPUT "    MINIMUM VALUE OF Kc";MINKC
510 PRINT:INPUT "    MAXIMUM VALUE OF Kc";MAXKC
520 PRINT: INPUT "CONVERGENCE CRITERION FOR Kc";EPS
530 DELKC=(MAXKC-MINKC)/2
540 KC=MINKC
```

```
550 IF KC>MAXKC THEN 910
560 FOR I = M + 1  TO 1 STEP -1
570 XCOF(I)=XCOFD(I) + XCOFN(I)*KC
580 NEXT I
590 IFIT = 0
600 N=M
610 IF (XCOF(N+1)) = 0  THEN 880
620 GOSUB 2000
630 PRINT
640 GOSUB 1000
650 IF PROBLEM=1 THEN 830
660 STABLE$="SYSTEM IS STABLE"
670 UNSTABLE$="SYSTEM IS UNSTABLE"
680 ST$=STABLE$
690 FOR I=1 TO M
700 IF ROOTR(I)<=0 THEN 720
710 ST$=UNSTABLE$
720 NEXT I
730 PRINT:PRINT"Kc=";KC; ": ";ST$:PRINT
740 IF ST$=STABLE$ THEN 790
750 MAXKC=KC
760 MINKC=KC-DELKC
770 KC=MINKC
780 DELKC=(MAXKC-MINKC)/2
790 IF KC=0 THEN 800 ELSE IF ST$=STABLE$ AND ABS(DELKC/KC)<EPS THEN 820
800 KC=KC+DELKC
810 GOTO 550
820 PRINT:PRINT"          ********  SEARCH HAS CONVERGED **********":PRINT
830 END
840 '
850 'Print error messages
860 '
870 PRINT:PRINT "DEGREE OF POLYNOMIAL CANNOT BE LESS THAN ONE":END
880 PRINT:PRINT "HIGH ORDER COEFFICIENT CANNOT BE ZERO":END
890 PRINT:PRINT "DEGREE OF POLYNOMIAL CANNOT EXCEED 36":END
900 PRINT:PRINT "UNABLE TO DETERMINE ROOT":END
910 PRINT:PRINT "CRITICAL VALUE OF KC NOT FOUND IN THIS RANGE":END
920 '
1000 '****************** Subroutine 1: Print the roots ********************
1010 '
1020 PRINT"*****************************************************************"
1030 PRINT:PRINT "THE ROOTS OF THE EQUATION ARE":PRINT
1040 FOR I = 1 TO M
1050 YY=1000
1060 IF ABS(ROOTR(I)*YY)>32000 THEN YY=YY/10:GOTO 1060
1070 IF ROOTR(I)=0 THEN 1100
1080 IF ABS(ROOTR(I)*YY)<1000 THEN YY=YY*10:GOTO 1080
1090 ROOTR(I)=CINT(ROOTR(I)*YY)/YY
1100 YY=1000
1110 IF ABS(ROOTI(I)*YY)>32000 THEN YY=YY/10:GOTO 1110
1120 IF ROOTI(I)=0 THEN 1150
1130 IF ABS(ROOTI(I)*YY)<1000 THEN YY=YY*10:GOTO 1130
1140 ROOTI(I)=CINT(ROOTI(I)*YY)/YY
1150 IF ROOTI(I) = 0  THEN 1160 ELSE 1170
1160 PRINT"X(";I;") = ";ROOTR(I):GOTO 1230
1170 IF ROOTR(I) = 0  THEN 1180 ELSE 1190
1180 PRINT"X(";I;") = ";ROOTI(I);"i":GOTO 1230
1190 IF ROOTI(I) > 0  THEN 1200 ELSE 1220
1200 PRINT"X(";I;") = ";ROOTR(I);"+";ROOTI(I);"i"
```

```
1210 GOTO 1230
1220 PRINT"X(";I;") = ";ROOTR(I);"-";ABS(ROOTI(I));"i"
1230 NEXT I
1240 RETURN
2000 '****************** Subroutine 2: Newton's method *********************
2010 '
2020 NX = N
2030 NXX = N + 1
2040 N2 = 1
2050 KJ1 = N + 1
2060 FOR L = 1 TO KJ1
2070 MT = KJ1 - L + 1
2080 COF(MT) = XCOF(L)
2090 NEXT L
2100 '
2110 'Set initial values
2120 '
2130 XO = .005
2140 YO = .01
2150 '
2160 'Zero initial value counter
2170 '
2180 IN = 0
2190 X = XO
2200 '
2210 'Increment initial values and counter
2220 '
2230 XO = -10 * YO
2240 YO = -10 * X
2250 '
2260 'Set X and Y to current value
2270 '
2280 X = XO
2290 Y = YO
2300 IN = IN + 1
2310 GOTO 2380
2320 IFIT = 1
2330 XPR = X
2340 YPR = Y
2350 '
2360 'Evaluate polynomials and derivarives
2370 '
2380 ICT = 0
2390 UX = 0
2400 UY = 0
2410 V = 0
2420 YT = 0
2430 XT = 1
2440 U = COF(N+1)
2450 IF U = 0  THEN 2880 ELSE 2460
2460 FOR I = 1 TO N
2470 L = N - I + 1
2480 TEMP = COF(L)
2490 XT2 = X * XT - Y * YT
2500 YT2 = X * YT + Y * XT
2510 U = U + TEMP * XT2
2520 V = V + TEMP * YT2
2530 FI = I
2540 UX = UX + FI * XT * TEMP
```

```
2550 UY = UY - FI * YT * TEMP
2560 XT = XT2
2570 YT = YT2
2580 NEXT I
2590 SUMSQ = UX * UX + UY * UY
2600 IF SUMSQ = 0  THEN 2790
2610 DX = (V * UY - U * UX) / SUMSQ
2620 X = X + DX
2630 DY = -(U * UY + V * UX) / SUMSQ
2640 Y = Y + DY
2650 IF (ABS(DY) + ABS(DX) - .00001) < 0  THEN 2730
2660 '
2670 'Step iteration counter
2680 '
2690 ICT = ICT + 1
2700 IF ICT < 500  THEN 2390
2710 IF IFIT = 0  THEN 2720 ELSE 2730
2720 IF IN < 5  THEN 2190 ELSE 900
2730 FOR L = 1 TO NXX
2740 MT = KJ1 - L + 1
2750 SWAP  XCOF(MT),COF(L)
2760 NEXT L
2770 SWAP  N ,NX
2780 IF IFIT = 0  THEN 2320 ELSE 2820
2790 IF IFIT = 0 THEN 2190
2800 X = XPR
2810 Y = YPR
2820 IFIT = 0
2830 IF (ABS(Y) - .0001 * ABS(X)) < 0  THEN 2910
2840 ALPHA = X + X
2850 SUMSQ = X * X + Y * Y
2860 N = N - 2
2870 GOTO 2950
2880 X = 0
2890 NX = NX - 1
2900 NXX = NXX - 1
2910 Y = 0
2920 SUMSQ = 0
2930 ALPHA = X
2940 N = N - 1
2950 COF(2) = COF(2) + ALPHA * COF(1)
2960 FOR L = 2 TO N
2970 COF(L+1) = COF(L+1) + ALPHA * COF(L) - SUMSQ * COF(L-1)
2980 NEXT L
2990 ROOTI(N2) = Y
3000 ROOTR(N2) = X
3010 N2 = N2 + 1
3020 IF SUMSQ = 0  THEN 3060
3030 Y = -Y
3040 SUMSQ = 0
3050 GOTO 2990
3060 IF N > 0  THEN 2130
3070 RETURN
```

RESULTS

```
******************************************************************
*                                                                *
*                        EXAMPLE 2-4                             *
*                                                                *
*         NEWTON'S METHOD FOR REAL AND COMPLEX ROOTS             *
*                                                                *
*           OF POLYNOMIALS AND TRANSFER FUNCTIONS                *
*                                                                *
*                        (NEWTON.BAS)                            *
*                                                                *
******************************************************************

PROBLEM TO SOLVE:

     1. nth DEGREE POLYNOMIAL

     2. TRANSFER FUNCTION

YOUR CHOICE? 2

DEGREE OF POLYNOMIAL IN THE DENOMINATOR OF THE FUNCTION? 4

   COEFFICIENT OF X TO THE POWER  4 ? 1

   COEFFICIENT OF X TO THE POWER  3 ? 11.5

   COEFFICIENT OF X TO THE POWER  2 ? 47.49

   COEFFICIENT OF X TO THE POWER  1 ? 83.06326

   COEFFICIENT OF X TO THE POWER  0 ? 51.23268

DEGREE OF POLYNOMIAL IN THE NUMERATOR OF THE FUNCTION? 1

   COEFFICIENT OF X TO THE POWER  1 ? 2.98

   COEFFICIENT OF X TO THE POWER  0 ? 6.705

THE RANGE OF SEARCH FOR THE GAIN (Kc):

   MINIMUM VALUE OF Kc? 0

   MAXIMUM VALUE OF Kc? 100

CONVERGENCE CRITERION FOR Kc? 0.001
```

```
*****************************************************************

THE ROOTS OF THE EQUATION ARE

X( 1 ) = -1.45
X( 2 ) = -2.85 - .002447 i
X( 3 ) = -2.85 + .002447 i
X( 4 ) = -4.35

Kc= 0 : SYSTEM IS STABLE

*****************************************************************

THE ROOTS OF THE EQUATION ARE

X( 1 ) = -2.246
X( 2 ) = -8.495
X( 3 ) = -.3796 - 4.485 i
X( 4 ) = -.3796 + 4.485 i

Kc= 50 : SYSTEM IS STABLE

*****************************************************************

THE ROOTS OF THE EQUATION ARE

X( 1 ) = -2.248
X( 2 ) = -9.851
X( 3 ) =  .2995 - 5.701 i
X( 4 ) =  .2995 + 5.701 i

Kc= 100 : SYSTEM IS UNSTABLE

*****************************************************************

THE ROOTS OF THE EQUATION ARE

X( 1 ) = -2.247
X( 2 ) = -9.248999
X( 3 ) = -.001993 - 5.163 i
X( 4 ) = -.001993 + 5.163 i

Kc= 75 : SYSTEM IS STABLE

*****************************************************************

THE ROOTS OF THE EQUATION ARE

X( 1 ) = -2.248
X( 2 ) = -9.851
X( 3 ) =  .2995 - 5.701 i
X( 4 ) =  .2995 + 5.701 i

Kc= 100 : SYSTEM IS UNSTABLE
```

```
*************************************************************

THE ROOTS OF THE EQUATION ARE

X( 1 ) = -2.248
X( 2 ) = -9.564
X( 3 ) =  .1559 - 5.445 i
X( 4 ) =  .1559 + 5.445 i

Kc= 87.5 : SYSTEM IS UNSTABLE

*************************************************************

THE ROOTS OF THE EQUATION ARE

X( 1 ) = -2.247
X( 2 ) = -9.41
X( 3 ) =  .07893 - 5.308 i
X( 4 ) =  .07893 + 5.308 i

Kc= 81.25 : SYSTEM IS UNSTABLE

*************************************************************

THE ROOTS OF THE EQUATION ARE

X( 1 ) = -2.247
X( 2 ) = -9.331001
X( 3 ) =  .039 - 5.237 i
X( 4 ) =  .039 + 5.237 i

Kc= 78.125 : SYSTEM IS UNSTABLE

*************************************************************

THE ROOTS OF THE EQUATION ARE

X( 1 ) = -2.247
X( 2 ) = -9.29
X( 3 ) =  .01864 - 5.2 i
X( 4 ) =  .01864 + 5.2 i

Kc= 76.5625 : SYSTEM IS UNSTABLE

*************************************************************

THE ROOTS OF THE EQUATION ARE

X( 1 ) = -2.247
X( 2 ) = -9.269
X( 3 ) =  8.360001E-03 - 5.182 i
X( 4 ) =  8.360001E-03 + 5.182 i

Kc= 75.78125 : SYSTEM IS UNSTABLE
```

```
******************************************************************

THE ROOTS OF THE EQUATION ARE

X( 1 ) = -2.247
X( 2 ) = -9.259
X( 3 ) =  .003192 - 5.173 i
X( 4 ) =  .003192 + 5.173 i

Kc= 75.39063 : SYSTEM IS UNSTABLE

******************************************************************

THE ROOTS OF THE EQUATION ARE

X( 1 ) = -2.247
X( 2 ) = -9.254
X( 3 ) =  .0006016 - 5.168 i
X( 4 ) =  .0006016 + 5.168 i

Kc= 75.19531 : SYSTEM IS UNSTABLE

******************************************************************

THE ROOTS OF THE EQUATION ARE

X( 1 ) = -2.247
X( 2 ) = -9.250999
X( 3 ) = -.0006954 - 5.166 i
X( 4 ) = -.0006954 + 5.166 i

Kc= 75.09766 : SYSTEM IS STABLE

******************************************************************

THE ROOTS OF THE EQUATION ARE

X( 1 ) = -2.247
X( 2 ) = -9.254
X( 3 ) =  .0006016 - 5.168 i
X( 4 ) =  .0006016 + 5.168 i

Kc= 75.19531 : SYSTEM IS UNSTABLE

******************************************************************

THE ROOTS OF THE EQUATION ARE

X( 1 ) = -2.247
X( 2 ) = -9.253
X( 3 ) = -4.685E-05 - 5.167 i
X( 4 ) = -4.685E-05 + 5.167 i

Kc= 75.14649 : SYSTEM IS STABLE

        ********  SEARCH HAS CONVERGED **********

Ok
```

DISCUSSION OF RESULTS The range of search for the proportional gain ($K_c$) is chosen to be between 0 and 100. A convergence criterion of 0.001 is used. The bisection method evaluates the roots at the low end of the range ($K_c = 0$) and finds them to have the predicted values of

$$-1.45 \quad -2.85 \quad -2.85 \quad \text{and} \quad -4.35$$

The small complex part of the two middle roots is due to accumulated error. It is very small in comparison to the real part and it can be ignored.

At the midrange ($K_c = 50$) the roots are

$$-2.246 \quad -8.495 \quad -0.3796 \pm 4.485i$$

The system is still stable since all the real parts of the roots are negative.

At the upper end of the range ($K_c = 100$) the system is unstable because of the positive real components of the roots.

The bisection method continues its search in the range 50 to 100. In a total of 15 evaluations, the algorithm arrives at the critical value of $K_c$ in the range

$$75.1465 < K_c < 75.1953$$

In the event that the critical value of the gain was outside the limits of the original range of search, the program would have detected this early in the search and would have issued a warning.

## PROBLEMS

**2-1** Evaluate all the roots of the following polynomial equations:

(*a*) $x^4 - 16x^3 + 96x^2 - 256x + 256 = 0$
(*b*) $x^4 - 32x^2 + 256 = 0$
(*c*) $x^4 + 3x^3 + 12x - 16 = 0$
(*d*) $x^4 + 4x^3 + 18x^2 - 20x + 125 = 0$
(*e*) $x^5 - 8x^4 + 35x^3 - 106x^2 + 170x - 200 = 0$
(*f*) $x^4 - 10x^3 + 35x^2 - 5x + 24 = 0$

First, use Descartes' rule to predict how many positive and how many negative roots each polynomial may have. Use Graeffe's root-squaring method to calculate the numerical values of the roots. Classify these polynomials according to the four categories described in Sec. 2.6.

**2-2** Using Newton's method for real and complex roots, evaluate the following:

(*a*) Verify the values of the roots calculated in Prob. 2-1.
(*b*) Calculate the roots of the following sixth-degree polynomials:
   (*i*) $x^6 - 8x^5 + 11x^4 + 78x^3 - 382x^2 + 800x - 800 = 0$
   (*ii*) $x^6 - 8x^5 + 10x^4 + 80x^3 - 371x^2 + 768x - 720 = 0$

**2-3** $F$ moles per hour of an $n$-component natural gas stream is introduced as feed to the flash vaporization tank shown in Fig. P2-3. The resulting vapor and liquid streams are withdrawn at the rate of $V$ and $L$ moles per hour, respectively. The mole fractions of the components in the feed, vapor, and liquid streams are designated by $z_i$, $y_i$, and $x_i$, respectively ($i = 1, 2, \ldots, n$) [Ref. 7, p. 204].

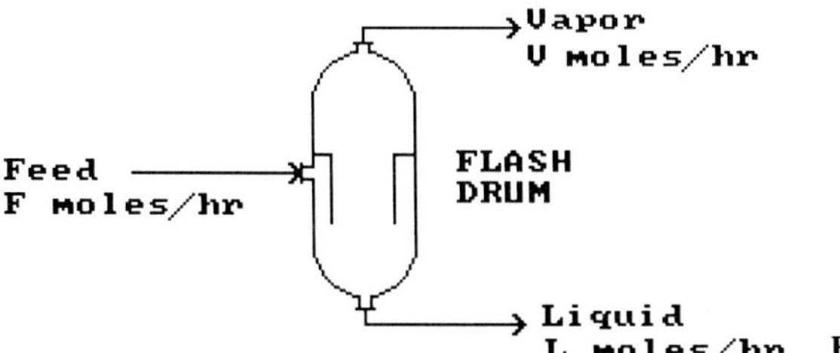


**Figure P2-3** Flash drum.

Assuming vapor/liquid equilibrium and steady-state operation, we have

| | | |
|---|---|---|
| Overall balance | $F = L + V$ | |
| Individual balance | $z_i F = x_i L + y_i V$ | $i = 1, 2, \ldots, n$ |
| Equilibrium relation | $K_i = \dfrac{y_i}{x_i}$ | |

Here, $K_i$ is the equilibrium constant for the $i$th component at the prevailing temperature and pressure in the tank. From these equations and the fact that

$$\sum_{i=1}^{n} x_i = \sum_{i=1}^{n} y_i = 1$$

show that

$$\sum_{i=1}^{n} \frac{z_i K_i F}{V(K_i - 1) + F} = 1.0$$

Write a program that reads values for $F$, the $z_i$, and the $K_i$ as data, and then uses the Newton-Raphson method to solve the last equation above for $V$. The program should also compute the values of $L$, the $x_i$, and the $y_i$ by using the first three equations given above.

The test data, shown in Table P2-3, relates to the flashing of a natural gas stream at 1600 psia and 120°F.

Assume that $F = 1000$ mol/h. A small tolerance $\varepsilon$ and an upper limit on the number of iterations should also be specified. What would be a good value $V_1$ for starting the iteration?

**Table P2-3**

| Component | $i$ | $z_i$ | $K_i$ |
|---|---|---|---|
| Carbon dioxide | 1 | 0.0046 | 1.65 |
| Methane | 2 | 0.8345 | 3.09 |
| Ethane | 3 | 0.0381 | 0.72 |
| Propane | 4 | 0.0163 | 0.39 |
| Isobutane | 5 | 0.0050 | 0.21 |
| $n$-Butane | 6 | 0.0074 | 0.175 |
| Pentanes | 7 | 0.0287 | 0.093 |
| Hexanes | 8 | 0.0220 | 0.065 |
| Heptanes+ | 9 | 0.0434 | 0.036 |
| | | 1.0000 | |

**2-4** The Underwood [2] equation for multicomponent distillation is given as

$$\left(\sum_{j=1}^{n} \frac{\alpha_j z_{jF} F}{\alpha_j - \phi}\right) - F(1-q) = 0$$

where $F$ = molar feed flow rate
$n$ = number of components in the feed
$z_{jF}$ = mole fraction of each component in the feed
$q$ = quality of the feed
$\alpha_j$ = relative volatility of each component at average column conditions
$\phi$ = root of the equation

It has been shown by Underwood that $n-1$ of the roots of this equation lie between the values of the relative volatilities as shown below:

$$\alpha_n < \phi_{n-1} < \alpha_{n-1} < \phi_{n-2} < \cdots < \alpha_3 < \phi_2 < \alpha_2 < \phi_1 < \alpha_1$$

Using the Newton-Raphson method, evaluate the $n-1$ roots of this equation for the case shown in Table P2-4.

**Table P2-4**

| Component in feed | Mole fraction, $z_{jF}$ | Relative volatility, $\alpha_j$ |
|---|---|---|
| $C_1$ | 0.05 | 10 |
| $C_2$ | 0.05 | 5 |
| $C_3$ | 0.10 | 2.05 |
| $C_4$ | 0.30 | 2.0 |
| $C_5$ | 0.05 | 1.5 |
| $C_6$ | 0.30 | 1.0 |
| $C_7$ | 0.10 | 0.9 |
| $C_8$ | 0.05 | 0.1 |
| | 1.00 | |

$F = 1000$ mol/h  $q = 1.0$ (saturated liquid)

**2-5** Carbon monoxide from a water gas plant is burned with air in an adiabatic reactor. Both the carbon monoxide and air are being fed to the reactor at 70°F and atmospheric pressure. For the reaction

$$CO + \tfrac{1}{2}O_2 = CO_2$$

the following standard free energy change (25°C) has been determined:

$$\Delta G^0_{T_0} = -62{,}000 \text{ cal/(g mol of CO)}$$

The standard enthalpy change at 25°C has been measured as

$$\Delta H^0_{T_0} = -67{,}611 \text{ cal/(g mol of CO)}$$

The standard states for all components are the pure gases at 1 atm.

Calculate the adiabatic flame temperature and the conversion of CO for the following two cases:

(*a*) 0.4 mol of oxygen per mole of CO is provided for the reaction.

(*b*) 0.8 mol of oxygen per mole of CO is provided for the reaction.

The constant pressure heat capacities for the various constituents in cal/(K·g·mol) with $T$ in kelvins are all of the form

$$C_{P_i} = A_i + B_i T_K + C_i T_K^2$$

For the gases involved here, the constants are as shown in Table P2-5.

**Table P2-5**

| Gas | A | B | C |
|---|---|---|---|
| CO | 6.25 | $2.091 \times 10^{-3}$ | $-0.459 \times 10^{-6}$ |
| $O_2$ | 6.13 | $2.990 \times 10^{-3}$ | $-0.806 \times 10^{-6}$ |
| $CO_2$ | 6.85 | $8.533 \times 10^{-3}$ | $-2.475 \times 10^{-6}$ |
| $N_2$ | 6.30 | $1.819 \times 10^{-3}$ | $-0.345 \times 10^{-6}$ |

*Hint* Combine the material balance, enthalpy balance, and equilibrium relationship to form two nonlinear algebraic equations in two unknowns: the temperature and conversion.

**2-6** Consider the three-mode feedback control of a stirred-tank heater system (Fig. P2-6). The measured output variable is the tank temperature and the input load variable is the feedstream temperature [11]. Using classical methods (i.e., deviation variables, linearization, and Laplace transforms) the overall closed-loop transfer function for the control system is given by

$$\frac{\bar{T}}{\bar{T}_i} = \frac{(\tau_I s)(\tau_v s + 1)(\tau_m s + 1)}{(\tau_I s)(\tau_P s + 1)(\tau_v s + 1)(\tau_m s + 1) + K(\tau_I s + 1 + \tau_D \tau_I s^2)}$$

where $\tau_I$ = reset time constant

$\tau_D$ = derivative time constant

$K = K_p K_v K_m K_c$

$K_P$ = first-order process static gain

$K_v$ = first-order valve constant

$K_m$ = first-order measurement constant

$K_c$ = proportional gain for the three-mode controller

$\bar{T}$ = Laplace transform of the output temperature deviation

$\bar{T}_i$ = Laplace transform of the input load temperature deviation

$\tau_P, \tau_m, \tau_v$ = first-order time constants for the process, measurement device, and process valve, respectively

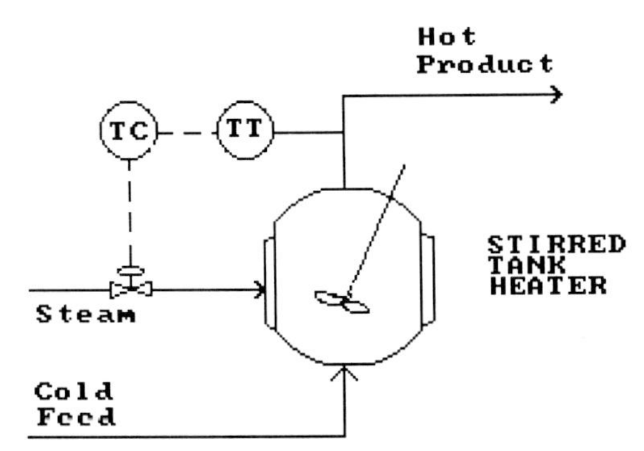


**Figure P2-6** Stirred tank heater.

For a given set of values, the stability of the system can be determined from the roots of the characteristic polynomial (i.e., the polynomial in the denominator of the overall transfer function). Thus,

$$\tau_I\tau_P\tau_m\tau_v s^4 + (\tau_I\tau_P\tau_m + \tau_I\tau_P\tau_v + \tau_I\tau_m\tau_v)s^3 + (K\tau_I\tau_D + \tau_I\tau_P + \tau_I\tau_v + \tau_I\tau_m)s^2 + (\tau_I + K\tau_I)s + K = 0$$

For the following set of parameter values, find the four roots to the characteristic polynomial when $K_c$ is equal to its "critical" value.

$$\tau_I = 10 \qquad \tau_D = 1$$

$$K_P = 10 \qquad K_v = 2 \qquad K_m = 0.09 \qquad K = 1.8K_c$$

$$\tau_P = 10 \qquad \tau_m = 5 \qquad \tau_v = 5$$

**2-7** Develop a computer program which uses the Benedict-Webb-Rubin equation of state to obtain the specific volume of a pure gas, given the temperature and pressure.

$$P = \frac{RT}{V} + \frac{B_0RT - A_0 - (C_0/T^2)}{V^2} + \frac{bRT - a}{V^3} + \frac{a\alpha}{V^6} + \frac{c}{V^3T^2}\left(1 + \frac{\gamma}{V^2}\right)e^{-\gamma/V^2}$$

where $A_0$, $B_0$, $C_0$, $a$, $b$, $c$, $\alpha$, and $\gamma$ are constants.

Use the Newton-Raphson technique to arrive at the correct root of the BWR equation.

Calculate the specific volume of $n$-butane at 425 K and at pressures of 1, 10, 20, 30, and 40 atm. Compare these results with the specific volumes obtained from the ideal gas law and the Beattie-Bridgeman equations of state (Example 2-2).

The values of the constants for $n$-butane are

$$A_0 = 10.0847 \qquad B_0 = 0.124361 \qquad C_0 = 0.992830 \times 10^6$$

$$a = 1.88231 \qquad b = 0.0399983 \qquad c = 0.316400 \times 10^6$$

$$\alpha = 1.10132 \times 10^{-3} \qquad \gamma = 3.400 \times 10^{-2} \qquad R = 0.08206$$

*Note* $P$ is in atmospheres, $V$ is in liters per mole, and $T$ is in kelvins.

**2-8** A direct-fired tubular reactor is used in the thermal cracking of light hydrocarbons or naphthas for the production of olefins, such as ethylene (see Fig. P2-8). The reactants are preheated in the

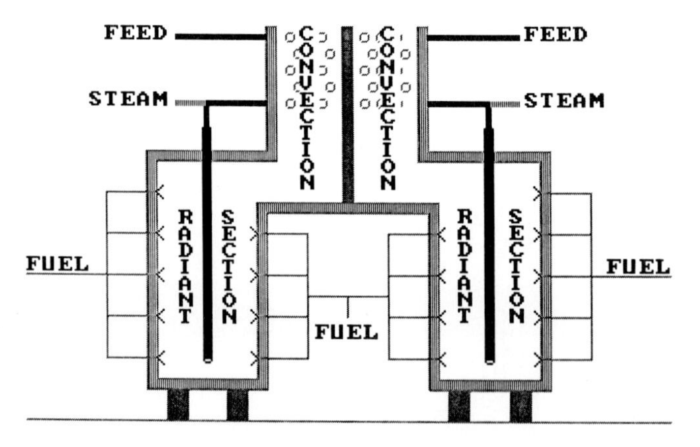


**Figure P2-8** Pyrolysis reactor.

convection section of the furnace, mixed with steam, and then subjected to high temperatures in the radiant section of the furnace. Heat transfer in the radiant section of the furnace takes place through three mechanisms: radiation, conduction, and convection. Heat is transferred by radiation from the walls of the furnace to the surface of the tubes which carry the reactants and it is transferred through the walls of the tubes by conduction and finally to the fluid inside the tubes by convection.

The three heat-transfer mechanisms are quantified as follows:

1. *Radiation* The Stefan-Boltzmann law of radiation may be written as

$$\frac{dQ}{dA_o} = \sigma\phi(T_R^4 - T_o^4)$$

where $dQ/dA_o$ is the rate of heat transfer per unit outside surface area of the tubes, $T_R$ is the "effective" furnace radiation temperature, and $T_o$ is the temperature on the outside surface of the tube. In furnaces with tube banks irradiated from both sides, a reasonable approximation is

$$T_R = T_G$$

where $T_G$ is the temperature of the flue gas in the reactor. Therefore, the Stefan-Boltzmann equation is revised to

$$\frac{dQ}{dA_o} = \sigma\phi(T_G^4 - T_o^4)$$

$\sigma$ is the Stefan-Boltzmann constant and $\phi$ is the tube geometry emissivity factor, which depends on the tube arrangement and tube surface emissivity. For single rows of tubes irradiated from both sides

$$\frac{1}{\phi} = \frac{1}{\varepsilon} - 1 + \frac{\pi}{2\Omega}$$

where $\varepsilon$ is the emissivity of the outside surface of the tube and

$$\Omega = \frac{S}{D_o} + \arctan\sqrt{\left(\frac{S}{D_o}\right)^2 - 1} - \sqrt{\left(\frac{S}{D_o}\right)^2 - 1}$$

S is the spacing (pitch) of the tubes (center-to-center) and $D_o$ is the outside diameter of the tubes.

2. *Conduction* Conduction through the tube wall is given by Fourier's equation

$$\frac{dQ}{dA_o} = \frac{k_t}{t_t}(T_o - T_i)$$

where $T_i$ is the temperature on the inside surface of the tube, $k_t$ is the thermal conductivity of the tube material, and $t_t$ is the thickness of the tube wall.

3. *Convection* Convection through the fluid film inside the tube is expressed by

$$\frac{dQ}{dA_o} = h_i\left(\frac{D_i}{D_o}\right)(T_i - T_f)$$

where $D_i$ is the inside diameter of the tube, $T_f$ is the temperature of the fluid in the tube, and $h_i$ is the heat-transfer film coefficient on the inside of the tube. The film coefficient may be approximated from the Dittus-Boelter equation [12]:

$$h_i = 0.023\left(\frac{k_f}{D_i}\right)(\mathrm{Re}_f)^{0.8}(\mathrm{Pr}_f)^{0.4}$$

where $Re_f$ is the Reynolds number, $Pr_f$ is the Prandtl number, and $k_f$ is the thermal conductivity of the fluid.

Conditions vary drastically along the length of the tube, as the temperature of the fluid inside the tube rises rapidly. The rate of heat transfer is the highest at the entrance conditions and lowest at the exit conditions of the fluids.

Calculate the rate of heat transfer $(dQ/dA_o)$, the temperature on the outside surface of the tube $(T_o)$, and the temperature on the inside surface of the tube $(T_i)$ at a point along the length of the tube where the following conditions exist:

$$T_G = 2200°\text{F} \qquad T_f = 1500°\text{F}$$

$$\varepsilon = 0.9 \qquad \sigma = 1.713 \times 10^{-9}\ \text{Btu}/(\text{h}\cdot\text{ft}^2\cdot°\text{R}^4)$$

$$S = 0.667\ \text{ft} \qquad D_i = 0.337\ \text{ft} \qquad D_o = 0.375\ \text{ft}$$

$$k_t = 12.5\ \text{Btu}/(\text{h}\cdot\text{ft}\cdot°\text{R}) \qquad t_t = 0.019\ \text{ft} \qquad k_f = 0.1012\ \text{Btu}/(\text{h}\cdot\text{ft}\cdot°\text{R})$$

$$\text{Re}_f = 388{,}000 \qquad \text{Pr}_f = 0.660$$

Explain in detail the method of solution used to arrive at your answers.

# REFERENCES

1. Himmelblau, D. M.: *Basic Principles and Calculations in Chemical Engineering*, Prentice-Hall, Inc., Englewood Cliffs, N.J., 1982, p. 223.
2. Underwood, A. J. V.: *Chem. Eng. Prog.*, vol. 44, 1948, p. 603.
3. Treybal, R. E.: *Mass-Transfer Operations*, 3d ed., McGraw-Hill Book Company, New York, 1980, p. 436.
4. Brodkey, R. S.: *The Phenomena of Fluid Motions*, Addison-Wesley Publishing Co., Reading, Mass., 1967, p. 254.
5. Salvadori, M. G., and Baron, M. L.: *Numerical Methods in Engineering*, Prentice-Hall Inc., Englewood Cliffs, N.J., 1961, p. 2.
6. Lapidus, L.: *Digital Computation for Chemical Engineering*, McGraw-Hill Book Company, New York, 1962, p. 289.
7. Carnahan, B., Luther, H. A., and Wilkes, J. O.: *Applied Numerical Methods*, John Wiley & Sons, Inc., New York, 1969, p. 142.
8. Jenson, V. G., and Jeffreys, G. V.: *Mathematical Methods in Chemical Engineering*, 2d ed., Academic Press, London, 1977, p. 126.
9. *System/360 Scientific Subroutine Package Version III Programmer's Manual*, IBM DOC. No. GH20-205-4, IBM Corp., Technical Publication Dept., White Plains, N.Y., 1970, p. 181.
10. Douglas, J. M.: *Process Dynamic and Control*, vol. 2, Prentice-Hall, Inc., Englewood Cliffs, N.J., 1972, p. 75.
11. Davidson, B. D.: Private communication, Department of Chemical and Biochemical Engineering, Rutgers University, New Brunswick, N.J., 1984.
12. Bennett, C. O., and Meyers, J. E.: *Momentum, Heat, and Mass Transfer*, McGraw-Hill Book Company, New York, 1973, p. 376.

CHAPTER

# THREE

# NUMERICAL SOLUTION OF SIMULTANEOUS LINEAR ALGEBRAIC EQUATIONS

## 3-1 INTRODUCTION

The mathematical analysis of linear physicochemical systems often results in models consisting of sets of linear algebraic equations. In addition, methods of solution of nonlinear systems and differential equations use the technique of linearization of the models, thus requiring the repetitive solution of sets of linear algebraic equations. These problems may range in complexity from a set of two simultaneous linear algebraic equations to a set involving 1000 or even 10,000 equations. The solution of a set of two to three linear algebraic equations can be obtained easily by the algebraic elimination of variables or by the application of Cramer's rule. However, for systems involving five or more equations the algebraic elimination method becomes too complex, and Cramer's rule requires a rapidly escalating number of arithmetic operations, too large even for today's high-speed digital computers.

In the remainder of this section, we give several examples of systems drawn from chemical engineering applications which yield sets of simultaneous linear algebraic equations. In the following sections of this chapter we discuss several methods for the numerical solution of such problems, and demonstrate the application of these methods on the personal computer.

Material and energy balances are the primary tools of chemical engineers. Such balances applied to multistage or multicomponent processes result in sets of equations which can be either differential or algebraic. Steady-state considerations simplify differential equations to algebraic ones. Often the systems under analysis are nonlinear, thus resulting in sets of nonlinear equations. However, many procedures have been developed which linearize the equations and apply iterative convergence techniques to arrive at the solutions of the nonlinear systems.

A classical example of the use of these techniques is in the analysis of distillation columns, such as the one shown in Fig. 3-1. Steady-state material

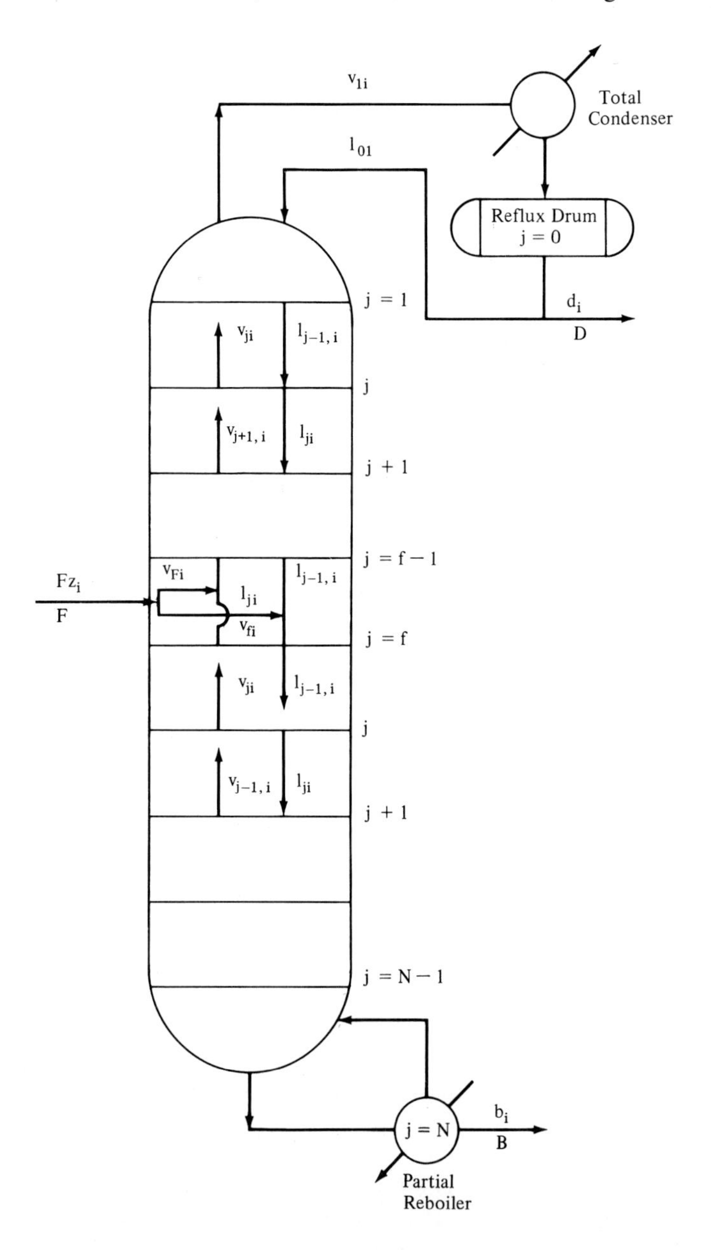


**Figure 3-1** Conventional distillation column.

balances applied to the rectifying section of the column yield the following equations [1]:

Balance around condenser: $V_1 y_{1i} = L_0 x_{0i} + D x_{Di}$ (3-1)

Balance above the $j$th stage: $V_j y_{ji} = L_{j-1} x_{j-1,\,i} + D x_{Di}$ (3-2)

Assuming that the stages are equilibrium stages and that the column uses a total condenser, the following equilibrium relations apply:

$$y_{ji} = K_{ji} x_{ji} \tag{3-3}$$

Substituting Eq. (3-3) in (3-1) and (3-2) and dividing through by $D x_{Di}$ we get

$$\frac{V_1 y_{1i}}{D x_{Di}} = \frac{L_0}{D x_{Di}} x_{Di} + 1 \tag{3-4}$$

$$\frac{V_j y_{ji}}{D x_{Di}} = \left(\frac{L_{j-1}}{K_{j-1,\,i} V_{j-1}}\right)\left(\frac{V_{j-1} y_{j-1,\,i}}{D x_{Di}}\right) + 1 \tag{3-5}$$

The molal flow rates of the individual components are defined as

$$v_{ji} = V_j y_{ji} \tag{3-6}$$

$$d_i = D x_{Di} \tag{3-7}$$

The adsorption ratio is defined as

$$A_{ji} = \frac{L_j}{K_{ji} V_j} \tag{3-8a}$$

for any stage $j$, and as

$$A_{0i} = \frac{L_0}{D} \tag{3-8b}$$

for the total condenser. Substitution of Eqs. (3-6) to (3-8) in (3-4) and (3-5) yields

$$\left(\frac{v_{1i}}{d_i}\right) = A_{0i} + 1 \tag{3-9}$$

$$\left(\frac{v_{ji}}{d_i}\right) = A_{j-1,\,i}\left(\frac{v_{j-1,i}}{d_i}\right) + 1 \qquad 2 \le j \le f-1 \tag{3-10}$$

For any given trial calculation the $A$'s are regarded as constants [1]. The unknowns in the above equations are the groups of terms $v_{ji}/d_i$. If these are replaced by $x_{ji}$ and the subscript $i$, designating component $i$, is dropped, the following set of equations can be written for a column containing five equilibrium stages above the feed stage:

$$
\begin{aligned}
x_1 \qquad\qquad\qquad\qquad\qquad\qquad &= A_0 + 1 \\
-A_1x_1 + x_2 \qquad\qquad\qquad\qquad &= 1 \\
-A_2x_2 + x_3 \qquad\qquad\qquad &= 1 \\
-A_3x_3 + x_4 \qquad\quad &= 1 \\
-A_4x_4 + x_5 &= 1
\end{aligned}
\tag{3-11}
$$

This is a set of simultaneous linear algebraic equations. It is actually a special set which has nonzero terms only on the diagonal and one adjacent element. It is a bidiagonal set.

The general formulation of a set of simultaneous linear algebraic equations is

$$
\begin{aligned}
a_{11}x_1 + a_{12}x_2 + \cdots + a_{1n}x_n &= c_1 \\
a_{21}x_1 + a_{22}x_2 + \cdots + a_{2n}x_n &= c_2 \\
\cdots\cdots\cdots\cdots\cdots\cdots\cdots\cdots \\
a_{n1}x_1 + a_{n2}x_2 + \cdots + a_{nn}x_n &= c_n
\end{aligned}
\tag{3-12}
$$

where all of the coefficients $a_{ij}$ could be nonzero. This set is usually condensed in vector-matrix notation as

$$\mathbf{Ax} = \mathbf{c} \tag{3-13}$$

where $\mathbf{A}$ is the coefficient matrix

$$\mathbf{A} = \begin{bmatrix} a_{11} & a_{12} & \cdots & a_{1n} \\ \cdots & \cdots & \cdots & \cdots \\ a_{n1} & a_{n2} & \cdots & a_{nn} \end{bmatrix} \tag{3-14}$$

$\mathbf{x}$ is the vector of unknown variables

$$\mathbf{x} = \begin{bmatrix} x_1 \\ x_2 \\ \vdots \\ x_n \end{bmatrix} \tag{3-15}$$

and $\mathbf{c}$ is the vector of constants

$$\mathbf{c} = \begin{bmatrix} c_1 \\ c_2 \\ \vdots \\ c_n \end{bmatrix} \tag{3-16}$$

When the vector $\mathbf{c}$ is the zero vector, the set of equations is called *homogeneous*.

Another example requiring the solution of linear algebraic equations comes from the analysis of complex reaction systems which have monomolecular kinetics. Figure 3-2 considers a chemical reaction between the three species, whose concentrations are designated by $Y_1$, $Y_2$, $Y_3$, taking place in a batch reactor [2, 3]. The equations describing the dynamics of this chemical reaction scheme are

$$\begin{aligned}\frac{dY_1}{dt} &= -(k_{21} + k_{31})Y_1 + k_{12}Y_2 + k_{13}Y_3 \\ \frac{dY_2}{dt} &= k_{21}Y_1 - (k_{12} + k_{32})Y_2 + k_{23}Y_3 \\ \frac{dY_3}{dt} &= k_{31}Y_1 + k_{32}Y_2 - (k_{13} + k_{23})Y_3\end{aligned} \tag{3-17}$$

The above set of linear ordinary differential equations may be condensed into matrix notation

$$\dot{\mathbf{y}} = \mathbf{K}\mathbf{y} \tag{3-18}$$

where $\dot{\mathbf{y}}$ is the vector of derivatives

$$\dot{\mathbf{y}} = \begin{bmatrix} \dfrac{dY_1}{dt} \\ \dfrac{dY_2}{dt} \\ \dfrac{dY_3}{dt} \end{bmatrix} \tag{3-19}$$

$\mathbf{y}$ is the vector of concentrations of the components

$$\mathbf{y} = \begin{bmatrix} Y_1 \\ Y_2 \\ Y_3 \end{bmatrix} \tag{3-20}$$

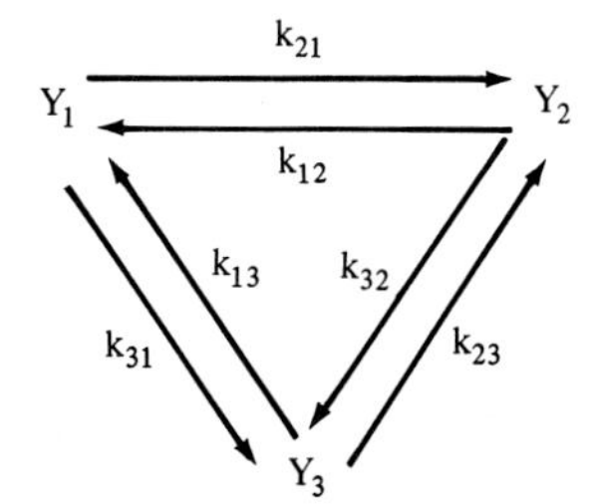


**Figure 3-2** System of chemical reactions.

and **K** is the matrix of kinetic rate constants

$$\mathbf{K} = \begin{bmatrix} -(k_{21} + k_{31}) & k_{12} & k_{13} \\ k_{21} & -(k_{12} + k_{32}) & k_{23} \\ k_{31} & k_{32} & -(k_{13} + k_{23}) \end{bmatrix} \tag{3-21}$$

The solution of the dynamic problem, which is modeled by Eq. (3-18), would require the evaluation of the characteristic values (eigenvalues) $\lambda_i$ and characteristic vectors (eigenvectors) $\mathbf{x}_i$ of the matrix **K**. It is shown in Chap. 5 that the solution of a set of linear ordinary differential equation can be obtained by using Eq. (5-68).

$$y = [\mathbf{X}\mathbf{e}^{\Lambda t}\mathbf{X}^{-1}]\mathbf{y}_0 \tag{5-68}$$

where **X** is a matrix whose columns are the eigenvectors $\mathbf{x}_i$ of **K**; $\mathbf{e}^{\Lambda t}$ is a matrix with $e^{\lambda i t}$ on the diagonal and zero elsewhere; $\mathbf{X}^{-1}$ is the inverse of **X**; and $\mathbf{y}_0$ is the vector of initial concentration of the variables $Y_i$. Methods for calculating eigenvalues and eigenvectors of matrices are developed in Sec. 3-8.

When a chemical reaction reaches steady state, the vector of derivatives in Eq. (3-18) becomes zero and Eq. (3-18) simplifies to

$$\mathbf{Ky} = 0 \tag{3-22}$$

This is a set of homogeneous linear algebraic equations whose solution describes the steady-state situation of the above chemical reaction problem.

Comparison of Eqs. (3-13) and (3-22) reveals that the difference between nonhomogeneous and homogeneous sets of equations is that in the latter the vector of constants **c** is the zero vector. The steady-state solution of the chemical reaction problem requires finding a unique solution to the set of homogeneous algebraic equations represented by Eq. (3-22).

In the analysis of fermentation processes, the prediction of the concentration of cell and metabolic products formed in the fermentor can be accomplished by the technique of material balancing [4]. For example, the stoichiometry of the fermentation of the microorganism *Brevibacterium flavum*, which produces glutamic acid, may be represented by the following two chemical reactions:

Biomass formation:

$$C_6H_{12}O_6 + b'O_2 + c'NH_3 \rightarrow d'C_wH_xO_yN_z + e'CO_2 + f'H_2O \tag{3-23}$$

Glutamic acid synthesis:

$$C_6H_{12}O_6 + 1.5O_2 + NH_3 \rightarrow C_5H_9O_4N + CO_2 + 3H_2O \tag{3-24}$$

$C_6H_{12}O_6$ is glucose, the basic nutrient in this fermentation, and $C_5H_9O_4N$ is glutamic acid. Biomass is represented by $C_wH_xO_yN_z$, where $w$, $x$, $y$, $z$ are the corresponding number of atoms of each element in the cell. This empirical formula can be determined by analyzing the carbon, hydrogen, oxygen, and nitrogen contents of the biomass. The inherent assumption is that C, H, O, and N are the only atoms of significant quantity in the cell biomass. The cellular composition is assumed to remain constant during growth.

Let $a$ be the number of moles of glucose used, $g$ be the fraction of glucose converted to product, and $1-g$ be the fraction of glucose converted to biomass. Then combining Eqs. (3-23) and (3-24) results in the overall reaction

$$\begin{aligned} aC_6H_{12}O_6 + [b(1-g) + 1.5ag]O_2 + [c(1-g) + ag]NH_3 \rightarrow \\ [d(1-g)]C_wH_xO_yN_z + agC_5H_9O_4N \\ + [e(1-g) + ag]CO_2 + [f(1-g) + 3ag]H_2O \end{aligned} \tag{3-25}$$

The primed coefficients ($b'$, $c'$, $d'$, $e'$, $f'$) have been replaced by unprimed coefficients ($b$, $c$, $d$, $e$, $f$). These are related to each other, as follows:

$$b = ab' \qquad c = ac' \qquad d = ad' \qquad e = ae' \qquad f = af'$$

Equation (3-25) contains seven unknown quantities: the stoichiometric coefficients ($a$, $b$, $c$, $d$, $e$, $f$) and the fraction of substrate used for product formation ($g$). Based on Eq. (3-25), four independent equations are derived from material balances on carbon, hydrogen, oxygen, and nitrogen, as shown in Table 3-1.

In order to determine all seven variables, three more independent relationships are required. This is accomplished by calculating total oxygen consumed (TOTAL $O_2$), total carbon dioxide released (TOTAL $CO_2$), and total glucose consumed ($a$).

**Table 3-1**

| | Elemental material balances from Eq. (3-25) | |
|---|---|---|
| Carbon: | $6a = wd(1-g) + 5ag + e(1-g) + ag$ | (3-26) |
| Hydrogen: | $12a + 3c(1-g) + 3ag = xd(1-g) + 9ag + 2f(1-g) + 6ag$ | (3-27) |
| Oxygen: | $6a + 2b(1-g) + 3ag = yd(1-g) + 4ag + 2e(1-g) + 2ag + f(1-g) + 3ag$ | (3-28) |
| Nitrogen: | $c(1-g) + ag = zd(1-g) + ag$ | (3-29) |
| | **Simplified elemental balances** | |
| Carbon: | $(wd + e - 6a)(1-g) = 0$ | (3-30) |
| Hydrogen: | $(xd + 2f - 3c - 12a)(1-g) = 0$ | (3-31) |
| Oxygen: | $(yd + 2e + f - 2b - 6a)(1-g) = 0$ | (3-32) |
| Nitrogen: | $(c - zd)(1-g) = 0$ | (3-33) |

Oxygen and carbon dioxide contents of inlet and exit gases are measured by means of a paramagnetic gaseous oxygen analyzer and an IR carbon dioxide analyzer, respectively. These quantities, combined with gas flow rates, complete the oxygen and carbon balances on the gas stream. The trapezoidal rule (Chap. 5) is used to calculate total oxygen consumed and total carbon dioxide released at any time during the fermentation. These quantities are related to the stoichiometric coefficients of Eq. (3-25) as follows:

$$\frac{\text{TOTAL } O_2}{M_{O_2}} = b(1-g) + 1.5ag \tag{3-34}$$

$$\frac{\text{TOTAL } CO_2}{M_{CO_2}} = e(1-g) + ag \tag{3-35}$$

where $M_{O_2}$ = molecular weight of $O_2$ (g/mol)
$M_{CO_2}$ = molecular weight of $CO_2$ (g/mol)

Since this system is a batch process with respect to glucose, the glucose consumption can be evaluated by subtracting the residual glucose concentration from the initial glucose concentration.

$$a = (\text{GLU}_0 - \text{GLU}_1)\frac{V}{M_G} \tag{3-36}$$

where $a$ = glucose consumption (mol)
$\text{GLU}_0$ = glucose concentration at time 0 (g/L)
$\text{GLU}_1$ = residual glucose concentration at time $t$ (g/L)
$V$ = liquid volume (L)
$M_G$ = molecular weight of glucose (g/mol)

The residual glucose concentration can be measured by an on-line enzymatic glucose analyzer.

The overall system consists of Eqs. (3-30) through (3-36). This is a set of nonlinear simultaneous algebraic equations which can be solved simultaneously using a combination of Newton's method (Sec. 2-7) and the Gauss-Jordan method to be developed in this chapter (Sec. 3-6).

The solution of complex boundary-value differential equations is sometimes carried out by solving sets of simultaneous finite difference equations (see Chap. 6). These equations are often linear in nature and can be solved by the methods which will be discussed in this chapter.

Optimization of a complex assembly of unit operations, such as a chemical plant, or of a cluster of interrelated assemblies, such as a group of refineries, can be accomplished by techniques of linear programming which handle large sets of simultaneous linear equations.

The application of linear and nonlinear regression analysis to fit mathematical models to experimental data and to evaluate the unknown parameters of these models (see Chap. 7) requires the repetitive solution of sets of linear algebraic equations. In addition, the ellipse formed by the correlation coefficient matrix in the parameter hyperspace of these systems, must be searched in the direction of the major and minor axes. The directions of these axes are defined by the eigenvectors of the correlation coefficient matrix, and the relative lengths of the axes are measured by the eigenvalues of the correlation coefficient matrix.

In developing systematic methods for the solution of linear algebraic equations and the evaluation of eigenvalues and eigenvectors of linear systems, we will make extensive use of matrix-vector notation. For this reason, and for the benefit of the reader, a review of selected matrix and vector operations is given in the next section.

## 3-2 REVIEW OF SELECTED MATRIX AND VECTOR OPERATIONS

### 3-2.1 Matrices and Determinants

A matrix is an array of elements arranged in rows and columns as

$$\mathbf{A} = \begin{bmatrix} a_{11} & a_{12} & \cdots & a_{1n} \\ a_{21} & a_{22} & \cdots & a_{2n} \\ \cdots & \cdots & \cdots & \cdots \\ a_{m1} & a_{m2} & \cdots & a_{mn} \end{bmatrix} \tag{3-37}$$

The elements $a_{ij}$ of the matrix may be real numbers, complex numbers, or functions of other variables. Matrix $\mathbf{A}$ has $m$ rows and $n$ columns and is said to be of order $m \times n$. If the number of rows of a matrix is equal to the number of columns, i.e., if $m = n$, then the matrix is a square matrix of $n$th order [2]. A special matrix containing only a single column is called a vector

$$\mathbf{x} = \begin{bmatrix} x_1 \\ x_2 \\ \vdots \\ x_n \end{bmatrix} \tag{3-38}$$

Define another matrix $\mathbf{B}$ with $k$ rows and $l$ columns:

$$\mathbf{B} = \begin{bmatrix} b_{11} & b_{12} & \cdots & b_{1l} \\ b_{21} & b_{22} & \cdots & b_{2l} \\ \cdots & \cdots & \cdots & \cdots \\ b_{k1} & b_{k2} & \cdots & b_{kl} \end{bmatrix} \tag{3-39}$$

The two matrices $\mathbf{A}$ and $\mathbf{B}$ can be added to (or subtracted from) each other if they have the same number of rows ($m = k$) and the same number of columns ($n = l$). For example, if both $\mathbf{A}$ and $\mathbf{B}$ are $3 \times 2$ matrices, their sum (or difference) can be written as

$$\mathbf{A} \pm \mathbf{B} = \begin{bmatrix} a_{11} \pm b_{11} & a_{12} \pm b_{12} \\ a_{21} \pm b_{21} & a_{22} \pm b_{22} \\ a_{31} \pm b_{31} & a_{32} \pm b_{32} \end{bmatrix} = \mathbf{C} \tag{3-40}$$

Matrix **C** is also a $3 \times 2$ matrix. The commutative and associative laws for addition and subtraction apply.

Two matrices can multiply each other, if they are conformable. Matrices **A** and **B** would be conformable in the order **AB** if **A** had the same number of columns as **B** has rows ($n = k$). If **A** is of order $4 \times 2$ and **B** is of order $2 \times 3$, then the product **AB** is

$$\mathbf{AB} = \begin{bmatrix} a_{11} & a_{12} \\ a_{21} & a_{22} \\ a_{31} & a_{32} \\ a_{41} & a_{42} \end{bmatrix} \begin{bmatrix} b_{11} & b_{12} & b_{13} \\ b_{21} & b_{22} & b_{23} \end{bmatrix}$$

$$= \begin{bmatrix} a_{11}b_{11} + a_{12}b_{21} & a_{11}b_{12} + a_{12}b_{22} & a_{11}b_{13} + a_{12}b_{23} \\ a_{21}b_{11} + a_{22}b_{21} & a_{21}b_{12} + a_{22}b_{22} & a_{21}b_{13} + a_{22}b_{23} \\ a_{31}b_{11} + a_{32}b_{21} & a_{31}b_{12} + a_{32}b_{22} & a_{31}b_{13} + a_{32}b_{23} \\ a_{41}b_{11} + a_{42}b_{21} & a_{41}b_{12} + a_{42}b_{22} & a_{41}b_{13} + a_{42}b_{23} \end{bmatrix} = \mathbf{E} \tag{3-41}$$

The resulting matrix **E** is of order $4 \times 3$. The general equation for performing matrix multiplication is

$$\mathbf{e}_{ij} = \sum_{p=1}^{n} a_{ip} b_{pj} \qquad \begin{cases} i = 1, 2, \ldots . m \\ j = 1, 2, \ldots, l \end{cases} \tag{3-42}$$

The resulting matrix would be of order $m \times l$.

The commutative law is not usually valid for matrix multiplication, i.e.,

$$\mathbf{AB} \neq \mathbf{BA} \tag{3-43}$$

even if the matrices are conformable. The distributive law for multiplication applies to matrices, provided that conformability exists

$$\mathbf{A}(\mathbf{B} + \mathbf{C}) = \mathbf{AB} + \mathbf{AC} \tag{3-44}$$

The associative law of multiplication is also valid for matrices

$$\mathbf{A}(\mathbf{BC}) = (\mathbf{AB})\mathbf{C} \tag{3-45}$$

If the rows of an $m \times n$ matrix are written as columns, a new matrix of order $n \times m$ is formed. This new matrix is called the *transpose* of the original matrix. For example, if matrix **A** is

$$\mathbf{A} = \begin{bmatrix} 1 & 2 \\ 3 & 4 \\ 5 & 6 \end{bmatrix} \tag{3-46}$$

then the *transpose* $\mathbf{A}^T$ is

$$\mathbf{A}^T = \begin{bmatrix} 1 & 3 & 5 \\ 2 & 4 & 6 \end{bmatrix} \tag{3-47}$$

The transpose of the sum of two matrices is given by

$$(\mathbf{A} + \mathbf{B})^T = \mathbf{A}^T + \mathbf{B}^T \tag{3-48}$$

The transpose of the product of two matrices is given by

$$(\mathbf{AB})^T = \mathbf{B}^T\mathbf{A}^T \tag{3-49}$$

The following definitions apply to square matrices only: A *symmetric* matrix is one which obeys the equation

$$\mathbf{A} = \mathbf{A}^T \tag{3-50}$$

If the symmetrically situated elements of a matrix are complex conjugates of each other, the matrix is called *Hermitian*. A *diagonal* matrix is one with nonzero elements on the principal diagonal and zero elements everywhere else:

$$\mathbf{D} = \begin{bmatrix} d_{11} & 0 & 0 & 0 \\ 0 & d_{22} & 0 & 0 \\ 0 & 0 & d_{33} & 0 \\ 0 & 0 & 0 & d_{44} \end{bmatrix} \tag{3-51}$$

A *unit* matrix (or *identity* matrix) is a diagonal matrix whose nonzero elements are unity:

$$\mathbf{I} = \begin{bmatrix} 1 & 0 & 0 & 0 \\ 0 & 1 & 0 & 0 \\ 0 & 0 & 1 & 0 \\ 0 & 0 & 0 & 1 \end{bmatrix} \tag{3-52}$$

Multiplication of a matrix (or a vector) by the identity matrix does not alter the matrix (or vector):

$$\mathbf{IA} = \mathbf{A} \qquad \mathbf{Ix} = \mathbf{x} \tag{3-53}$$

A *tridiagonal* matrix is one which has nonzero elements on the principal diagonal and its two adjacent "diagonals," which we will refer to as the subdiagonal (below) and superdiagonal (above), and zero elements everywhere else:

$$\mathbf{T} = \begin{bmatrix} a_{11} & a_{12} & 0 & 0 & 0 \\ a_{21} & a_{22} & a_{23} & 0 & 0 \\ 0 & a_{32} & a_{33} & a_{34} & 0 \\ 0 & 0 & a_{43} & a_{44} & a_{45} \\ 0 & 0 & 0 & a_{54} & a_{55} \end{bmatrix} \tag{3-54}$$

An *upper triangular* matrix is one which has all zero elements below the principal diagonal:

$$\mathbf{U} = \begin{bmatrix} a_{11} & a_{12} & a_{13} & a_{14} & a_{15} \\ 0 & a_{22} & a_{23} & a_{24} & a_{25} \\ 0 & 0 & a_{33} & a_{34} & a_{35} \\ 0 & 0 & 0 & a_{44} & a_{45} \\ 0 & 0 & 0 & 0 & a_{55} \end{bmatrix} \tag{3-55}$$

A *lower triangular* matrix is one which has all zero elements above the principal diagonal:

$$\mathbf{L} = \begin{bmatrix} a_{11} & 0 & 0 & 0 & 0 \\ a_{21} & a_{22} & 0 & 0 & 0 \\ a_{31} & a_{32} & a_{33} & 0 & 0 \\ a_{41} & a_{42} & a_{43} & a_{44} & 0 \\ a_{51} & a_{52} & a_{53} & a_{54} & a_{55} \end{bmatrix} \tag{3-56}$$

A *supertriangular* matrix, also called a *Hessenberg* matrix, is one which has all zero elements below the subdiagonal (upper Hessenberg) or above the superdiagonal (lower Hessenberg):

$$\mathbf{H}_U = \begin{bmatrix} a_{11} & a_{12} & a_{13} & a_{14} & a_{15} \\ a_{21} & a_{22} & a_{23} & a_{24} & a_{25} \\ 0 & a_{32} & a_{33} & a_{34} & a_{35} \\ 0 & 0 & a_{43} & a_{44} & a_{45} \\ 0 & 0 & 0 & a_{54} & a_{55} \end{bmatrix} \tag{3-57}$$

$$\mathbf{H}_L = \begin{bmatrix} a_{11} & a_{12} & 0 & 0 & 0 \\ a_{21} & a_{22} & a_{23} & 0 & 0 \\ a_{31} & a_{32} & a_{33} & a_{34} & 0 \\ a_{41} & a_{42} & a_{43} & a_{44} & a_{45} \\ a_{51} & a_{52} & a_{53} & a_{54} & a_{55} \end{bmatrix} \tag{3-58}$$

Tridiagonal, triangular, and Hessenberg matrices are called *banded* matrices.

Matrices can be divided into two general categories: *dense* and *sparse* matrices. The dense matrices are usually of low order and have no zero elements. The sparse matrices may be of high order with many zero elements. A special subcategory of sparse matrices is the group of banded matrices described above.

The sum of the elements on the main diagonal of a square matrix is called the *trace*:

$$\operatorname{tr} \mathbf{A} = \sum_{i=1}^{n} a_{ii} \tag{3-59}$$

The sum of the *eigenvalues* of a square matrix is equal to the trace of that matrix:

$$\sum_{i=1}^{n} \lambda_i = \operatorname{tr} \mathbf{A} \tag{3-60}$$

Matrix division is not defined in the normal algebraic sense. Instead, an inverse operation is defined which uses multiplication to achieve the same result [5]. If a square matrix $\mathbf{A}$ and another square matrix $\mathbf{B}$, of same order as $\mathbf{A}$, lead to the identity matrix $\mathbf{I}$ when multiplied together

$$\mathbf{AB} = \mathbf{I} \tag{3-61}$$

then $\mathbf{B}$ is called the *inverse* of $\mathbf{A}$ and is written as $\mathbf{A}^{-1}$. It follows then that

$$\mathbf{AA}^{-1} = \mathbf{A}^{-1}\mathbf{A} = \mathbf{I} \tag{3-62}$$

The inverse of the product of two matrices is the product of the inverses of these matrices multiplied in reverse order:

$$(\mathbf{AB})^{-1} = \mathbf{B}^{-1}\mathbf{A}^{-1} \tag{3-63}$$

This can be generalized to products of more than two matrices:

$$(\mathbf{ABC} \cdots \mathbf{KLM})^{-1} = \mathbf{M}^{-1}\mathbf{L}^{-1}\mathbf{K}^{-1} \cdots \mathbf{C}^{-1}\mathbf{B}^{-1}\mathbf{A}^{-1} \tag{3-64}$$

Only *nonsingular* matrices have inverses. A matrix is nonsingular if the determinant of the matrix is nonzero.

The value of the determinant, which exists for square matrices only, can be calculated from Laplace's expansion theorem which involves *minors* and *cofactors* of square matrices. If the row and column containing an element $a_{ij}$ in a square matrix $\mathbf{A}$ are deleted, the determinant of the remaining square array is called the *minor* of $a_{ij}$ and is denoted by $M_{ij}$ [6]. The *cofactor* of $a_{ij}$, denoted by $A_{ij}$, is given by

$$A_{ij} = (-1)^{i+j} M_{ij} \tag{3-65}$$

Laplace's expansion theorem states that the determinant of a square matrix is equal to the sum of the products of the elements of any row (or column) and their respective cofactors

$$|\mathbf{A}| = \sum_{k=1}^{n} a_{ik} A_{ik} \tag{3-66}†$$

for any row $i$, or

$$|\mathbf{A}| = \sum_{k=1}^{n} a_{kj} A_{kj} \tag{3-67}$$

for any column $j$.

Determinants have the following properties [5]:

*Property 1.* If all the elements of any row or column of a matrix are zero, its determinant is equal to zero.

*Property 2.* If the corresponding rows and columns of a matrix are interchanged, its determinant is unchanged.

*Property 3.* If two rows or two columns of a matrix are interchanged, the sign of the determinant changes.

*Property 4.* If the elements of two rows or two columns of a matrix are equal, the determinant of the matrix is zero.

*Property 5.* If the elements of any row or column of a matrix are multiplied by a scalar, this is equivalent to multiplying the determinant by the scalar.

*Property 6.* Adding the product of a scalar and any row (or column) to any other row (or column) of a matrix leaves the determinant unchanged.

*Property 7.* The determinant of a triangular matrix is equal to the product of its diagonal elements:

$$|\mathbf{U}| = \prod_{i=1}^{n} a_{ii} \tag{3-68}$$

Calculating determinants by the expansion of cofactors is a very time-intensive task. Each determinant has $n!$ groups of terms and each group is the product of $n$ elements; thus the total number of multiplications is $(n-1)(n!)$. Evaluating the determinant of a matrix of order $10 \times 10$ would require 32,659,200 multiplications. More efficient methods have been developed for evaluating determinants. It will be shown in Sec. 3-5 that the Gauss elimination method can be used to calculate the determinant of a matrix in addition to finding the solution of simultaneous linear algebraic equations.

The inverse of a matrix cannot always be determined accurately. There are many matrices which are ill-conditioned. An ill-conditioned matrix can be identified using the following criteria [5]:

*a.* When the determinant of a matrix is very small, the matrix is ill-conditioned.

*b.* When the ratio of the absolute values of the largest and smallest eigenvalues of the matrix is very large, the matrix is ill-conditioned.

† The straight brackets | | will be used to designate the determinant of a matrix.

The *rank* $r$ of a matrix $\mathbf{A}$ is defined as the order of the largest nonsingular square submatrix within $\mathbf{A}$. Consider the $m \times n$ matrix

$$\mathbf{A} = \begin{bmatrix} a_{11} & a_{12} & \cdots & a_{1n} \\ a_{21} & a_{22} & \cdots & a_{2n} \\ \cdots & \cdots & \cdots & \cdots \\ a_{m1} & a_{m2} & \cdots & a_{mn} \end{bmatrix} \tag{3-69}$$

where $n \geq m$. The largest square submatrix within $\mathbf{A}$ is of order $m \times m$. If the determinant of this $m \times m$ submatrix is nonzero, then the rank of $\mathbf{A}$ is $m$ ($r = m$). However, if the determinant of the $m \times m$ submatrix is equal to zero, then the rank of $\mathbf{A}$ is less than $m$ ($r < m$). The order of the next largest nonsingular submatrix that can be located within $\mathbf{A}$ would determine the value of the rank.

As an example, let us look at the following $3 \times 4$ matrix:

$$\mathbf{A} = \begin{bmatrix} 3 & 1 & 2 & -4 \\ 5 & 2 & 1 & 3 \\ 6 & 2 & 4 & -8 \end{bmatrix} \tag{3-70}$$

There are four submatrices of order $3 \times 3$, whose determinants are evaluated below using Laplace's expansion theorem:

$$\begin{vmatrix} 3 & 1 & 2 \\ 5 & 2 & 1 \\ 6 & 2 & 4 \end{vmatrix} = (3)(-1)^2 \begin{vmatrix} 2 & 1 \\ 2 & 4 \end{vmatrix} + (1)(-1)^3 \begin{vmatrix} 5 & 1 \\ 6 & 4 \end{vmatrix} + 2(-1)^4 \begin{vmatrix} 5 & 2 \\ 6 & 2 \end{vmatrix}$$

$$= (3)(8-2) - (1)(20-6) + (2)(10-12) = 0 \tag{3-71}$$

Similarly,

$$\begin{vmatrix} 3 & 1 & -4 \\ 5 & 2 & 3 \\ 6 & 2 & -8 \end{vmatrix} = 0$$

$$\begin{vmatrix} 3 & 2 & -4 \\ 5 & 1 & 3 \\ 6 & 4 & -8 \end{vmatrix} = 0 \tag{3-72}$$

$$\begin{vmatrix} 1 & 2 & -4 \\ 2 & 1 & 3 \\ 2 & 4 & -8 \end{vmatrix} = 0$$

Since all the above $3 \times 3$ submatrices are singular; the rank of $\mathbf{A}$ is less than 3. It is easy to find several $2 \times 2$ submatrices that are nonsingular; therefore $r = 2$.

The same conclusion, regarding the singularity of the $3 \times 3$ submatrices, could have been reached by the application of Properties 4 and 5, which were mentioned earlier in this section. Property 4 states, "If the elements of two rows or two columns of a matrix are equal, the determinant of the matrix is zero." Property 5 states, "If the elements of any row or column of a matrix are multiplied by a scalar, this is equivalent to multiplying the determinant by a scalar." Careful inspection of the four $3 \times 3$ submatrices shows that the first and third rows are multiples of each other. In accordance with Properties 4 and 5, the determinants are zero.

### 3-2.2 Matrix Transformations

It is often desirable to transform a matrix to a different form which is more amenable to solution. There are several such transformations which convert matrices without significantly changing their properties. We will divide these transformations into two categories: *elementary* transformations and *similarity* transformations.

*Elementary* transformations usually change the shape of the matrix but preserve the value of its determinant. In addition, if the matrix represents a set of linear algebraic equations, the solution of the set is not affected by the elementary transformation. The following series of matrix multiplications

$$\mathbf{L}_{n-1}\mathbf{L}_{n-2}\cdots\mathbf{L}_2\mathbf{L}_1\mathbf{A} = \mathbf{U} \tag{3-73}$$

represents an elementary transformation of matrix $\mathbf{A}$ to an upper triangular matrix $\mathbf{U}$. This operation can be shown in condensed form as

$$\mathbf{LA} = \mathbf{U} \tag{3-74}$$

where the transformation matrix $\mathbf{L}$ is the product of the lower triangular matrices $\mathbf{L}_i$. The form of the $\mathbf{L}_i$ matrices will be defined in Sec. 3-5, in conjunction with the development of the Gauss elementary transformation procedure.

*Similarity* transformations are of the form

$$\mathbf{Q}^{-1}\mathbf{AQ} = \mathbf{B} \tag{3-75}$$

where $\mathbf{Q}$ is a nonsingular square matrix. In this operation matrix $\mathbf{A}$ is transformed to matrix $\mathbf{B}$, which is said to be *similar* to $\mathbf{A}$. Similarity in this case implies that

1. The determinants of $\mathbf{A}$ and $\mathbf{B}$ are equal:

$$|\mathbf{A}| = |\mathbf{B}| \tag{3-76}$$

2. The traces of $\mathbf{A}$ and $\mathbf{B}$ are the same:

$$\operatorname{tr}\mathbf{A} = \operatorname{tr}\mathbf{B} \tag{3-77}$$

3. The eigenvalues of **A** and **B** are identical:

$$|\mathbf{A} - \lambda\mathbf{I}| = |\mathbf{B} - \lambda\mathbf{I}| \tag{3-78}$$

If the columns of matrix **Q** are real mutually orthogonal unit vectors, then **Q** is an *orthogonal* matrix, and the following relations are true:

$$\mathbf{Q}^T\mathbf{Q} = \mathbf{I} \tag{3-79}$$

and

$$\mathbf{Q}^T = \mathbf{Q}^{-1} \tag{3-80}$$

In this case the similarity transformation, represented by Eq. (3-75), can be written as

$$\mathbf{Q}^T\mathbf{A}\mathbf{Q} = \mathbf{B} \tag{3-81}$$

and is called an *orthogonal* transformation. Since an orthogonal transformation is a similarity transformation, the three identities [Eqs. (3-76) to (3-78)] pertaining to determinants, traces, and eigenvalues of **A** and **B** are equally valid.

### 3-2.3 Matrix Polynomials and Power Series

The definition of a scalar polynomial was given in Chap. 2 as

$$f(x) = a_n x^n + a_{n-1} x^{n-1} + \cdots + a_1 x + a_0 \tag{3-82}$$

Similarly, a matrix polynomial can be defined as

$$P(\mathbf{A}) = \alpha_n \mathbf{A}^n + \alpha_{n-1}\mathbf{A}^{n-1} + \cdots + \alpha_1\mathbf{A} + \alpha_0\mathbf{I} \tag{3-83}$$

where **A** is a square matrix, $\mathbf{A}^n$ is the product of **A** by itself $n$ times, and $\mathbf{A}^0 = \mathbf{I}$.

Matrices can be used in infinite series, such as the exponential, trigonometric, and logarithmic series. For example, the matrix exponential function is defined as

$$e^{\mathbf{A}} = \mathbf{I} + \mathbf{A} + \frac{\mathbf{A}^2}{2!} + \frac{\mathbf{A}^3}{3!} + \cdots \tag{3-84}$$

and the matrix trigonometric functions as

$$\sin\mathbf{A} = \mathbf{A} - \frac{\mathbf{A}^3}{3!} + \frac{\mathbf{A}^5}{5!} - \cdots \tag{3-85}$$

$$\cos\mathbf{A} = \mathbf{I} - \frac{\mathbf{A}^2}{2!} + \frac{\mathbf{A}^4}{4!} - \cdots \tag{3-86}$$

### 3-2.4 Vector Operations

Consider two vectors **x** and **y**:

$$\mathbf{x} = \begin{bmatrix} x_1 \\ x_2 \\ \vdots \\ x_n \end{bmatrix} \qquad \mathbf{y} = \begin{bmatrix} y_1 \\ y_2 \\ \vdots \\ y_n \end{bmatrix} \tag{3-87}$$

and their transpose

$$\mathbf{x}^T = [x_1 \quad x_2 \quad \cdots \quad x_n] \qquad \mathbf{y}^T = [y_1 \quad y_2 \quad \cdots \quad y_n] \tag{3-88}$$

The *scalar* product (or *inner* product) of these two vectors is defined as

$$\mathbf{x}^T\mathbf{y} = [x_1 \quad x_2 \quad \cdots \quad x_n] \begin{bmatrix} y_1 \\ y_2 \\ \vdots \\ y_n \end{bmatrix} = x_1y_1 + x_2y_2 + \cdots + x_ny_n \tag{3-89}$$

As the name implies, this is a scalar quantity. The scalar product is sometimes called the *dot* product. The *dyadic* product of these two vectors is defined as

$$\mathbf{x}\mathbf{y}^T = \begin{bmatrix} x_1 \\ x_2 \\ \vdots \\ x_n \end{bmatrix} [y_1 \quad y_2 \quad \cdots \quad y_n] = \begin{bmatrix} x_1y_1 & x_1y_2 & \cdots & x_1y_n \\ x_2y_1 & x_2y_2 & \cdots & x_2y_n \\ \cdots & \cdots & \cdots & \cdots \\ x_ny_1 & x_ny_2 & \cdots & x_ny_n \end{bmatrix} \tag{3-90}$$

This is a matrix of order $n \times n$. The *cross* product of two vectors is a vector:

$$\mathbf{x}\mathbf{y} = \begin{bmatrix} x_1 \\ x_2 \\ \vdots \\ x_n \end{bmatrix} \begin{bmatrix} y_1 \\ y_2 \\ \vdots \\ y_n \end{bmatrix} = \begin{bmatrix} x_1y_1 \\ x_2y_2 \\ \vdots \\ x_ny_n \end{bmatrix} \tag{3-91}$$

Two nonzero vectors are *orthogonal* if their scalar product is zero:

$$\mathbf{x}^T\mathbf{y} = 0 \tag{3-92}$$

The length of a vector can be calculated from

$$|\mathbf{x}| = \sqrt{\mathbf{x}^T\mathbf{x}} \qquad \text{(3-93)†}$$

A *unit vector* is a vector whose length is unity.

A set of vectors **x**, **y**, **z**, . . . is *linearly dependent* if there exists a set of

† In this case, the straight brackets | | are used to designate the length of the vector.

scalars $c_1, c_2, c_3, \ldots$ so that

$$c_1\mathbf{x} + c_2\mathbf{y} + c_3\mathbf{z} + \cdots = 0 \tag{3-94}$$

Otherwise the vectors are *linearly independent.*

## 3-3 CONSISTENCY OF EQUATIONS AND EXISTENCE OF SOLUTIONS

Consider the set of simultaneous linear algebraic equations represented by

$$\begin{aligned} a_{11}x_1 + a_{12}x_2 + \cdots + a_{1n}x_n &= c_1 \\ a_{21}x_1 + a_{22}x_2 + \cdots + a_{2n}x_n &= c_2 \\ \cdots\cdots\cdots\cdots\cdots\cdots\cdots\cdots \\ a_{n1}x_1 + a_{n2}x_2 + \cdots + a_{nn}x_n &= c_n \end{aligned} \tag{3-95}$$

The *coefficient matrix* is $\mathbf{A}$, the *vector of unknowns* is $\mathbf{x}$, and the *vector of constants* is $\mathbf{c}$.

The *augmented* matrix $\mathbf{A}'$ is defined as the matrix resulting from joining the vector $\mathbf{c}$ to the columns of matrix $\mathbf{A}$ as shown below:

$$\mathbf{A}' = \begin{bmatrix} a_{11} & a_{12} & \cdots & a_{1n} & c_1 \\ a_{21} & a_{22} & \cdots & a_{2n} & c_2 \\ \cdots & \cdots & \cdots & \cdots & \cdots \\ a_{n1} & a_{n2} & \cdots & a_{nn} & c_n \end{bmatrix} \tag{3-96}$$

The set of linear equations has a solution if, and only if, the rank of the augmented matrix is equal to the rank of the coefficient matrix. If, in addition, the rank is equal to $n$ $(r = n)$, the solution is unique. If the rank is less than $n$ $(r < n)$, there are more unknowns in the set than there are independent equations. In that case, the set of equations can be reduced to $r$ independent equations. The remaining $n - r$ unknowns must be assigned arbitrary values. This implies that the system of $n$ equations has an infinite number of possible solutions, since the values of $n - r$ unknowns are given arbitrary values, and the rest of unknowns depend on these $n - r$ values.

A special subcategory of linear algebraic equations is the set whose vector of constants $\mathbf{c}$ is the zero vector:

$$\mathbf{A}\mathbf{x} = 0 \tag{3-97}$$

This is called the *homogeneous* set of linear algebraic equations. This set always has the solution

$$x_1 = x_2 = \cdots = x_n = 0 \tag{3-98}$$

It is called the *trivial* solution, because it is not of any particular interest. The coefficient matrix and the augmented matrix of a homogeneous set always have the same rank, since the vector **c** is the zero vector. As stated earlier, if the rank of **A** is equal to $n$ ($r = n$), then the set of equations has a unique solution. However, in the case of the homogeneous equations, this unique solution is none other than the trivial one. In order for a homogeneous set to have nontrivial solutions the determinant of **A** must be zero, i.e., **A** must be singular.

In summary, the nonhomogeneous set has a *unique nontrivial* solution if the matrix of coefficients **A** is nonsingular. It has an infinite number of solutions if the matrix **A** is singular and the ranks of **A** and **A′** are equal to each other. It has no solution at all if the rank of **A′** is higher than the rank of **A**.

The homogeneous set has a *unique*, but *trivial*, solution if the matrix of coefficients **A** is nonsingular. It has an infinite number of solutions if the matrix **A** is singular. The rank of **A′** is always equal to the rank of **A** for a homogeneous system, since the vector of constants is the zero vector.

## 3-4 CRAMER'S RULE

Cramer's rule calculates the solution of nonhomogeneous linear algebraic equations of the form

$$\mathbf{A}\mathbf{x} = \mathbf{c} \tag{3-99}$$

using the determinants of the coefficient matrix **A** and the *substituted* matrix $\mathbf{A}_j$ as follows:

$$x_j = \frac{|\mathbf{A}_j|}{|\mathbf{A}|} \qquad j = 1, 2, \ldots, n \tag{3-100}$$

The substituted matrix $\mathbf{A}_j$ is obtained by replacing column $j$ of matrix **A** with the vector **c**:

$$\mathbf{A}_j = \begin{bmatrix} a_{11} & a_{12} & \cdots & c_1 & \cdots & a_{1n} \\ a_{21} & a_{22} & \cdots & c_2 & \cdots & a_{2n} \\ \cdots & \cdots & \cdots & \cdots & \cdots & \cdots \\ a_{n1} & a_{n2} & \cdots & c_n & \cdots & a_{nn} \end{bmatrix} \tag{3-101}$$

The set of equations must be nonhomogeneous, since the determinant of **A** appears in the denominator of Eq. (3-100); the determinant cannot be zero, i.e., matrix **A** must be nonsingular.

For a system of $n$ equations, Cramer's rule evaluates $n + 1$ determinants and performs $n$ divisions. Since the calculation of each determinant requires $(n - 1)(n!)$ multiplications, the total number of multiplications and divisions is

$$(n + 1)(n - 1)(n!) + n \tag{3-102}$$

Table 3-2 illustrates how the number of operations required by Cramer's rule increases as the value of $n$ increases.

For $n = 3$, a total of 51 multiplications and divisions are needed. However, when $n = 10$ this number climbs to 359,251,210. For this reason Cramer's rule is rarely used for systems with $n > 3$. The Gauss elimination, Gauss-Jordan reduction, and Gauss-Seidel methods, to be described in the next three sections of this chapter, are much more efficient methods of solution of linear equations than Cramer's rule.

## 3-5 THE GAUSS ELIMINATION METHOD

The most widely used method for the solution of simultaneous linear algebraic equations is the Gauss elimination method. This is based on the principle of converting the set of $n$ equations in $n$ unknowns

$$\begin{array}{c} a_{11}x_1 + a_{12}x_2 + \cdots + a_{1n}x_n = c_1 \\ a_{21}x_1 + a_{22}x_2 + \cdots + a_{2n}x_n = c_2 \\ \cdots\cdots\cdots\cdots\cdots\cdots\cdots\cdots \\ a_{n1}x_1 + a_{n2}x_2 + \cdots + a_{nn}x_n = c_n \end{array} \tag{3-103}$$

to a triangular set of the form

$$\begin{array}{rl} a_{11}x_1 + a_{12}x_2 + a_{13}x_3 + a_{14}x_4 + \cdots + a_{1n}x_n &= c_1 \\ a'_{22}x_2 + a'_{23}x_3 + a'_{24}x_4 + \cdots + a'_{2n}x_n &= c'_2 \\ a'_{33}x_3 + a'_{34}x_4 + \cdots + a'_{3n}x_n &= c'_3 \\ \vdots \quad & \quad \vdots \\ + a'_{n-1,\,n-1}x_{n-1} + a'_{n-1,\,n}x_n &= c'_{n-1} \\ a'_{nn}x_n &= c'_n \end{array} \tag{3-104}$$

whose solution is the same as that of the original set of equations.

**Table 3-2 Number of operations needed by Cramer's rule**

| $n$ | $(n+1)(n-1)(n!) + n$ |
|---|---|
| 3 | 51 |
| 4 | 364 |
| 5 | 2,885 |
| 6 | 25,206 |
| 7 | 241,927 |
| 8 | 2,540,168 |
| 9 | 29,030,409 |
| 10 | 359,251,210 |

The process is essentially that of converting the set

$$\mathbf{Ax} = \mathbf{c} \tag{3-105}$$

to the equivalent triangular set

$$\mathbf{Ux} = \mathbf{c}' \tag{3-106}$$

where **U** is an upper triangular matrix. Once triangularization is achieved, the solution of the set can be obtained easily by back substitution starting with variable $n$ and working backward to variable 1.

### 3-5.1 Gauss Elimination in Formula Form

The Gauss elimination is accomplished by a series of elementary operations which do not alter the solution of the equations. These operations are

1. Any equation in the set can be multiplied (or divided) by a nonzero scalar, without affecting the solution.
2. Any equation in the set can be added to (or subtracted from) another equation, without affecting the solution.
3. Any two equations can interchange positions within the set, without affecting the solution.

Two matrices which can be obtained from each other by successive application of the above elementary operations are said to be *equivalent* matrices. The rank and determinant of these matrices are unaltered by the application of elementary operations.

In order to demonstrate the application of the Gauss elimination method, let's apply the triangularization procedure to obtain the solution of the following set of three equations:

$$\begin{aligned} 3x_1 + 18x_2 + 9x_3 &= 18 \\ 2x_1 + \;\; 3x_2 + 3x_3 &= 117 \\ 4x_1 + \;\;\;\; x_2 + 2x_3 &= 283 \end{aligned} \tag{3-107}$$

First, form the $3 \times 4$ augmented matrix of coefficients and constants

$$\left[\begin{array}{ccc|c} 3 & 18 & 9 & 18 \\ 2 & 3 & 3 & 117 \\ 4 & 1 & 2 & 283 \end{array}\right] \tag{3-108}$$

Each complete row of the augmented matrix represents one of the equations of the linear set (3-107). Therefore, any operations performed on a row of the augmented matrix are automatically performed on the corresponding equation.

To begin the solution, divide the first row by 3, multiply it by 2, and subtract it from the second row to obtain

$$\left[\begin{array}{ccc:c} 3 & 18 & 9 & 18 \\ 0 & -9 & -3 & 105 \\ 4 & 1 & 2 & 283 \end{array}\right] \tag{3-109}$$

Divide the first row by 3, multiply it by 4, and subtract it from the third row to obtain

$$\left[\begin{array}{ccc:c} 3 & 18 & 9 & 18 \\ 0 & -9 & -3 & 105 \\ 0 & -23 & -10 & 259 \end{array}\right] \tag{3-110}$$

Note that the coefficients in the first column below the diagonal have become zero. Continue the elimination by dividing the second row by $-9$, multiplying it by $-23$, and subtracting it from the third row to obtain

$$\left[\begin{array}{ccc:c} 3 & 18 & 9 & 18 \\ 0 & -9 & -3 & 105 \\ 0 & 0 & -\frac{7}{3} & -\frac{28}{3} \end{array}\right] \tag{3-111}$$

The triangularization of the coefficient part of the augmented matrix is complete, and matrix (3-111) represents the triangular set of equations

$$3x_1 + 18x_2 + 9x_3 = 18 \tag{3-112$a$}$$

$$-\ 9x_2 - 3x_3 = 105 \tag{3-112$b$}$$

$$-\tfrac{7}{3}x_3 = -\tfrac{28}{3} \tag{3-112$c$}$$

whose solution is identical to that of the original set (3-107). The solution is obtained by back substitution. Rearrangement of (3-112*c*) yields

$$x_3 = 4$$

Substitution of the value of $x_3$ in (3-112*b*) and rearrangement gives

$$x_2 = -13$$

Substitution of the values of $x_3$ and $x_2$ in (3-112*a*) and rearrangement yields

$$x_1 = 72$$

The overall Gauss elimination procedure applied on the $n \times (n + 1)$ augmented matrix is condensed into a three-part mathematical formula for initialization, elimination, and back substitution [7, 8] as shown below:

Initialization formula:

$$\left.\begin{array}{ll} a_{ij}^{(0)} = a_{ij} & j = 1, 2, \ldots, n \\ a_{ij}^{(0)} = c_i & j = n + 1 \end{array}\right\} \quad i = 1, 2, \ldots, n \tag{3-113}$$

Elimination formula:

$$a_{ij}^{(k)} = a_{ij}^{(k-1)} - \frac{a_{ik}^{(k-1)}}{a_{kk}^{(k-1)}} a_{kj}^{(k-1)} \begin{cases} j = n+1, n, \ldots, k \\ i = k+1, k+2, \ldots, n \end{cases} \begin{cases} k = 1, 2, \ldots, n-1 \\ a_{kk}^{(k-1)} \neq 0 \end{cases} \tag{3-114}$$

where the initialization step places the elements of the coefficient matrix and the vector of constants into the augmented matrix, and the elimination formula reduces to zero the elements below the diagonal. The counter $k$ is the iteration counter of the outside loop in a set of nested loops that perform the elimination.

It should be noted that the element $a_{kk}$ in the denominator of Eq. (3-114) is always the diagonal element. It is called the *pivot* element. This pivot element must not be zero; otherwise the computer program will result in overflow. The computer program can be written so that it rearranges the equations at each step to attain *diagonal dominance* in the coefficient matrix, i.e., the row with the largest pivot element is chosen. This strategy is called *partial pivoting*, and it serves two purposes in the gaussian elimination procedure: it reduces the possibility of division by zero and it increases the accuracy of the Gauss elimination method by using the largest pivot element. If, in addition to rows, the columns are also searched for the maximum available pivot element, then the strategy is called *complete pivoting*. If pivoting cannot locate a nonzero element to place on the diagonal, the matrix must be singular. When two columns are interchanged, the corresponding variables must also be interchanged. A program which performs complete pivoting must keep track of the column interchanges in order to interchange the corresponding variables.

When triangularization of the coefficient matrix has been completed, the algorithm transfers the calculation to the back-substitution formula:

$$\begin{aligned} x_n &= \frac{a_{n,n+1}}{a_{nn}} \\ x_i &= \frac{a_{i,n+1} - \sum_{j=i+1}^{n} a_{ij} x_j}{a_{ii}} \qquad i = n-1, n-2, \ldots, 1 \end{aligned} \tag{3-115}$$

The above formulas complete the solution of the equations by the Gauss elimination method by calculating all the unknowns from $x_n$ to $x_1$. The Gauss elimination algorithm requires $n^3/3$ multiplications to evaluate the vector $x$.

### 3-5.2 Gauss Elimination in Matrix Form

The Gauss elimination procedure, which was described above in formula form, can also be accomplished by a series of matrix multiplications. Two types of special matrices are involved in this operation [8]. Both these matrices are modifications of the identity matrix. The first type, which we designate as $\mathbf{P}_{ij}$, is the identity matrix with the following changes: the unity at position $ii$ switches places with the zero at position $ij$, and the unity at position $jj$ switches places with the zero at position $ji$. For example, $\mathbf{P}_{23}$ for a fifth-order system is

$$\mathbf{P}_{23} = \begin{bmatrix} 1 & 0 & 0 & 0 & 0 \\ 0 & 0 & 1 & 0 & 0 \\ 0 & 1 & 0 & 0 & 0 \\ 0 & 0 & 0 & 1 & 0 \\ 0 & 0 & 0 & 0 & 1 \end{bmatrix} \tag{3-116}$$

Premultiplication of matrix $\mathbf{A}$ by $\mathbf{P}_{ij}$ has the effect of interchanging rows $i$ and $j$. Postmultiplication causes interchange of columns $i$ and $j$. By definition $\mathbf{P}_{ii} = \mathbf{I}$, and multiplication of $\mathbf{A}$ by $\mathbf{P}_{ii}$ causes no interchanges. The inverse of $\mathbf{P}_{ij}$ is identical to $\mathbf{P}_{ij}$.

The second type of matrices used by the Gauss elimination method are unit lower triangular matrices of the form

$$\mathbf{L}_1 = \begin{bmatrix} 1 & 0 & 0 & 0 & 0 \\ -\dfrac{a_{21}^{(0)}}{a_{11}^{(0)}} & 1 & 0 & 0 & 0 \\ -\dfrac{a_{31}^{(0)}}{a_{11}^{(0)}} & 0 & 1 & 0 & 0 \\ -\dfrac{a_{41}^{(0)}}{a_{11}^{(0)}} & 0 & 0 & 1 & 0 \\ -\dfrac{a_{51}^{(0)}}{a_{11}^{(0)}} & 0 & 0 & 0 & 1 \end{bmatrix} \tag{3-117}$$

where the superscript (0) indicates that each $\mathbf{L}_k$ matrix uses the elements $a_{ik}^{(k-1)}$ of the previous transformation step. Premultiplication of matrix $\mathbf{A}$ by $\mathbf{L}_k$ has the effect of reducing to zero the elements below the diagonal in column $k$. The inverse of $\mathbf{L}_k$ has the same form as $\mathbf{L}_k$ but with the signs of the off-diagonal elements reversed.

Therefore, the entire Gauss elimination method, which reduces a nonsingular matrix $\mathbf{A}$ to an upper triangular matrix $\mathbf{U}$, can be represented by the following series of matrix multiplications:

$$\mathbf{L}_{n-1}\mathbf{L}_{n-2} \ldots \mathbf{P}_{ij} \ldots \mathbf{L}_2\mathbf{L}_1\mathbf{P}_{ij}\mathbf{A} = \mathbf{U} \tag{3-118}$$

where the multiplications by $\mathbf{P}_{ij}$ cause pivoting, if and when needed, and the multiplications by $\mathbf{L}_k$ cause elimination. If pivoting is not performed, Eq. (3-118) simplifies to

$$\mathbf{L}_{n-1}\mathbf{L}_{n-2} \ldots \mathbf{L}_2\mathbf{L}_1\mathbf{A} = \mathbf{U} \tag{3-119}$$

The matrices $\mathbf{L}_k$ are unit lower triangular and their product, defined by matrix $\mathbf{L}$, is also unit lower triangular. With this definition of $\mathbf{L}$, Eq. (3-119) condenses to

$$\mathbf{LA} = \mathbf{U} \tag{3-120}$$

Since matrix $\mathbf{L}$ is unit lower triangular, it is nonsingular. Its inverse exists and is also a unit lower triangular matrix. If we premultiply both sides of Eq. (3-120) by $\mathbf{L}^{-1}$, we obtain

$$\mathbf{A} = \mathbf{L}^{-1}\mathbf{U} \tag{3-121}$$

This equation represents the *decomposition* of a nonsingular matrix $\mathbf{A}$ into a unit lower triangular matrix and an upper triangular matrix. Furthermore, this decomposition is unique [8]. Therefore, the matrix operation of Eq. (3-120) when applied to the augmented matrix $[\mathbf{A} \,\vdots\, \mathbf{c}]$ yields the unique solution

$$\mathbf{L}[\mathbf{A} \,\vdots\, \mathbf{c}] \Rightarrow [\mathbf{U} \,\vdots\, \mathbf{c}'] \tag{3-122}$$

of the system of linear algebraic equations

$$\mathbf{Ax} = \mathbf{c} \tag{3-105}$$

whose matrix of coefficients $\mathbf{A}$ is nonsingular.

### 3-5.3 Calculation of Determinants by the Gauss Method

The Gauss elimination method is also very useful in the calculation of determinants of matrices. The elementary operations used in the Gauss method are consistent with the properties of determinants listed in Sec. 3-2. Therefore, the reduction of a matrix to the equivalent triangular matrix by the Gauss elimination procedure would not alter the value of the determinant of the matrix. The determinant of a triangular matrix is equal to the product of its diagonal elements:

$$|\mathbf{U}| = \prod_{i=1}^{n} a_{ii} \tag{3-68}$$

Therefore, a matrix whose determinant is to be evaluated should first be converted to the triangular form using the Gauss method and then its determinant should be calculated from the product of the diagonal elements of the triangular matrix.

Example 3-1 demonstrates the Gauss elimination method in solving a set of simultaneous linear algebraic equations and in calculating the determinant of the matrix of coefficients. The program uses complete pivoting strategy and it is written in such a way that it gives a step-by-step explanation of the complete Gauss elimination algorithm.

**Example 3-1 The Gauss elimination method for simultaneous linear algebraic equations** Write a general computer program which implements the Gauss elimination method for the solution of nonhomogeneous linear algebraic equations. The program should have the following features:

1. Give a step-by-step explanation of the Gauss algorithm.
2. Use complete pivoting strategy.
3. Calculate the determinant of the matrix of coefficients.
4. Identify singular matrices and give their rank.

Use this program to calculate the following:

*a*. The solution to the set of equations in (3-107). Compare the pathway to solution developed by the program, which utilizes complete pivoting, to that performed manually without pivoting.
*b*. Find the solution to the set of equations shown below:

$$2x_1 - 3x_2 - 3x_3 + 6x_4 = 15$$

$$4x_1 + 2x_2 + 3x_3 - 4x_4 = 9.75$$

$$5x_1 + 6x_2 + x_3 - 12x_4 = 5$$

$$3x_1 - x_2 + 2x_3 + 2x_4 = 13$$

METHOD OF SOLUTION The program uses the initialization formula [Eq. (3-113)] to construct the augmented matrix. It applies complete pivoting strategy by searching rows and columns for the maximum pivot element. It keeps track of column interchanges, which affect the positions of the unknown variables. The program uses the elimination formula [Eq. (3-114)] to eliminate the elements below the diagonal of the coefficient segment of the augmented matrix. It applies the back-substitution formula [Eq. (3-115)] to calculate the unknown variables and interchanges their order to correct for column pivoting.

The determinant of the matrix is calculated from the triangularized matrix using Eq. (3-68). Since the sign of the determinant is affected by row and column interchanges, the program performs a sign change on the determinant for each row or column interchange.

If during pivoting, the program cannot find a nonzero pivot element, the matrix must be singular. The program detects this, identifies the rank of the singular matrix, sets ($n$ − rank) unknowns to unity, and reduces the problem to finding (rank) unknowns.

PROGRAM DESCRIPTION This program, which we call GAUSS.BAS, consists of the sections listed in Table E3-1.

*Main program* Lines 120–420 are the input part of the program in which the user enters the number of equations, the coefficients and constants of

**Table E3-1**

| Program section | Line numbers |
|---|---|
| Title | 1–90 |
| Main program | 100–1320 |
| Input | 120–420 |
| Complete pivoting | 440–760 |
| Elimination | 770–880 |
| Back substitution | 900–1030 |
| Interchange of unknowns | 1050–1080 |
| Evaluation of determinant | 1100–1130 |
| Printing of results | 1140–1210 |
| Rerun | 1220–1290 |
| Subroutine | |
| Print the matrix | 2000–2090 |

each equation, the minimum allowable value of the pivot element, and any corrections, if needed.

Lines 440–760 implement the complete pivoting strategy and keep track of rows and columns that have been interchanged. The sign of the determinant is adjusted according to the number of interchanges that occur.

Lines 770–880 perform the elimination step according to formula [Eq. (3-114)].

Lines 900–1030 apply the back-substitution formula [Eq. (3-115)] to evaluate all the unknowns. If the matrix is singular, the program sets ($n$ − rank) unknowns to unity and determines the (rank) unknowns.

Lines 1050–1080 interchange the order of the unknowns to correct for any column interchanges that took place during complete pivoting.

Lines 1100–1130 evaluate the determinant using the elements on the diagonal of the triangularized matrix.

Lines 1140–1210 print out the values of all the variables calculated by back substitution.

Lines 1220–1290 give the user the opportunity to rerun the program with modifications.

Lines 1300–1320 terminate the program.

*Subroutine 1* This subroutine prints out the $n \times (n + 1)$ augmented matrix and the reduced matrices at each step of the elimination procedure, if needed. A brief pause is generated by line 2080 to enable the user to observe the step-by-step results. The length of the pause may be adjusted by the user by changing the upper limit of the iteration counter IPAUSE, or the pause may be eliminated entirely by removing line 2080 from the program.

## PROGRAM

```
1 SCREEN 0:WIDTH 80:CLS:KEY OFF
5 PRINT "**********************************************************************"
10 PRINT"*                                                                    *"
15 PRINT"*                            EXAMPLE 3-1                             *"
20 PRINT"*                                                                    *"
25 PRINT"*                    THE GAUSS ELIMINATION METHOD                    *"
30 PRINT"*                                                                    *"
35 PRINT"*          FOR SIMULTANEOUS LINEAR ALGEBRAIC EQUATIONS               *"
40 PRINT"*                                                                    *"
45 PRINT"*                            (GAUSS.BAS)                             *"
50 PRINT"*                                                                    *"
90 PRINT"**********************************************************************"
100 '*************************** Main Program *****************************
110 '
120 'Enter the number of equations, the coefficients, and constants
130 '
140 PRINT:PRINT "NUMBER OF EQUATIONS";:INPUT N
150 DIM A(N,N+1), B(N,N+1), X(N), NPIVROW(N,2),NPIVCOL(N,2)
160 PRINT:PRINT "ENTER COEFFICIENTS AND CONSTANT FOR EACH EQUATION"
170 FOR K = 1 TO N
180 PRINT : PRINT "EQUATION ";K: PRINT
190 FOR J = 1 TO N
200 PRINT "   COEFFICIENT (";K;",";J;") =";: INPUT B(K,J)
210 NEXT J
220 PRINT:PRINT "   CONSTANT ";K;" =";: INPUT B(K,N+1)
230 NEXT K
240 NC=N+1
250 PRINT
260 PRINT"GIVE THE MINIMUM ALLOWABLE VALUE OF THE PIVOT ELEMENT";:INPUT EPS
270 PRINT CHR$(12)
280 DET=1
290 FOR K = 1 TO N
300 FOR J = 1 TO NC
310 A(K,J)=B(K,J)
320 NEXT J : NEXT K
330 PRINT:PRINT
340 PRINT"**********************************************************************"
350 PRINT"AUGMENTED MATRIX:"
360 GOSUB 2000
370 PRINT: INPUT "IS THE AUGMENTED MATRIX CORRECT(Y/N)"; Q$:PRINT
380 IF Q$ = "Y" OR Q$ = "y" THEN 450
390 PRINT"GIVE THE POSITION OF THE ELEMENT TO BE CORRECTED:":PRINT
400 INPUT "   ROW NUMBER";NROW :INPUT "   COLUMN NUMBER";NCOL
410 PRINT:INPUT "   CORRECT VALUE OF THE ELEMENT"; B(NROW,NCOL):PRINT
420 GOTO 270
430 '
440 'Beginning of the Gauss elimination procedure
450 INPUT "DO YOU WANT TO SEE STEP-BY-STEP RESULTS(Y/N)";Q2$:PRINT
460 PRINT"**********************************************************************"
470 FOR K = 1 TO N
480 'Apply complete pivoting strategy
490 MAXPIVOT = ABS(A(K,K))
500 NPIVROW(K,1)=K: NPIVROW(K,2)=K
510 NPIVCOL(K,1)=K: NPIVCOL(K,2)=K
520 FOR I = K TO N
530 FOR J = K TO N
540 IF MAXPIVOT > = ABS(A(I,J)) GOTO 580
```

```
550 MAXPIVOT=ABS(A(I,J))
560 NPIVROW(K,1)=K: NPIVROW(K,2)=I
570 NPIVCOL(K,1)=K: NPIVCOL(K,2)=J
580 NEXT J:NEXT I
590 IF MAXPIVOT > = EPS GOTO 610
600 PRINT"PIVOT ELEMENT SMALLER THAN";EPS;". MATRIX MAY BE SINGULAR.":GOTO 910
610 IF NPIVROW(K,2)=K GOTO 690
620 IF Q2$="Y" OR Q2$="y" THEN PRINT"PIVOTING ROWS:"
630 IF Q2$="Y" OR Q2$="y" THEN PRINT "INTERCHANGE ROWS ";NPIVROW(K,2);" AND ";K
640 FOR J = K TO NC
650 SWAP  A(NPIVROW(K,2),J),A(K,J)
660 NEXT J
670 DET=DET*(-1)
680 IF Q2$="Y" OR Q2$="y" THEN GOSUB 2000
690 IF NPIVCOL(K,2)=K GOTO 770
700 IF Q2$="Y" OR Q2$="y" THEN PRINT"PIVOTING COLUMNS:"
710 IF Q2$="Y" OR Q2$="y" THEN PRINT "INTERCHANGE COLUMNS ";NPIVCOL(K,2);" AND "
;K
720 FOR I = 1 TO N
730 SWAP  A(I,NPIVCOL(K,2)),A(I,K)
740 NEXT I
750 DET=DET*(-1)
760 IF Q2$="Y" OR Q2$="y" THEN GOSUB 2000
770 IF K=N THEN GOTO 880
780 IF Q2$="Y" OR Q2$="y" THEN PRINT "PERFORM ELIMINATION:"
790 FOR I = K+1 TO N
800 IF Q2$="Y" OR Q2$="y" THEN PRINT "DIVIDE ROW ";K;" BY ";A(K,K)
810 IF Q2$="Y" OR Q2$="y" THEN PRINT "MULTIPLY ROW ";K;" BY ";A(I,K) ;"AND SUBTR
ACT FROM ROW ";I
820 MULT = - A(I,K)/A(K,K)
830 FOR J = NC TO K STEP -1
840 A(I,J) = A(I,J) + MULT * A(K,J)
850 NEXT J
860 IF Q2$="Y" OR Q2$="y" THEN GOSUB 2000
870 NEXT I
880 NEXT K
890 '
900 'Apply the back-substitution formulas
910 RANK=K-1 :PRINT"RANK =";RANK:NMR=N-RANK
920 IF RANK=N THEN X(N) = A(N,N+1) / A(N,N) :NCOUNT=N-1: GOTO 970
925 IF ABS(A(N,N+1)) > EPS THEN PRINT "THE RANK OF THE AUGMENTED MATRIX IS GREAT
ER THAN THE RANK OF THE":PRINT "COEFFICIENT MATRIX.  THE SYSTEM HAS NO SOLUTIONS
.":GOTO 1200
930 PRINT"THE PROGRAM SETS ";NMR;" UNKNOWN(S) TO UNITY,"
940 PRINT"AND REDUCES THE PROBLEM TO FINDING OTHER ";RANK;" UNKNOWN(S)."
950 FOR JJ=1 TO NMR : X(N+1-JJ) = 1: NEXT JJ
960 NCOUNT=RANK
970 FOR I = NCOUNT TO 1 STEP -1
980 SUM = 0
990 FOR J = I+1 TO N
1000 SUM = SUM + A(I,J) * X(J)
1010 NEXT J
1020 X(I) = (A(I,N+1) - SUM) / A(I,I)
1030 NEXT I
1040 '
1050 'Interchange the order of the unknowns to correct for the column pivoting
1060 FOR K=N TO 1 STEP -1
1070 SWAP X(NPIVCOL(K,2)), X(NPIVCOL(K,1))
1080 NEXT K
```

```
1090 '
1100 'Evaluate the determinant of the matrix
1110 FOR I=1 TO N
1120 DET=DET*A(I,I)
1130 NEXT I
1140 PRINT"*******************************************************************"
1150 PRINT:PRINT "RESULTS BY BACK SUBSTITUTION:": PRINT
1160 FOR J = 1 TO N
1170 PRINT "X(";J;") =";X(J)
1180 NEXT J
1190 PRINT:PRINT"VALUE OF DETERMINANT=";DET:PRINT
1200 PRINT"*******************************************************************"
1210 PRINT"*******************************************************************"
1220 PRINT:PRINT "DO YOU WANT TO REPEAT THE CALCULATIONS":PRINT"WITH MINOR CHANG
ES TO THE COEFFICIENTS(Y/N)";:INPUT V$
1230 IF V$ = "Y" OR V$ = "y" THEN 1240 ELSE 1250
1240 CLS : GOTO 270
1250 PRINT:INPUT "DO YOU WANT TO RESET ALL THE COEFFICIENTS(Y/N)";W$
1260 IF W$ = "Y" OR W$ = "y" THEN 1270 ELSE 1300
1270 PRINT:INPUT "IS THE NEW SET OF THE SAME ORDER AS THE PREVIOUS SET";WW$
1280 IF WW$="N" OR WW$="n" THEN PRINT CHR$(12):RUN 100
1290 CLS : GOTO 160
1300 PRINT:PRINT
1310 PRINT"*********************** END OF PROGRAM ****************************"
1320 END
2000 '**************** Subroutine 1: Print the matrix ************************
2010 '
2020 FOR KA = 1 TO N
2030 PRINT
2040 FOR J = 1 TO NC
2050 PRINT A(KA,J),
2060 NEXT J:PRINT: NEXT KA:PRINT
2070 PRINT"*******************************************************************"
2080 FOR
```

RESULTS

```
***************************************************************************
*                                                                         *
*                              EXAMPLE 3-1                                *
*                                                                         *
*                    THE GAUSS ELIMINATION METHOD                         *
*                                                                         *
*          FOR SIMULTANEOUS LINEAR ALGEBRAIC EQUATIONS                    *
*                                                                         *
*                              (GAUSS.BAS)                                *
*                                                                         *
***************************************************************************

NUMBER OF EQUATIONS? 3
ENTER COEFFICIENTS AND CONSTANT FOR EACH EQUATION
EQUATION  1
   COEFFICIENT ( 1 , 1 ) =? 3
   COEFFICIENT ( 1 , 2 ) =? 18
   COEFFICIENT ( 1 , 3 ) =? 9
   CONSTANT  1  =? 18
EQUATION  2
   COEFFICIENT ( 2 , 1 ) =? 2
   COEFFICIENT ( 2 , 2 ) =? 3
   COEFFICIENT ( 2 , 3 ) =? 3
   CONSTANT  2  =? 117
EQUATION  3
   COEFFICIENT ( 3 , 1 ) =? 4
   COEFFICIENT ( 3 , 2 ) =? 1
   COEFFICIENT ( 3 , 3 ) =? 2
   CONSTANT  3  =? 283
GIVE THE MINIMUM ALLOWABLE VALUE OF THE PIVOT ELEMENT? 0.00001
***************************************************************************
AUGMENTED MATRIX:
 3              18              9               18
 2              3               3               117
 4              1               2               283
***************************************************************************
IS THE AUGMENTED MATRIX CORRECT(Y/N)? Y

DO YOU WANT TO SEE STEP-BY-STEP RESULTS(Y/N)? Y

***************************************************************************
PIVOTING COLUMNS:
INTERCHANGE COLUMNS  2  AND  1
 18             3               9               18
 3              2               3               117
 1              4               2               283
```

```
******************************************************************
PERFORM ELIMINATION:
DIVIDE ROW  1  BY  18
MULTIPLY ROW  1  BY  3 AND SUBTRACT FROM ROW  2

 18            3              9              18

 0             1.5            1.5            114

 1             4              2              283

******************************************************************
DIVIDE ROW  1  BY  18
MULTIPLY ROW  1  BY  1 AND SUBTRACT FROM ROW  3

 18            3              9              18

 0             1.5            1.5            114

 0             3.833333       1.5            282

******************************************************************
PIVOTING ROWS:
INTERCHANGE ROWS  3  AND  2

 18            3              9              18

 0             3.833333       1.5            282

 0             1.5            1.5            114

******************************************************************
PERFORM ELIMINATION:
DIVIDE ROW  2  BY  3.833333
MULTIPLY ROW  2  BY  1.5 AND SUBTRACT FROM ROW  3

 18            3              9              18

 0             3.833333       1.5            282

 0             0              .9130435       3.652176

******************************************************************
RANK = 3
******************************************************************

RESULTS BY BACK SUBSTITUTION:

X( 1 ) = 72
X( 2 ) =-13
X( 3 ) = 4.000002

VALUE OF DETERMINANT= 63

******************************************************************
******************************************************************

DO YOU WANT TO REPEAT THE CALCULATIONS
WITH MINOR CHANGES TO THE COEFFICIENTS(Y/N)? N
```

```
DO YOU WANT TO RESET ALL THE COEFFICIENTS(Y/N)? Y

IS THE NEW SET OF THE SAME ORDER AS THE PREVIOUS SET? N
NUMBER OF EQUATIONS? 4

ENTER COEFFICIENTS AND CONSTANT FOR EACH EQUATION

EQUATION  1

   COEFFICIENT ( 1 , 1 ) =? 2
   COEFFICIENT ( 1 , 2 ) =? -3
   COEFFICIENT ( 1 , 3 ) =? -3
   COEFFICIENT ( 1 , 4 ) =? 6

   CONSTANT  1  =? 15

EQUATION  2

   COEFFICIENT ( 2 , 1 ) =? 4
   COEFFICIENT ( 2 , 2 ) =? 2
   COEFFICIENT ( 2 , 3 ) =? 3
   COEFFICIENT ( 2 , 4 ) =? -4

   CONSTANT  2  =? 9.75

EQUATION  3

   COEFFICIENT ( 3 , 1 ) =? 5
   COEFFICIENT ( 3 , 2 ) =? 6
   COEFFICIENT ( 3 , 3 ) =? 1
   COEFFICIENT ( 3 , 4 ) =? -12

   CONSTANT  3  =? 5

EQUATION  4

   COEFFICIENT ( 4 , 1 ) =? 3
   COEFFICIENT ( 4 , 2 ) =? -1
   COEFFICIENT ( 4 , 3 ) =? 2
   COEFFICIENT ( 4 , 4 ) =? 2

   CONSTANT  4  =? 13

GIVE THE MINIMUM ALLOWABLE VALUE OF THE PIVOT ELEMENT? 0.00001

*********************************************************************
AUGMENTED MATRIX:

 2          -3          -3          6          15


 4           2           3         -4          9.75


 5           6           1        -12          5


 3          -1           2          2          13


*************************************************************
```

```
IS THE AUGMENTED MATRIX CORRECT(Y/N)? Y

DO YOU WANT TO SEE STEP-BY-STEP RESULTS(Y/N)? Y

*****************************************************************
PIVOTING ROWS:
INTERCHANGE ROWS  3  AND  1

 5          6           1           -12          5

 4          2           3           -4           9.75

 2         -3          -3            6           15

 3         -1           2            2           13

*****************************************************************
PIVOTING COLUMNS:
INTERCHANGE COLUMNS  4  AND  1

-12         6           1            5           5

-4          2           3            4           9.75

 6         -3          -3            2           15

 2         -1           2            3           13

*****************************************************************
PERFORM ELIMINATION:
DIVIDE ROW  1  BY -12
MULTIPLY ROW  1  BY -4 AND SUBTRACT FROM ROW  2

-12         6           1            5           5

 0          0           2.666667     2.333333    8.083333

 6         -3          -3            2           15

 2         -1           2            3           13

*****************************************************************


 2         -1           2            3           13
```

```
DIVIDE ROW  1  BY -12
MULTIPLY ROW  1  BY  6 AND SUBTRACT FROM ROW  3

-12          6          1          5          5

 0           0          2.666667   2.333333   8.083333

 0           0         -2.5        4.5        17.5

 2          -1          2          3          13

*******************************************************************
DIVIDE ROW  1  BY -12
MULTIPLY ROW  1  BY  2 AND SUBTRACT FROM ROW  4

-12          6          1          5          5

 0           0          2.666667   2.333333   8.083333

 0           0         -2.5        4.5        17.5

 0           0          2.166667   3.833334   13.83333

*******************************************************************
PIVOTING ROWS:
INTERCHANGE ROWS  3  AND  2

-12          6          1          5          5

 0           0         -2.5        4.5        17.5

 0           0          2.666667   2.333333   8.083333

 0           0          2.166667   3.833334   13.83333
*******************************************************************
PIVOTING COLUMNS:
INTERCHANGE COLUMNS  4  AND  2

-12          5          1          6          5

 0           4.5       -2.5        0          17.5

 0           2.333333   2.666667   0          8.083333

 0           3.833334   2.166667   0          13.83333

*******************************************************************
```

```
PERFORM ELIMINATION:
DIVIDE ROW  2  BY  4.5
MULTIPLY ROW  2  BY  2.333333 AND SUBTRACT FROM ROW  3

-12             5             1             6             5

 0              4.5          -2.5           0             17.5

 0              0             3.962963      0            -.9907408

 0              3.833334      2.166667      0             13.83333

******************************************************************
DIVIDE ROW  2  BY  4.5
MULTIPLY ROW  2  BY  3.833334 AND SUBTRACT FROM ROW  4

-12             5             1             6             5

 0              4.5          -2.5           0             17.5

 0              0             3.962963      0            -.9907408

 0              0             4.296296      0            -1.074075
******************************************************************
PIVOTING ROWS:
INTERCHANGE ROWS  4  AND  3

-12             5             1             6             5

 0              4.5          -2.5           0             17.5

 0              0             4.296296      0            -1.074075

 0              0             3.962963      0            -.9907408

******************************************************************
PERFORM ELIMINATION:
DIVIDE ROW  3  BY  4.296296
MULTIPLY ROW  3  BY  3.962963 AND SUBTRACT FROM ROW  4

-12             5             1             6             5

 0              4.5          -2.5           0             17.5

 0              0             4.296296      0            -1.074075

 0              0             0             0             6.556511E-07
```

```
*****************************************************************
PIVOT ELEMENT SMALLER THAN .00001 . MATRIX MAY BE SINGULAR.
RANK = 3
THE PROGRAM SETS  1  UNKNOWN(S) TO UNITY,
AND REDUCES THE PROBLEM TO FINDING OTHER  3  UNKNOWN(S).
*****************************************************************

RESULTS BY BACK SUBSTITUTION:

X( 1 ) = 3.75
X( 2 ) = 1
X( 3 ) =-.2500002
X( 4 ) = 1.625

VALUE OF DETERMINANT= 0

*****************************************************************
*****************************************************************

DO YOU WANT TO REPEAT THE CALCULATIONS
WITH MINOR CHANGES TO THE COEFFICIENTS(Y/N)? N

DO YOU WANT TO RESET ALL THE COEFFICIENTS(Y/N)? N

********************** END OF PROGRAM ***************************
Ok
```

### Discussion of results

*Part a* The complete pivoting strategy finds the maximum pivot element in column 2 and interchanges columns 2 and 1; it performs the elimination to get zeros below the diagonal in column 1; it interchanges rows 3 and 2 and eliminates the elements below the diagonal in column 2. Finally, it evaluates the unknowns by back substitution. These are identical to the values obtained manually:

$$x_1 = 72$$

$$x_2 = -13$$

$$x_3 = 4$$

*Part b* This example was chosen specifically because its matrix of coefficients is singular. Columns 2 and 4 are multiples of each other. Note that with complete pivoting, the Gauss elimination method obtains zeros in all the coefficient positions in one row of the augmented matrix and shifts this row to the bottom of the matrix. The program detects the fact that it cannot locate a nonzero pivot element and identifies the matrix as singular of rank = 3. The program reduces the set of four equations to a set of three independent equations, assumes the value of one of the unknowns to be equal to unity, and calculates the other three unknowns.

## 3-6 THE GAUSS-JORDAN REDUCTION METHOD

The Gauss-Jordan reduction method is an extension of the Gauss elimination method. It reduces a set of $n$ equations from its canonical form of

$$\mathbf{Ax} = \mathbf{c} \tag{3-105}$$

to the diagonal set of the form

$$\mathbf{Ix} = \mathbf{c}' \tag{3-123}$$

where $\mathbf{I}$ is the unit matrix. Equation (3-123) is identical to

$$\mathbf{x} = \mathbf{c}' \tag{3-124}$$

i.e., the solution vector is given by the $\mathbf{c}'$ vector.

The Gauss-Jordan reduction method applies the same series of elementary operations that are used by the Gauss elimination method. It applies these operations both below and above the diagonal in order to reduce all the off-diagonal elements of the matrix to zero.

### 3-6.1 Gauss-Jordan Reduction in Formula Form

We will apply the Gauss-Jordan procedure, without pivoting, to the set of Eqs. (3-107) shown in Sec. 3-5.1 in order to observe the differences between the Gauss-Jordan and the Gauss methods. Starting with the augmented matrix,

$$\left[\begin{array}{ccc|c} 3 & 18 & 9 & 18 \\ 2 & 3 & 3 & 117 \\ 4 & 1 & 2 & 283 \end{array}\right] \tag{3-108}$$

normalize the first row by dividing it by 3:

$$\left[\begin{array}{ccc|c} 1 & 6 & 3 & 6 \\ 2 & 3 & 3 & 117 \\ 4 & 1 & 2 & 283 \end{array}\right] \tag{3-125}$$

Multiply the normalized first row by 2 and subtract it from the second row:

$$\left[\begin{array}{ccc|c} 1 & 6 & 3 & 6 \\ 0 & -9 & -3 & 105 \\ 4 & 1 & 2 & 283 \end{array}\right] \tag{3-126}$$

Multiply the normalized first row by 4 and subtract it from the third row:

$$\left[\begin{array}{ccc|c} 1 & 6 & 3 & 6 \\ 0 & -9 & -3 & 105 \\ 0 & -23 & -10 & 259 \end{array}\right] \tag{3-127}$$

Normalize the second row by dividing it by $-9$:

$$\left[\begin{array}{ccc|c} 1 & 6 & 3 & 6 \\ 0 & 1 & \frac{1}{3} & -\frac{35}{3} \\ 0 & -23 & -10 & 259 \end{array}\right] \tag{3-128}$$

Multiply the normalized second row by 6 and subtract it from the first row:

$$\left[\begin{array}{ccc|c} 1 & 0 & 1 & 76 \\ 0 & 1 & \frac{1}{3} & -\frac{35}{3} \\ 0 & -23 & -10 & 259 \end{array}\right] \tag{3-129}$$

Multiply the normalized second row by $-23$ and subtract it from the third row:

$$\left[\begin{array}{ccc|c} 1 & 0 & 1 & 76 \\ 0 & 1 & \frac{1}{3} & -\frac{35}{3} \\ 0 & 0 & -\frac{7}{3} & -\frac{28}{3} \end{array}\right] \tag{3-130}$$

Normalize the third row by dividing by $-\frac{7}{3}$:

$$\left[\begin{array}{ccc|c} 1 & 0 & 1 & 76 \\ 0 & 1 & \frac{1}{3} & -\frac{35}{3} \\ 0 & 0 & 1 & 4 \end{array}\right] \tag{3-131}$$

Multiply the third row by 1 and subtract it from the first row:

$$\left[\begin{array}{ccc|c} 1 & 0 & 0 & 72 \\ 0 & 1 & \frac{1}{3} & -\frac{35}{3} \\ 0 & 0 & 1 & 4 \end{array}\right] \tag{3-132}$$

Finally, multiply the third row by $\frac{1}{3}$ and subtract it from the second row:

$$\left[\begin{array}{ccc|c} 1 & 0 & 0 & 72 \\ 0 & 1 & 0 & -13 \\ 0 & 0 & 1 & 4 \end{array}\right] \tag{3-133}$$

This reduced matrix [Eq. (3-133)] is equivalent to the set of equations

$$\mathbf{I}\mathbf{x} = \mathbf{c}' \tag{3-123}$$

The vector $\mathbf{c}'$, which is the last column of the reduced matrix, is the solution of the original set of equations (3-107). There is no need for back substitution since the solution is obtained in its final form in vector $\mathbf{c}'$.

The Gauss-Jordan reduction procedure applied to the $(n) \times (n+1)$ augmented matrix can be given in a three-part mathematical formula for the initialization, normalization, and reduction steps as shown below:

Initialization formula:

$$a_{ij}^{(0)} = a_{ij} \quad \{j = 1, 2, \ldots, n \qquad a_{ij}^{(0)} = c_i \quad \{j = n+1 \qquad \Big\{ i = 1, 2, \ldots, n \tag{3-134}$$

Normalization formula:

$$a_{kj}^{(k)} = \frac{a_{kj}^{(k-1)}}{a_{kk}^{(k-1)}} \quad \{ j = n+1, n, \ldots, k \tag{3-135}$$

Reduction formula:

$$a_{ij}^{(k)} = a_{ij}^{(k-1)} - a_{ik}^{(k-1)} a_{kj}^{(k)} \quad \{ j = n+1, n, \ldots, k \quad \begin{cases} i = 1, 2, \ldots, n \\ i \neq k \end{cases} \tag{3-136}$$

$$\text{(3-135), (3-136):} \quad \begin{cases} k = 1, 2, \ldots, n \\ a_{kk}^{(k-1)} \neq 0 \end{cases}$$

The initialization formula places the elements of the coefficient matrix in columns 1 to $n$ and the vector of constants in column $n + 1$ of the augmented matrix. The normalization formula divides each row of the augmented matrix by its pivot element and makes this change permanent, thus causing the diagonal elements of the coefficient segment of the augmented matrix to become unity. Finally, the reduction formula reduces to zero the off-diagonal elements in each row and column in the coefficient segment of the augmented matrix, and converts column $n + 1$ to the solution vector.

## 3-6.2 Gauss-Jordan Reduction in Matrix Form

The Gauss-Jordan reduction procedure can also be accomplished by a series of matrix multiplications, similar to those performed in the Gauss elimination method (Sec. 3-5.2). The matrix $\mathbf{P}_{ij}$, which causes pivoting, is identical to that defined by Eq. (3-116). The matrix $\mathbf{L}_k$ must have additional terms above the diagonal to cause the reduction to zero of elements above, as well as below, the diagonal. We will designate this matrix as $\tilde{\mathbf{L}}_k$ and give an example for a fifth-order system with $k = 2$.

$$\tilde{\mathbf{L}}_2 = \begin{bmatrix} 1 & -\dfrac{a_{12}^{(1)}}{a_{22}^{(1)}} & 0 & 0 & 0 \\ 0 & -\dfrac{1}{a_{22}^{(1)}} & 0 & 0 & 0 \\ 0 & -\dfrac{a_{32}^{(1)}}{a_{22}^{(1)}} & 1 & 0 & 0 \\ 0 & -\dfrac{a_{42}^{(1)}}{a_{22}^{(1)}} & 0 & 1 & 0 \\ 0 & -\dfrac{a_{52}^{(1)}}{a_{22}^{(1)}} & 0 & 0 & 1 \end{bmatrix} \tag{3-137}$$

where the superscript (1) indicates that each $\tilde{\mathbf{L}}_k$ matrix uses the elements $a_{ik}^{(k-1)}$ of the previous transformation step.

The Gauss-Jordan algorithm reduces a nonsingular matrix **A** to the identity matrix **I** by the following series of matrix multiplications:

$$\tilde{\mathbf{L}}_n\tilde{\mathbf{L}}_{n-1}\ldots\mathbf{P}_{ij}\ldots\tilde{\mathbf{L}}_2\tilde{\mathbf{L}}_1\mathbf{P}_{ij}\mathbf{A}=\mathbf{I} \tag{3-138}$$

where the multiplications by $\mathbf{P}_{ij}$ cause pivoting, if and when needed, and the multiplications by $\mathbf{L}_k$ cause normalization and reduction. If pivoting is not performed, Eq. (3-138) simplifies to

$$\tilde{\mathbf{L}}_n\tilde{\mathbf{L}}_{n-1}\ldots\tilde{\mathbf{L}}_2\tilde{\mathbf{L}}_1\mathbf{A}=\mathbf{I} \tag{3-139}$$

By defining the product of all the $\tilde{\mathbf{L}}_j$ matrixes as $\tilde{\mathbf{L}}$, we can condense Eq. (3-139) to

$$\tilde{\mathbf{L}}\mathbf{A}=\mathbf{I} \tag{3-140}$$

The matrix operation of Eq. (3-140), when applied to the augmented matrix $[\mathbf{A} \mid \mathbf{c}]$, yields the unique solution

$$\tilde{\mathbf{L}}[\mathbf{A} \mid \mathbf{c}]\Rightarrow[\mathbf{I} \mid \mathbf{c}'] \tag{3-141}$$

of the system of linear algebraic equations

$$\mathbf{A}\mathbf{x}=\mathbf{c} \tag{3-105}$$

whose matrix of coefficients **A** is nonsingular.

### 3-6.3 Gauss-Jordan Reduction with Matrix Inversion

Matrix $\tilde{\mathbf{L}}$, in Eq. (3-140), is a nonsingular matrix; therefore its inverse exists. Premultiplying both sides of Eq. (3-140) by $\tilde{\mathbf{L}}^{-1}$, we obtain

$$\mathbf{A}=\tilde{\mathbf{L}}^{-1}\mathbf{I} \tag{3-142}$$

Taking the inverse of both sides of Eq. (3-142) results in

$$\mathbf{A}^{-1}=\tilde{\mathbf{L}}\mathbf{I} \tag{3-143}$$

This simply states that the inverse of **A** is equal to $\tilde{\mathbf{L}}$. This has very important implications in numerical methods because it shows that the Gauss-Jordan reduction method is essentially a matrix inversion algorithm. The application of the reduction operation $\tilde{\mathbf{L}}$ on the identity matrix yields the inverse of **A**:

$$\tilde{\mathbf{L}}\mathbf{I}=\mathbf{A}^{-1} \tag{3-143}$$

This observation can be used to extend the formula form of the Gauss-Jordan algorithm to give the inverse of matrix **A** every time it calculates the solution to the set of equations

$$\mathbf{A}\mathbf{x} = \mathbf{c} \tag{3-105}$$

This is done by forming the augmented matrix of order $n \times (2n + 1)$:

$$[\mathbf{A} \mid \mathbf{c} \mid \mathbf{I}] \tag{3-144}$$

and applying the Gauss-Jordan reduction to the augmented matrix. In this case, the three-part mathematical formula for the initialization, normalization, and reduction steps is the following:

Initialization formula:

$$\left.\begin{aligned} a_{ij}^{(0)} &= a_{ij} && \{j = 1, 2, \ldots, n \\ a_{ij}^{(0)} &= c_i && \{j = n + 1 \\ a_{ij}^{(0)} &= 0 \quad \{i \neq j && \multirow{2}{*}{$\Big\{ j = n + 2, \ldots, 2n + 1$} \\ a_{ij}^{(0)} &= 1 \quad \{i = j && \end{aligned}\right\} i = 1, 2, \ldots, n \tag{3-145}$$

Normalization formula:

$$a_{kj}^{(k)} = \frac{a_{kj}^{(k-1)}}{a_{kk}^{(k-1)}} \quad \{j = 2n + 1, 2n, \ldots, k \tag{3-146}$$

Reduction formula:

$$a_{ij}^{(k)} = a_{ij}^{(k-1)} - a_{ik}^{(k-1)} a_{kj}^{(k)} \quad \{j = 2n + 1, 2n, \ldots, k \quad \begin{cases} i = 1, 2, \ldots, n \\ i \neq k \end{cases} \tag{3-147}$$

for (3-146) and (3-147): $k = 1, 2, \ldots, n$, $\quad a_{kk}^{(k-1)} \neq 0$

The first two parts of the initialization formula place the elements of the coefficient matrix in columns 1 to $n$ and the vector of constants in column $n + 1$ of the augmented matrix. The last two parts of the initialization step expand the augmented matrix to include the identity matrix in columns $n + 2$ to $2n + 1$. The normalization formula divides each row of the entire matrix by its pivot element, thus causing the diagonal elements of the coefficient segment of the augmented matrix to become unity. Finally, the reduction formula reduces to zero the off-diagonal elements in each row and column in the coefficient segment of the augmented matrix, converts column $n + 1$ to the solution vector, and converts the identity matrix in columns $(n + 2)$ to $(2n + 1)$ to the inverse of **A**.

Example 3-2 demonstrates the use of the Gauss-Jordan reduction method for the solution of simultaneous linear algebraic equations and for matrix inversion.

**Example 3-2 The Gauss-Jordan reduction method for simultaneous linear algebraic equations and matrix inversion** Figure E3-2 represents the steam distribution system of a chemical plant. The material and energy balances of this system are given below.

$$181.60 - x_3 - 132.57 - x_4 - x_5 = -y_1 - y_2 + y_5 + y_4 = 5.1 \qquad (1)$$

$$1.17x_3 - x_6 = 0 \qquad (2)$$

$$132.57 - 0.745x_7 = 61.2 \qquad (3)$$

$$x_5 + x_7 - x_8 - x_9 - x_{10} + x_{15} = y_7 + y_8 - y_3 = 99.1 \qquad (4)$$

$$x_8 + x_9 + x_{10} + x_{11} - x_{12} - x_{13} = -y_7 = -8.4 \qquad (5)$$

$$x_6 - x_{15} = y_6 - y_5 = 24.2 \qquad (6)$$

$$-1.15(181.60) + x_3 - x_6 + x_{12} + x_{16} = 1.15y_1 - y_9 + 0.4 = -19.7 \qquad (7)$$

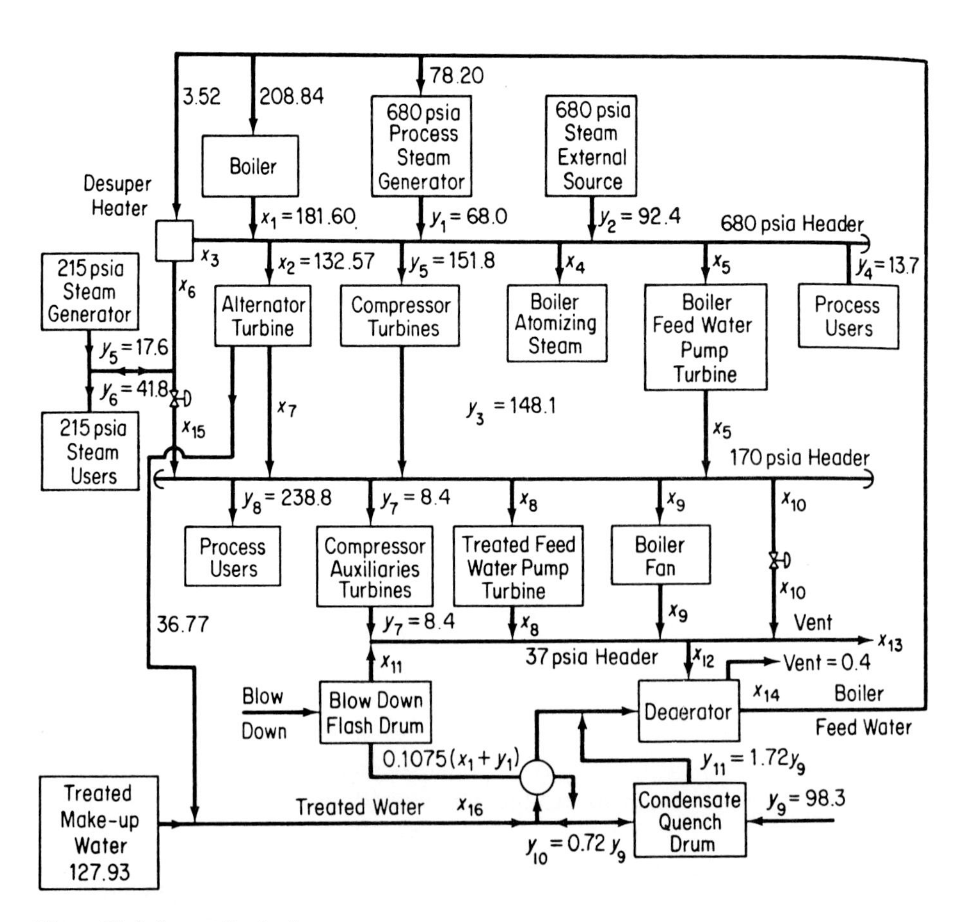


**Figure E3-2** Steam distribution system.

$$181.60 - 4.594x_{12} - 0.11x_{16} = -y_1 + 1.0235y_9 + 2.45 = 35.05 \qquad (8)$$

$$-0.0423(181.60) + x_{11} = 0.0423y_1 = 2.88 \qquad (9)$$

$$-0.016(181.60) + x_4 = 0 \qquad (10)$$

$$x_8 - 0.0147x_{16} = 0 \qquad (11)$$

$$x_5 - 0.07x_{14} = 0 \qquad (12)$$

$$-0.0805(181.60) + x_9 = 0 \qquad (13)$$

$$x_{12} - x_{14} + x_{16} = 0.4 - y_9 = -97.9 \qquad (14)$$

There are four levels of steam in this plant: 680, 215, 170, and 37 psia. The fourteen $x_i$, $i = 3, \ldots, 16$ are the unknowns and the $y_i$ are given parameters for the system. Both $x_i$ and $y_i$ have the units of 1000 lb/h.

Use the Gauss-Jordan method to determine the values of the fourteen unknown quantities $x_i$, $i = 3, \ldots, 16$.†

METHOD OF SOLUTION The fourteen equations of this problem represent balances around the following fourteen units, respectively:

| | |
|---|---|
| Equation (1) | 680-psia header |
| Equation (2) | Desuper heater |
| Equation (3) | Alternator turbine |
| Equation (4) | 170-psia header |
| Equation (5) | 37-psia header |
| Equation (6) | 215-psia steam |
| Equation (7) | BFW balance |
| Equation (8) | Condensate quench drum |
| Equation (9) | Blow down flash drum |
| Equation (10) | Boiler atomizing |
| Equation (11) | Treated feedwater pump |
| Equation (12) | Boiler feedwater pump |
| Equation (13) | Boiler fan |
| Equation (14) | Deaerator-quench |

The set of equations given in the statement of the problem simplifies to a set containing fourteen unknowns ($x_3$ to $x_{16}$):

$$x_3 + x_4 + x_5 = 43.93 \qquad (1)$$

$$1.17x_3 - x_6 = 0.0 \qquad (2)$$

$$x_7 = 95.798 \qquad (3)$$

$$x_5 + x_7 - x_8 - x_9 - x_{10} + x_{15} = 99.1 \qquad (4)$$

† This problem was adapted from Ref. 9 by permission of the author.

$$x_8 + x_9 + x_{10} + x_{11} - x_{12} - x_{13} = -8.4 \qquad (5)$$

$$x_6 - x_{15} = 24.2 \qquad (6)$$

$$x_3 - x_6 + x_{12} + x_{16} = 189.14 \qquad (7)$$

$$4.594x_{12} + 0.11x_{16} = 146.55 \qquad (8)$$

$$x_{11} = 10.56 \qquad (9)$$

$$x_4 = 2.9056 \qquad (10)$$

$$x_8 - 0.0147x_{16} = 0.0 \qquad (11)$$

$$x_5 - 0.07x_{14} = 0.0 \qquad (12)$$

$$x_9 = 14.6188 \qquad (13)$$

$$x_{12} - x_{14} + x_{16} = -97.9 \qquad (14)$$

For convenience in programming, we normalize the equations so that $x_3$ becomes $x_1$, $x_4$ becomes $x_2$, etc.

$$x_1 + x_2 + x_3 = 43.93 \qquad (1)$$

$$1.17x_1 - x_4 = 0.0 \qquad (2)$$

$$x_5 = 95.798 \qquad (3)$$

$$x_3 + x_5 - x_6 - x_7 - x_8 + x_{13} = 99.1 \qquad (4)$$

$$x_6 + x_7 + x_8 + x_9 - x_{10} - x_{11} = -8.4 \qquad (5)$$

$$x_4 - x_{13} = 24.2 \qquad (6)$$

$$x_1 - x_4 + x_{10} + x_{14} = 189.14 \qquad (7)$$

$$4.594x_{10} + 0.11x_{14} = 146.55 \qquad (8)$$

$$x_9 = 10.56 \qquad (9)$$

$$x_2 = 2.9056 \qquad (10)$$

$$x_6 - 0.0147x_{14} = 0.0 \qquad (11)$$

$$x_3 - 0.07x_{12} = 0.0 \qquad (12)$$

$$x_7 = 14.6188 \qquad (13)$$

$$x_{10} - x_{12} + x_{14} = -97.9 \qquad (14)$$

The above set of simultaneous linear algebraic equations can be represented by

$$\mathbf{Ax} = \mathbf{c}$$

where **A** is a sparse matrix containing many zeros. It is not a banded set or a predominantly diagonal set. The Gauss-Jordan reduction is an appropriate method for the solution of this problem. The computer program, which is described in the next section, implements the extended Gauss-Jordan algorithm [Eqs. (3-145) to (3-147)] which solves the linear equations and obtains the inverse of matrix **A** at the same time. The program uses a partial pivoting strategy.

PROGRAM DESCRIPTION This program, which we call JORDAN.BAS, consists of the sections listed in Table E3-2*a*.

*Main program* Lines 120–160 provide the user of this program with three options:

1. To solve a set of linear algebraic equations
2. To find the inverse of a matrix
3. To solve a set of equations and also find the inverse of the coefficient matrix

Lines 180–340 are the input part of the program in which the user enters the set of equations or the matrix, depending on which option was selected.

Lines 360–440 implement the initialization formula [Eq. (3-145)] of the extended Gauss-Jordan algorithm.

**Table E3-2*a***

| Program section | Line numbers |
|---|---|
| Title | 1–9 |
| Main program | 100–1110 |
| Options | 120–160 |
| Input | 180–340 |
| Initialization | 360–440 |
| Corrections | 450–540 |
| Partial pivoting | 550–730 |
| Normalization | 740–800 |
| Reduction | 810–900 |
| Printing of results | 910–1000 |
| Rerun | 1010–1080 |
| Subroutines | |
| 1. Print the matrix | 2000–2090 |
| 2. Print the inverse of the matrix | 3000–3080 |
| 3. Check the product of matrix and inverse | 4000–4100 |

Lines 450–540 allow the user to make corrections to the input.

Lines 550–730 implement the partial pivoting strategy where only rows are pivoted.

Lines 740–800 perform the normalization step according to formula (3-146).

Lines 810–900 apply the reduction step according to formula (3-147).

Lines 910–960 print the values of all the unknowns calculated by the program.

Line 990, in conjunction with subroutines 2 and 3, prints the inverse of the matrix and the product $\mathbf{AA}^{-1}$.

Lines 1010–1080 give the user the opportunity to rerun the program with modifications.

*Subroutine 1* This subroutine prints out the $n \times (n+1)$ augmented matrix and reduced matrices at each step of the reduction procedure, if needed.

*Subroutine 2* This subroutine prints out the inverse of the coefficient matrix.

*Subroutine 3* This subroutine obtains the product $\mathbf{AA}^{-1}$ and prints it out. The user must verify the accuracy of the inverse by ascertaining whether $\mathbf{AA}^{-1} = \mathbf{I}$ is satisfied.

PROGRAM

```
1 SCREEN 0:WIDTH 80:CLS:KEY OFF
5 PRINT "***********************************************************************"
10 PRINT"*                                                                      *"
15 PRINT"*                           EXAMPLE 3-2                                *"
20 PRINT"*                                                                      *"
25 PRINT"*                THE GAUSS-JORDAN REDUCTION METHOD                     *"
30 PRINT"*                                                                      *"
35 PRINT"*           FOR SIMULTANEOUS LINEAR ALGEBRAIC EQUATIONS                *"
40 PRINT"*                                                                      *"
45 PRINT"*                      AND MATRIX INVERSION                            *"
50 PRINT"*                                                                      *"
55 PRINT"*                          (JORDAN.BAS)                                *"
60 PRINT"*                                                                      *"
90 PRINT"***********************************************************************"
100 '************************** Main Program ******************************
110 '
120 PRINT:PRINT"YOU MAY USE THIS PROGRAM TO:"
130 PRINT:PRINT"                1. SOLVE LINEAR ALGEBRAIC EQUATIONS "
140 PRINT:PRINT"                2. FIND THE INVERSE OF A MATRIX     "
150 PRINT:PRINT"                3. DO BOTH OF THE ABOVE             "
160 PRINT:INPUT"THE NUMBER OF YOUR SELECTION";SEL
170 '
180 'Enter the number of equations, the coefficients, and constants
190 '
200 PRINT: IF SEL =2 THEN INPUT"NUMBER OF ROWS OF THE MATRIX";N
210 IF SEL <> 2 THEN INPUT"NUMBER OF EQUATIONS";N
220 DIM A(N,2*N+1), B(N,N+1), C(N,N),X(N)
230 PRINT:IF SEL = 2 THEN PRINT "ENTER ELEMENTS OF MATRIX" ELSE
     PRINT "ENTER COEFFICIENTS AND CONSTANT FOR EACH EQUATION"
240 FOR K = 1 TO N
250 PRINT : IF SEL = 2 THEN PRINT "ROW ";K ELSE PRINT"EQUATION ";K
260 PRINT
270 FOR J = 1 TO N
280 IF SEL = 2 THEN PRINT "   ELEMENT (";K;",";J;") =";: INPUT B(K,J)
290 IF SEL <> 2 THEN PRINT "   COEFFICIENT (";K;",";J;") =";: INPUT B(K,J)
300 NEXT J
310 IF SEL <> 2 THEN PRINT:PRINT "   CONSTANT ";K;" =";: INPUT B(K,N+1)
320 NEXT K
330 PRINT
340 PRINT"GIVE THE MINIMUM ALLOWABLE VALUE OF THE PIVOT ELEMENT";:INPUT EPS
350 PRINT CHR$(12)
360 FOR K = 1 TO N
370 FOR J = 1 TO N + 1
380 A(K,J) = B(K,J)
390 NEXT J
400 FOR J = N+2 TO 2*N + 1
410 A(K,J) = 0
420 NEXT J
430 A(K,K-1+N+2) = 1
440 NEXT K
450 PRINT:PRINT:PRINT:PRINT
460 PRINT"***********************************************************************"
470 PRINT"AUGMENTED MATRIX:"
480 GOSUB 2000
490 PRINT: INPUT "IS THE AUGMENTED MATRIX CORRECT(Y/N)"; Q$:PRINT
500 IF Q$ = "Y" OR Q$ = "y" THEN 560
510 PRINT"GIVE THE POSITION OF THE ELEMENT TO BE CORRECTED:":PRINT
```

```
520 INPUT "   ROW NUMBER";NROW :INPUT "   COLUMN NUMBER";NCOL
530 PRINT:INPUT "   CORRECT VALUE OF THE ELEMENT"; B(NROW,NCOL):PRINT
540 GOTO 360
550 'Beginning of the Gauss-Jordan reduction procedure
560 INPUT "DO YOU WANT TO SEE STEP-BY-STEP RESULTS(Y/N)";Q2$:PRINT
570 PRINT"*******************************************************************"
580 FOR K = 1 TO N
590 'Apply partial pivoting strategy
600 MAXPIVOT = ABS(A(K,K)):NPIVOT=K
610 FOR I = K TO N
620 IF MAXPIVOT > = ABS(A(I,K)) GOTO 640
630 MAXPIVOT=ABS(A(I,K)) : NPIVOT=I
640 NEXT I
650 IF MAXPIVOT > = EPS GOTO 670
660 PRINT"PIVOT ELEMENT SMALLER THAN";EPS;". MATRIX MAY BE SINGULAR. RANK=";K-1
:GOTO 1090
670 IF NPIVOT = K GOTO 740
680 IF Q2$="Y" OR Q2$="y" THEN PRINT"PARTIAL PIVOTING:"
690 IF Q2$="Y" OR Q2$="y" THEN PRINT "INTERCHANGE ROWS ";NPIVOT;" AND ";K
700 FOR J = K TO 2*N + 1
710 SWAP  A(NPIVOT,J),A(K,J)
720 NEXT J
730 IF Q2$="Y" OR Q2$="y" THEN GOSUB 2000
740 IF Q2$="Y" OR Q2$="y" THEN PRINT "PERFORM NORMALIZATION:"
750 IF Q2$="Y" OR Q2$="y" THEN PRINT "DIVIDE ROW ";K;" BY ";A(K,K)
760 D=A(K,K)
770 FOR J = 2*N+1 TO K STEP -1
780 A(K,J) = A(K,J)/D
790 NEXT J
800 IF Q2$="Y" OR Q2$="y" THEN GOSUB 2000
810 IF Q2$="Y" OR Q2$="y" THEN PRINT "PERFORM REDUCTION:"
820 FOR I = 1 TO N
830 IF I=K GOTO 900
840 MULT=A(I,K)
850 IF Q2$="Y" OR Q2$="y" THEN PRINT "MULTIPLY ROW ";K;" BY ";A(I,K) ;"AND SUBTR
ACT FROM ROW ";I
860 FOR J = 2*N+1 TO K STEP -1
870 A(I,J) = A(I,J) - MULT* A(K,J)
880 NEXT J
890 IF Q2$="Y" OR Q2$="y" THEN GOSUB 2000
900 NEXT I:NEXT K:IF SEL=2 THEN GOTO 990
910 PRINT"*******************************************************************"
920 PRINT:PRINT "RESULTS:": PRINT
930 FOR J = 1 TO N
940 X(J)=A(J,N+1)
950 PRINT "X(";J;") =";X(J)
960 NEXT J
970 PRINT
980 PRINT"*******************************************************************"
990 IF SEL > 1 THEN GOSUB 3000: GOSUB 4000
1000 PRINT"*******************************************************************"
1010 PRINT:PRINT "DO YOU WANT TO REPEAT THE CALCULATIONS":PRINT"WITH MINOR CHANG
ES TO THE COEFFICIENTS(Y/N)";:INPUT V$
1020 IF V$ = "Y" OR V$ = "y" THEN 1030 ELSE 1040
1030 CLS : GOTO 350
1040 PRINT:INPUT "DO YOU WANT TO RESET ALL THE COEFFICIENTS(Y/N)";W$
1050 IF W$ = "Y" OR W$ = "y" THEN 1060 ELSE 1090
1060 PRINT:INPUT "IS THE NEW SET OF THE SAME ORDER AS THE PREVIOUS SET";WW$
1070 IF WW$="N" OR WW$="n" THEN PRINT CHR$(12):RUN 100
```

```
1080 CLS : GOTO 230
1090 PRINT:PRINT
1100 PRINT"*********************** END OF PROGRAM ****************************"
1110 END
2000 '***************** Subroutine 1: Print the matrix ***********************
2010 '
2020 FOR KA = 1 TO N
2030 PRINT
2040 FOR J = 1 TO N + 1
2050 PRINT A(KA,J),
2060 NEXT J:PRINT: NEXT KA:PRINT
2070 PRINT"*******************************************************************"
2080 FOR IPAUSE = 1 TO 3000 : NEXT
2090 RETURN
3000 '********** Subroutine 2: Print the inverse of the matrix ***************
3010 PRINT"INVERSE OF MATRIX:"
3020 FOR KA = 1 TO N
3030 PRINT
3040 FOR J = N+2 TO 2*N + 1
3050 PRINT A(KA,J),
3060 NEXT J:PRINT:NEXT KA:PRINT
3070 PRINT"*******************************************************************"
3080 RETURN
4000 '********* Subroutine 3: Check the product of matrix and inverse ********
4010 PRINT"PRODUCT OF MATRIX AND INVERSE SHOULD BE THE IDENTITY MATRIX:":PRINT
4020 FOR I = 1 TO N
4030 FOR J = 1 TO N
4040 C(I,J)=0
4050 FOR K = 1 TO N
4060 C(I,J)=C(I,J) + B(I,K)*A(K,J+N+1)
4070 NEXT K
4080 PRINT USING "##.####       ";C(I,J),
4082 IF I=J AND ABS(C(I,J)-1)<EPS THEN GOTO 4090
4084 IF I<>J AND ABS(C(I,J))<EPS THEN GOTO 4090
4086 PRINT "CAUTION: INVERSE MAY NOT BE CORRECT."
4090 NEXT J:PRINT:NEXT I:PRINT
4100 RETURN
```

RESULTS

```
*******************************************************************
*                                                                 *
*                         EXAMPLE 3-2                             *
*                                                                 *
*            THE GAUSS-JORDAN REDUCTION METHOD                    *
*                                                                 *
*       FOR SIMULTANEOUS LINEAR ALGEBRAIC EQUATIONS               *
*                                                                 *
*                  AND MATRIX INVERSION                           *
*                                                                 *
*                      (JORDAN.BAS)                               *
*                                                                 *
*******************************************************************

YOU MAY USE THIS PROGRAM TO:

            1. SOLVE LINEAR ALGEBRAIC EQUATIONS

            2. FIND THE INVERSE OF A MATRIX

            3. DO BOTH OF THE ABOVE

THE NUMBER OF YOUR SELECTION? 1

NUMBER OF EQUATIONS? 14

ENTER COEFFICIENTS AND CONSTANT FOR EACH EQUATION

EQUATION  1

   COEFFICIENT ( 1 , 1 ) =? 1
   COEFFICIENT ( 1 , 2 ) =? 1
   COEFFICIENT ( 1 , 3 ) =? 1
   COEFFICIENT ( 1 , 4 ) =? 0
   COEFFICIENT ( 1 , 5 ) =? 0
   COEFFICIENT ( 1 , 6 ) =? 0
   COEFFICIENT ( 1 , 7 ) =? 0
   COEFFICIENT ( 1 , 8 ) =? 0
   COEFFICIENT ( 1 , 9 ) =? 0
   COEFFICIENT ( 1 , 10 ) =? 0
   COEFFICIENT ( 1 , 11 ) =? 0
   COEFFICIENT ( 1 , 12 ) =? 0
   COEFFICIENT ( 1 , 13 ) =? 0
   COEFFICIENT ( 1 , 14 ) =? 0

   CONSTANT  1  =? 43.93
```

```
EQUATION  2

   COEFFICIENT ( 2 , 1 ) =? 1.17
   COEFFICIENT ( 2 , 2 ) =? 0
   COEFFICIENT ( 2 , 3 ) =? 0
   COEFFICIENT ( 2 , 4 ) =? -1
   COEFFICIENT ( 2 , 5 ) =? 0
   COEFFICIENT ( 2 , 6 ) =? 0
   COEFFICIENT ( 2 , 7 ) =? 0
   COEFFICIENT ( 2 , 8 ) =? 0
   COEFFICIENT ( 2 , 9 ) =? 0
   COEFFICIENT ( 2 , 10 ) =? 0
   COEFFICIENT ( 2 , 11 ) =? 0
   COEFFICIENT ( 2 , 12 ) =? 0
   COEFFICIENT ( 2 , 13 ) =? 0
   COEFFICIENT ( 2 , 14 ) =? 0

   CONSTANT  2  =? 0


EQUATION  3

   COEFFICIENT ( 3 , 1 ) =? 0
   COEFFICIENT ( 3 , 2 ) =? 0
   COEFFICIENT ( 3 , 3 ) =? 0
   COEFFICIENT ( 3 , 4 ) =? 0
   COEFFICIENT ( 3 , 5 ) =? 1
   COEFFICIENT ( 3 , 6 ) =? 0
   COEFFICIENT ( 3 , 7 ) =? 0
   COEFFICIENT ( 3 , 8 ) =? 0
   COEFFICIENT ( 3 , 9 ) =? 0
   COEFFICIENT ( 3 , 10 ) =? 0
   COEFFICIENT ( 3 , 11 ) =? 0
   COEFFICIENT ( 3 , 12 ) =? 0
   COEFFICIENT ( 3 , 13 ) =? 0
   COEFFICIENT ( 3 , 14 ) =? 0

   CONSTANT  3  =? 95.798


EQUATION  4

   COEFFICIENT ( 4 , 1 ) =? 0
   COEFFICIENT ( 4 , 2 ) =? 0
   COEFFICIENT ( 4 , 3 ) =? 1
   COEFFICIENT ( 4 , 4 ) =? 0
   COEFFICIENT ( 4 , 5 ) =? 1
   COEFFICIENT ( 4 , 6 ) =? -1
   COEFFICIENT ( 4 , 7 ) =? -1
   COEFFICIENT ( 4 , 8 ) =? -1
   COEFFICIENT ( 4 , 9 ) =? 0
   COEFFICIENT ( 4 , 10 ) =? 0
   COEFFICIENT ( 4 , 11 ) =? 0
   COEFFICIENT ( 4 , 12 ) =? 0
   COEFFICIENT ( 4 , 13 ) =? 1
   COEFFICIENT ( 4 , 14 ) =? 0

   CONSTANT  4  =? 99.1
```

```
EQUATION  5

    COEFFICIENT ( 5 , 1 ) =? 0
    COEFFICIENT ( 5 , 2 ) =? 0
    COEFFICIENT ( 5 , 3 ) =? 0
    COEFFICIENT ( 5 , 4 ) =? 0
    COEFFICIENT ( 5 , 5 ) =? 0
    COEFFICIENT ( 5 , 6 ) =? 1
    COEFFICIENT ( 5 , 7 ) =? 1
    COEFFICIENT ( 5 , 8 ) =? 1
    COEFFICIENT ( 5 , 9 ) =? 1
    COEFFICIENT ( 5 , 10 ) =? -1
    COEFFICIENT ( 5 , 11 ) =? -1
    COEFFICIENT ( 5 , 12 ) =? 0
    COEFFICIENT ( 5 , 13 ) =? 0
    COEFFICIENT ( 5 , 14 ) =? 0

    CONSTANT  5  =? -8.4


EQUATION  6

    COEFFICIENT ( 6 , 1 ) =? 0
    COEFFICIENT ( 6 , 2 ) =? 0
    COEFFICIENT ( 6 , 3 ) =? 0
    COEFFICIENT ( 6 , 4 ) =? 1
    COEFFICIENT ( 6 , 5 ) =? 0
    COEFFICIENT ( 6 , 6 ) =? 0
    COEFFICIENT ( 6 , 7 ) =? 0
    COEFFICIENT ( 6 , 8 ) =? 0
    COEFFICIENT ( 6 , 9 ) =? 0
    COEFFICIENT ( 6 , 10 ) =? 0
    COEFFICIENT ( 6 , 11 ) =? 0
    COEFFICIENT ( 6 , 12 ) =? 0
    COEFFICIENT ( 6 , 13 ) =? -1
    COEFFICIENT ( 6 , 14 ) =? 0

    CONSTANT  6  =? 24.2


EQUATION  7

    COEFFICIENT ( 7 , 1 ) =? 1
    COEFFICIENT ( 7 , 2 ) =? 0
    COEFFICIENT ( 7 , 3 ) =? 0
    COEFFICIENT ( 7 , 4 ) =? -1
    COEFFICIENT ( 7 , 5 ) =? 0
    COEFFICIENT ( 7 , 6 ) =? 0
    COEFFICIENT ( 7 , 7 ) =? 0
    COEFFICIENT ( 7 , 8 ) =? 0
    COEFFICIENT ( 7 , 9 ) =? 0
    COEFFICIENT ( 7 , 10 ) =? 1
    COEFFICIENT ( 7 , 11 ) =? 0
    COEFFICIENT ( 7 , 12 ) =? 0
    COEFFICIENT ( 7 , 13 ) =? 0
    COEFFICIENT ( 7 , 14 ) =? 1

    CONSTANT  7  =? 189.14
```

```
EQUATION  8

   COEFFICIENT ( 8 , 1 ) =? 0
   COEFFICIENT ( 8 , 2 ) =? 0
   COEFFICIENT ( 8 , 3 ) =? 0
   COEFFICIENT ( 8 , 4 ) =? 0
   COEFFICIENT ( 8 , 5 ) =? 0
   COEFFICIENT ( 8 , 6 ) =? 0
   COEFFICIENT ( 8 , 7 ) =? 0
   COEFFICIENT ( 8 , 8 ) =? 0
   COEFFICIENT ( 8 , 9 ) =? 0
   COEFFICIENT ( 8 , 10 ) =? 4.594
   COEFFICIENT ( 8 , 11 ) =? 0
   COEFFICIENT ( 8 , 12 ) =? 0
   COEFFICIENT ( 8 , 13 ) =? 0
   COEFFICIENT ( 8 , 14 ) =? 0.11

   CONSTANT  8  =? 146.55


EQUATION  9

   COEFFICIENT ( 9 , 1 ) =? 0
   COEFFICIENT ( 9 , 2 ) =? 0
   COEFFICIENT ( 9 , 3 ) =? 0
   COEFFICIENT ( 9 , 4 ) =? 0
   COEFFICIENT ( 9 , 5 ) =? 0
   COEFFICIENT ( 9 , 6 ) =? 0
   COEFFICIENT ( 9 , 7 ) =? 0
   COEFFICIENT ( 9 , 8 ) =? 0
   COEFFICIENT ( 9 , 9 ) =? 1
   COEFFICIENT ( 9 , 10 ) =? 0
   COEFFICIENT ( 9 , 11 ) =? 0
   COEFFICIENT ( 9 , 12 ) =? 0
   COEFFICIENT ( 9 , 13 ) =? 0
   COEFFICIENT ( 9 , 14 ) =? 0

   CONSTANT  9  =? 10.56


EQUATION  10

   COEFFICIENT ( 10 , 1 ) =? 0
   COEFFICIENT ( 10 , 2 ) =? 1
   COEFFICIENT ( 10 , 3 ) =? 0
   COEFFICIENT ( 10 , 4 ) =? 0
   COEFFICIENT ( 10 , 5 ) =? 0
   COEFFICIENT ( 10 , 6 ) =? 0
   COEFFICIENT ( 10 , 7 ) =? 0
   COEFFICIENT ( 10 , 8 ) =? 0
   COEFFICIENT ( 10 , 9 ) =? 0
   COEFFICIENT ( 10 , 10 ) =? 0
   COEFFICIENT ( 10 , 11 ) =? 0
   COEFFICIENT ( 10 , 12 ) =? 0
   COEFFICIENT ( 10 , 13 ) =? 0
   COEFFICIENT ( 10 , 14 ) =? 0

   CONSTANT  10  =? 2.9056
```

```
EQUATION  11

   COEFFICIENT ( 11 , 1 ) =? 0
   COEFFICIENT ( 11 , 2 ) =? 0
   COEFFICIENT ( 11 , 3 ) =? 0
   COEFFICIENT ( 11 , 4 ) =? 0
   COEFFICIENT ( 11 , 5 ) =? 0
   COEFFICIENT ( 11 , 6 ) =? 1
   COEFFICIENT ( 11 , 7 ) =? 0
   COEFFICIENT ( 11 , 8 ) =? 0
   COEFFICIENT ( 11 , 9 ) =? 0
   COEFFICIENT ( 11 , 10 ) =? 0
   COEFFICIENT ( 11 , 11 ) =? 0
   COEFFICIENT ( 11 , 12 ) =? 0
   COEFFICIENT ( 11 , 13 ) =? 0
   COEFFICIENT ( 11 , 14 ) =? -0.0147

   CONSTANT  11  =? 0


EQUATION  12

   COEFFICIENT ( 12 , 1 ) =? 0
   COEFFICIENT ( 12 , 2 ) =? 0
   COEFFICIENT ( 12 , 3 ) =? 1
   COEFFICIENT ( 12 , 4 ) =? 0
   COEFFICIENT ( 12 , 5 ) =? 0
   COEFFICIENT ( 12 , 6 ) =? 0
   COEFFICIENT ( 12 , 7 ) =? 0
   COEFFICIENT ( 12 , 8 ) =? 0
   COEFFICIENT ( 12 , 9 ) =? 0
   COEFFICIENT ( 12 , 10 ) =? 0
   COEFFICIENT ( 12 , 11 ) =? 0
   COEFFICIENT ( 12 , 12 ) =? -0.07
   COEFFICIENT ( 12 , 13 ) =? 0
   COEFFICIENT ( 12 , 14 ) =? 0

   CONSTANT  12  =? 0


EQUATION  13

   COEFFICIENT ( 13 , 1 ) =? 0
   COEFFICIENT ( 13 , 2 ) =? 0
   COEFFICIENT ( 13 , 3 ) =? 0
   COEFFICIENT ( 13 , 4 ) =? 0
   COEFFICIENT ( 13 , 5 ) =? 0
   COEFFICIENT ( 13 , 6 ) =? 0
   COEFFICIENT ( 13 , 7 ) =? 1
   COEFFICIENT ( 13 , 8 ) =? 0
   COEFFICIENT ( 13 , 9 ) =? 0
   COEFFICIENT ( 13 , 10 ) =? 0
   COEFFICIENT ( 13 , 11 ) =? 0
   COEFFICIENT ( 13 , 12 ) =? 0
   COEFFICIENT ( 13 , 13 ) =? 0
   COEFFICIENT ( 13 , 14 ) =? 0

   CONSTANT  13  =? 14.6188
```

```
EQUATION  14

   COEFFICIENT ( 14 , 1 ) =? 0
   COEFFICIENT ( 14 , 2 ) =? 0
   COEFFICIENT ( 14 , 3 ) =? 0
   COEFFICIENT ( 14 , 4 ) =? 0
   COEFFICIENT ( 14 , 5 ) =? 0
   COEFFICIENT ( 14 , 6 ) =? 0
   COEFFICIENT ( 14 , 7 ) =? 0
   COEFFICIENT ( 14 , 8 ) =? 0
   COEFFICIENT ( 14 , 9 ) =? 0
   COEFFICIENT ( 14 , 10 ) =? 1
   COEFFICIENT ( 14 , 11 ) =? 0
   COEFFICIENT ( 14 , 12 ) =? -1
   COEFFICIENT ( 14 , 13 ) =? 0
   COEFFICIENT ( 14 , 14 ) =? 1

   CONSTANT  14  =? -97.9

GIVE THE MINIMUM ALLOWABLE VALUE OF THE PIVOT ELEMENT? 0.00001


******************************************************************
AUGMENTED MATRIX:

 1              1              1              0              0
 0              0              0              0              0
 0              0              0              0              43.93

 1.17           0              0             -1              0
 0              0              0              0              0
 0              0              0              0              0

 0              0              0              0              1
 0              0              0              0              0
 0              0              0              0              95.79799

 0              0              1              0              1
-1             -1             -1              0              0
 0              0              1              0              99.1

 0              0              0              0              0
 1              1              1              1             -1
-1              0              0              0             -8.399999

 0              0              0              1              0
 0              0              0              0              0
 0              0             -1              0              24.2

 1              0              0             -1              0
 0              0              0              0              1
 0              0              0              1              189.14

 0              0              0              0              0
 0              0              0              0              4.594
 0              0              0              .11            146.55
```

```
0              0              0              0              0
0              0              0              1              0
0              0              0              0              10.56

0              1              0              0              0
0              0              0              0              0
0              0              0              0              2.9056

0              0              0              0              0
1              0              0              0              0
0              0              0             -.0147          0

0              0              1              0              0
0              0              0              0              0
0             -.07            0              0              0

0              0              0              0              0
0              1              0              0              0
0              0              0              0              14.6188

0              0              0              0              0
0              0              0              0              1
0             -1              0              1             -97.9
*****************************************************************
IS THE AUGMENTED MATRIX CORRECT(Y/N)? Y

DO YOU WANT TO SEE STEP-BY-STEP RESULTS(Y/N)? N
*****************************************************************
*****************************************************************

RESULTS:

X( 1 ) = 20.68544
X( 2 ) = 2.905602
X( 3 ) = 20.33896
X( 4 ) = 24.20197
X( 5 ) = 95.79799
X( 6 ) = 2.421088
X( 7 ) = 14.6188
X( 8 ) =-9.679794E-04
X( 9 ) = 10.56
X( 10 ) = 27.95669
X( 11 ) = 8.042231
X( 12 ) = 290.5565
X( 13 ) = 1.966477E-03
X( 14 ) = 164.6998

*****************************************************************
*****************************************************************

DO YOU WANT TO REPEAT THE CALCULATIONS
WITH MINOR CHANGES TO THE COEFFICIENTS(Y/N)? N

DO YOU WANT TO RESET ALL THE COEFFICIENTS(Y/N)? N

********************** END OF PROGRAM ***************************
Ok
```

DISCUSSION OF RESULTS The results are listed in Table E3-2*b* and show the correspondence between the program-variable numbering sequence and the problem-variable numbering sequence.

**Table E3-2*b***

| Program variable | Value | Problem variable |
|---|---|---|
| X(1) | 20.7 | $x_3$ |
| X(2) | 2.9 | $x_4$ |
| X(3) | 20.3 | $x_5$ |
| X(4) | 24.2 | $x_6$ |
| X(5) | 95.8 | $x_7$ |
| X(6) | 2.4 | $x_8$ |
| X(7) | 14.6 | $x_9$ |
| X(8) | 0.0 | $x_{10}$ |
| X(9) | 10.6 | $x_{11}$ |
| X(10) | 28.0 | $x_{12}$ |
| X(11) | 8.0 | $x_{13}$ |
| X(12) | 290.6 | $x_{14}$ |
| X(13) | 0.0 | $x_{15}$ |
| X(14) | 164.7 | $x_{16}$ |

The units of the above quantities are 1000 lb/h. The values of variables $x_{10}$ and $x_{15}$ are zero, as may be expected from the flow diagram of Fig. E3-2.

## 3-7 The GAUSS-SEIDEL SUBSTITUTION METHOD

Certain engineering problems yield sets of simultaneous linear algebraic equations which are *predominantly diagonal* systems. One such example is the solution of finite difference equations resulting from the approximation of partial differential equations (Chap. 6). A *predominantly diagonal* system of linear equations has coefficients on the diagonal which are larger in absolute value than the sum of the absolute values of the other coefficients. For example, the set of equations

$$\begin{aligned} -10x_1 + 2x_2 + 3x_3 &= 6 \\ x_1 + 8x_2 - 2x_3 &= 9 \\ -3x_1 - x_2 - 7x_3 &= -33 \end{aligned} \tag{3-148}$$

is a predominantly diagonal set because

$$|-10| > |2| + |3| \tag{3-149}$$

and

$$|8| > |1| + |-2| \tag{3-150}$$

and

$$|-7| > |-3| + |1| \tag{3-151}$$

Each equation in set (3-148) can be solved for the unknown on its diagonal:

$$\begin{aligned} x_1 &= \frac{6 - (2x_2 + 3x_3)}{-10} \\ x_2 &= \frac{9 - (x_1 - 2x_3)}{8} \\ x_3 &= \frac{-33 - (-3x_1 - x_2)}{-7} \end{aligned} \tag{3-152}$$

For the general set of $n$ equations in $n$ unknowns

$$\mathbf{Ax} = \mathbf{c} \tag{3-105}$$

the above operation corresponds to the formula

$$x_i = \frac{\left[c_i - \sum_{\substack{j=1 \\ j \neq i}}^{n} a_{ij} x_j\right]}{a_{ii}} \qquad i = 1, 2, \ldots, n; \; k = 1, \ldots \tag{3-153}$$

The Gauss-Seidel substitution method requires an initial guess of the values of the unknowns $x_2$ to $x_n$. These values are used in Eq. (3-153) to begin evaluation of new estimates of the $x$'s. Each newly calculated $x_i$ replaces its previous value in subsequent calculations. The iteration continues until *all* the newly calculated $x$'s converge to within a convergence criterion $\varepsilon$ of their previous values.

The Gauss-Seidel method converges to the correct solution, no matter what the initial estimate is, provided that the system of equations is *predominantly diagonal*. On the other hand, if the system is not predominantly diagonal, the correct solution may still be obtained if the initial estimate of the values of $x_2$ to $x_n$ is close with the correct set. The Gauss-Seidel method is a very simple algorithm to program and it is computationally very efficient, in comparison with the other methods described in this chapter, provided, of course, that the system is predominantly diagonal. These advantages account for this method's wide use in the solution of engineering problems.

**Example 3-3 The Gauss-Seidel substitution method for diagonal systems of linear algebraic equations** A chemical reaction takes place in a series of four continuous stirred tank reactors arranged as shown in Fig. E3-3.

The chemical reaction is a first-order irreversible reaction of the type

$$A \xrightarrow{k_i} B$$

The conditions of temperature in each reactor are such that the value of $k_i$ is different in each reactor. Also, the volume of each reactor is different. The values of $k_i$ and $V_i$ are given in Table E3-3*a*.

**Table E3-3*a***

| Reactor | $V_i$, L | $k_i$, $h^{-1}$ |
|---|---|---|
| 1 | 1000 | 0.1 |
| 2 | 1500 | 0.2 |
| 3 | 100 | 0.4 |
| 4 | 500 | 0.3 |

The following assumptions can be made regarding this system:

1. The system is at steady state.
2. The reactions are in the liquid phase.
3. There is no change in volume or density of the liquid.
4. The rate of disappearance of component A in each reactor is given by

$$R_i = V_i k_i c_{A_i} \qquad \text{mol/h}$$

Respond to the following questions:

*a*. Set up the material balance equation for each of the four reactors. What type of equations do you have in this set of material balances?
*b*. What method do you recommend as the best one to use to solve for the exit concentration from each reactor ($c_{A_i}$)?
*c*. Write a computer program to solve this set of equations and find the exit concentration from each reactor.

METHOD OF SOLUTION

*Part a* The general unsteady-state material balance for each reactor is

$$\text{Input} = \text{output} + \text{disappearance by reaction} + \text{accumulation}$$

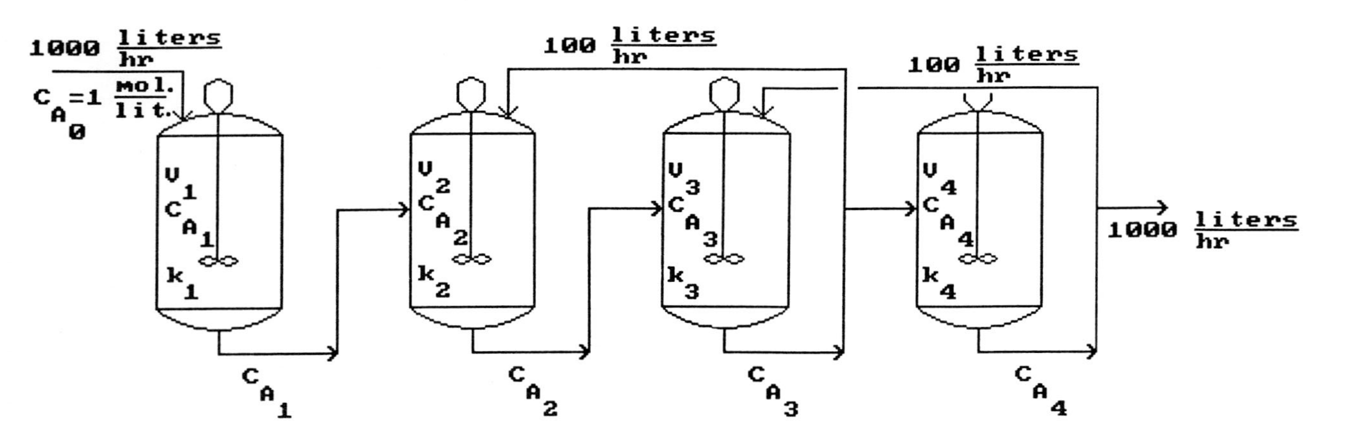

1000 liters/hr
C A0 =1 mol./lit.
V 1
C A 1
k 1
C A 1
V 2
C A 2
k 2
C A 2
100 liters/hr
V 3
C A 3
k 3
C A 3
100 liters/hr
V 4
C A 4
k 4
C A 4
1000 liters/hr


Figure E3-3

Since the system is at steady state, the accumulation term is zero; therefore, the material balance simplifies to

$$\text{Input} = \text{output} + \text{disappearance by reaction}$$

This balance applied to each of the four reactors yields the following set of equations:

$$(1000)(1) = 1000c_{A_1} + V_1 k_1 c_{A_1}$$
$$1000c_{A_1} + 100c_{A_3} = 1100c_{A_2} + V_2 k_2 c_{A_2}$$
$$1100c_{A_2} + 100c_{A_4} = 1200c_{A_3} + V_3 k_3 c_{A_3}$$
$$1100c_{A_3} = 1100c_{A_4} + V_4 k_4 c_{A_4}$$

Substituting the values of $V_i$ and $k_i$ and rearranging,

$$1100c_{A_1} = 1000$$
$$1000c_{A_1} - 1400c_{A_2} + 100c_{A_3} = 0$$
$$1100c_{A_2} - 1240c_{A_3} + 100c_{A_4} = 0$$
$$1100c_{A_3} - 1250c_{A_4} = 0$$

The above is a set of four simultaneous linear algebraic equations. It appears to be a predominantly diagonal system of equations since the coefficients on the diagonal are larger in absolute value than the sum of the absolute values of the other coefficients.

*Part b* The best method of solution for a predominantly diagonal set is the Gauss-Seidel method.

*Part c* The four material balance equations are rearranged to solve for the unknown on the diagonal position of each equation:

$$c_{A_1} = \frac{1000}{1100}$$

$$c_{A_2} = \frac{1000c_{A_1} + 100c_{A_3}}{1400}$$

$$c_{A_3} = \frac{1100c_{A_2} + 100c_{A_4}}{1240}$$

$$c_{A_4} = \frac{1100c_{A_3}}{1250}$$

These equations are of the form of Eq. (3-153):

$$x_i = \frac{\left[c_i - \sum\limits_{\substack{j=1 \\ i \neq j}}^{n} a_{ij} x_j\right]}{a_{ii}} \qquad i = 1, 2, \ldots, n;\ k = 1, \ldots \tag{3-153}$$

The general program, which uses the Gauss-Seidel method, is described in the next section.

An initial guess of the unknowns $c_{A_2}$ to $c_{A_4}$ is needed to start the Gauss-Seidel algorithm. Since this system of equations is a predominantly diagonal set, any initial guess for unknowns $c_{A_2}$ to $c_{A_4}$ will yield convergence. However, the initial guess of 0.6 for all three unknowns seems to be an appropriate choice based on the fact that $c_{A_0} = 1.0$. Two cases will be run in order to test the ability of the Gauss-Seidel method to converge. The first case will use 0.6 as the initial values and the second will use 100 as the starting values.

PROGRAM DESCRIPTION This program, which we call SEIDEL.BAS, consists of the sections listed in Table E3-3*b*.

**Table E3-3*b***

| Program section | Line numbers |
|---|---|
| Title | 1–90 |
| Main program | 100–740 |
| Input | 120–430 |
| Gauss-Seidel method | 450–620 |
| Rerun | 640–710 |
| Subroutines | |
| 1. Print the matrix | 2000–2100 |
| 2. Print the results | 3000–3090 |

*Main program* Lines 120–430 constitute the input part of the program in which the user enters the number of equations, the coefficients and constants of each equation, the convergence criterion $\varepsilon$ for all the unknowns, the initial estimates for the unknowns, and any corrections, if needed.

Lines 450–620 implement the Gauss-Seidel algorithm by programming Eq. (3-153). The convergence criterion, which must be met by all the unknowns, is

$$\left| \frac{\text{Old value of } x - \text{new value of } x}{\text{New value of } x} \right| < \varepsilon$$

This is tested in lines 560–600.

Lines 640–710 give the user the opportunity to rerun the program with modifications.

Lines 720–740 terminate the program.

*Subroutine 1* This subroutine prints the matrix of coefficients and constants.

*Subroutine 2* This subroutine prints the results.

PROGRAM

```
1 SCREEN 0:WIDTH 80:CLS:KEY OFF
5 PRINT "*********************************************************************"
10 PRINT"*                                                                   *"
15 PRINT"*                            EXAMPLE 3-3                            *"
20 PRINT"*                                                                   *"
25 PRINT"*                THE GAUSS-SEIDEL SUBSTITUTION METHOD               *"
30 PRINT"*                                                                   *"
35 PRINT"*       FOR DIAGONAL SYSTEMS OF LINEAR ALGEBRAIC EQUATIONS          *"
40 PRINT"*                                                                   *"
45 PRINT"*                            (SEIDEL.BAS)                           *"
50 PRINT"*                                                                   *"
90 PRINT"*******************************************************************"
100 '************************** Main Program *****************************
110 '
120 'Enter the number of equations, the coefficients, and constants
130 '
140 PRINT:PRINT "NUMBER OF EQUATIONS";:INPUT N
150 DIM A(N,N+1), B(N,N+1), X(N),OLDX(N)
160 PRINT:PRINT "ENTER COEFFICIENTS AND CONSTANT FOR EACH EQUATION"
170 FOR K = 1 TO N
180 PRINT : PRINT "EQUATION ";K: PRINT
190 FOR J = 1 TO N
200 PRINT "   COEFFICIENT (";K;",";J;") =";: INPUT B(K,J)
210 NEXT J
220 PRINT:PRINT "   CONSTANT ";K;" =";: INPUT B(K,N+1)
230 NEXT K
240 NC=N+1
250 PRINT
260 PRINT"GIVE THE CONVERGENCE CRITERION FOR THE UNKNOWNS";:INPUT EPS
270 PRINT CHR$(12)
280 FOR K = 1 TO N
290 FOR J = 1 TO NC
300 A(K,J) = B(K,J)
310 NEXT J : NEXT K
320 PRINT"GIVE INITIAL ESTIMATES OF THE UNKNOWNS:":PRINT
330 FOR I=2 TO  N:PRINT "   UNKNOWN";I;:INPUT X(I):NEXT I
340 PRINT
350 PRINT"*******************************************************************"
360 PRINT"AUGMENTED MATRIX:"
370 GOSUB 2000
380 PRINT: INPUT "IS THE AUGMENTED MATRIX CORRECT(Y/N)"; Q$:PRINT
390 IF Q$ = "Y" OR Q$ = "y" THEN 460
400 PRINT"GIVE THE POSITION OF THE ELEMENT TO BE CORRECTED:":PRINT
410 INPUT "   ROW NUMBER";NROW :INPUT "   COLUMN NUMBER";NCOL
420 PRINT:INPUT "   CORRECT VALUE OF THE ELEMENT"; B(NROW,NCOL):PRINT
430 GOTO 270
440 '
450 'Beginning of the Gauss-Seidel substitution procedure
460 INPUT "DO YOU WANT TO SEE STEP-BY-STEP RESULTS(Y/N)";Q2$:PRINT
470 COUNT=0
480 FOR I=1 TO N
490 OLDX(I)=X(I)
500 SUM=0
510 FOR J=1 TO N
520 IF J=I THEN GOTO 540
530 SUM=SUM + A(I,J)*X(J)
540 NEXT J
```

```
550 X(I)=(A(I,NC) - SUM)/A(I,I)
560 DEL=(OLDX(I)-X(I))/X(I)
570 IF ABS(DEL) < EPS THEN COUNT=COUNT + 1
580 NEXT I
590 IF Q2$="Y" OR Q2$="y" THEN GOSUB 3000
600 IF COUNT=N GOTO 620
610 GOTO 470
620 GOSUB 3000
630 '
640 PRINT:PRINT "DO YOU WANT TO REPEAT THE CALCULATIONS":PRINT"WITH MINOR CHANGE
S TO THE COEFFICIENTS(Y/N)";:INPUT V$
650 IF V$ = "Y" OR V$ = "y" THEN 660 ELSE 670
660 CLS : GOTO 270
670 PRINT:INPUT "DO YOU WANT TO RESET ALL THE COEFFICIENTS(Y/N)";W$
680 IF W$ = "Y" OR W$ = "y" THEN 690 ELSE 720
690 PRINT:INPUT "IS THE NEW SET OF THE SAME ORDER AS THE PREVIOUS SET";WW$
700 IF WW$="N" OR WW$="n" THEN PRINT CHR$(12):RUN 100
710 CLS : GOTO 160
720 PRINT:PRINT
730 PRINT"************************ END OF PROGRAM ****************************"
740 END
2000 '**************** Subroutine 1: Print the matrix ************************
2010 '
2020 FOR KA = 1 TO N
2030 PRINT
2040 FOR J = 1 TO NC
2050 PRINT A(KA,J),
2060 NEXT J:PRINT
2070 NEXT KA:PRINT
2080 PRINT"*******************************************************************"
2090 RETURN
2100 '
3000 '**************** Subroutine 2: Print the results ***********************
3010 '
3020 PRINT"*******************************************************************"
3030 PRINT:PRINT "RESULTS:": PRINT
3040 FOR J = 1 TO N
3050 PRINT "X(";J;") =";X(J)
3060 NEXT J
3070 PRINT
3080 PRINT"*******************************************************************"
3090 RETURN
```

RESULTS

```
*******************************************************************
*                                                                 *
*                           EXAMPLE 3-3                           *
*                                                                 *
*               THE GAUSS-SEIDEL SUBSTITUTION METHOD              *
*                                                                 *
*     FOR DIAGONAL SYSTEMS OF LINEAR ALGEBRAIC EQUATIONS          *
*                                                                 *
*                           (SEIDEL.BAS)                          *
*                                                                 *
*******************************************************************
NUMBER OF EQUATIONS? 4
ENTER COEFFICIENTS AND CONSTANT FOR EACH EQUATION
EQUATION  1

   COEFFICIENT ( 1 , 1 ) =? 1100
   COEFFICIENT ( 1 , 2 ) =? 0
   COEFFICIENT ( 1 , 3 ) =? 0
   COEFFICIENT ( 1 , 4 ) =? 0

   CCNSTANT  1  =? 1000

EQUATION  2

   COEFFICIENT ( 2 , 1 ) =? 1000
   COEFFICIENT ( 2 , 2 ) =? -1400
   COEFFICIENT ( 2 , 3 ) =? 100
   COEFFICIENT ( 2 , 4 ) =? 0

   CONSTANT  2  =? 0

EQUATION  3

   COEFFICIENT ( 3 , 1 ) =? 0
   COEFFICIENT ( 3 , 2 ) =? 1100
   COEFFICIENT ( 3 , 3 ) =? -1240
   COEFFICIENT ( 3 , 4 ) =? 100

   CONSTANT  3  =? 0

EQUATION  4

   COEFFICIENT ( 4 , 1 ) =? 0
   COEFFICIENT ( 4 , 2 ) =? 0
   COEFFICIENT ( 4 , 3 ) =? 1100
   COEFFICIENT ( 4 , 4 ) =? -1250

   CONSTANT  4  =? 0
GIVE THE CONVERGENCE CRITERION FOR THE UNKNOWNS? 0.00001
GIVE INITIAL ESTIMATES OF THE UNKNOWNS:

   UNKNOWN 2 ? 0.6
   UNKNOWN 3 ? 0.6
   UNKNOWN 4 ? 0.6
```

```
*********************************************************************
AUGMENTED MATRIX:

 1100          0             0             0             1000

 1000          -1400         100           0             0

 0             1100          -1240         100           0

 0             0             1100          -1250         0
*********************************************************************

IS THE AUGMENTED MATRIX CORRECT(Y/N)? Y

DO YOU WANT TO SEE STEP-BY-STEP RESULTS(Y/N)? Y

*********************************************************************

RESULTS:

X( 1 ) = .9090909
X( 2 ) = .6922078
X( 3 ) = .6624423
X( 4 ) = .5829493

*********************************************************************
*********************************************************************

RESULTS:

X( 1 ) = .9090909
X( 2 ) = .696668
X( 3 ) = .665024
X( 4 ) = .5852211

*********************************************************************
*********************************************************************

RESULTS:

X( 1 ) = .9090909
X( 2 ) = .6968523
X( 3 ) = .6653706
X( 4 ) = .5855262

*********************************************************************
*********************************************************************


RESULTS:

X( 1 ) = .9090909
X( 2 ) = .6968771
X( 3 ) = .6654173
X( 4 ) = .5855672

*********************************************************************
*********************************************************************
```

```
RESULTS:

X( 1 ) = .9090909
X( 2 ) = .6968805
X( 3 ) = .6654235
X( 4 ) = .5855728

*******************************************************************
*******************************************************************

RESULTS:

X( 1 ) = .9090909
X( 2 ) = .6968805
X( 3 ) = .6654235
X( 4 ) = .5855728

*******************************************************************

DO YOU WANT TO REPEAT THE CALCULATIONS
WITH MINOR CHANGES TO THE COEFFICIENTS(Y/N)? Y

GIVE INITIAL ESTIMATES OF THE UNKNOWNS:

   UNKNOWN 2 ? 100
   UNKNOWN 3 ? 100
   UNKNOWN 4 ? 100

*******************************************************************
AUGMENTED MATRIX:

 1100          0             0             0             1000

 1000         -1400          100           0             0

 0             1100         -1240          100           0

 0             0             1100         -1250          0

*******************************************************************
IS THE AUGMENTED MATRIX CORRECT(Y/N)? Y

DO YOU WANT TO SEE STEP-BY-STEP RESULTS(Y/N)? Y

*******************************************************************
RESULTS:

X( 1 ) = .9090909
X( 2 ) = 7.792208
X( 3 ) = 14.97696
X( 4 ) = 13.17972

*******************************************************************
*******************************************************************
RESULTS:

X( 1 ) = .9090909
X( 2 ) = 1.719134
X( 3 ) = 2.587919
X( 4 ) = 2.277369

*******************************************************************
*******************************************************************
```

```
RESULTS:

X( 1 ) = .9090909
X( 2 ) = .834202
X( 3 ) = .9236766
X( 4 ) = .8128354

*******************************************************************
*******************************************************************

RESULTS:

X( 1 ) = .9090909
X( 2 ) = .7153276
X( 3 ) = .700116
X( 4 ) = .6161021

*******************************************************************
*******************************************************************

RESULTS:

X( 1 ) = .9090909
X( 2 ) = .699359
X( 3 ) = .6700847
X( 4 ) = .5896746

*******************************************************************
*******************************************************************

RESULTS:

X( 1 ) = .9090909
X( 2 ) = .6972138
X( 3 ) = .6660505
X( 4 ) = .5861245

*******************************************************************
*******************************************************************

RESULTS:

X( 1 ) = .9090909
X( 2 ) = .6969256
X( 3 ) = .6655086
X( 4 ) = .5856476

*******************************************************************
*******************************************************************

RESULTS:

X( 1 ) = .9090909
X( 2 ) = .696887
X( 3 ) = .6654358
X( 4 ) = .5855835

*******************************************************************
*******************************************************************
```

```
RESULTS:

X( 1 ) = .9090909
X( 2 ) = .6968818
X( 3 ) = .665426
X( 4 ) = .5855749

*******************************************************************
*******************************************************************

RESULTS:

X( 1 ) = .9090909
X( 2 ) = .6968811
X( 3 ) = .6654247
X( 4 ) = .5855738

*******************************************************************
*******************************************************************

RESULTS:

X( 1 ) = .9090909
X( 2 ) = .6968811
X( 3 ) = .6654247
X( 4 ) = .5855738

*******************************************************************


DO YOU WANT TO REPEAT THE CALCULATIONS
WITH MINOR CHANGES TO THE COEFFICIENTS(Y/N)? N

DO YOU WANT TO RESET ALL THE COEFFICIENTS(Y/N)? N


************************ END OF PROGRAM ***************************
Ok
```

DISCUSSION OF RESULTS The first case uses the value of 0.6 as the initial guess for the values of the unknowns $c_{A_2}$ to $c_{A_4}$. The Gauss-Seidel method converges to the solution

$$c_{A_1} = 0.9091 \qquad c_{A_2} = 0.6969 \qquad c_{A_3} = 0.6654 \qquad c_{A_4} = 0.5856$$

in six iterations. The convergence criterion, which is satisfied by all the unknowns, is 0.00001.

In the second case, the value of 100 is used as the initial guess for each of the unknowns $c_{A_2}$ to $c_{A_4}$. Convergence to exactly the same answer as in the first case is accomplished in 11 iterations.

## 3-8 HOMOGENEOUS ALGEBRAIC EQUATIONS AND THE CHARACTERISTIC-VALUE PROBLEM

We mentioned earlier that a homogeneous set of equations

$$\mathbf{Ax} = 0 \tag{3-97}$$

has a nontrivial solution, if and only if the matrix **A** is singular, i.e., if the rank $r$ of **A** is less than $n$. The system of equations would consist of $r$ independent equations, $r$ unknowns that can be evaluated independently, and $n - r$ unknowns which must be chosen arbitrarily in order to complete the solution. Choosing nonzero values for the $n - r$ unknowns transforms the homogeneous set to a nonhomogeneous set of order $r$. The Gauss and Gauss-Jordan methods, which are applicable to nonhomogeneous systems, can then be used to obtain the complete solution of the problem. In fact, these methods can be used first on the homogeneous system to determine the number of independent equations (or the rank of **A**) and then applied to the set of $r$ nonhomogeneous independent equations to evaluate the $r$ unknowns. This concept will be demonstrated later in this section in conjunction with the calculation of eigenvectors.

A special class of homogeneous linear algebraic equations arises in the study of vibrating systems, structure analysis, and electric circuit system analysis, and in the solution and stability analysis of linear ordinary differential equations (Chap. 5). This system of equations has the form

$$\mathbf{Ax} = \lambda\mathbf{x} \tag{3-154}$$

which can be alternatively expressed as

$$(\mathbf{A} - \lambda\mathbf{I})\mathbf{x} = 0 \tag{3-155}$$

where the scalar $\lambda$ is called an *eigenvalue* (or a *characteristic value*) of matrix **A**. The vector **x** is called the *eigenvector* (or *characteristic vector*) corresponding to $\lambda$. The matrix **I** is the identity matrix. The problem often requires the solution of the homogeneous set of equations, represented by Eq. (3-155), to determine the values of $\lambda$ and **x** which satisfy this set. However, before we proceed with developing methods of solution, let's examine Eq. (3-154) from a geometric perspective.

The multiplication of a vector by a matrix is a linear transformation of the original vector to a new vector of different direction and length. For example, matrix **A** transforms the vector **y** to the vector **z** in the operation

$$\mathbf{Ay} = \mathbf{z} \tag{3-156}$$

In contrast to this, if **x** is the eigenvector of **A**, then the multiplication of the eigenvector **x** by matrix **A** yields the same vector **x** multiplied by a scalar $\lambda$, that is, the same vector but of different length:

$$\mathbf{Ax} = \lambda \mathbf{x} \tag{3-154}$$

It can be stated that for a nonsingular matrix **A** of order $n$ there are $n$ characteristic directions in which the operation by **A** does not change the direction of the vector, but only changes its length. More simply stated, matrix **A** has $n$ eigenvectors and $n$ eigenvalues.

The types of eigenvalues that exist for a set of special matrices are listed in Table 3-3.

The homogeneous problem

$$(\mathbf{A} - \lambda \mathbf{I})\mathbf{x} = 0 \tag{3-155}$$

possesses nontrivial solutions if the determinant of the matrix $(\mathbf{A} - \lambda \mathbf{I})$, called the *characteristic matrix* of **A**, vanishes:

$$|\mathbf{A} - \lambda \mathbf{I}| = \begin{vmatrix} a_{11} - \lambda & a_{12} & \cdots & a_{1n} \\ a_{21} & a_{22} - \lambda & \cdots & a_{2n} \\ \cdots & \cdots & \cdots & \cdots \\ a_{n1} & a_{n2} & \cdots & a_{nn} - \lambda \end{vmatrix} = 0 \tag{3-157}$$

The determinant can be expanded by minors to yield a polynomial of $n$th degree

$$\lambda^n - \alpha_1 \lambda^{n-1} - \alpha_2 \lambda^{n-2} - \cdots - \alpha_n = 0 \tag{3-158}$$

This polynomial, which is called the *characteristic equation* of matrix **A**, has $n$ roots which are the eigenvalues of **A**. These roots may be real distinct, real

**Table 3-3†**

| Matrix | Eigenvalues |
|---|---|
| Singular, $\lvert\mathbf{A}\rvert = 0$ | At least one 0 eigenvalue |
| Nonsingular, $\lvert\mathbf{A}\rvert \neq 0$ | No zero eigenvalues |
| Symmetric, $\mathbf{A} = \mathbf{A}^T$ | All real eigenvalues |
| Hermitian | All real eigenvalues |
| Zero matrix, $\mathbf{A} = 0$ | All zero eigenvalues |
| Identity, $\mathbf{A} = \mathbf{I}$ | All unity eigenvalues |
| Diagonal, $\mathbf{A} = \mathbf{D}$ | Equal to diagonal elements of **A** |
| Inverse, $\mathbf{A}^{-1}$ | Inverse eigenvalues of **A** |
| Transformed, $\mathbf{B} = \mathbf{Q}^{-1}\mathbf{AQ}$ | Eigenvalues of **B** = eigenvalues of **A** |

† Adapted from a similar table in Ref. 5, p. 218.

repeated, or complex depending on matrix **A** (see Table 3-3). A nonsingular real symmetric matrix of order $n$ has $n$ real distinct eigenvalues and $n$ linearly independent eigenvectors. The eigenvectors of a real symmetric matrix are orthogonal to each other. The coefficients $\alpha_i$ of the characteristic polynomial are functions of the matrix elements $a_{ij}$, and must be determined before the polynomial can be used.

The well-known Cayley-Hamilton theorem states that a square matrix satisfies its own characteristic equation, i.e.,

$$\mathbf{A}^n - \alpha_1\mathbf{A}^{n-1} - \alpha_2\mathbf{A}^{n-2} - \cdots - \alpha_n\mathbf{I} = 0 \tag{3-159}$$

The problem of evaluating the eigenvalues and eigenvectors of matrices is a complex multipstep procedure. Several methods have been developed for this purpose. Some of these apply to symmetric matrices, others to tridiagonal matrices, and a few can be used for general matrices. We can classify these methods into two categories:

1. The methods in this category work with the original matrix **A** and its characteristic polynomial [Eq. (3-158)] to evaluate the coefficients $\alpha_i$ of the polynomial. One such method is the Faddeev-Leverrier procedure, which will be described later. Once the coefficients of the polynomial are known, the methods use root-finding techniques, such as the Newton-Raphson or Graeffe's method, to determine the eigenvalues. Finally, the algorithms employ a reduction method, such as Gauss elimination, to calculate the eigenvectors.
2. The methods in this category reduce the original matrix **A** to tridiagonal form (when **A** is symmetric) or to Hessenberg form (when **A** is nonsymmetric) by orthogonal transformations or elementary similarity transformations. They apply successive factorization procedures, such as the LR or QR algorithms, to extract the eigenvalues, and, finally, they use a reduction method to calculate the eigenvectors.

In the remaining part of this chapter we will discuss the following methods: (*a*) the Faddeev-Leverrier procedure for calculating the coefficients of the characteristic polynomial, (*b*) the elementary similarity transformation for converting a matrix to Hessenberg form, (*c*) the QR algorithm of successive factorization for the determination of the eigenvalues, and, finally, (*d*) the Gauss elimination method applied for the evaluation of the eigenvectors. These methods were chosen for their general applicability to both symmetric and nonsymmetric matrices. For a complete discussion of these and other methods the reader is referred to Ralston and Rabinowitz [8].

### 3-8.1 The Faddeev-Leverrier method

The Faddeev-Leverrier method [7, 10] calculates the coefficients $\alpha_1$ to $\alpha_n$ of the characteristic polynomial [Eq. (3-158)] by generating a series of matrices $\mathbf{A}_k$

whose traces yield the coefficients of the polynomial. The starting matrix and first coefficient are

$$\mathbf{A}_1 = \mathbf{A} \qquad \alpha_1 = \operatorname{tr} \mathbf{A}_1 \tag{3-160}$$

and the subsequent matrices are evaluated from the recursive equations:

$$\left.\begin{aligned} \mathbf{A}_k &= \mathbf{A}(\mathbf{A}_{k-1} - \alpha_{k-1}\mathbf{I}) \\ \alpha_k &= \frac{1}{k} \operatorname{tr} \mathbf{A}_k \end{aligned}\right\} \qquad k = 2, 3, \ldots, n \tag{3-161}$$

In addition to this, the Faddeev-Leverrier method yields the inverse of the matrix $\mathbf{A}$ by

$$\mathbf{A}^{-1} = \frac{1}{\alpha_n} (\mathbf{A}_{n-1} - \alpha_{n-1}\mathbf{I}) \tag{3-162}$$

To elucidate this method we will determine the coefficients of the characteristic polynomial of the following set of homogeneous equations:

$$\begin{aligned} (1-\lambda)x_1 + 2x_2 + x_3 &= 0 \\ 3x_1 + (1-\lambda)x_2 + 2x_3 &= 0 \\ 4x_1 + 2x_2 + (3-\lambda)x_3 &= 0 \end{aligned} \tag{3-163}$$

The characteristic polynomial for this third-order system is

$$\lambda^3 - \alpha_1\lambda^2 - \alpha_2\lambda - \alpha_3 = 0 \tag{3-164}$$

The matrix $\mathbf{A}$ is

$$\mathbf{A} = \begin{bmatrix} 1 & 2 & 1 \\ 3 & 1 & 2 \\ 4 & 2 & 3 \end{bmatrix} \tag{3-165}$$

Application of Eq. (3-160) gives

$$\mathbf{A}_1 = \mathbf{A} \qquad \text{and} \qquad \alpha_1 = \operatorname{tr} \mathbf{A}_1 = 5 \tag{3-166}$$

Application of Eq. (3-161), with $k = 2$, yields

$$\mathbf{A}_2 = \mathbf{A}(\mathbf{A}_1 - \alpha_1 \mathbf{I})$$

$$= \begin{bmatrix} 1 & 2 & 1 \\ 3 & 1 & 2 \\ 4 & 2 & 3 \end{bmatrix} \left\{ \begin{bmatrix} 1 & 2 & 1 \\ 3 & 1 & 2 \\ 4 & 2 & 3 \end{bmatrix} - \begin{bmatrix} 5 & 0 & 0 \\ 0 & 5 & 0 \\ 0 & 0 & 5 \end{bmatrix} \right\}$$

$$= \begin{bmatrix} 6 & -4 & 3 \\ -1 & 6 & 1 \\ 2 & 6 & 2 \end{bmatrix} \tag{3-167}$$

$$\alpha_2 = \tfrac{1}{2} \operatorname{tr} \mathbf{A}_2 = 7 \tag{3-168}$$

Repetition of Eq. (3-161), with $k = 3$, results in

$$\mathbf{A}_3 = \mathbf{A}(\mathbf{A}_2 - \alpha_2 \mathbf{I})$$

$$= \begin{bmatrix} 1 & 2 & 1 \\ 3 & 1 & 2 \\ 4 & 2 & 3 \end{bmatrix} \left\{ \begin{bmatrix} 6 & -4 & 3 \\ -1 & 6 & 1 \\ 2 & 6 & 2 \end{bmatrix} - \begin{bmatrix} 7 & 0 & 0 \\ 0 & 7 & 0 \\ 0 & 0 & 7 \end{bmatrix} \right\} \tag{3-169}$$

$$= \begin{bmatrix} -1 & 0 & 0 \\ 0 & -1 & 0 \\ 0 & 0 & -1 \end{bmatrix}$$

$$\alpha_3 = \tfrac{1}{3} \operatorname{tr} \mathbf{A}_3 = -1 \tag{3-170}$$

Therefore the characteristic polynomial is

$$\lambda^3 - 5\lambda^2 - 7\lambda + 1 = 0 \tag{3-171}$$

The root-finding techniques described in Chap. 2 can be used to determine the $\lambda$'s of this polynomial. The eigenvectors corresponding to each eigenvalue can be calculated using the Gauss elimination method, as will be described later in this chapter. The Faddeev-Leverrier method, the Newton-Raphson method with synthetic division, and the Gauss elimination method constitute a complete algorithm for the evaluation of all the eigenvalues and eigenvectors of this characteristic-value problem. The combined use of these techniques is demonstrated in Example 3-4.

**Example 3-4 Calculation of eigenvalues and eigenvectors using the Faddeev-Leverrier/Newton-Raphson/Gauss methods** Consider the following chemical reaction system:

$$\mathrm{A} \xrightarrow{k_1} \mathrm{B} \underset{k_3}{\overset{k_2}{\rightleftharpoons}} \mathrm{C}$$

Assume that all steps are first-order reactions and write the set of linear ordinary differential equations which describe the kinetics of these reactions. The values of the kinetic rate constants are

$$k_1 = 1 \qquad k_2 = 2 \qquad k_3 = 3$$

The initial concentrations of the three components are

$$A_0 = 1 \qquad B_0 = 0 \qquad C_0 = 0$$

Determine the eigenvalues and eigenvectors of the matrix of kinetic rate constants. Verify the eigenvalues and eigenvectors.

METHOD OF SOLUTION Assuming that all steps are first-order reactions, the set of differential equations which give the rate of formation of each compound is the following:

$$\frac{dA}{dt} = -k_1 A$$

$$\frac{dB}{dt} = k_1 A - k_2 B + k_3 C$$

$$\frac{dC}{dt} = k_2 B - k_3 C$$

In matrix form these equations reduce to

$$\dot{\mathbf{y}} = \mathbf{K}\mathbf{y}$$

where

$$\dot{\mathbf{y}} = \begin{bmatrix} \dfrac{dA}{dt} \\ \dfrac{dB}{dt} \\ \dfrac{dC}{dt} \end{bmatrix} \qquad \mathbf{y} = \begin{bmatrix} A \\ B \\ C \end{bmatrix}$$

and

$$\mathbf{K} = \begin{bmatrix} -k_1 & 0 & 0 \\ k_1 & -k_2 & k_3 \\ 0 & k_2 & -k_3 \end{bmatrix}$$

Using the values of the kinetic rate constants given in the problem, matrix **K** becomes

$$\mathbf{K} = \begin{bmatrix} -1 & 0 & 0 \\ 1 & -2 & 3 \\ 0 & 2 & -3 \end{bmatrix}$$

In order to be consistent with the terminology used in the rest of this chapter, we will redesignate matrix **K** as matrix **A**:

$$\mathbf{A} = \mathbf{K} = \begin{bmatrix} -1 & 0 & 0 \\ 1 & -2 & 3 \\ 0 & 2 & -3 \end{bmatrix}$$

The eigenvalues and eigenvectors of matrix **A** are related by Eq. (3-154):

$$\mathbf{Ax} = \lambda \mathbf{x}$$

which rearranges to

$$(\mathbf{A} - \lambda \mathbf{I})\mathbf{x} = 0$$

The above set of homogeneous equations has a nontrivial solution when the determinant of the matrix is zero:

$$|\mathbf{A} - \lambda \mathbf{I}| = 0$$

This results in a characteristic polynomial of the form given by Eq. (3-158):

$$\lambda^3 - \alpha_1 \lambda^2 - \alpha_2 \lambda - \alpha_3 = 0$$

The coefficients of this polynomial are determined by the Faddeev-Leverrier method. The roots of the polynomial, which are the eigenvalues of matrix **A**, are calculated using the Newton-Raphson method. Finally, the eigenvectors are obtained using a modification of the Gauss elimination method.

The program verifies the eigenvalues by applying Eq. (3-60), which states that the sum of the eigenvalues of a square matrix is equal to the trace of that matrix:

$$\sum_{i=1}^{n} \lambda_i = \text{tr}\, \mathbf{A} \tag{3-60}$$

The program also verifies the eigenvectors by performing the multiplication

$$\mathbf{Ax} = \lambda \mathbf{x}$$

and printing out both sides of the equation for the user to compare.

**Program description** This program, which we call EIGEN.BAS, consists of the sections listed in Table E3-4.

**Table E3-4**

| Program section | Line numbers |
|---|---|
| Title | 1–90 |
| Main program | 100–1900 |
| Input | 120–400 |
| Faddeev-Leverrier method | 410–830 |
| Newton-Raphson method | 840–1390 |
| Gauss elimination method | 1560–1790 |
| Rerun | 1800–1870 |
| Subroutines | |
| 1. Print and matrix | 2000–2090 |
| 2. Newton-Raphson method | 3000–3120 |
| 3. Evaluation of polynomial and its derivative | 4000–4080 |
| 4. Gauss elimination method | 9000–9820 |

*Main program and subroutines* Lines 120–400 constitute the input part of the program in which the user enters the order of the matrix, the elements of the matrix, and any corrections, if needed.

Lines 410–830 implement the Faddeev-Leverrier method. Equations (3-160) and (3-161) are used to evaluate the coefficients $\alpha_i$ of the characteristic polynomial.

Lines 840–1390, together with subroutines 2 and 3, implement the Newton-Raphson method for finding the roots of the characteristic polynomial. This method is identical to the one described in Example 2-2, without the graphical demonstration.

Lines 1320–1360 verify the eigenvalues using Eq. (3-60).

Lines 1560–1790, together with subroutine 4, implement the Gauss elimination procedure. The matrix $\mathbf{A} - \lambda\mathbf{I}$ is set up in lines 1670–1750 and the subroutine is called in line 1760. The Gauss elimination procedure in subroutine 4 is essentially the same as that described in Example 3-1 with one basic difference: The elimination procedure is applied to the $\mathbf{A} - \lambda\mathbf{I}$ matrix and the rank $r$ of this matrix is determined. Then the program sets $n - r$ elements of the eigenvector to unity and reduces the problem to finding the remaining $r$ elements. This is repeated for each eigenvalue in order to determine all eigenvectors. Finally, the Gauss subroutine verifies the eigenvector by performing the multiplication

$$\mathbf{Ax} = \lambda\mathbf{I}$$

and printing out the results for comparison (lines 9680–9780).

Lines 1800–1870 give the user the opportunity to rerun the program with modifications.

Lines 1880–1900 terminate the program.

PROGRAM

```
1 SCREEN 0:WIDTH 80:CLS:KEY OFF
5 PRINT "********************************************************************"
10 PRINT"*                                                                  *"
15 PRINT"*                         EXAMPLE 3-4                              *"
20 PRINT"*                                                                  *"
25 PRINT"*      CALCULATION OF EIGENVALUES AND EIGENVECTORS  USING          *"
30 PRINT"*                                                                  *"
35 PRINT"*    THE FADDEEV-LEVERRIER / NEWTON-RAPHSON / GAUSS METHODS        *"
40 PRINT"*                                                                  *"
45 PRINT"*                         (EIGEN.BAS)                              *"
50 PRINT"*                                                                  *"
90 PRINT"*********************************************************************"
100 '*************************** Main Program ******************************
110 '
120 'Enter the order of the matrix and its elements
130 '
140 PRINT: INPUT"ENTER THE ORDER OF THE MATRIX";N
150 DIM A(N,N+1),B(N,N),C(N,N),X(N),ALPHA(N),AA(N),BB(N), XW(N) ,NPIVROW(N-1,2),
NPIVCOL(N-1,2)
160 PRINT:PRINT "ENTER ELEMENTS OF MATRIX:"
170 FOR K = 1 TO N
180 PRINT : PRINT "ROW ";K :PRINT
190 FOR J = 1 TO N
200 PRINT "   ELEMENT (";K;",";J;") =";: INPUT B(K,J)
210 NEXT J
220 NEXT K
230 NN=N: NC=N
240 TRC=0
250 FOR K = 1 TO N
260 FOR J = 1 TO N
270 A(K,J) = B(K,J)
280 NEXT J
290 TRC=TRC + B(K,K)
300 NEXT K
310 PRINT:PRINT
320 PRINT"*********************************************************************"
330 PRINT"ORIGINAL MATRIX:"
340 GOSUB 2000 'Print matrix
350 PRINT: INPUT "IS THE MATRIX CORRECT(Y/N)"; Q$:PRINT
360 IF Q$ = "Y" OR Q$ = "y" THEN 410
370 PRINT"GIVE THE POSITION OF THE ELEMENT TO BE CORRECTED:":PRINT
380 INPUT "   ROW NUMBER";NROW :INPUT "   COLUMN NUMBER";NCOL
390 PRINT:INPUT "   CORRECT VALUE OF THE ELEMENT"; B(NROW,NCOL):PRINT
400 GOTO 240
410 'Beginning of the Faddeev-Leverrier procedure
420 PRINT CHR$(12)
430 PRINT"*****************  FADDEEV-LEVERRIER METHOD  ********************"
440 PRINT
450 PRINT "DO YOU WANT TO SEE STEP-BY-STEP RESULTS OF THE "
460 PRINT"FADDEEV-LEVERRIER PROCEDURE(Y/N)";:INPUT Q2$:PRINT
470 PRINT"*********************************************************************"
480 FOR K = 2 TO N + 1
490 TRACE=0
500 FOR I = 1 TO N
510 TRACE=TRACE + A(I,I)
520 NEXT I
530 ALPHA(K-1)=TRACE/(K-1)
540 IF K=N+1 GOTO 750
```

```
550 FOR I = 1 TO N
560 FOR J = 1 TO N
570 IF I=J THEN FACT=1 ELSE FACT=0
580 A(I,J)=A(I,J) - FACT*ALPHA(K-1)
590 NEXT J: NEXT I
600 IF Q2$="Y" OR Q2$="y" THEN GOSUB 2000
610 FOR I = 1 TO N
620 FOR J = 1 TO N
630 SUM=0
640 FOR L = 1 TO N
650 SUM = SUM + B(I,L)*A(L,J)
660 NEXT L
670 C(I,J)=SUM
680 NEXT J:NEXT I
690 FOR I = 1 TO N
700 FOR J = 1 TO N
710 A(I,J)=C(I,J)
720 NEXT J:NEXT I
730 IF Q2$="Y" OR Q2$="y" THEN GOSUB 2000
740 NEXT K
750 PRINT"*******************************************************************"
760 PRINT:PRINT "COEFFICIENTS OF CHARACTERISTIC POLYNOMIAL:": PRINT
770 FOR J = 1 TO N
780 PRINT "ALPHA(";J;") =";ALPHA(J)
790 NEXT J
800 PRINT
810 PRINT"*******************************************************************"
820 PRINT"*******************************************************************"
830 PRINT:INPUT"PRESS THE ENTER KEY TO CONTINUE",KK$
840 PRINT CHR$(12)
850 PRINT"*******************  NEWTON-RAPHSON METHOD  **********************"
860 AA(N)=1
870 FOR I=1 TO N
880 AA(N-I)=-ALPHA(I)
890 NEXT I
900 NW=N
910 PRINT:PRINT"GIVE YOUR BEST GUESS OF THE EIGENVALUE";:INPUT X1
920 PRINT:PRINT "GIVE THE CONVERGENCE VALUE OF POLYNOMIAL";: INPUT EPS
930 GOSUB 3000 'Root-finding routine
940 XW(N) = X1
950 '
960 'Synthetic division
970 IF N = 1 GOTO 1260
980 BB(N-1)=AA(N)
990 FOR R = 1 TO N - 1
1000 BB(N - 1 - R) = AA(N - R) + BB(N - R) * X1
1010 NEXT
1020 N = N - 1
1030 FOR JJ = N TO 0 STEP  - 1
1040 AA(JJ) = BB(JJ)
1050 NEXT
1060 '
1070 'Check for complex roots
1080 IF N<>2 GOTO 1220
1090 WW=AA(1)^2-4*AA(2)*AA(0)
1100 IF WW>=0 GOTO 1220
1110 XW(2)=-AA(1)/(2*AA(2))
1120 XW(1)=SQR(-WW)/(2*AA(2))
1130 PRINT "THE ";NW;" EIGENVALUES ARE:":PRINT
1140 FOR JK = NW TO 3 STEP -1
1150 PRINT XW(JK)
```

```
1160 NEXT
1170 '
1180 'Print complex roots
1190 PRINT XW(2);"+";XW(1);"i"
1200 PRINT XW(2);"-";XW(1);"i"
1210 PRINT"EIGENVALUES ARE COMPLEX.":GOTO 1890
1220 FOR PAUSE=1 TO 1000:NEXT PAUSE
1230 IF N > 0 GOTO 930
1240 '
1250 'Print the roots
1260 PRINT:PRINT"THE ";NW;" EIGENVALUES ARE:":PRINT
1270 EIGENSUM=0
1280 FOR I=NW TO 1 STEP -1
1290 PRINT"   ";XW(I)
1300 EIGENSUM=EIGENSUM + XW(I)
1310 NEXT I
1320 PRINT:PRINT"VERIFICATION:"
1330 PRINT:PRINT"THE SUM OF THE EIGENVALUES = ";EIGENSUM
1340 PRINT"THE TRACE OF THE MATRIX    = ";TRC
1350 PRINT"THE DIFFERENCE IS          = ";EIGENSUM-TRC : PRINT
1360 IF ABS(EIGENSUM-TRC) > EPS THEN PRINT "THE DIFFERENCE IS LARGER THAN THE CO
NVERGENCE CRITERION" : PRINT
1370 PRINT"*******************************************************************"
1380 N=NW
1390 PRINT:PRINT:INPUT"PRESS THE ENTER KEY TO CONTINUE",KK$
1560 PRINT CHR$(12)
1570 PRINT"******************  GAUSS ELIMINATION METHOD  *********************"
1580 PRINT
1590 PRINT"GIVE THE MINIMUM ALLOWABLE VALUE OF THE PIVOT ELEMENT FOR THE"
1600 PRINT"GAUSS ELIMINATION PROCEDURE ";:INPUT EPS:PRINT
1610 PRINT"DO YOU WANT TO SEE STEP-BY-STEP RESULTS OF THE "
1620 PRINT"GAUSS ELIMINATION PROCEDURE(Y/N)";:INPUT Q2$:PRINT
1630 FOR KK = N TO 1 STEP -1
1640 PRINT"*******************************************************************"
1650 PRINT"HOMOGENEOUS SET: (A - EIGENVALUE*I)*X = 0, FOR EIGENVALUE =";XW(KK)
1660 NC=N+1
1670 'Subtract eigenvalue from the matrix and use the Gauss method
1680 FOR I = 1 TO N
1690 FOR J = 1 TO N
1700 A(I,J) = B(I,J)
1710 NEXT J
1720 A(I,I)= B(I,I) - XW(KK)
1730 A(I,NC)=0
1740 NEXT I
1750 GOSUB 2000 'Print matrix
1760 GOSUB 9000 'Perform elimination
1770 NEXT KK
1780 PRINT:PRINT
1790 PRINT"*******************************************************************"
1800 PRINT:PRINT "DO YOU WANT TO REPEAT THE CALCULATIONS":PRINT"WITH MINOR CHANG
ES TO THE COEFFICIENTS(Y/N)";:INPUT V$
1810 IF V$ = "Y" OR V$ = "y" THEN 1820 ELSE 1830
1820 CLS : GOTO 230
1830 PRINT:INPUT "DO YOU WANT TO RESET ALL THE COEFFICIENTS(Y/N)";W$
1840 IF W$ = "Y" OR W$ = "y" THEN 1850 ELSE 1880
1850 PRINT:INPUT "IS THE NEW SET OF THE SAME ORDER AS THE PREVIOUS SET";WW$
1860 IF WW$="N" OR WW$="n" THEN PRINT CHR$(12):RUN 100
1870 CLS : GOTO 160
1880 PRINT:PRINT
1890 PRINT"*********************** END OF PROGRAM ****************************"
1900 END
```

```
2000 '**************** Subroutine 1: Print the matrix ***********************
2010 '
2020 FOR KA = 1 TO N
2030 PRINT
2040 FOR J = 1 TO NC
2050 PRINT A(KA,J),
2060 NEXT J:PRINT:NEXT KA:PRINT
2070 PRINT"******************************************************************"
2090 RETURN
3000 '**************** Subroutine 2: Newton-Raphson method ******************
3010 '
3020 PRINT:PRINT"CONVERGENCE TO EIGENVALUE";(NW-N+1):PRINT
3030 FOR ITER=0 TO 40 STEP 2
3040 X=X1:GOSUB 4000
3050 PRINT TAB(3) "x=";X1 TAB(20) "f(x)=";F
3080 IF  ABS (F)< EPS GOTO 3120
3090 X1=X1-F/FP
3100 NEXT ITER
3110 PRINT"EXCEEDS ITERATION LIMIT:NO CONVERGENCE":
      PRINT"ITER=";ITER:PRINT"EPS=";EPS:PRINT"F=";F:
      PRINT"FUNCTION KEY F5 WILL CONTINUE PROGRAM":STOP
3120 RETURN
4000 '***** Subroutine 3: Evaluation of the polynomial and its derivative ****
4010 '
4020 F = 0:FP=0
4030 FOR KK = N TO 0 STEP  - 1
4040 F = F + AA(KK) * X ^ KK
4050 IF KK=0 GOTO 4080
4060 FP= FP+ KK*AA(KK) * X ^ (KK-1)
4070 NEXT KK
4080 RETURN
9000 '**************** Subroutine 4: Gauss elimination method ***************
9010 '
9020 FOR K = 1 TO N-1
9030 'Apply complete pivoting strategy
9040 MAXPIVOT = ABS(A(K,K))
9050 NPIVROW(K,1)=K: NPIVROW(K,2)=K
9060 NPIVCOL(K,1)=K: NPIVCOL(K,2)=K
9070 FOR I = K TO N
9080 FOR J = K TO N
9090 IF MAXPIVOT > = ABS(A(I,J)) GOTO 9130
9100 MAXPIVOT=ABS(A(I,J))
9110 NPIVROW(K,1)=K: NPIVROW(K,2)=I
9120 NPIVCOL(K,1)=K: NPIVCOL(K,2)=J
9130 NEXT J:NEXT I
9140 IF MAXPIVOT > = EPS GOTO 9160
9150 GOTO 9420
9160 IF NPIVROW(K,2)=K GOTO 9230
9170 IF Q2$="Y" OR Q2$="y" THEN PRINT"PIVOTING ROWS:"
9180 IF Q2$="Y" OR Q2$="y" THEN PRINT "INTERCHANGE ROWS ";NPIVROW(K,2);" AND ";K

9190 FOR J = K TO NC
9200 SWAP  A(NPIVROW(K,2),J),A(K,J)
9210 NEXT J
9220 IF Q2$="Y" OR Q2$="y" THEN GOSUB 2000 'Print matrix
9230 IF NPIVCOL(K,2)=K GOTO 9300
9240 IF Q2$="Y" OR Q2$="y" THEN PRINT"PIVOTING COLUMNS:"
9250 IF Q2$="Y" OR Q2$="y" THEN PRINT "INTERCHANGE COLUMNS ";NPIVCOL(K,2);" AND
";K
```

```
9260 FOR I = 1 TO N
9270 SWAP  A(I,NPIVCOL(K,2)),A(I,K)
9280 NEXT I
9290 IF Q2$="Y" OR Q2$="y" THEN GOSUB 2000 'Print matrix
9300 IF Q2$="Y" OR Q2$="y" THEN PRINT "PERFORM ELIMINATION:"
9310 FOR I = K+1 TO N
9320 IF Q2$="Y" OR Q2$="y" THEN PRINT "DIVIDE ROW ";K;" BY ";A(K,K)
9330 IF Q2$="Y" OR Q2$="y" THEN PRINT "MULTIPLY ROW ";K;" BY ";A(I,K) ;"AND SUBT
RACT FROM ROW ";I
9340 MULT = - A(I,K)/A(K,K)
9350 FOR J = NC TO K STEP -1
9360 A(I,J) = A(I,J) + MULT * A(K,J)
9370 NEXT J
9380 IF Q2$="Y" OR Q2$="y" THEN GOSUB 2000 'Print matrix
9390 NEXT I: NEXT K
9400 '
9410 'Apply the back-substitution formulas
9420 RANK=K-1 :PRINT"RANK =";RANK:NMR=N-RANK
9430 PRINT"THE PROGRAM SETS ";NMR;" ELEMENT(S) OF THE EIGENVECTOR TO UNITY,"
9440 PRINT"AND REDUCES THE PROBLEM TO FINDING ";RANK;" ELEMENT(S)."
9450 FOR JJ=1 TO NMR : X(N+1-JJ) = 1: NEXT JJ
9460 FOR I = RANK TO 1 STEP -1
9470 SUM = 0
9480 FOR J = I+1 TO N
9490 SUM = SUM + A(I,J) * X(J)
9500 NEXT J
9510 X(I) = (A(I,NC) - SUM) / A(I,I)
9520 IF ABS(X(I)) < EPS THEN X(I)=0
9530 NEXT I
9540 '
9550 'Interchange the order of the unknowns to correct for the column pivoting
9560 FOR K=N-1 TO 1 STEP -1
9570 SWAP X(NPIVCOL(K,2)), X(NPIVCOL(K,1))
9580 NEXT K
9590 '
9600 PRINT"********************************************************************"
9610 PRINT
9620 PRINT "EIGENVECTOR CORRESPONDING TO EIGENVALUE =";XW(KK): PRINT
9630 FOR J = 1 TO N
9640 PRINT "X(";J;") =";X(J)
9650 NEXT J
9660 PRINT
9670 PRINT"********************************************************************"
9680 PRINT:PRINT"VERIFICATION:"
9690 PRINT:PRINT"A*X              =   EIGENVALUE*X":PRINT
9700 FOR I=1 TO N
9710 AX=0
9720 FOR J=1 TO N
9730 AX=AX + B(I,J)*X(J)
9740 NEXT J
9750 EX=XW(KK)*X(I)
9760 PRINT AX TAB(20) EX
9770 NEXT I
9780 PRINT
9790 PRINT"********************************************************************"
9800 PRINT:INPUT"PRESS THE ENTER KEY TO CONTINUE",KK$
9810 PRINT CHR$(12)
9820 RETURN
```

## RESULTS

```
*******************************************************************
*                                                                 *
*                          EXAMPLE 3-4                            *
*                                                                 *
*      CALCULATION OF EIGENVALUES AND EIGENVECTORS  USING         *
*                                                                 *
*    THE FADDEEV-LEVERRIER / NEWTON-RAPHSON / GAUSS METHODS       *
*                                                                 *
*                          (EIGEN.BAS)                            *
*                                                                 *
*******************************************************************
ENTER THE ORDER OF THE MATRIX? 3

ENTER ELEMENTS OF MATRIX:

ROW  1

   ELEMENT ( 1 , 1 ) =? -1
   ELEMENT ( 1 , 2 ) =? 0
   ELEMENT ( 1 , 3 ) =? 0

ROW  2

   ELEMENT ( 2 , 1 ) =? 1
   ELEMENT ( 2 , 2 ) =? -2
   ELEMENT ( 2 , 3 ) =? 3

ROW  3

   ELEMENT ( 3 , 1 ) =? 0
   ELEMENT ( 3 , 2 ) =? 2
   ELEMENT ( 3 , 3 ) =? -3

*******************************************************************
ORIGINAL MATRIX:

-1              0               0

 1             -2               3

 0              2              -3

*******************************************************************

 IS THE MATRIX CORRECT(Y/N)? Y

*****************  FADDEEV-LEVERRIER METHOD  ********************

DO YOU WANT TO SEE STEP-BY-STEP RESULTS OF THE
FADDEEV-LEVERRIER PROCEDURE(Y/N)? N

*******************************************************************
*******************************************************************
```

```
COEFFICIENTS OF CHARACTERISTIC POLYNOMIAL:

ALPHA( 1 ) =-6
ALPHA( 2 ) =-5
ALPHA( 3 ) = 0

******************************************************************
******************************************************************

PRESS THE ENTER KEY TO CONTINUE

*******************  NEWTON-RAPHSON METHOD  **********************

GIVE YOUR BEST GUESS OF THE EIGENVALUE? 0

GIVE THE CONVERGENCE VALUE OF POLYNOMIAL? 0.00001

CONVERGENCE TO EIGENVALUE 1

  x= 0               f(x)= 0

CONVERGENCE TO EIGENVALUE 2

  x= 0               f(x)= 5
  x=-.8333333        f(x)= .6944447
  x=-.9935898        f(x)= 2.568245E-02
  x=-.9999899        f(x)= 4.1008E-05
  x=-1               f(x)=-9.536743E-07

CONVERGENCE TO EIGENVALUE 3

  x=-1               f(x)= 4
  x=-5               f(x)= 0

THE  3  EIGENVALUES ARE:

    0
   -1
   -5

VERIFICATION:

THE SUM OF THE EIGENVALUES = -6
THE TRACE OF THE MATRIX    = -6
THE DIFFERENCE IS          =  0

******************************************************************

PRESS THE ENTER KEY TO CONTINUE

******************  GAUSS ELIMINATION METHOD  ********************

GIVE THE MINIMUM ALLOWABLE VALUE OF THE PIVOT ELEMENT FOR THE
GAUSS ELIMINATION PROCEDURE ? 0.00001

DO YOU WANT TO SEE STEP-BY-STEP RESULTS OF THE
GAUSS ELIMINATION PROCEDURE(Y/N)? N
```

```
*******************************************************************
HOMOGENEOUS SET: (A - EIGENVALUE*I)*X = 0, FOR EIGENVALUE = 0

-1              0              0              0

 1             -2              3              0

 0              2             -3              0

*******************************************************************

RANK = 2
THE PROGRAM SETS  1  ELEMENT(S) OF THE EIGENVECTOR TO UNITY,
AND REDUCES THE PROBLEM TO FINDING  2  ELEMENT(S).

*******************************************************************

EIGENVECTOR CORRESPONDING TO EIGENVALUE = 0

X( 1 ) = 0
X( 2 ) = 1
X( 3 ) = .6666667

*******************************************************************

VERIFICATION:

A*X              =   EIGENVALUE*X

 0                    0
 0                    0
 0                    0

*******************************************************************

PRESS THE ENTER KEY TO CONTINUE

*******************************************************************
HOMOGENEOUS SET: (A - EIGENVALUE*I)*X = 0, FOR EIGENVALUE =-1

 1.192093E-07                  0              0              0


 1             -.9999999      3              0

 0              2             -2              0

*******************************************************************

RANK = 2
THE PROGRAM SETS  1  ELEMENT(S) OF THE EIGENVECTOR TO UNITY,
AND REDUCES THE PROBLEM TO FINDING  2  ELEMENT(S).

*******************************************************************

EIGENVECTOR CORRESPONDING TO EIGENVALUE =-1

X( 1 ) = 1
X( 2 ) =-.4999999
X( 3 ) =-.5
```

```
*******************************************************************

VERIFICATION:

A*X                =    EIGENVALUE*X

-1                     -1
 .4999999               .5
 .5000001               .5

*******************************************************************

PRESS THE ENTER KEY TO CONTINUE

*******************************************************************
HOMOGENEOUS SET: (A - EIGENVALUE*I)*X = 0, FOR EIGENVALUE =-5

 4              0              0              0

 1              3              3              0

 0              2              2              0

*******************************************************************
RANK = 2
THE PROGRAM SETS  1  ELEMENT(S) OF THE EIGENVECTOR TO UNITY,
AND REDUCES THE PROBLEM TO FINDING  2  ELEMENT(S).
*******************************************************************

EIGENVECTOR CORRESPONDING TO EIGENVALUE =-5

X( 1 ) = 0
X( 2 ) =-1
X( 3 ) = 1

*******************************************************************

VERIFICATION:

A*X                =    EIGENVALUE*X

 0                      0
 5                      5
-5                     -5

*******************************************************************

PRESS THE ENTER KEY TO CONTINUE

*******************************************************************

DO YOU WANT TO REPEAT THE CALCULATIONS
WITH MINOR CHANGES TO THE COEFFICIENTS(Y/N)? N

DO YOU WANT TO RESET ALL THE COEFFICIENTS(Y/N)? N


********************** END OF PROGRAM ***************************
Ok
```

DISCUSSION OF RESULTS The eigenvalues of **K** are calculated by the Newton-Raphson method to be 0, −1, and −5. The zero eigenvalue is obtained because the matrix **K** is singular. The rank of this $3 \times 3$ matrix is 2. This implies that only two of the three differential equations are independent of each other. This is a direct consequence of the law of conservation of mass, which states that mass cannot be created or destroyed in a reaction of this type. Therefore, the sum of the concentrations of the three components, $A$, $B$, and $C$, must be equal to the sum of the initial concentrations:

$$A + B + C = A_0 + B_0 + C_0$$

Taking the derivative with respect to time of both sides of the equation we have

$$\frac{dA}{dt} + \frac{dB}{dt} + \frac{dC}{dt} = 0$$

Rearranging to solve for $dC/dt$,

$$\frac{dC}{dt} = -\frac{dA}{dt} - \frac{dB}{dt}$$

which proves that the third differential equation in this set is a linear combination of the other two equations.

The eigenvectors for each eigenvalue of **K** are determined by the Gauss method to be the following:

$$\mathbf{x}_1 = \begin{bmatrix} 0 \\ 1 \\ 0.6667 \end{bmatrix} \qquad \text{for } \lambda_1 = 0$$

$$\mathbf{x}_2 = \begin{bmatrix} 1 \\ -0.5 \\ -0.5 \end{bmatrix} \qquad \text{for } \lambda_2 = -1$$

$$\mathbf{x}_3 = \begin{bmatrix} 0 \\ -1 \\ 1 \end{bmatrix} \qquad \text{for } \lambda_3 = -5$$

It was mentioned in the introduction to this chapter (Sec. 3-1) that the solution of a set of linear ordinary differential equations can be obtained by Eq. (5-68). Thus, the complete solution of the chemical reaction system described in this problem can be obtained from the above eigenvalues, eigenvectors, and the initial concentration vector. This is discussed in more detail in Chap. 5.

### 3-8.2 Elementary Similarity Transformations

In Sec. 3-5.2, we showed that the Gauss elimination method can be represented in matrix form as

$$\mathbf{LA} = \mathbf{U} \tag{3-120}$$

Matrix **A** is nonsingular, matrix **L** is unit lower triangular, and matrix **U** is upper triangular. The inverse of **L** is also a unit lower triangular matrix. Postmultiplying both sides of Eq. (3-120) by $\mathbf{L}^{-1}$ we obtain

$$\mathbf{LAL}^{-1} = \mathbf{UL}^{-1} = \mathbf{B} \tag{3-172}$$

This is a similarity transformation of the type described in Sec. 3-2.2. The transformation converts matrix **A** to a *similar* matrix **B**. The two matrices, **A** and **B**, have identical eigenvalues, determinants, and traces.

We, therefore, conclude that if the Gauss elimination method is extended so that matrix **A** is postmultiplied by $\mathbf{L}^{-1}$, at each step of the operation, in addition to being premultiplied by **L**, the resulting matrix **B** is similar to **A**. This operation is called the *elementary similarity transformation*.

In the determination of eigenvalues, it is desirable to reduce matrix **A** to a supertriangular matrix of upper Hessenberg form:

$$\mathbf{H} = \begin{bmatrix} h_{11} & h_{12} & h_{13} & h_{14} & h_{15} \\ h_{21} & h_{22} & h_{23} & h_{24} & h_{25} \\ 0 & h_{32} & h_{33} & h_{34} & h_{35} \\ 0 & 0 & h_{43} & h_{44} & h_{45} \\ 0 & 0 & 0 & h_{54} & h_{55} \end{bmatrix} \tag{3-173}$$

This can be done by using the $(k+1)$st row to eliminate the elements $k+2$ to $n$ of column $k$. Consequently, the elements of the subdiagonal do not vanish. The transformation matrices that perform this elimination are unit lower triangular of the form shown in Eq. (3-174). For a $5 \times 5$ matrix, the elimination matrix $\bar{\mathbf{L}}_1$ that would eliminate the elements of column 1 below the subdiagonal is

$$\bar{\mathbf{L}}_1 = \begin{bmatrix} 1 & 0 & 0 & 0 & 0 \\ 0 & 1 & 0 & 0 & 0 \\ 0 & -\dfrac{h_{31}^{(0)}}{h_{21}^{(0)}} & 1 & 0 & 0 \\ 0 & -\dfrac{h_{41}^{(0)}}{h_{21}} & 0 & 1 & 0 \\ 0 & -\dfrac{h_{51}^{(0)}}{h_{21}^{(0)}} & 0 & 0 & 1 \end{bmatrix} \tag{3-174}$$

where the superscript (0) indicates that each $\bar{\mathbf{L}}_k$ matrix uses the elements $h_{ik}^{(k-1)}$ of the previous transformation step. The reader is encouraged to compare $\bar{\mathbf{L}}_1$ with $\mathbf{L}_1$ of Eq. (3-117).

The inverse of $\bar{\mathbf{L}}_1$ is given by Eq. (3-175):

$$\bar{\mathbf{L}}_1^{-1} = \begin{bmatrix} 1 & 0 & 0 & 0 & 0 \\ 0 & 1 & 0 & 0 & 0 \\ 0 & \dfrac{h_{31}^{(0)}}{h_{21}^{(0)}} & 1 & 0 & 0 \\ 0 & \dfrac{h_{41}^{(0)}}{h_{21}^{(0)}} & 0 & 1 & 0 \\ 0 & \dfrac{h_{51}^{(0)}}{h_{21}^{(0)}} & 0 & 0 & 1 \end{bmatrix} \tag{3-175}$$

The complete elementary similarity transformation which converts matrix $\mathbf{A}$ to the upper Hessenberg matrix $\mathbf{H}$ is shown by

$$\bar{\mathbf{L}}_{n-1}\bar{\mathbf{L}}_{n-2}\cdots\bar{\mathbf{L}}_2\bar{\mathbf{L}}_1\mathbf{A}\bar{\mathbf{L}}_1^{-1}\bar{\mathbf{L}}_2^{-1}\cdots\bar{\mathbf{L}}_{n-2}^{-1}\bar{\mathbf{L}}_{n-1}^{-1} = \mathbf{H} \tag{3-176}$$

Each postmultiplication step by the inverse $\bar{\mathbf{L}}_i^{-1}$ preserves the zeros previously obtained in the premultiplication step by $\mathbf{L}_i$ [11].

For simplicity in the above discussion, the partial pivoting matrices $\mathbf{P}_{ij}$ were not applied. However, use of partial pivoting is strongly recommended in order to reduce roundoff errors. Premultiplication by $\mathbf{P}_{ij}$ interchanges two rows and causes the sign of the determinant to change. Postmultiplication by $\mathbf{P}_{ij}^{-1}$ (which is identical to $\mathbf{P}_{ij}$) interchanges the corresponding two columns and causes the sign of the determinant to change again. The premultiplication step must be followed immediately by the postmultiplication step in order to balance the symmetry of the transformation and to preserve the form of the transformed matrix.

For the sake of programming the elementary similarity transformation to produce an upper Hessenberg matrix, we give the formula form of the operation as follows:

Initialization step:

$$h_{ij}^{(0)} = a_{ij} \qquad \{j = 1, 2, \ldots, n \qquad \{i = 1, 2, \ldots, n \tag{3-177a}$$

Transformation formula:

$$m_{i,k+1}^{(k)} = \frac{h_{ij}^{(k-1)}}{h_{k+1,k}^{(k-1)}} \qquad \text{(3-177}b\text{)}$$

Premultiplication step:

$$h_{ij}^{(k-1/2)} = h_{ij}^{(k-1)} - m_{i,k+1}^{(k)} h_{k+1,j}^{(k-1)} \qquad \left\{ \begin{array}{l} j = n, n-1, \ldots, k \\ i = k+2, \ldots, n \end{array} \right. \qquad \text{(3-178)}$$

Postmultiplication step:

$$h_{i,k+1}^{(k)} = h_{i,k+1}^{(k-1/2)} + h_{ij}^{(k-1/2)} m_{j,k+1}^{(k)} \qquad \left\{ \begin{array}{l} j = k+2, \ldots, n \\ i = n, n-1, \ldots, 1 \end{array} \right. \qquad \text{(3-179)}$$

for all three equations: $k = 1, 2, \ldots, n-2$, $h_{k+1,k} \neq 0$

where the superscript $(k - 1/2)$ means that only half the complete transformation (i.e., only premultiplication) has been completed at that point.

The QR algorithm, which will be discussed next, utilizes the upper Hessenberg matrix **H** to determine its eigenvalues, which are equivalent to the eigenvalues of matrix **A**.

### 3-8.3 The QR Algorithm of Successive Factorization

The QR algorithm is based on the possible decomposition of a matrix **A** into a product of two matrices

$$\mathbf{A} = \mathbf{QR} \qquad \text{(3-180)}$$

where **Q** is orthogonal and **R** is upper triangular with nonnegative diagonal elements. This decomposition always exists, and when **A** is nonsingular, the decomposition is unique [8].

The above decomposition can be used to form a series of successive matrices $\mathbf{A}_k$ which are *similar* to the original matrix **A**; therefore, their eigenvalues are the same. To do this, let us first define $\mathbf{A}_1 = \mathbf{A}$ and convert Eq. (3-180) to

$$\mathbf{A}_1 = \mathbf{Q}_1\mathbf{R}_1 \qquad \text{(3-181)}$$

Premultiply each side by $\mathbf{Q}_1^{-1}$ and rearrange to obtain

$$\mathbf{R}_1 = \mathbf{Q}_1^{-1}\mathbf{A}_1 \qquad \text{(3-182)}$$

Form a second matrix $\mathbf{A}_2$ from the product of $\mathbf{R}_1$ with $\mathbf{Q}_1$

$$\mathbf{A}_2 = \mathbf{R}_1\mathbf{Q}_1 \qquad \text{(3-183)}$$

and use Eq. (3-182) to eliminate $\mathbf{R}_1$ from (3-183)

$$\mathbf{A}_2 = \mathbf{Q}_1^{-1}\mathbf{A}_1\mathbf{Q}_1 \qquad \text{(3-184)}$$

Since $\mathbf{Q}_1$ is an orthogonal matrix this is an orthogonal transformation of $\mathbf{A}_1$ to $\mathbf{A}_2$; therefore, these two matrices are similar. They have the same eigenvalues. The inverse of an orthogonal matrix is equal to its transpose, thus Eq. (3-184) can also be written as

$$\mathbf{A}_2 = \mathbf{Q}_1^T \mathbf{A}_1 \mathbf{Q}_1 \tag{3-185}$$

In the particular case where matrix $\mathbf{A}$ is symmetric, an orthogonal transformation of $\mathbf{A}$ can be found which yields a diagonal matrix $\mathbf{D}$

$$\mathbf{D} = \mathbf{Q}^T \mathbf{A} \mathbf{Q} \tag{3-186}$$

whose diagonal elements are the eigenvalues of $\mathbf{A}$. Our discussion, however, will focus on nonsymmetric matrices which transform to triangular matrices.

The orthogonal matrix $\mathbf{Q}_1$ is determined by finding a series of $\mathbf{S}_{ij}^T$ orthogonal transformation matrices, each of which eliminates one element, in position $ij$, below the diagonal of the matrix it is postmultiplying. The complete set of transformations converts matrix $\mathbf{A}_1$ to upper triangular form with nonnegative diagonal elements:

$$\mathbf{S}_{n,\,n-1}^T \cdots \mathbf{S}_{ij}^T \cdots \mathbf{S}_{n1}^T \cdots \mathbf{S}_{31}^T \mathbf{S}_{21}^T \mathbf{A}_1 = \mathbf{R}_1 \tag{3-187}$$

where the counter $i$ increases from $j+1$ to $n$, and the counter $j$ increases from 1 to $n-1$.

Each of the $\mathbf{S}_{ij}^T$ matrices is orthogonal and the product of orthogonal matrices is also orthogonal. Direct comparison of Eq. (3-187) with Eq. (3-182) reveals that $\mathbf{Q}_1^{-1}$ is equal to the product of the $\mathbf{S}_{ij}^T$ matrices

$$\mathbf{Q}_1^{-1} = \mathbf{S}_{n,\,n-1}^T \cdots \mathbf{S}_{ij}^T \cdots \mathbf{S}_{31}^T \mathbf{S}_{21}^T \tag{3-188}$$

Since the transpose of an orthogonal matrix is equal to its inverse, it follows that

$$\mathbf{Q}_1^T = \mathbf{S}_{n,\,n-1}^T \cdots \mathbf{S}_{ij}^T \cdots \mathbf{S}_{31}^T \mathbf{S}_{21}^T \tag{3-189}$$

and

$$\mathbf{Q}_1 = \mathbf{S}_{21} \mathbf{S}_{31} \cdots \mathbf{S}_{ij} \cdots \mathbf{S}_{n,\,n-1} \tag{3-190}$$

Therefore, Eq. (3-184) can be rewritten in terms of the $\mathbf{S}_{ij}$ matrices:

$$\mathbf{A}_2 = \mathbf{S}_{n,\,n-1}^T \cdots \mathbf{S}_{ij}^T \cdots \mathbf{S}_{31}^T \mathbf{S}_{21}^T \mathbf{A}_1 \mathbf{S}_{21} \mathbf{S}_{31} \cdots \mathbf{S}_{ij} \cdots \mathbf{S}_{n,\,n-1} \tag{3-191}$$

As an example of the orthogonal transformation matrices $\mathbf{S}_{ij}$ we give the $\mathbf{S}_{pq}$ matrix for a $6 \times 6$-order system, with $p = 6$ and $q = 3$:

$$\mathbf{S}_{6,3} = \begin{bmatrix} 1 & 0 & 0 & 0 & 0 & 0 \\ 0 & 1 & 0 & 0 & 0 & 0 \\ 0 & 0 & s_{33} & 0 & 0 & s_{36} \\ 0 & 0 & 0 & 1 & 0 & 0 \\ 0 & 0 & 0 & 0 & 1 & 0 \\ 0 & 0 & s_{63} & 0 & 0 & s_{66} \end{bmatrix} \tag{3-192}$$

where the diagonal elements of this matrix are specified as

$$s_{pp} = s_{qq} = \cos\theta \tag{3-193}$$

$$s_{ii} = 1 \qquad \text{for } i \neq p \text{ or } q \tag{3-194}$$

and the off-diagonal elements as

$$s_{pq} = -s_{qp} = \sin\theta \tag{3-195}$$

$$s_{ij} = 0 \qquad \text{everywhere else} \tag{3-196}$$

Premultiplication of matrix **A** by $\mathbf{S}_{pq}^T$ eliminates the element $pq$ and causes a rotation of axes in the $(p, q)$ plane. The $\mathbf{S}_{ij}$ matrices clearly satisfy the orthogonality requirement that

$$\mathbf{S}_{ij}^T\mathbf{S}_{ij} = \mathbf{I} \tag{3-197}$$

This is left as an exercise for the reader to verify.

The angle of axis rotation $\theta$, in Eqs. (3-193) and (3-195), is chosen so that the element $pq$, of the matrix being transformed, vanishes. It has been shown by Givens [12] that it is not necessary to actually calculate the value of $\theta$ itself. The trigonometric terms $\cos\theta$ and $\sin\theta$ can be obtained from the values of the elements of the matrix being transformed. Givens has determined that the elements of the matrix $\mathbf{S}_{pq}$ are calculated as follows:
Diagonal elements:

$$s_{pp}^{(k)} = s_{qq}^{(k)} = \frac{a_{qq}^{(k-1)}}{\sqrt{(a_{qq}^{(k-1)})^2 + (a_{pq}^{(k-1)})^2}} \tag{3-198}$$

$$s_{ii}^{(k)} = 1 \qquad \text{for } i \neq p \text{ or } q \tag{3-199}$$

Off-diagonal elements:

$$s_{pq}^{(k)} = \frac{a_{pq}^{(k-1)}}{\sqrt{(a_{qq}^{(k-1)})^2 + (a_{pq}^{(k-1)})^2}} \qquad s_{qp} = -s_{pq} \tag{3-200}$$

$$s_{ij}^{(k)} = 0 \qquad \text{everywhere else} \tag{3-201}$$

The superscripts $(k-1)$ have been used in the above equations to remind the reader that the elements $a_{pq}^{(k-1)}$ and $a_{qq}^{(k-1)}$ are those of the matrix from the previous transformation step and not those of the original matrix.

Givens' method of plane rotations can reduce a nonsymmetric matrix to upper triangular form and a symmetric matrix to tridiagonal form. However, a large number of computations is required. It is computationally more efficient to apply first the elementary similarity transformation to reduce the matrix to upper Hessenberg form, as was described in Sec. 3-8.2, and then to use plane rotations to reduce it to triangular form. In the rest of this section we will assume that the matrix $\mathbf{A}$ has been already reduced to upper Hessenberg form, $\mathbf{H}_1$, and we will show how the QR algorithm further reduces the matrix to obtain its eigenvalues.

If the eigenvalues of matrix $\mathbf{H}_1$ are $\lambda$, then the eigenvalues of matrix $(\mathbf{H}_1 - \gamma_1\mathbf{I})$ are $(\lambda - \gamma_1)$, where $\gamma_1$ is called the *shift factor*. The orthogonal transformation applied to $\mathbf{A}_1$ above can also be applied to the shifted matrix $(\mathbf{H}_1 - \gamma_1\mathbf{I})$ as follows: Decompose the matrix $(\mathbf{H}_1 - \gamma_1\mathbf{I})$ into $\mathbf{Q}_1$ and $\mathbf{R}_1$ matrices:

$$\mathbf{H}_1 - \gamma_1\mathbf{I} = \mathbf{Q}_1\mathbf{R}_1 \tag{3-202}$$

Rearrange the above equation to obtain $\mathbf{R}_1$:

$$\mathbf{R}_1 = \mathbf{Q}_1^{-1}(\mathbf{H}_1 - \gamma_1\mathbf{I}) \tag{3-203}$$

Form a new matrix $(\mathbf{H}_2 - \gamma_1\mathbf{I})$ from the product of $\mathbf{R}_1$ and $\mathbf{Q}_1$:

$$\mathbf{H}_2 - \gamma_1\mathbf{I} = \mathbf{R}_1\mathbf{Q}_1 \tag{3-204}$$

Eliminate $\mathbf{R}_1$ using Eq. (3-203):

$$\mathbf{H}_2 - \gamma_1\mathbf{I} = \mathbf{Q}_1^{-1}(\mathbf{H}_1 - \gamma_1\mathbf{I})\mathbf{Q}_1 \tag{3-205}$$

Solve for $\mathbf{H}_2$:

$$\mathbf{H}_2 = \mathbf{Q}_1^{-1}(\mathbf{H}_1 - \gamma_1\mathbf{I})\mathbf{Q}_1 + \gamma_1\mathbf{I} \tag{3-206}$$

It has been shown [8] that if the shift factor $\gamma_1$ is chosen to be a good estimate of one of the eigenvalues and that if the magnitudes of the eigenvalues are

$$|\lambda_1| > |\lambda_2| > \cdots > |\lambda_n| \tag{3-207}$$

then the matrix $\mathbf{H}_k$ will converge to an almost triangular form with the elements $h_{n,n-1} \rightarrow 0$ and $h_{nn} \rightarrow \lambda_n$.

Estimation of the shift factor $\gamma_1$ is relatively easy when the matrix has been reduced to upper Hessenberg form:

$$\mathbf{H}_1 = \begin{bmatrix} h_{11} & h_{12} & h_{13} & \cdots & h_{1,n-1} & h_{1,n} \\ h_{21} & h_{22} & h_{23} & \cdots & \vdots & \vdots \\ 0 & h_{32} & h_{33} & \cdots & \vdots & \vdots \\ 0 & 0 & h_{43} & \ddots & \vdots & \vdots \\ & & & \ddots & h_{n-1,n-1} & h_{n-1,n} \\ 0 & 0 & 0 & \cdot & h_{n,n-1} & h_{nn} \end{bmatrix} \tag{3-208}$$

The eigenvalues of the lower $2 \times 2$ submatrix

$$\begin{bmatrix} h_{n-1,n-1} & h_{n-1,n} \\ h_{n,n-1} & h_{nn} \end{bmatrix} \tag{3-209}$$

can be used to determine the shift factor. The two eigenvalues of this matrix are obtained from the quadratic characteristic equation

$$\gamma^2 - (h_{n-1,n-1} + h_{nn})\gamma + (h_{n-1,n-1}h_{nn} - h_{n,n-1}h_{n-1,n}) = 0 \tag{3-210}$$

whose solution is given by the quadratic formula

$$\gamma_{+,-} = \frac{(h_{n-1,n-1} + h_{nn}) \pm \sqrt{(h_{n-1,n-1} + h_{nn})^2 - 4(h_{n-1,n-1}h_{nn} - h_{n,n-1}h_{n-1,n})}}{2} \tag{3-211}$$

The value of $\gamma$ which is closest to $h_{nn}$ is chosen from the two roots. In the case where the roots are complex conjugates, the real part of the root is chosen as the shift factor.

In the QR iteration procedure the subsequent values of the shift factor, $\gamma_2 \ldots \gamma_k$, are similarly chosen from matrices $\mathbf{H}_2 \ldots \mathbf{H}_k$.

The steps of the QR algorithm for calculating the eigenvalues and eigenvectors of a nonsingular nonsymmetric matrix **A** with real eigenvalues are the following:

1. Use the elementary similarity transformations [Eqs. (3-176) to (3-179)] to transform matrix **A** to the upper Hessenberg matrix $\mathbf{H}_1$.
2. Utilize the lower $2 \times 2$ submatrix of $\mathbf{H}_1$ [Eq. (3-209)] to estimate the shift factor $\gamma_1$ from Eq. (3-211).
3. Construct the shifted matrix $(\mathbf{H}_1 - \gamma_1\mathbf{I})$.
4. Calculate the elements of the transformation matrix $\mathbf{S}_{21}$ from the elements of the shifted matrix $(\mathbf{H}_1 - \gamma_1\mathbf{I})$ using Eqs. (3-198) to (3-201).
5. Perform the premultiplication $\mathbf{S}_{21}^T(\mathbf{H}_1 - \gamma_1\mathbf{I})$ which eliminates the elements in position (2, 1) of the matrix $(\mathbf{H}_1 - \gamma_1\mathbf{I})$.

6. Repeat steps 4 and 5, calculating the transformation matrix $\mathbf{S}_{pq}$ and eliminating one element on the subdiagonal in each set of steps. The application of steps 4 and 5 for $n-1$ times, with the counter $q$ increasing from 1 to $n-1$ and the counter $p$ set at $q+1$, will convert the Hessenberg matrix $\mathbf{H}_1$ to a triangular matrix $\mathbf{R}_1$:

$$\mathbf{S}^T_{n,\,n-1} \cdots \mathbf{S}^T_{32}\mathbf{S}^T_{21}\mathbf{H}_1 = \mathbf{R}_1 \tag{3-212}$$

7. Perform the postmultiplication of $\mathbf{R}_1$ by $\mathbf{S}_{pq}$ to obtain the transformed shifted matrix $(\mathbf{H}_2 - \gamma_1\mathbf{I})$:

$$\mathbf{H}_2 - \gamma_1\mathbf{I} = \mathbf{R}_1\mathbf{S}_{21}\mathbf{S}_{32} \cdots \mathbf{S}_{n,\,n-1} \tag{3-213}$$

8. Solve Eq. (3-213) for the transformed Hessenberg matrix $\mathbf{H}_2$:

$$\mathbf{H}_2 = \mathbf{R}_1\mathbf{S}_{21}\mathbf{S}_{32} \cdots \mathbf{S}_{n,\,n-1} + \gamma_1\mathbf{I} \tag{3-214}$$

9. Use $\mathbf{H}_2$ as the new Hessenberg matrix and repeat steps 2 to 8 until $|h_{n,\,n-1}| \le \varepsilon$, where $\varepsilon$ is a small convergence criterion. At this point the element $h_{nn}$ will give one eigenvalue $\lambda_n$.
10. Deflate the $\mathbf{H}_k$ matrix to order $n-1$ by eliminating the $n$th row and $n$th column, and repeat steps 2 to 10 until all the eigenvalues are calculated.
11. Apply the Gauss elimination method with complete pivoting to the matrix $(\mathbf{A} - \lambda\mathbf{I})$ to evaluate the eigenvectors corresponding to each eigenvalue. Several different possibilities exist when the eigenvalues are real:

   *a*. *Distinct nonzero eigenvalues*: Matrix $\mathbf{A}$ is nonsingular and matrix $(\mathbf{A} - \lambda\mathbf{I})$ is singular of rank $n-1$. Application of the Gauss elimination method with complete pivoting on the matrix $(\mathbf{A} - \lambda\mathbf{I})$ triangularizes the matrix and causes the last row to contain all zero values, because the rank is $n-1$. Assume the value of the $n$th element of the eigenvector to be equal to unity and reduce the problem to finding the remaining $n-1$ elements.

   *b*. *One zero eigenvalue*: Matrix $\mathbf{A}$ is singular of rank $n-1$ and matrix $(\mathbf{A} - \lambda\mathbf{I})$ is singular of rank $n-1$. Application of the Gauss elimination method proceeds as in *a*. One element of each eigenvector will be found to be a zero element.

   *c*. *One pair of repeated eigenvalues*: Matrix $\mathbf{A}$ is nonsingular and matrix $(\mathbf{A} - \lambda\mathbf{I})$ is of rank $n-2$. Application of the Gauss elimination method with complete pivoting on the matrix $(\mathbf{A} - \lambda\mathbf{I})$ triangularizes the matrix and causes the last two rows to contain all zero values, because the rank is $n-2$. Assume the values of the last two elements in the eigenvector to be equal to unity and reduce the problem to finding the remaining $n-2$ elements.

The QR algorithm described in this section applies well to both symmetric and nonsymmetric matrices with real eigenvalues. A more general method, called the double QR algorithm, which can evaluate complex eigenvalues, is described by Ralston and Rabinowitz [8].

Example 3-5 demonstrates the use of the QR algorithm in locating the eigenvalues and eigenvectors of the variance/covariance matrix in a nonlinear regression problem.

**Example 3-5 Calculation of eigenvalues and eigenvectors using elementary similarity transformations to convert matrix to Hessenberg form and the QR algorithm with plane rotations** The variance/covariance matrix, obtained in the application of nonlinear regression methods for evaluating the parameters of mathematical models from experimental data, is given by $\sigma^2\mathbf{I}(\mathbf{A}^T\mathbf{A})^{-1}$. The correlation coefficient matrix **R** is defined as

$$\mathbf{R} = r_{ij} = \left(\frac{a_{ij}}{\sqrt{a_{ii}a_{jj}}}\right)$$

where $a_{ij}$ are the elements of the variance/covariance matrix (see Chap. 7).

The correlation coefficient matrix measures the correlation between pairs of parameters and defines the shape and nature of the confidence region of the estimates of these parameters. The eigenvectors of the correlation coefficient matrix represent the directions of the major and minor axes of the confidence hyperellipsoid, and the eigenvalues give the relative magnitudes of these axes. Therefore, knowledge of the eigenvectors and eigenvalues of the correlation coefficient matrix are extremely useful in studying the shape of the confidence region of the parameter estimates.

The following correlation coefficient matrix was obtained in the application of the Marquardt nonlinear regression method for evaluating four parameters in a mathematical model describing the dynamics of the penicillin fermentation process [13]:

Correlation coefficient matrix:

$$\begin{bmatrix} 1 & -0.5484 & -0.8134 & -0.7757 \\ -0.5484 & 1 & 0.4339 & 0.5557 \\ -0.8134 & 0.4339 & 1 & 0.9791 \\ -0.7757 & 0.5557 & 0.9791 & 1 \end{bmatrix}$$

Determine the eigenvalues and eigenvectors of this correlation matrix. Use elementary similarity transformations to convert the matrix to Hessenberg form and then apply the QR algorithm of successive factorization to get the eigenvalues. Apply the Gauss elimination method (as in Example 3-4) to get the eigenvectors.

METHOD OF SOLUTION The method of solution used in this example is identical to that described in steps 1 to 11 in Sec. 3-8.3.

PROGRAM DESCRIPTION This computer program, which we call QR.BAS, consists of the sections listed in Table E3-5.

*Main program and subroutines* Lines 120–400 constitute the input part of the program in which the user enters the order of the matrix, the elements of the matrix, and any corrections, if needed.

Lines 420–920 implement the elementary similarity transformations [Eqs. (3-177) to (3-179)] to transform matrix **A** to the upper Hessenberg matrix $\mathbf{H}_1$. Partial pivoting strategy is used with this transformation.

Lines 930–990 input the convergence criterion and option for the successive factorization procedure.

Lines 1020–1330 constitute a WHILE . . . END loop which calls subroutines 3 to 7 to implement the successive factorization algorithm (steps 2 to 9 of the QR algorithm described in Sec. 3-8.3). The sequence of steps is as follows: Subroutine 3 is called to estimate the shift factor from Eq. (3-211). The main program (line 1070) constructs the shifted matrix. Subroutine 4 is called to calculate the transpose of the transformation matrix using Eqs. (3-198) to (3-201). Subroutine 5 premultiplies the shifted

**Table E3-5**

| Program section | Line numbers |
|---|---|
| Title | 1–90 |
| Main program | 100–1900 |
| Input | 120–400 |
| Elementary similarity transformation | 420–920 |
| QR successive factorization | 930–1550 |
| Gauss elimination method | 1560–1790 |
| Rerun | 1800–1870 |
| Subroutines | |
| 1. Print matrix **A** | 2000–2080 |
| 2. Print matrix **S** | 3000–3080 |
| 3. Calculate the shift factor | 4000–4080 |
| 4. Calculate the transpose of the transformation matrix | 5000–5150 |
| 5. Perform premultiplication | 6000–6120 |
| 6. Calculate the transformation matrix | 7000–7120 |
| 7. Perform postmultiplication | 8000–8120 |
| 8. Gauss elimination method | 9000–9820 |

matrix by the transpose of the transformation matrix to obtain the triangular matrix as in Eq. (3-212). Subroutine 6 calculates the transformation matrix. Subroutine 7 postmultiplies the triangular matrix by the transformation matrix to obtain the transformed shifted matrix as in Eq. (3-213). The main program (line 1270) solves for the transformed matrix using Eq. (3-214). Finally, the main program (line 1310) checks for convergence $|h_{n,\,n-1}| < \varepsilon$.

Lines 1360–1370 obtain the $n$th eigenvalue $\lambda_n$ from the $h_{nn}$ element of the transformed matrix.

Lines 1380–1400 deflate the matrix to order $n-1$ and repeat the successive factorization until all eigenvalues are calculated.

Lines 1410–1420 reset the order of the matrix to the original value.

Lines 1430–1530 print the eigenvalues and verify them using Eq. (3-60).

Lines 1560–1790, together with subroutine 8, implement the Gauss elimination procedure. These are identical to the corresponding lines in Example 3-4.

Lines 1800–1870 give the user the opportunity to rerun the program with modifications.

Lines 1880–1900 terminate the program.

PROGRAM

```
1 SCREEN 0:WIDTH 80:CLS:KEY OFF
10 PRINT"*********************************************************************"
15 PRINT"*                                                                   *"
20 PRINT"*                            EXAMPLE 3-5                            *"
25 PRINT"*                                                                   *"
30 PRINT"*   CALCULATION OF EIGENVALUES AND EIGENVECTORS USING ELEMENTARY    *"
35 PRINT"*                                                                   *"
40 PRINT"* SIMILARITY TRANSFORMATIONS TO CONVERT MATRIX TO HESSENBERG FORM *"
45 PRINT"*                                                                   *"
50 PRINT"*             AND THE QR ALGORITHM WITH PLANE ROTATIONS             *"
55 PRINT"*                                                                   *"
60 PRINT"*                              (QR.BAS)                             *"
65 PRINT"*                                                                   *"
90 PRINT"*********************************************************************"
100 '************************** Main Program ******************************
110 '
120 'Enter the order of the matrix and its elements
130 '
140 PRINT:PRINT "ENTER THE ORDER OF THE MATRIX";:INPUT N
150 DIM A(N,N+1),B(N,N),C(N,N),S(N,N),SAVED(N-1,2),NPIVROW(N-1,2),NPIVCOL(N-1,2)
,X(N),XW(N)
160 PRINT:PRINT "ENTER ELEMENTS OF MATRIX:"
170 FOR K = 1 TO N
180 PRINT : PRINT "ROW ";K: PRINT
190 FOR J = 1 TO N
200 PRINT "   ELEMENT (";K;",";J;") =";: INPUT B(K,J)
210 NEXT J
220 NEXT K
230 NN=N : NC=N
240 TRC=0
250 FOR K = 1 TO N
260 FOR J = 1 TO N
270 A(K,J) = B(K,J)
280 NEXT J
290 TRC=TRC + B(K,K)
300 NEXT K
310 PRINT
320 PRINT"*********************************************************************"
330 PRINT"ORIGINAL MATRIX:"
340 GOSUB 2000 'Print matrix
350 INPUT "IS THE MATRIX CORRECT(Y/N)"; Q$:PRINT
360 IF Q$ = "Y" OR Q$ = "y" THEN 420
370 PRINT"GIVE THE POSITION OF THE ELEMENT TO BE CORRECTED:":PRINT
380 INPUT "   ROW NUMBER";NROW :INPUT "   COLUMN NUMBER";NCOL
390 PRINT:INPUT "   CORRECT VALUE OF THE ELEMENT"; B(NROW,NCOL):PRINT
400 GOTO 240
410 '
420 'Beginning of the elementary similarity transformation procedure
430 PRINT CHR$(12)
440 PRINT"************* ELEMENTARY SIMILARITY TRANSFORMATION ***************"
450 PRINT
460 PRINT"GIVE THE MINIMUM ALLOWABLE VALUE OF THE PIVOT ELEMENT FOR THE"
470 PRINT"ELEMENTARY SIMILARITY TRANSFORMATION ";:INPUT EPS:PRINT
480 PRINT"DO YOU WANT TO SEE STEP-BY-STEP RESULTS OF THE "
490 PRINT"ELEMENTARY SIMILARITY TRANSFORMATION(Y/N)";:INPUT Q2$:PRINT
500 PRINT"*********************************************************************"
510 FOR K = 1 TO N - 2
520 'Apply partial pivoting strategy
```

```
530 MAXPIVOT = ABS(A(K+1,K)):NPIVOT=K+1
540 FOR I = K+1 TO N
550 IF MAXPIVOT > = ABS(A(I,K)) GOTO 570
560 MAXPIVOT=ABS(A(I,K)) : NPIVOT=I
570 NEXT I
580 IF MAXPIVOT > = EPS GOTO 600
590 PRINT"PIVOT ELEMENT SMALLER THAN";EPS".  MATRIX MAY BE SINGULAR":GOTO 1880
600 IF NPIVOT = K+1 GOTO 710
610 IF Q2$="Y" OR Q2$="y" THEN PRINT"PARTIAL PIVOTING:"
620 IF Q2$="Y" OR Q2$="y" THEN PRINT "INTERCHANGE ROWS ";NPIVOT;" AND ";K+1
630 FOR J = K TO N
640 SWAP  A(NPIVOT,J),A(K+1,J)
650 NEXT J
660 IF Q2$="Y" OR Q2$="y" THEN PRINT "INTERCHANGE COLUMNS ";NPIVOT;" AND ";K+1
670 FOR I = 1 TO N
680 SWAP  A(I,NPIVOT),A(I,K+1)
690 NEXT I
700 IF Q2$="Y" OR Q2$="y" THEN GOSUB 2000 'Print matrix
710 IF Q2$="Y" OR Q2$="y" THEN PRINT "PERFORM PREMULTIPLICATION:"
720 FOR I = K+2 TO N
730 IF Q2$="Y" OR Q2$="y" THEN PRINT "DIVIDE ROW ";K+1;" BY ";A(K+1,K)
740 IF Q2$="Y" OR Q2$="y" THEN PRINT "MULTIPLY ROW ";K+1;" BY ";A(I,K) ;"AND SUB
TRACT FROM ROW ";I
750 S(I,K+1) = A(I,K)/A(K+1,K)
760 FOR J = N TO K STEP -1
770 A(I,J) = A(I,J) - S(I,K+1) * A(K+1,J)
780 NEXT J
790 IF Q2$="Y" OR Q2$="y" THEN GOSUB 2000 'Print matrix
800 NEXT I
810 IF Q2$="Y" OR Q2$="y" THEN PRINT "PERFORM POSTMULTIPLICATION:"
820 FOR I = N TO 1 STEP -1
830 FOR J = K+2 TO N
840 A(I,K+1) = A(I,K+1) +  A(I,J) * S(J,K+1)
850 NEXT J
860 NEXT I
870 IF Q2$="Y" OR Q2$="y" THEN GOSUB 2000 'Print matrix
880 NEXT K
890 PRINT"*****************************************************************"
900 PRINT"MATRIX REDUCED TO HESSENBERG FORM:"
910 GOSUB 2000 'Print matrix
920 PRINT:INPUT"PRESS THE ENTER KEY TO CONTINUE",KK$
930 'Beginning of the QR successive factorization procedure
940 PRINT CHR$(12)
950 PRINT"************* SUCCESSIVE FACTORIZATION PROCEDURE *****************"
960 PRINT: PRINT"GIVE THE CONVERGENCE CRITERION FOR THE"
970 PRINT"SUCCESSIVE FACTORIZATION PROCEDURE";:INPUT EPS:PRINT
980 PRINT:PRINT"DO YOU WANT TO SEE STEP-BY-STEP RESULTS OF THE "
990 PRINT"SUCCESSIVE FACTORIZATION PROCEDURE(Y/N)";:INPUT Q2$:PRINT
1000 IF Q2$="N" OR Q2$="n" THEN PRINT"CALCULATING RESULTS.  PLEASE WAIT.":PRINT
1010 K=1
1020 WHILE K < 31
1030 IF Q2$="Y" OR Q2$="y" THEN PRINT "ITERATION=";K:PRINT
1040 GOSUB 4000 'Calculate the shift factor
1050 IF Q2$="Y" OR Q2$="y" THEN PRINT "SHIFT FACTOR=";GAMMA:PRINT
1060 'Construct the shifted matrix
1070 FOR I=1 TO N : A(I,I)=A(I,I) - GAMMA :NEXT I
1080 IF Q2$="Y" OR Q2$="Y" THEN PRINT"******************************************
************************"
1090 IF Q2$="Y" OR Q2$="y" THEN PRINT "SHIFTED MATRIX:" :GOSUB 2000
1100 FOR Q=1 TO N-1
```

```
1110 P=Q + 1
1120 GOSUB 5000 'Calculate the transpose of the transformation matrix
1130 IF Q2$="Y" OR Q2$="y" THEN  PRINT "TRANSPOSE OF TRANSFORMATION MATRIX:" :GO
SUB 3000
1140 GOSUB 6000 'Perform premultiplication
1150 IF Q2$="Y" OR Q2$="y" THEN PRINT "INTERMEDIATE MATRIX:" :GOSUB 2000
1160 NEXT Q
1170 IF Q2$="Y" OR Q2$="y" THEN PRINT "TRIANGULARIZED MATRIX:" :GOSUB 2000
1180 FOR Q=1 TO N-1
1190 P=Q + 1
1200 GOSUB 7000 'Calculate the transformation matrix
1210 IF Q2$="Y" OR Q2$="y" THEN PRINT "TRANSFORMATION MATRIX:" :GOSUB 3000
1220 GOSUB 8000 'Perform postmultiplication
1230 IF Q2$="Y" OR Q2$="y" THEN PRINT "INTERMEDIATE MATRIX:" :GOSUB 2000
1240 NEXT Q
1250 IF Q2$="Y" OR Q2$="y" THEN PRINT "TRANSFORMED SHIFTED MATRIX:" :GOSUB 2000
1260 'Solve for the transformed matrix
1270 FOR I=1 TO N : A(I,I)=A(I,I) + GAMMA :NEXT I
1280 IF Q2$="Y" OR Q2$="y" THEN PRINT "TRANSFORMED MATRIX:" :GOSUB 2000
1290 'Check for convergence
1300 IF Q2$="Y" OR Q2$="y" THEN PRINT CHR$(12)
1310 IF ABS(A(N,N-1)) < EPS THEN GOTO 1360
1320 K=K + 1
1330 WEND
1340 PRINT:PRINT"****** SUCCESSIVE FACTORIZATION DID NOT CONVERGE ******"
1350 PRINT"ITERATION=";K;" VALUE OF A(N,N)=";A(N,N);"  VALUE OF EPS=";EPS :GOTO
1880
1360 IF ABS(A(N,N)) < EPS THEN A(N,N)=0
1370 XW(N)=A(N,N)
1380 'Deflate matrix
1390 N=N-1: NC=NC-1
1400 IF N>0 THEN GOTO 1010
1410 'Reset the order of the matrix
1420 N=NN: NC=NN
1430 PRINT:PRINT"THE ";N;" EIGENVALUES ARE:":PRINT
1440 EIGENSUM=0
1450 FOR I=N TO 1 STEP -1
1460 PRINT"   ";XW(I)
1470 EIGENSUM=EIGENSUM + XW(I)
1480 NEXT I
1490 PRINT:PRINT"VERIFICATION:"
1500 PRINT:PRINT"THE SUM OF THE EIGENVALUES = ";EIGENSUM
1510 PRINT"THE TRACE OF THE MATRIX    = ";TRC
1520 PRINT"THE DIFFERENCE IS          = ";EIGENSUM-TRC : PRINT
1530 IF ABS(EIGENSUM-TRC) > EPS THEN PRINT "THE DIFFERENCE IS LARGER THAN THE CO
NVERGENCE CRITERION" : PRINT
1540 PRINT"*****************************************************************"
1550 PRINT:INPUT"PRESS THE ENTER KEY TO CONTINUE",KK$
1560 PRINT CHR$(12)
1570 PRINT"*****************  GAUSS ELIMINATION METHOD  ********************"
1580 PRINT
1590 PRINT"GIVE THE MINIMUM ALLOWABLE VALUE OF THE PIVOT ELEMENT FOR THE"
1600 PRINT"GAUSS ELIMINATION PROCEDURE ";:INPUT EPS:PRINT
1610 PRINT"DO YOU WANT TO SEE STEP-BY-STEP RESULTS OF THE "
1620 PRINT"GAUSS ELIMINATION PROCEDURE(Y/N)";:INPUT Q2$:PRINT
1630 FOR KK = N TO 1 STEP -1
1640 PRINT"*****************************************************************"
1650 PRINT"HOMOGENEOUS SET: (A - EIGENVALUE*I)*X = 0, FOR EIGENVALUE =";XW(KK)
1660 NC=N+1
1670 'Subtract the eigenvalue from the matrix and use the Gauss method
```

```
1680 FOR I = 1 TO N
1690 FOR J = 1 TO N
1700 A(I,J)  = B(I,J)
1710 NEXT J
1720 A(I,I)= B(I,I) - XW(KK)
1730 A(I,NC)=0
1740 NEXT I
1750 GOSUB 2000 'Print matrix
1760 GOSUB 9000 'Perform elimination
1770 NEXT KK
1780 PRINT:PRINT
1790 PRINT"*******************************************************************"
1800 PRINT:PRINT "DO YOU WANT TO REPEAT THE CALCULATIONS":PRINT"WITH MINOR CHANG
ES TO THE COEFFICIENTS(Y/N)";:INPUT V$
1810 IF V$ = "Y" OR V$ = "y" THEN 1820 ELSE 1830
1820 CLS : GOTO 230
1830 PRINT:INPUT "DO YOU WANT TO RESET ALL THE COEFFICIENTS(Y/N)";W$
1840 IF W$ = "Y" OR W$ = "y" THEN 1850 ELSE 1880
1850 PRINT:INPUT "IS THE NEW SET OF THE SAME ORDER AS THE PREVIOUS SET";WW$
1860 IF WW$="N" OR WW$="n" THEN PRINT CHR$(12):RUN 100
1870 CLS : GOTO 160
1880 PRINT:PRINT
1890 PRINT"*********************** END OF PROGRAM ****************************"
1900 END
2000 '****************** Subroutine 1: Print matrix A ************************
2010 '
2020 FOR KA = 1 TO N
2030 PRINT
2040 FOR J = 1 TO NC
2050 PRINT A(KA,J),
2060 NEXT J:PRINT: NEXT KA:PRINT
2070 PRINT"*******************************************************************"
2080 RETURN
3000 '****************** Subroutine 2: Print matrix S ************************
3010 '
3020 FOR KA = 1 TO N
3030 PRINT
3040 FOR J = 1 TO N
3050 PRINT S(KA,J),
3060 NEXT J:PRINT: NEXT KA:PRINT
3070 PRINT"*******************************************************************"
3080 RETURN
4000 '************** Subroutine 3: Calculate the shift factor ****************
4010 '
4020 APA=A(N-1,N-1) + A(N,N)
4030 GR=APA^2 - 4*(A(N-1,N-1)*A(N,N) - A(N,N-1)*A(N-1,N))
4040 IF GR<0 THEN GAMMA=APA/2: GOTO 4080
4050 GPLUS=(APA + SQR(GR))/2
4060 GMINUS=(APA - SQR(GR))/2
4070 IF ABS(GPLUS - A(N,N)) <= ABS(GMINUS-A(N,N)) THEN  GAMMA=GPLUS ELSE GAMMA=G
MINUS
4080 RETURN
5000 '** Subroutine 4: Calculate the transpose of the transformation matrix **
5010 '
5020 FOR I=1 TO N
5030 FOR J=1 TO N
5040 S(I,J)=0
5050 NEXT J
5060 S(I,I)=1
5070 NEXT I
```

```
5080 S(P,P)=A(Q,Q)/SQR(A(Q,Q)^2 + A(P,Q)^2)
5090 S(Q,Q)=S(P,P)
5100 S(Q,P)=A(P,Q)/SQR(A(Q,Q)^2 + A(P,Q)^2)
5110 S(P,Q)=-S(Q,P)
5120 'Save the elements
5130 SAVED(Q,1)=S(P,P)
5140 SAVED(Q,2)=S(Q,P)
5150 RETURN
6000 '************** Subroutine 5: Perform premultiplication *****************
6010 '
6020 FOR I=1 TO N
6030 FOR J=1 TO N
6040 C(I,J)=0
6050 FOR L=1 TO N
6060 C(I,J)=C(I,J) + S(I,L)*A(L,J)
6070 NEXT L:NEXT J:NEXT I
6080 FOR I=1 TO N
6090 FOR J=1 TO N
6100 A(I,J)=C(I,J)
6110 NEXT J:NEXT I
6120 RETURN
7000 '*********** Subroutine 6: Calculate the transformation matrix **********
7010 '
7020 FOR I=1 TO N
7030 FOR J=1 TO N
7040 S(I,J)=0
7050 NEXT J
7060 S(I,I)=1
7070 NEXT I
7080 S(P,P)=SAVED(Q,1)
7090 S(Q,Q)=S(P,P)
7100 S(P,Q)=SAVED(Q,2)
7110 S(Q,P)=-S(P,Q)
7120 RETURN
8000 '************** Subroutine 7: Perform postmultiplication *****************
8010 '
8020 FOR I=1 TO N
8030 FOR J=1 TO N
8040 C(I,J)=0
8050 FOR L=1 TO N
8060 C(I,J)=C(I,J) + A(I,L)*S(L,J)
8070 NEXT L:NEXT J:NEXT I
8080 FOR I=1 TO N
8090 FOR J=1 TO N
8100 A(I,J)=C(I,J)
8110 NEXT J:NEXT I
8120 RETURN
9000 '**************** Subroutine 8: Gauss elimination method ****************
9010 '
9020 FOR K = 1 TO N-1
9030 'Apply complete pivoting strategy
9040 MAXPIVOT = ABS(A(K,K))
9050 NPIVROW(K,1)=K: NPIVROW(K,2)=K
9060 NPIVCOL(K,1)=K: NPIVCOL(K,2)=K
9070 FOR I = K TO N
9080 FOR J = K TO N
9090 IF MAXPIVOT > = ABS(A(I,J)) GOTO 9130
9100 MAXPIVOT=ABS(A(I,J))
9110 NPIVROW(K,1)=K: NPIVROW(K,2)=I
9120 NPIVCOL(K,1)=K: NPIVCOL(K,2)=J
```

```
9130 NEXT J:NEXT I
9140 IF MAXPIVOT > = EPS GOTO 9160
9150 GOTO 9420
9160 IF NPIVROW(K,2)=K GOTO 9230
9170 IF Q2$="Y" OR Q2$="y" THEN PRINT"PIVOTING ROWS:"
9180 IF Q2$="Y" OR Q2$="y" THEN PRINT "INTERCHANGE ROWS ";NPIVROW(K,2);" AND ";K

9190 FOR J = K TO NC
9200 SWAP  A(NPIVROW(K,2),J),A(K,J)
9210 NEXT J
9220 IF Q2$="Y" OR Q2$="y" THEN GOSUB 2000 'Print matrix
9230 IF NPIVCOL(K,2)=K GOTO 9300
9240 IF Q2$="Y" OR Q2$="y" THEN PRINT"PIVOTING COLUMNS:"
9250 IF Q2$="Y" OR Q2$="y" THEN PRINT "INTERCHANGE COLUMNS ";NPIVCOL(K,2);" AND
";K
9260 FOR I = 1 TO N
9270 SWAP  A(I,NPIVCOL(K,2)),A(I,K)
9280 NEXT I
9290 IF Q2$="Y" OR Q2$="y" THEN GOSUB 2000 'Print matrix
9300 IF Q2$="Y" OR Q2$="y" THEN PRINT "PERFORM ELIMINATION:"
9310 FOR I = K+1 TO N
9320 IF Q2$="Y" OR Q2$="y" THEN PRINT "DIVIDE ROW ";K;" BY ";A(K,K)
9330 IF Q2$="Y" OR Q2$="y" THEN PRINT "MULTIPLY ROW ";K;" BY ";A(I,K) ;"AND SUBT
RACT FROM ROW ";I
9340 MULT = - A(I,K)/A(K,K)
9350 FOR J = NC TO K STEP -1
9360 A(I,J) = A(I,J) + MULT * A(K,J)
9370 NEXT J
9380 IF Q2$="Y" OR Q2$="y" THEN GOSUB 2000 'Print matrix
9390 NEXT I: NEXT K
9400 '
9410 'Apply the back-substitution formulas
9420 RANK=K-1 :PRINT"RANK =";RANK:NMR=N-RANK
9430 PRINT"THE PROGRAM SETS ";NMR;" ELEMENT(S) OF THE EIGENVECTOR TO UNITY,"
9440 PRINT"AND REDUCES THE PROBLEM TO FINDING ";RANK;" ELEMENT(S)."
9450 FOR JJ=1 TO NMR : X(N+1-JJ) = 1: NEXT JJ
9460 FOR I = RANK TO 1 STEP -1
9470 SUM = 0
9480 FOR J = I+1 TO N
9490 SUM = SUM + A(I,J) * X(J)
9500 NEXT J
9510 X(I) = (A(I,NC) - SUM) / A(I,I)
9520 IF ABS(X(I)) < EPS THEN X(I)=0
9530 NEXT I
9540 '
9550 'Interchange the order of the unknowns to correct for the column pivoting
9560 FOR K=N-1 TO 1 STEP -1
9570 SWAP X(NPIVCOL(K,2)), X(NPIVCOL(K,1))
9580 NEXT K
9590 '
9600 PRINT"**************************************************************"
9610 PRINT
9620 PRINT "EIGENVECTOR CORRESPONDING TO EIGENVALUE =";XW(KK): PRINT
9630 FOR J = 1 TO N
9640 PRINT "X(";J;") =";X(J)
9650 NEXT J
9660 PRINT
9670 PRINT"**************************************************************"
9680 PRINT:PRINT"VERIFICATION:"
```

```
9690 PRINT:PRINT"A*X              =    EIGENVALUE*X":PRINT
9700 FOR I=1 TO N
9710 AX=0
9720 FOR J=1 TO N
9730 AX=AX + B(I,J)*X(J)
9740 NEXT J
9750 EX=XW(KK)*X(I)
9760 PRINT AX TAB(20) EX
9770 NEXT I
9780 PRINT
9790 PRINT"********************************************************************"
9800 PRINT:INPUT"PRESS THE ENTER KEY TO CONTINUE",KK$
9810 PRINT CHR$(12)
9820 RETURN
```

RESULTS

```
*******************************************************************
*                                                                 *
*                          EXAMPLE 3-5                            *
*                                                                 *
*   CALCULATION OF EIGENVALUES AND EIGENVECTORS USING ELEMENTARY  *
*                                                                 *
* SIMILARITY TRANSFORMATIONS TO CONVERT MATRIX TO HESSENBERG FORM *
*                                                                 *
*           AND THE QR ALGORITHM WITH PLANE ROTATIONS             *
*                                                                 *
*                           (QR.BAS)                              *
*                                                                 *
*******************************************************************
ENTER THE ORDER OF THE MATRIX? 4
ENTER ELEMENTS OF MATRIX:

ROW  1

   ELEMENT ( 1 , 1 ) =? 1
   ELEMENT ( 1 , 2 ) =? -0.5484
   ELEMENT ( 1 , 3 ) =? -0.8134
   ELEMENT ( 1 , 4 ) =? -0.7757

ROW  2

   ELEMENT ( 2 , 1 ) =? -0.5484
   ELEMENT ( 2 , 2 ) =? 1
   ELEMENT ( 2 , 3 ) =? 0.4339
   ELEMENT ( 2 , 4 ) =? 0.5557

ROW  3

   ELEMENT ( 3 , 1 ) =? -0.8134
   ELEMENT ( 3 , 2 ) =? 0.4339
   ELEMENT ( 3 , 3 ) =? 1
   ELEMENT ( 3 , 4 ) =? 0.9791

ROW  4

   ELEMENT ( 4 , 1 ) =? -0.7757
   ELEMENT ( 4 , 2 ) =? 0.5557
   ELEMENT ( 4 , 3 ) =? 0.9791
   ELEMENT ( 4 , 4 ) =? 1
*******************************************************************
ORIGINAL MATRIX:

 1              -.5484          -.8134          -.7757

-.5484           1               .4339           .5557

-.8134           .4339          1                .9791

-.7757           .5557           .9791          1

*******************************************************************
IS THE MATRIX CORRECT(Y/N)? Y
```

```
************* ELEMENTARY SIMILARITY TRANSFORMATION ****************

GIVE THE MINIMUM ALLOWABLE VALUE OF THE PIVOT ELEMENT FOR THE
ELEMENTARY SIMILARITY TRANSFORMATION ? 0.00001

DO YOU WANT TO SEE STEP-BY-STEP RESULTS OF THE
ELEMENTARY SIMILARITY TRANSFORMATION(Y/N)? N

******************************************************************
******************************************************************
MATRIX REDUCED TO HESSENBERG FORM:

 1            -1.922882     -1.183554     -.5484

-.8134         2.226258      1.301798      .4339

 0             .1843339      .1718212      .1419107

 0             0             .2939481      .6019204

******************************************************************

PRESS THE ENTER KEY TO CONTINUE

************* SUCCESSIVE FACTORIZATION PROCEDURE *****************

GIVE THE CONVERGENCE CRITERION FOR THE
SUCCESSIVE FACTORIZATION PROCEDURE? 0.00001


DO YOU WANT TO SEE STEP-BY-STEP RESULTS OF THE
SUCCESSIVE FACTORIZATION PROCEDURE(Y/N)? N

CALCULATING RESULTS.  PLEASE WAIT.


THE  4  EIGENVALUES ARE:

    .6476151
    5.024747E-03
    .2563368
    3.091024

VERIFICATION:

THE SUM OF THE EIGENVALUES =  4
THE TRACE OF THE MATRIX    =  4
THE DIFFERENCE IS          =  0

******************************************************************

PRESS THE ENTER KEY TO CONTINUE

*****************  GAUSS ELIMINATION METHOD  ********************

GIVE THE MINIMUM ALLOWABLE VALUE OF THE PIVOT ELEMENT FOR THE
GAUSS ELIMINATION PROCEDURE ? 0.00001

DO YOU WANT TO SEE STEP-BY-STEP RESULTS OF THE
GAUSS ELIMINATION PROCEDURE(Y/N)? N
```

```
*******************************************************************
HOMOGENEOUS SET: (A - EIGENVALUE*I)*X = 0, FOR EIGENVALUE = .6476151

 .3523849     -.5484         -.8134         -.7757          0


-.5484         .3523849      .4339          .5557          0


-.8134         .4339         .3523849       .9791          0


-.7757         .5557         .9791          .3523849       0

*******************************************************************
RANK = 3
THE PROGRAM SETS  1  ELEMENT(S) OF THE EIGENVECTOR TO UNITY,
AND REDUCES THE PROBLEM TO FINDING  3  ELEMENT(S).
*******************************************************************


EIGENVECTOR CORRESPONDING TO EIGENVALUE = .6476151

X( 1 ) = 7.700096E-02
X( 2 ) = 1
X( 3 ) =-.425155
X( 4 ) =-.2261763


*******************************************************************

VERIFICATION:

A*X                =   EIGENVALUE*X

 4.986699E-02         4.986698E-02
 .6476118             .6476151
-.2753368            -.2753368
-.1464752            -.1464752

*******************************************************************

PRESS THE ENTER KEY TO CONTINUE


*******************************************************************
HOMOGENEOUS SET: (A - EIGENVALUE*I)*X = 0, FOR EIGENVALUE = 5.024747E-03

 .9949752     -.5484         -.8134         -.7757          0


-.5484         .9949752      .4339          .5557          0


-.8134         .4339         .9949752       .9791          0


-.7757         .5557         .9791          .9949752       0
```

```
*******************************************************************
RANK = 3
THE PROGRAM SETS  1  ELEMENT(S) OF THE EIGENVECTOR TO UNITY,
AND REDUCES THE PROBLEM TO FINDING  3  ELEMENT(S).
*******************************************************************


EIGENVECTOR CORRESPONDING TO EIGENVALUE = 5.024747E-03

X( 1 ) = .1862585
X( 2 ) = .1962924
X( 3 ) = 1
X( 4 ) =-.9484648

*******************************************************************

VERIFICATION:

A*X              =    EIGENVALUE*X

 9.358525E-04         9.359017E-04
 9.863377E-04         9.863194E-04
 5.026818E-03         5.024747E-03
-4.765809E-03        -4.765795E-03

*******************************************************************

PRESS THE ENTER KEY TO CONTINUE

*******************************************************************
HOMOGENEOUS SET: (A - EIGENVALUE*I)*X = 0, FOR EIGENVALUE = .2563368

 .7436633     -.5484          -.8134          -.7757          0


-.5484         .7436633        .4339           .5557          0


-.8134         .4339           .7436633        .9791          0


-.7757         .5557           .9791           .7436633       0

*******************************************************************
RANK = 3
THE PROGRAM SETS  1  ELEMENT(S) OF THE EIGENVECTOR TO UNITY,
AND REDUCES THE PROBLEM TO FINDING  3  ELEMENT(S).
*******************************************************************


EIGENVECTOR CORRESPONDING TO EIGENVALUE = .2563368

X( 1 ) = 1
X( 2 ) = .1675103
X( 3 ) = .289702
X( 4 ) = .5364892
```

```
*******************************************************************
VERIFICATION:

A*X                =   EIGENVALUE*X

 .2563391              .2563368
 4.293907E-02          4.293905E-02
 7.426131E-02          7.426127E-02
 .1375219              .1375219
*******************************************************************
PRESS THE ENTER KEY TO CONTINUE

*******************************************************************
HOMOGENEOUS SET: (A - EIGENVALUE*I)*X = 0, FOR EIGENVALUE = 3.091024

-2.091024      -.5484         -.8134         -.7757          0

-.5484         -2.091024       .4339          .5557          0

-.8134          .4339         -2.091024       .9791          0

-.7757          .5557          .9791         -2.091024       0

*******************************************************************
RANK = 3
THE PROGRAM SETS  1  ELEMENT(S) OF THE EIGENVECTOR TO UNITY,
AND REDUCES THE PROBLEM TO FINDING  3  ELEMENT(S).
*******************************************************************
EIGENVECTOR CORRESPONDING TO EIGENVALUE = 3.091024

X( 1 ) =-.941349
X( 2 ) = .7166419
X( 3 ) = .9831282
X( 4 ) = 1
*******************************************************************
VERIFICATION:

A*X                =   EIGENVALUE*X

-2.909732            -2.909732
 2.215157             2.215157
 3.038873             3.038872
 3.091023             3.091024
*******************************************************************

PRESS THE ENTER KEY TO CONTINUE

*******************************************************************

DO YOU WANT TO REPEAT THE CALCULATIONS
WITH MINOR CHANGES TO THE COEFFICIENTS(Y/N)? N

DO YOU WANT TO RESET ALL THE COEFFICIENTS(Y/N)? N


*********************** END OF PROGRAM ****************************
Ok
```

DISCUSSION OF RESULTS It should be noted that the correlation coefficient matrix of this problem is a symmetric matrix. The QR algorithm works equally well with symmetric and nonsymmetric matrices.

The elementary similarity transformation reduces the correlation matrix to the upper Hessenberg matrix, as expected. The successive factorization procedure obtains the following four eigenvalues for the correlation matrix:

$$0.6476$$

$$0.0050$$

$$0.2563$$

$$3.0910$$

and verifies that their sum is equal to the trace of the matrix.

The eigenvectors, corresponding to each eigenvalue, are calculated by the Gauss elimination method as shown on the printout.

It is discussed in Chap. 7 that if the estimated parameters are completely uncorrelated with each other, the correlation coefficient matrix is the identity matrix. Therefore, in such a case, all the eigenvalues would be unity and all the eigenvectors would be unit vectors. Such a matrix defines a spherical hyperspace. This situation, however, is rarely possible in nonlinear parameter estimation.

The results obtained in this problem are typical of a real physical situation where the experimental data and the mathematical models are such that the extent of correlation between parameter estimates is considerable. These four eigenvalues and four eigenvectors can be used to construct (conceptually) a fourth-dimensional hyperellipse which describes the shape of the confidence region of the four parameter estimates. Because of the unequal magnitude of the eigenvalues, the axes of the hyperellipse would vary quite drastically in length, i.e., the hyperellipse will be elongated.

## PROBLEMS

**3-1** When a pure sample of gas is bombarded by low-energy electrons in a mass spectrometer the galvanometer shows peak heights that correspond to individual $m/e$ (mass-to-charge) ratios for the resulting mixture of ions. For the $i$th peak produced by a pure sample $j$, one can then assign a sensitivity $S_{ij}$ [peak height per micron ($\mu$m) of Hg sample pressure]. These coefficients are unique for each type of gas.

A distribution of peak heights may also be obtained for an $n$-component gas mixture that is to be analyzed for the partial pressures $p_1, p_2, \ldots, p_n$ of each of its constituents. The height $h_i$ of a certain peak is a linear combination of the products of the individual sensitivities and partial pressures:

$$\sum_{j=1}^{n} S_{ij} p_j = h_i$$

In general, more than $n$ peaks may be available. However, if the $n$ most distinct ones are chosen we have $i = 1, 2, \ldots, n$, so that the individual partial pressures are given by the solution of $n$ simultaneous linear equations.

Write a program that will accept values for the number of components $N$, the sensitivities $S(1,1) \cdots S(N,N)$, and the peak heights $H(1) \cdots H(N)$. The program should then compute and print values for the individual partial pressures $P(1) \cdots P(N)$. The Gauss-Seidel method should be used in the computations.

*Suggested test data* The sensitivities given in Table P3-1 were reported by Carnahan [see Ref. 14] in connection with the analysis of a hydrogen gas mixture.

A particular gas mixture produced the following peak heights: $h_1 = 17.1$, $h_2 = 65.1$, $h_3 = 186.0$, $h_4 = 82.7$, $h_5 = 84.2$, $h_6 = 63.7$, and $h_7 = 119.7$. The measured total pressure of the mixture was 38.78 μm of Hg, which can be compared with the sum of the computed partial pressures.

**3-2** Aniline is being removed from water by solvent extraction using toluene [15]. The unit is a 10-stage countercurrent tower, shown in Fig. P3-2. The equilibrium relationship valid at each stage is, to a first approximation,

$$m = \frac{Y_i}{X_i} = 9$$

where $Y_i$ = (lb of aniline in the toluene phase)/(lb of toluene in the toluene phase)
$X_i$ = (lb of aniline in the water phase)/(lb of water in the water phase)

(*a*) The solution to this problem is a set of 10 simultaneous equations. Derive these equations from material balances around each stage. Present these equations using compact notation.

(*b*) Solve the above set of equations to find the concentration in both the aqueous and organic phases leaving each stage of the system ($X_i$ and $Y_i$).

(*c*) If the slope of the equilibrium relationship is replaced by the expression $m = 9 + 20X_i$, the solution becomes a set of simultaneous nonlinear equations. Describe a procedure which would solve this problem.

(*d*) Solve the problem described in *c* above.

**Table P3-1**

| Peak index, $i$ | $m/e$ | Component index, $j$ | | | | | | |
|---|---|---|---|---|---|---|---|---|
| | | 1 Hydrogen | 2 Methane | 3 Ethylene | 4 Ethane | 5 Propylene | 6 Propane | 7 $n$-Pentane |
| 1 | 2 | 16.87 | 0.1650 | 0.2019 | 0.3170 | 0.2340 | 0.1820 | 0.1100 |
| 2 | 16 | 0.0 | 27.70 | 0.8620 | 0.0620 | 0.0730 | 0.1310 | 0.1200 |
| 3 | 26 | 0.0 | 0.0 | 22.35 | 13.05 | 4.420 | 6.001 | 3.043 |
| 4 | 30 | 0.0 | 0.0 | 0.0 | 11.28 | 0.0 | 1.110 | 0.3710 |
| 5 | 40 | 0.0 | 0.0 | 0.0 | 0.0 | 9.850 | 1.1684 | 2.108 |
| 6 | 44 | 0.0 | 0.0 | 0.0 | 0.0 | 0.2990 | 15.98 | 2.107 |
| 7 | 72 | 0.0 | 0.0 | 0.0 | 0.0 | 0.0 | 0.0 | 4.670 |

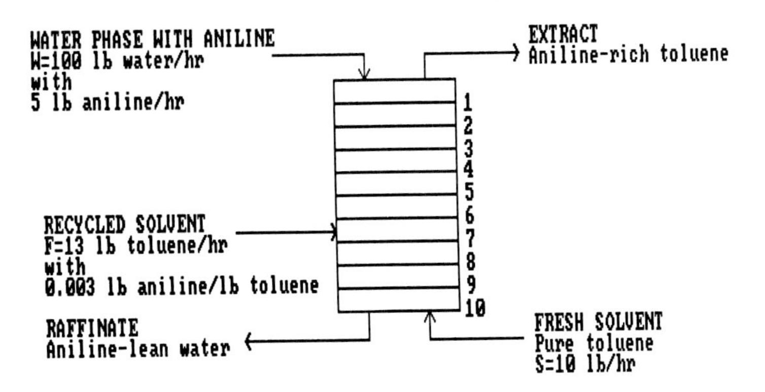


**Figure P3-2**

**3-3** In the study of chemical reaction, Aris [16] developed a technique of writing simultaneous chemical reactions in the form of linear algebraic equations. For example, the following two simultaneous chemical equations

$$C_2H_6 \rightleftarrows C_2H_4 + H_2$$

$$2C_2H_6 \rightleftarrows C_2H_4 + 2CH_4$$

can be rearranged in the form

$$C_2H_4 + H_2 - C_2H_6 = 0$$

$$C_2H_4 + 2CH_4 - 2C_2H_6 = 0$$

If we identify $A_1$ with $C_2H_4$, $A_2$ with $H_2$, $A_3$ with $CH_4$, and $A_4$ with $C_2H_6$, the set of equations becomes

$$A_1 + A_2 - A_4 = 0$$

$$A_1 + 2A_3 - 2A_4 = 0$$

This can be generalized to a system of $R$ reactions between $S$ chemical species by the set of equations represented by

$$\sum_{j=1}^{S} \alpha_{ij} A_j = 0 \qquad i = 1, 2, \ldots, R$$

where $\alpha_{ij}$ are the stoichiometric coefficients of each species $A_j$ in each reaction $i$.

Aris demonstrated that the number of independent chemical reactions in a set of $R$ reactions is equal to the rank of the matrix of stoichiometric coefficients $\alpha_{ij}$. Using Aris' method, and the techniques developed in this chapter, determine the number of independent chemical reactions in the following reaction system:

$$4NH_3 + 5O_2 \rightleftarrows 4NO + 6H_2O$$

$$4NH_3 + 3O_2 \rightleftarrows 2N_2 + 6H_2O$$

$$4NH_3 + 6NO \rightleftarrows 5N_2 + 6H_2O$$

$$2NO + O_2 \rightleftarrows 2NO_2$$

$$2NO \rightleftarrows N_2 + O_2$$

$$N_2 + 2O_2 \rightleftarrows 2NO_2$$

**3-4** The multistage distillation tower shown in Fig. P3-4 is equipped with a total condenser and a partial reboiler. This tower is to be used for the separation of a multicomponent mixture. Assume that for this particular mixture the tower contains the equivalent of 10 equilibrium stages including the reboiler; that is, $N = 10$ and $j = 11$.

The feed to the column has a flow rate $F = 1000$ mol/h. It is a saturated liquid and it enters the column on equilibrium stage 5 ($j = 6$). It contains five components ($n = 5$) whose mole

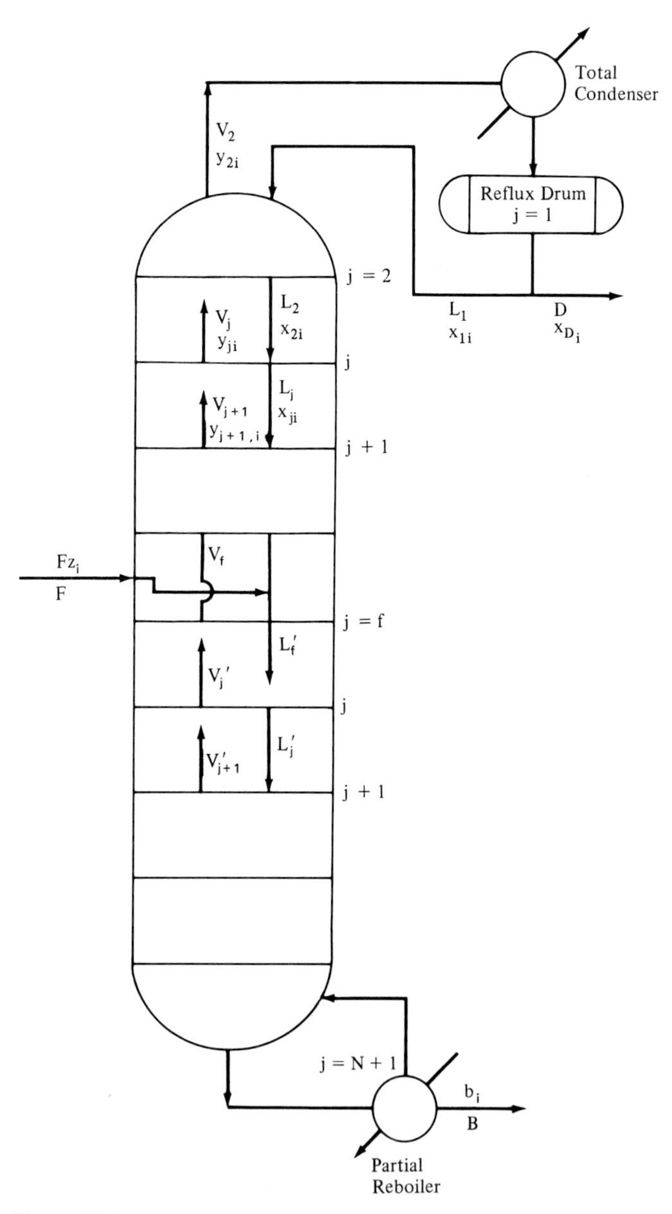


**Figure P3-4**

fractions are

$$z_1 = 0.06 \qquad z_2 = 0.17 \qquad z_3 = 0.22 \qquad z_4 = 0.20 \qquad z_5 = 0.35$$

It is required to recover a distillate product at a rate of 500 mol/h.

Develop all the material balances for component $i$ for all 10 equilibrium stages and for the condenser. For this problem make the following assumptions:

1. The external reflux ratio is

$$\frac{L_1}{D} = 2.5$$

2. Constant molal overflow occurs in each section of the tower.
3. The initial guesses of the temperatures corresponding to the equilibrium stages are $T_2 = 140°$, $T_3 = 150°F$, $T_4 = 160°F$, $T_5 = 170°F$, $T_6 = 180°F$, $T_7 = 190°F$, $T_8 = 200°F$, $T_9 = 210°F$, $T_{10} = 220°F$, and $T_{11} = 230°F$.
4. The equilibrium constants $K_{j,i}$ can be approximated by the following equation:

$$K_{j,i} = \alpha_i + \beta_i T_j + \gamma_i T_j^2$$

where the temperatures are in degrees Fahrenheit and the coefficients for each individual component are listed in Table P3-4 [17].

Solve the resulting set of equations in order to determine the following:

(*a*) The molal flow rates of all vapor and liquid streams in the tower.

(*b*) The mole fraction of each component in the vapor and liquid streams. Note that the mole fractions in each stage do not add up to unity, because the above solution is only a single step in the solution of multicomponent distillation problems. Assumptions 2 and 3 are only initial guesses which must be subsequently corrected from energy balances and bubble-point calculations.

**3-5** The following equations can be shown to relate the temperatures and pressures on either side of a detonation wave that is moving into a zone of unburned gas [14]:

$$\frac{\gamma_2 m_2 T_1}{m_1 T_2}\left(\frac{P_2}{P_1}\right)^2 - (\gamma_2 + 1)\frac{P_2}{P_1} + 1 = 0$$

$$\frac{\Delta H_{R1}}{c_{p2} T_1} + \frac{T_2}{T_1} - 1 = \frac{(\gamma_2 - 1) m_2}{2\gamma_2 m_1}\left(\frac{P_2}{P_1} - 1\right)\left(1 + \frac{m_1 T_2 P_1}{m_2 T_1 P_2}\right)$$

Here, $T$ = absolute temperature, $P$ = absolute pressure, $\gamma$ = ratio of specific heat at constant pressure to that at constant volume, $m$ = mean molecular weight, $\Delta H_{R1}$ = heat of reaction, $c_p$ = specific heat, and the subscripts 1 and 2 refer to the unburned and burned gas, respectively.

Write a program that will accept values for $m_1$, $m_2$, $\gamma_2$, $\Delta H_{R1}$, $c_{p2}$, $T_1$, and $P_1$ as data, and

**Table P3-4**

| Component $i$ | $\alpha_i$ | $\beta_i$ | $\gamma_i$ |
|---|---|---|---|
| 1 | 0.70 | $0.30 \times 10^{-2}$ | $0.60 \times 10^{-4}$ |
| 2 | 2.21 | $1.95 \times 10^{-2}$ | $0.90 \times 10^{-4}$ |
| 3 | 1.50 | $-1.60 \times 10^{-2}$ | $0.80 \times 10^{-4}$ |
| 4 | 0.86 | $-0.97 \times 10^{-2}$ | $0.46 \times 10^{-4}$ |
| 5 | 0.71 | $-0.87 \times 10^{-2}$ | $0.42 \times 10^{-4}$ |

that will proceed to compute and print values for $T_2$ and $P_2$. Run the program with the following data, which apply to the detonation of a mixture of hydrogen and oxygen. $m_1 = 12$ g/g mol, $m_2 = 18$ g/g mol, $\gamma_2 = 1.31$, $\Delta H_{R1} = -58{,}300$ cal/g mol, $c_{p2} = 9.86$ cal/(g mol · K), $T_1 = 300$ K, and $P_1 = 1$ atm.

**3-6** The system of highly coupled chemical reactions shown in Fig. P3-6 takes place in a batch reactor. The conditions of temperature and pressure in the reactor are such that the kinetic rate constants attain the following values:

$$k_{21} = 0.2 \qquad k_{34} = 0.1$$
$$k_{12} = 0.1 \qquad k_{54} = 0.05$$
$$k_{31} = 0.1 \qquad k_{45} = 0.1$$
$$k_{13} = 0.05 \qquad k_{64} = 0.2$$
$$k_{32} = 0.1 \qquad k_{46} = 0.2$$
$$k_{23} = 0.05 \qquad k_{65} = 0.1$$
$$k_{43} = 0.2 \qquad k_{56} = 0.1$$

If the chemical reaction starts with the following initial concentrations

$$A_0 = 1.0 \text{ mol/L} \qquad D_0 = 0$$
$$B_0 = 0 \qquad E_0 = 1.0 \text{ mol/L}$$
$$C_0 = 0 \qquad F_0 = 0$$

calculate the steady-state concentration of all components. Assume that all reactions are of first order.

**3-7** A linear mathematical model which has three independent variables, $X_1$, $X_2$, and $X_3$, may be written as

$$Y = b_1X_1 + b_2X_2 + b_3X_3$$

where $b_1$, $b_2$, and $b_3$ are parameter constants to be determined from experimental observations. It can be shown that the vector of parameters **b** may be calculated from

$$\mathbf{b} = (\mathbf{X}^T\mathbf{X})^{-1}\mathbf{X}^T\mathbf{Y}$$

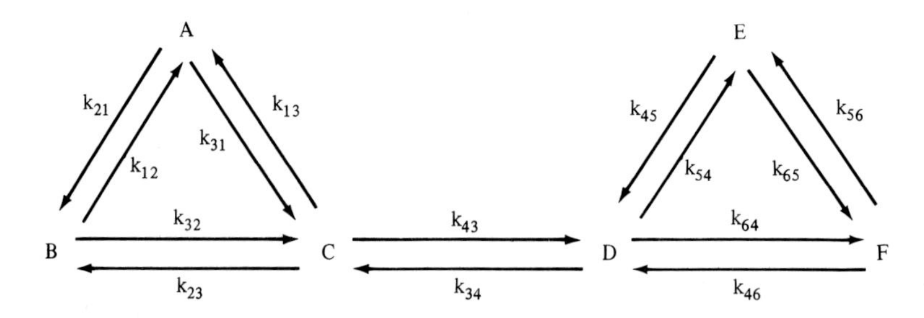


**Figure P3-6** System of coupled chemical reactions.

**Table P3-7**

| $X_1$ | $X_2$ | $X_3$ | $Y$ |
|---|---|---|---|
| 1 | 0.2 | 5.0 | 1.0 |
| 2 | 0.6 | 4.1 | 5.0 |
| 3 | 0.7 | 3.0 | 7.0 |
| 4 | 1.0 | 2.0 | 10.0 |
| 5 | 1.5 | 1.2 | 12.5 |
| 6 | 2.0 | 0.5 | 15.0 |

where $\mathbf{X}$ is the matrix which contains the vectors of independent variable observations, $\mathbf{X}_1$, $\mathbf{X}_2$, $\mathbf{X}_3$, as columns

$$\mathbf{X} = [\mathbf{X}_1 \quad \mathbf{X}_2 \quad \mathbf{X}_3]$$

and $\mathbf{Y}$ is the vector of dependent variable observations (see Chap. 7).

Using the experimental observations shown in Table P3-7 determine the values of the parameters $b_1$, $b_2$, and $b_3$ for this linear model.

## REFERENCES


1. Holland, C. D., and Liapis, A. I.: *Computer Methods for Solving Dynamic Separation Problems*, McGraw-Hill Book Co., New York, 1983.
2. Amundson, N. R.: *Mathematical Methods in Chemical Engineering, Matrices and Their Application*, Prentice-Hall, Inc., Englewood Cliffs, N.J., 1966, p. 225.
3. Wei, J., and Prater, C. D.: "Analysis of Complex Reaction Systems," in *Advances in Catalysis*, vol. 13, Academic Press, New York, 1962, p. 203.
4. Constantinides, A., and Shao, P.: "Material Balancing Applied to the Prediction of Glutamic Acid Production and Cell Mass Formation," in A. Constantinides, W. R. Vieth, and K. Venkatasubramanian (eds.), *Biochemical Engineering II*, New York Academy of Sciences Annals, vol. 369, 1981, p. 167.
5. Lapidus, L.: *Digital Computation for Chemical Engineers*, McGraw-Hill Book Co., New York, 1962, pp. 199–201.
6. Hildebrand, F. B.: *Methods of Applied Mathematics*, 2d ed., Prentice-Hall, Inc., Englewood Cliffs, N.J., 1965, p. 11.
7. James, M. L., Smith, G. M., and Wolford, J. C.: *Applied Numerical Methods for Digital Computation with FORTRAN and CSMP*, 2d ed., Harper & Row Publishers Incorporated, New York, 1977, p. 174.
8. Ralston, A., and Rabinowitz, P.: *A First Course in Numerical Analysis*, 2d ed., McGraw-Hill Book Co., New York, 1978.
9. Himmelblau, D. M.: *Basic Principles and Calculations in Chemical Engineering*, 4th ed., Prentice-Hall, Inc., Englewood Cliffs, N.J., 1982, p. 529.
10. Faddeev, D. K., and Faddeeva, U. N.: *Computational Methods in Linear Algebra*, translated by R. C. Williams, W. H. Freeman and Company, New York, 1963.
11. Johnson, L. W., and Riess, R. D.: *Numerical Analysis*, 2d ed., Addison-Wesley Publishing Co., Inc., Reading, Mass., 1982.
12. Givens, M.: "Computation of Plane Unitary Rotation Transforming a General Matrix to Triangular Form," *J. Soc. Ind. Appl. Math.*, vol. 6, 1958, pp. 26–50.

13. Constantinides, A., Spencer, J. L., and Gaden E. L. Jr.: "Optimization of Batch Fermentation Processes, I. Development of Mathematical Models for Batch Penicillin Fermentations," *Biotech. and Bioeng.*, vol. 12, 1970, p. 803.
14. Carnahan, B., Luther, H. A., and Wilkes, J. O.: *Applied Numerical Methods*, John Wiley & Sons, Inc., New York, 1969, p. 331.
15. Freeman, R.: Private communication, Department of Chemical and Biochemical Engineering, Rutgers University, New Brunswick, N.J., 1984.
16. Aris, R.: *Introduction to the Analysis of Chemical Reactors*, Prentice-Hall, Inc., Englewood Cliffs, N.J., 1965, p. 12.
17. Chang, H-Y., and Over, I. E.: *Selected Numerical Methods and Computer Programs for Chemical Engineers*, Sterling Swift Publishing Co., Manchaca, Tex, 1981, p. 93.

CHAPTER
# FOUR

# FINITE DIFFERENCE METHODS

## 4-1 INTRODUCTION

The most commonly encountered mathematical models in engineering and science are in the form of differential equations. The dynamics of physical systems that have one independent variable can be modeled by ordinary differential equations, while systems with two, or more, independent variables require the use of partial differential equations. Several types of ordinary differential equations, and a few partial differential equations, render themselves to analytical (closed form) solutions. These methods have been developed thoroughly in differential calculus. However, the great majority of differential equations, especially the nonlinear ones and those which involve large sets of simultaneous differential equations, do not have analytical solutions but require the application of numerical techniques for their solution.

Several numerical methods for the solution of ordinary and partial differential equations are discussed in Chaps. 5 and 6 of this book. These methods are based on the concept of *finite differences*. Therefore, the purpose of this chapter is to develop the systematic terminology used in the *calculus of finite differences* and to derive the relationships between finite differences and differential operators, which are needed in the numerical solution of ordinary and partial differential equations.

The calculus of finite differences may be characterized as a "two-way street" which enables the user to take a differential equation and integrate it numerically by calculating the values of the function at a discrete (finite) number of points. Or, conversely, if a set of finite values is available, such as experimental data, these may be differentiated, or integrated, using the calculus of finite differences. It has been pointed out, however, that numerical differentiation is inherently less accurate than numerical integration [1].

Another very useful application of the calculus of finite differences is in the derivation of interpolation/extrapolation formulas, the so-called *interpolating polynomials*, which can be used to represent experimental data when the actual functionality of these data is not well known. The discussion of several interpolating polynomials is given in Sec. 4-7.

## 4-2 SYMBOLIC OPERATORS

In differential calculus, the definition of the derivative is given as

$$\left.\frac{df(x)}{dx}\right|_{x_0} = f'(x_0) = \lim_{x \to x_0} \frac{f(x) - f(x_0)}{x - x_0} \tag{4-1}$$

In the calculus of finite differences, the value of $x - x_0$ does not approach zero but remains a finite quantity. If we represent this quantity by $h$

$$h = x - x_0 \tag{4-2}$$

then the derivative may be *approximated* by

$$f'(x_0) \cong \frac{f(x) - f(x_0)}{h} \tag{4-3}$$

Under certain circumstances, there is a point, $\xi$, in the interval $(a, b)$ for which the derivative can be calculated *exactly* from Eq. (4-3). This is confirmed by the mean-value theorem of differential calculus [2]:

**Mean-value theorem** Let $f(x)$ be continuous for $a \le x \le b$ and differentiable for $a < x < b$; then there exists at least one $\xi$, $a < \xi < b$, for which

$$f'(\xi) = \frac{f(b) - f(a)}{b - a} \tag{4-4}$$

This theorem forms the basis for both the differential calculus and the finite difference calculus.

A function $f(x)$, which is continuous and differentiable in the interval $[x_0, x]$, can be represented by a Taylor series

$$f(x) = f(x_0) + (x - x_0)f'(x_0) + \frac{(x - x_0)^2 f''(x_0)}{2!} + \frac{(x - x_0)^3 f'''(x_0)}{3!} + \cdots + \frac{(x - x_0)^n f^n(x_0)}{n!} + R_n(x) \tag{4-5}$$

where $R_n(x)$ is called the *remainder*. This term lumps together the remaining terms in the infinite series from $n + 1$ to infinity; it, therefore, represents the *truncation* error, when the function is evaluated using the terms up to, and including, the $n$th-order term of the infinite series.

The mean-value theorem can be used to show that there exists a point $\xi$ in the interval $(x_0, x)$ so that the remainder term is given by

$$R_n(x) = \frac{(x - x_0)^{n+1} f^{(n+1)}(\xi)}{(n+1)!} \tag{4-6}$$

The value of $\xi$ is an unknown function of $x$; therefore, it is impossible to evaluate the remainder, or truncation error, term exactly [2]. The remainder is a term of order $(n + 1)$, since it is a function of $(x - x_0)^{n+1}$ and of the $(n + 1)$st derivative. For this reason, in our discussion of truncation errors we will always specify the order of the remainder term, and will usually abbreviate it using the notation $O(h^{n+1})$.

The calculus of finite differences is used in conjunction with a series of discrete values, which can be either experimental data, such as

$$y_{i-3} \quad y_{i-2} \quad y_{i-1} \quad y_i \quad y_{i+1} \quad y_{i+2} \quad y_{i+3}$$

or discrete values of a continuous function $y(x)$:

$$y(x-3h) \quad y(x-2h) \quad y(x-h) \quad y(x) \quad y(x+h) \quad y(x+2h) \quad y(x+3h)$$

or, equivalently, values of a function $f(x)$:

$$f(x-3h) \quad f(x-2h) \quad f(x-h) \quad f(x) \quad f(x+h) \quad f(x+2h) \quad f(x+3h)$$

In all the above cases, the values of the dependent variable, $y$ or $f$, are those corresponding to *equally spaced* values of the independent variable $x$. This concept is demonstrated in Fig. 4-1 for a smooth function $y(x)$.

A set of *linear symbolic operators* drawn from differential calculus and from finite difference calculus will be defined in conjunction with the above

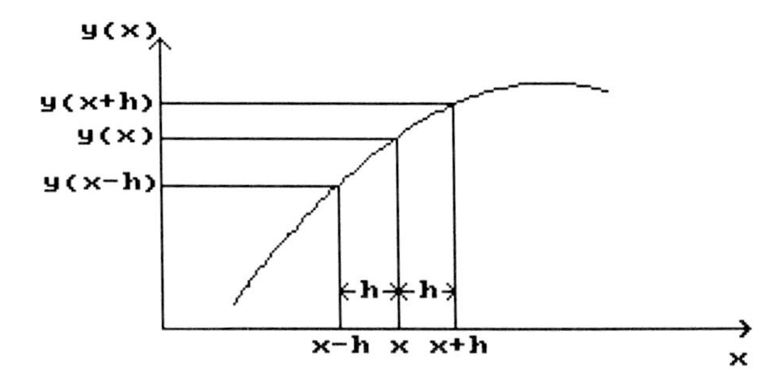


**Figure 4-1** Values of the function $y(x)$ at equally spaced points of the independent variable $x$.

series of discrete values. These definitions will then be used to derive the interrelationships between the operators. The linear symbolic operators are

$D$ = differential operator

$I$ = integral operator

$E$ = shift operator

$\Delta$ = forward difference operator

$\nabla$ = backward difference operator

$\delta$ = central difference operator

$\mu$ = averager operator

All these operators may be treated as algebraic variables since they satisfy the distributive, commutative, and associative laws of algebra [3].

The first two operators are well known from differential calculus. The *differential operator* $D$ has the following effect when applied to the function $y(x)$:

$$Dy(x) = \frac{dy(x)}{dx} = y'(x) \tag{4-7}$$

while the *integral operator* is

$$Iy(x) = \int_x^{x+h} y(x)\, dx \tag{4-8}$$

The integral operator is equivalent to the inverse of the differential operator

$$I = D^{-1} \tag{4-9}$$

The *shift operator* causes the function to shift to the next successive value of the independent variable:

$$Ey(x) = y(x + h) \tag{4-10}$$

The inverse of the shift operator $E^{-1}$ causes the function to shift in the negative direction of the independent variable:

$$E^{-1}y(x) = y(x - h) \tag{4-11}$$

Higher powers of the shift operator are defined as

$$E^n y(x) = y(x + nh) \tag{4-12}$$

The shift operator can be expressed in terms of the differential operator by expanding the function $y(x+h)$ into a Taylor series about $x$:

$$y(x+h) = y(x) + \frac{h}{1!}\, y'(x) + \frac{h^2}{2!}\, y''(x) + \frac{h^3}{3!}\, y'''(x) + \cdots \tag{4-13}$$

Using the differential operator $D$ to indicate the derivatives of $y$, we obtain

$$y(x+h) = y(x) + \frac{h}{1!}\, Dy(x) + \frac{h^2}{2!}\, D^2y(x) + \frac{h^3}{3!}\, D^3y(x) + \cdots \tag{4-14}$$

Factoring out the term $y(x)$ from the right-hand side of Eq. (4-14),

$$y(x+h) = \left(1 + \frac{h}{1!}\, D + \frac{h^2}{2!}\, D^2 + \frac{h^3}{3!}\, D^3 + \cdots\right) y(x) \tag{4-15}$$

The terms in the parentheses are equivalent to the series expansion

$$e^{hD} = 1 + \frac{hD}{1!} + \frac{h^2D^2}{2!} + \frac{h^3D^3}{3!} + \cdots \tag{4-16}$$

Therefore, Eq. (4-15) can be written as

$$y(x+h) = e^{hD}y(x) \tag{4-17}$$

Comparing Eq. (4-10) with (4-17), we conclude that the shift operator can be expressed in terms of the differential operator by the relation

$$E = e^{hD} \tag{4-18}$$

Similarly, the inverse of the shift operator can be related to the differential operator by expanding the function $y(x-h)$ into a Taylor series about $x$:

$$y(x-h) = y(x) - \frac{h}{1!}\, y'(x) + \frac{h^2}{2!}\, y''(x) - \frac{h^3}{3!}\, y'''(x) + \cdots \tag{4-19}$$

Replacing the derivatives with the differential operators and rearranging, we obtain

$$y(x-h) = \left(1 - \frac{h}{1!}\, D + \frac{h^2}{2!}\, D^2 - \frac{h^3}{3!}\, D^3 + \cdots\right) y(x) \tag{4-20}$$

The terms in the parentheses are equivalent to the series expansion

$$e^{-hD} = 1 - \frac{hD}{1!} + \frac{h^2D^2}{2!} - \frac{h^3D^3}{3!} + \cdots \tag{4-21}$$

Therefore, Eq. (4-19) can be written as

$$y(x-h) = e^{-hD}y(x) \tag{4-22}$$

It follows from a comparison of Eq. (4-11) with (4-22) that

$$E^{-1} = e^{-hD} \tag{4-23}$$

With these introductory concepts in mind, let us proceed to develop the backward, forward, and central difference operators and the relationships between these and the differential operators.

## 4-3 BACKWARD FINITE DIFFERENCES

Consider the set of values

$$y_{i-3} \qquad y_{i-2} \qquad y_{i-1} \qquad y_i \qquad y_{i+1} \qquad y_{i+2} \qquad y_{i+3}$$

or the equivalent set

$$y(x-3h) \quad y(x-2h) \quad y(x-h) \quad y(x) \quad y(x+h) \quad y(x+2h) \quad y(x+3h)$$

The *first backward difference* of $y$ at $i$ (or $x$) is defined as

$$\nabla y_i = y_i - y_{i-1}$$

or

$$\nabla y(x) = y(x) - y(x-h) \tag{4-24}$$

The *second backward difference* of $y$ at $i$ (or $x$) is defined as

$$\nabla^2 y_i = \nabla(\nabla y_i) = \nabla(y_i - y_{i-1}) = \nabla y_i - \nabla y_{i-1}$$

$$= (y_i - y_{i-1}) - (y_{i-1} - y_{i-2})$$

$$\nabla^2 y_i = y_i - 2y_{i-1} + y_{i-2}$$

or

$$\nabla^2 y(x) = y(x) - 2y(x-h) + y(x-2h) \tag{4-25}$$

The *third backward difference* of $y$ at $i$ defined as

$$\nabla^3 y_i = \nabla(\nabla^2 y_i) = \nabla(y_i - 2y_{i-1} + y_{i-2})$$

$$= \nabla y_i - 2\nabla y_{i-1} + \nabla y_{i-2}$$

$$= (y_i - y_{i-1}) - 2(y_{i-1} - y_{i-2}) + (y_{i-2} - y_{i-3})$$

$$\nabla^3 y_i = y_i - 3y_{i-1} + 3y_{i-2} - y_{i-3} \tag{4-26}$$

Higher-order backward differences are similarly derived:

$$\nabla^4 y_i = y_i - 4y_{i-1} + 6y_{i-2} - 4y_{i-3} + y_{i-4} \tag{4-27}$$

$$\nabla^5 y_i = y_i - 5y_{i-1} + 10y_{i-2} - 10y_{i-3} + 5y_{i-4} - y_{i-5} \tag{4-28}$$

$$\vdots$$

The coefficients of the terms in each of the above finite differences correspond to those of the binomial expansion $(a - b)^n$, where $n$ is the order of the finite difference. It should also be noted that the sum of the coefficients of this binomial expansion is always equal to zero. This can be used as a check to ensure that higher-order differences have been expanded correctly.

The relationship between backward difference operators and differential operators can now be established. Combine Eqs. (4-22) and (4-24) to obtain

$$\begin{aligned} \nabla y(x) &= y(x) - y(x-h) = y(x) - e^{-hD} y(x) \\ &= (1 - e^{-hD}) y(x) \end{aligned} \tag{4-29}$$

which shows that the backward difference operator is given by

$$\nabla = 1 - e^{-hD} \tag{4-30}$$

Using the infinite series expression of $e^{-hD}$ [Eq. (4-21)], Eq. (4-30) becomes

$$\nabla = hD - \frac{h^2 D^2}{2} + \frac{h^3 D^3}{6} - \cdots \tag{4-31}$$

The higher-order backward difference operator, $\nabla^2$, $\nabla^3 \ldots$, can be obtained by raising the first backward difference operator to higher powers†:

$$\nabla^2 = (1 - e^{-hD})^2 = (1 - 2e^{-hD} + e^{-2hD}) \tag{4-32}$$

$$\nabla^3 = (1 - e^{-hD})^3 = (1 - 3e^{-hD} + 3e^{-2hD} - e^{-3hD}) \tag{4-33}$$

$$\vdots$$

$$\nabla^n = (1 - e^{-hD})^n \tag{4-34}$$

Expansion of the exponential terms and rearrangement yields the following equations for the second and third backward difference operators:

$$\nabla^2 = h^2 D^2 - h^3 D^3 + \tfrac{7}{12} h^4 D^4 - \cdots \tag{4-35}$$

$$\nabla^3 = h^3 D^3 - \tfrac{3}{2} h^4 D^4 + \tfrac{5}{4} h^5 D^5 - \cdots \tag{4-36}$$

† These relationships can also be obtained by combining the definitions of the backward differences [Eqs. (4-25) and (4-26)] with the definition of the inverse shift operator [Eqs. (4-11) and (4-23)].

Equations (4-31), (4-35), and (4-36) express the backward difference operators in terms of infinite series of differential operators. In order to complete the set of relationships, equations that express the differential operators in terms of backward difference operators will also be derived [4]. To do so, first rearrange Eq. (4-30) to solve for $e^{-hD}$:

$$e^{-hD} = 1 - \nabla \tag{4-37}$$

Take the natural logarithm of both sides of this equation:

$$\ln e^{-hD} = -hD = \ln(1 - \nabla) \tag{4-38}$$

Utilize the infinite series expansion:

$$\ln(1 - \nabla) = -\nabla - \frac{\nabla^2}{2} - \frac{\nabla^3}{3} - \frac{\nabla^4}{4} - \frac{\nabla^5}{5} - \cdots \tag{4-39}$$

Combine (4-38) with (4-39) to obtain:

$$hD = \nabla + \frac{\nabla^2}{2} + \frac{\nabla^3}{3} + \frac{\nabla^4}{4} + \cdots \tag{4-40}$$

The higher-order differential operators can be obtained by simply raising both sides of Eq. (4-40) to higher powers:

$$h^2D^2 = \nabla^2 + \nabla^3 + \tfrac{11}{12}\nabla^4 + \tfrac{5}{6}\nabla^5 + \cdots \tag{4-41}$$

$$h^3D^3 = \nabla^3 + \tfrac{3}{2}\nabla^4 + \tfrac{7}{4}\nabla^5 + \cdots \tag{4-42}$$

$$\vdots$$

$$h^nD^n = \left(\nabla + \frac{\nabla^2}{2} + \frac{\nabla^3}{3} + \frac{\nabla^4}{4} + \cdots\right)^n \tag{4-43}$$

The complete set of relationships between backward difference operators and differential operators is summarized in Table 4-1. These equations enable us to develop a variety of formulas expressing derivatives of functions in terms of backward finite differences, and vice versa. In addition, these formulas may have any degree of accuracy desired, provided that a sufficient number of terms is retained in the manipulation of these infinite series. This concept will be demonstrated with several examples in the remainder of this section.

**Table 4-1 Backward finite differences**

| Backward difference operators | |
|---|---|
| $\nabla = hD - \frac{h^2D^2}{2} + \frac{h^3D^3}{6} - \cdots$ | (4-31) |
| $\nabla^2 = h^2D^2 - h^3D^3 + \frac{7}{12}h^4D^4 - \cdots$ | (4-35) |
| $\nabla^3 = h^3D^3 - \frac{3}{2}h^4D^4 + \frac{5}{4}h^5D^5 - \cdots$ | (4-36) |
| $\nabla^n = (1 - e^{-hD})^n$ | (4-34) |
| **Differential operators** | |
| $hD = \nabla + \frac{\nabla^2}{2} + \frac{\nabla^3}{3} + \frac{\nabla^4}{4} + \cdots$ | (4-40) |
| $h^2D^2 = \nabla^2 + \nabla^3 + \frac{11}{12}\nabla^4 + \frac{5}{6}\nabla^5 + \cdots$ | (4-41) |
| $h^3D^3 = \nabla^3 + \frac{3}{2}\nabla^4 + \frac{7}{4}\nabla^5 + \cdots$ | (4-42) |
| $h^nD^n = \left(\nabla + \frac{\nabla^2}{2} + \frac{\nabla^3}{3} + \frac{\nabla^4}{4} + \cdots\right)^n$ | (4-43) |

**Example 4-1** Express the first-order derivative of $y$ in terms of backward finite differences with error of order $h$.

SOLUTION Rearrange Eq. (4-31) to solve for the differential operator $D$:

$$D = \frac{1}{h}\nabla + \frac{hD^2}{2} - \frac{h^2D^3}{6} + \cdots$$

Apply this operator to the function $y$ at $i$:

$$Dy_i = \frac{1}{h}\nabla y_i + \frac{hD^2y_i}{2} - \frac{h^2D^3y_i}{6} + \cdots$$

Truncate the series, retaining only the first term, and show the order of the truncation error:

$$Dy_i = \frac{1}{h}\nabla y_i + O(h)$$

Express the differential and backward operators in terms of their respective definitions:

$$\frac{dy_i}{dx} = \frac{1}{h}(y_i - y_{i-1}) + O(h) \tag{4-44}$$

Equation (4-44), therefore, enables us to evaluate the first-order derivative of $y$ at position $i$ in terms of backward finite differences.

The term $O(h)$ is used to represent the order of the first term in the truncated portion of the series. When $h < 1.0$ and the function is smooth and continuous, the first term in the truncated portion of the series is the predominant term. It should be emphasized that for $h < 1.0$,

$$h > h^2 > h^3 > h^4 > \cdots > h^n$$

Therefore, when $h < 1.0$, formulas with higher-order error terms, $O(h^n)$, have smaller truncation errors, i.e., they are more accurate approximations of derivatives.

On the other hand, when $h > 1.0$

$$h < h^2 < h^3 < h^4 < \cdots < h^n$$

therefore, formulas with higher-order error terms have larger truncation errors and are less accurate approximations of derivatives.

It is obvious then, that the choice of step size $h$ is very important in determining the accuracy and stability of numerical integration and differentiation. This concept will be discussed in detail in Chaps. 5 and 6.

**Example 4-2** Express the second-order derivative of $y$ in terms of backward finite differences with error of order $h$.

SOLUTION Rearrange Eq. (4-35) to solve for $D^2$:

$$D^2 = \frac{1}{h^2} \nabla^2 + hD^3 - \tfrac{7}{12} h^2 D^4 + \cdots$$

Apply this operator to the function $y$ at $i$:

$$D^2 y_i = \frac{1}{h^2} \nabla^2 y_i + hD^3 y_i - \tfrac{7}{12} h^2 D^4 y_i + \cdots$$

Truncate the series, retaining only the first term, and express the operators in terms of their respective definitions:

$$\frac{d^2 y_i}{dx^2} = \frac{1}{h^2} (y_i - 2y_{i-1} + y_{i-2}) + O(h) \tag{4-45}$$

This equation evaluates the second-order derivative of $y$ at position $i$, in terms of backward finite differences, with error of order $h$.

**Example 4-3** Express the first-order derivative of $y$ in terms of backward finite differences with error of order $h^2$.

SOLUTION Rearrange Eq. (4-31) to solve for $hD$:

$$hD = \nabla + \frac{h^2D^2}{2} - \frac{h^3D^3}{6} + \cdots$$

Rearrange Eq. (4-35) to solve for $h^2D^2$:

$$h^2D^2 = \nabla^2 + h^3D^3 - \tfrac{7}{12}h^4D^4 + \cdots$$

Combine these two equations to eliminate $h^2D^2$:

$$hD = \nabla + \tfrac{1}{2}[\nabla^2 + h^3D^3 - \tfrac{7}{12}h^4D^4 \cdots] - \frac{h^3D^3}{6} + \cdots$$

$$= \nabla + \tfrac{1}{2}\nabla^2 + \frac{h^3D^3}{3} - \cdots$$

Divide through by $h$, and apply this operator to the function $y$ at $i$:

$$Dy_i = \frac{1}{h}\nabla y_i + \frac{1}{2h}\nabla^2 y_i + \frac{h^2D^3y_i}{3} - \cdots$$

Truncate the series, retaining only the first *two* terms, and express the operators in terms of their respective definitions:

$$\frac{dy_i}{dx} = \frac{1}{h}(y_i - y_{i-1}) + \frac{1}{2h}(y_i - 2y_{i-1} + y_{i-2}) + O(h^2)$$

$$= \frac{1}{2h}(3y_i - 4y_{i-1} + y_{i-2}) + O(h^2) \tag{4-46}$$

In this example, the first derivative of $y$ is obtained with error of order $h^2$. For the case where $h < 1.0$, Eq. (4-46) is a more accurate approximation of the first derivative than Eq. (4-44). To obtain the higher accuracy, however, a larger number of terms is involved in the calculation.

**Example 4-4** Express the second-order derivative of $y$ in terms of backward finite differences with error of order $h^2$.

SOLUTION Rearrange Eq. (4-35) to solve for $h^2D^2$:

$$h^2D^2 = \nabla^2 + h^3D^3 - \tfrac{7}{12}h^4D^4 + \cdots$$

Rearrange Eq. (4-36) to solve for $h^3D^3$:

$$h^3D^3 = \nabla^3 + \tfrac{3}{2}h^4D^4 - \tfrac{5}{4}h^5D^5 + \cdots$$

Combine these two equations to eliminate $h^3D^3$:

$$h^2D^2 = \nabla^2 + (\nabla^3 + \tfrac{3}{2}h^4D^4 - \tfrac{5}{4}h^5D^5 + \cdots) - \tfrac{7}{12}h^4D^4 + \cdots$$
$$= \nabla^2 + \nabla^3 + \tfrac{11}{12}h^4D^4 - \cdots$$

Divide through by $h^2$ and apply the operator to the function $y$ at $i$:

$$D^2y_i = \frac{1}{h^2}\nabla^2 y_i + \frac{1}{h^2}\nabla^3 y_i + \tfrac{11}{12}h^2D^4y_i - \cdots$$

Truncate the series, retaining only the first two terms, and express the operators in terms of their respective definitions:

$$\frac{d^2y_i}{dx^2} = \frac{1}{h^2}(y_i - 2y_{i-1} + y_{i-2}) + \frac{1}{h^2}(y_i - 3y_{i-1} + 3y_{i-2} - y_{i-3}) + O(h^2)$$
$$= \frac{1}{h^2}(2y_i - 5y_{i-1} + 4y_{i-2} - y_{i-3}) + O(h^2) \qquad (4\text{-}47)$$

It should be noted that this same equation could have been derived using Eq. (4-41) and an equation for $\nabla^4$ (not shown here). This statement applies to all these examples, which can be solved utilizing both sets of equations shown in Table 4-1.

The formulas for the first- and second-order derivatives, developed in the above four examples, together with those of the third-order derivative [4], are summarized in Table 4-2. It can be concluded, from these examples, that any derivative can be expressed in terms of finite differences with any degree of accuracy desired. These formulas may be used to differentiate the function $y(x)$ given a set of values of this function at equally spaced intervals of $x$, such as a set of experimental data. Conversely, these same formulas may be used in the numerical integration of differential equations, as shown in Chaps. 5 and 6.

**Table 4-2 Derivatives in terms of backward finite differences**

|  |  |
|---|---|
| **Error of order $h$** | |
| $\frac{dy_i}{dx} = \frac{1}{h}(y_i - y_{i-1}) + O(h)$ | (4-44) |
| $\frac{d^2y_i}{dx^2} = \frac{1}{h^2}(y_i - 2y_{i-1} + y_{i-2}) + O(h)$ | (4-45) |
| $\frac{d^3y_i}{dx^3} = \frac{1}{h^3}(y_i - 3y_{i-1} + 3y_{i-2} - y_{i-3}) + O(h)$ | (4-48) |
| **Error of order $h^2$** | |
| $\frac{dy_i}{dx} = \frac{1}{2h}(3y_i - 4y_{i-1} + y_{i-2}) + O(h^2)$ | (4-46) |
| $\frac{d^2y_i}{dx^2} = \frac{1}{h^2}(2y_i - 5y_{i-1} + 4y_{i-2} - y_{i-3}) + O(h^2)$ | (4-47) |
| $\frac{d^3y_i}{dx^3} = \frac{1}{2h^3}(5y_i - 18y_{i-1} + 24y_{i-2} - 14y_{i-3} + 3y_{i-4}) + O(h^2)$ | (4-49) |

## 4-4 FORWARD FINITE DIFFERENCES

The development of forward finite differences follows a course parallel to that used in the development of backward differences.

Consider the set of values

$$y_{i-3} \qquad y_{i-2} \qquad y_{i-1} \qquad y_i \qquad y_{i+1} \qquad y_{i+2} \qquad y_{i+3}$$

or the equivalent set

$$y(x-3h) \quad y(x-2h) \quad y(x-h) \quad y(x) \quad y(x+h) \quad y(x+2h) \quad y(x+3h)$$

The *first forward difference* of $y$ at $i$ (or $x$) is defined as

$$\Delta y_i = y_{i+1} - y_i$$

or

$$\Delta y(x) = y(x+h) - y(x) \tag{4-50}$$

The *second forward difference* of $y$ at $i$ (or $x$) is defined as

$$\begin{aligned}
\Delta^2 y_i &= \Delta(\Delta y_i) = \Delta(y_{i+1} - y_i) = \Delta y_{i+1} - \Delta y_i \\
&= (y_{i+2} - y_{i+1}) - (y_{i+1} - y_i) \\
\Delta^2 y_i &= y_{i+2} - 2y_{i+1} + y_i
\end{aligned}$$

or

$$\Delta^2 y(x) = y(x+2h) - 2y(x+h) + y(x) \tag{4-51}$$

The *third forward difference* of $y$ at $i$ is defined as

$$\begin{aligned}
\Delta^3 y_i &= \Delta(\Delta^2 y_i) = \Delta(y_{i+2} - 2y_{i+1} + y_i) \\
&= \Delta y_{i+2} - 2\Delta y_{i+1} + \Delta y_i \\
&= (y_{i+3} - y_{i+2}) - 2(y_{i+2} - y_{i+1}) + (y_{i+1} - y_i) \\
\Delta^3 y_i &= y_{i+3} - 3y_{i+2} + 3y_{i+1} - y_i
\end{aligned} \tag{4-52}$$

Higher-order forward differences are similarly derived:

$$\Delta^4 y_i = y_{i+4} - 4y_{i+3} + 6y_{i+2} - 4y_{i+1} + y_i \tag{4-53}$$

$$\Delta^5 y_i = y_{i+5} - 5y_{i+4} + 10y_{i+3} - 10y_{i+2} + 5y_{i+1} - y_i \tag{4-54}$$

$$\vdots$$

In similarity to the backward finite differences, the forward finite differences also have coefficients which correspond to those of the binomial expansion $(a-b)^n$.

The relationships between forward difference operators and differential operators can now be developed. Combine Eqs. (4-50) and (4-17) to obtain

$$\Delta y(x) = y(x+h) - y(x) = e^{hD}y(x) - y(x)$$
$$= (e^{hD} - 1)y(x) \qquad (4\text{-}55)$$

which shows that the forward difference operator is given by

$$\Delta = e^{hD} - 1 \qquad (4\text{-}56)$$

Using the infinite series expansion of $e^{hD}$ [Eq. (4-16)], Eq. (4-56) becomes

$$\Delta = hD + \frac{h^2D^2}{2} + \frac{h^3D^3}{6} + \cdots \qquad (4\text{-}57)$$

The higher-order forward difference operators, $\Delta^2, \Delta^3 \ldots$, can be obtained by raising the first forward difference operator to higher powers:

$$\Delta^2 = (e^{hD} - 1)^2 = e^{2hD} - 2e^{hD} + 1 \qquad (4\text{-}58)$$

$$\Delta^3 = (e^{hD} - 1)^3 = e^{3hD} - 3e^{2hD} + 3e^{hD} - 1 \qquad (4\text{-}59)$$
$$\vdots$$
$$\Delta^n = (e^{hD} - 1)^n \qquad (4\text{-}60)$$

Expansion of the exponential terms and rearrangement yields the following equations for the second and third forward difference operators:

$$\Delta^2 = h^2D^2 + h^3D^3 + \tfrac{7}{12}h^4D^4 + \cdots \qquad (4\text{-}61)$$

$$\Delta^3 = h^3D^3 + \tfrac{3}{2}h^4D^4 + \tfrac{5}{4}h^5D^5 + \cdots \qquad (4\text{-}62)$$

Equations (4-57), (4-61), and (4-62) express the forward difference operators in terms of infinite series of differential operators. In order to complete the set of relationships, equations which express the differential operators in terms of forward difference operators will also be derived. To do this, first rearrange Eq. (4-56) to solve for $e^{hD}$:

$$e^{hD} = 1 + \Delta \qquad (4\text{-}63)$$

Take the natural logarithm of both sides of this equation:

$$\ln e^{hD} = hD = \ln(1 + \Delta) \qquad (4\text{-}64)$$

Utilize the infinite series expansion:

$$\ln(1 + \Delta) = \Delta - \frac{\Delta^2}{2} + \frac{\Delta^3}{3} - \frac{\Delta^4}{4} + \frac{\Delta^5}{5} \cdots \qquad (4\text{-}65)$$

Combine (4-64) with (4-65) to obtain

$$hD = \Delta - \frac{\Delta^2}{2} + \frac{\Delta^3}{3} - \frac{\Delta^4}{4} + \frac{\Delta^5}{5} - \cdots \tag{4-66}$$

The higher-order differential operators can be obtained by simply raising both sides of Eq. (4-66) to higher powers:

$$h^2D^2 = \Delta^2 - \Delta^3 + \tfrac{11}{12}\Delta^4 - \tfrac{5}{6}\Delta^5 + \cdots \tag{4-67}$$

$$h^3D^3 = \Delta^3 - \tfrac{3}{2}\Delta^4 + \tfrac{7}{4}\Delta^5 - \cdots \tag{4-68}$$

$$\vdots$$

$$h^nD^n = \left(\Delta - \frac{\Delta^2}{2} + \frac{\Delta^3}{3} - \frac{\Delta^4}{4} + \frac{\Delta^5}{5} - \cdots\right)^n \tag{4-69}$$

The complete set of relationships between forward difference operators and differential operators is summarized in Table 4-3. These equations enable us to develop a variety of formulas expressing derivatives of functions in terms of forward finite differences, and vice versa. As was demonstrated in Sec. 4-3, these formulas may have any degree of accuracy desired provided that a sufficient number of terms is retained in the manipulation of these infinite series. A set of examples, parallel to those of Sec. 4-3, will be worked out using the forward finite differences.

**Table 4-3 Forward finite differences**

| Forward difference operators | |
|---|---|
| $\Delta = hD + \frac{h^2D^2}{2} + \frac{h^3D^3}{6} + \cdots$ | (4-57) |
| $\Delta^2 = h^2D^2 + h^3D^3 + \frac{7}{12}h^4D^4 + \cdots$ | (4-61) |
| $\Delta^3 = h^3D^3 + \frac{3}{2}h^4D^4 + \frac{5}{4}h^5D^5 + \cdots$ | (4-62) |
| $\Delta^n = (e^{hD} - 1)^n$ | (4-60) |
| **Differential operators** | |
| $hD = \Delta - \frac{\Delta^2}{2} + \frac{\Delta^3}{3} - \frac{\Delta^4}{4} + \cdots$ | (4-66) |
| $h^2D^2 = \Delta^2 - \Delta^3 + \frac{11}{12}\Delta^4 - \frac{5}{6}\Delta^5 + \cdots$ | (4-67) |
| $h^3D^3 = \Delta^3 - \frac{3}{2}\Delta^4 + \frac{7}{4}\Delta^5 - \cdots$ | (4-68) |
| $h^nD^n = \left(\Delta - \frac{\Delta^2}{2} + \frac{\Delta^3}{3} - \frac{\Delta^4}{4} + \frac{\Delta^5}{5} - \cdots\right)^n$ | (4-69) |

**Example 4-5** Express the first-order derivative of $y$ in terms of forward finite differences with error of order $h$.

SOLUTION Rearrange Eq. (4-57) to solve for $D$:

$$D = \frac{1}{h}\Delta - \frac{hD^2}{2} - \frac{h^2D^3}{6} - \cdots$$

Apply this operator to the function $y$ at $i$:

$$Dy_i = \frac{1}{h}\Delta y_i - \frac{hD^2y_i}{2} - \frac{h^2D^3y_i}{6} - \cdots$$

Truncate the series retaining only the first term:

$$Dy_i = \frac{1}{h}\Delta y_i + O(h)$$

Express the differential and forward operators in terms of their respective definitions:

$$\frac{dy_i}{dx} = \frac{1}{h}(y_{i+1} - y_i) + O(h) \tag{4-70}$$

Equation (4-70) enables us to evaluate the first-order derivative of $y$ at position $i$ in terms of forward finite differences with error of order $h$.

**Example 4-6** Express the second-order derivative of $y$ in terms of forward finite differences with error of order $h$.

SOLUTION Rearrange Eq. (4-61) to solve for $D^2$:

$$D^2 = \frac{1}{h^2}\Delta^2 - hD^3 - \tfrac{7}{12}h^2D^4 - \cdots$$

Apply this operator to the function $y$ at $i$:

$$D^2y_i = \frac{1}{h^2}\Delta^2y_i - hD^3y_i - \tfrac{7}{12}h^2D^4y_i - \cdots$$

Truncate the series, retaining only the first term, and express the operators in terms of their respective definitions:

$$\frac{d^2y_i}{dx^2} = \frac{1}{h^2}(y_{i+2} - 2y_{i+1} + y_i) + O(h) \tag{4-71}$$

This equation evaluates the second-order derivative of $y$ at position $i$ in terms of forward finite differences with error of order $h$.

**Example 4-7** Express the first-order derivative of $y$ in terms of forward finite differences with error of order $h^2$.

SOLUTION Rearrange Eq. (4-57) to solve for $hD$:

$$hD = \Delta - \frac{h^2D^2}{2} - \frac{h^3D^3}{6} - \cdots$$

Rearrange Eq. (4-61) to solve for $h^2D^2$:

$$h^2D^2 = \Delta^2 - h^3D^3 - \tfrac{7}{12}h^4D^4 - \cdots$$

Combine these two equations to eliminate $h^2D^2$:

$$hD = \Delta - \tfrac{1}{2}(\Delta^2 - h^3D^3 - \tfrac{7}{12}h^4D^4 - \cdots) - \frac{h^3D^3}{6} - \cdots$$

$$= \Delta - \tfrac{1}{2}\Delta^2 + \frac{h^3D^3}{3} + \cdots$$

Divide through by $h$, and apply this operator to the function $y$ at $i$:

$$Dy_i = \frac{1}{h}\Delta y_i - \frac{1}{2h}\Delta^2 y_i + \frac{h^2D^3y_i}{3} + \cdots$$

Truncate the series, retaining only the first two terms, and express the operators in terms of their respective definitions:

$$\frac{dy_i}{dx} = \frac{1}{h}(y_{i+1} - y_i) - \frac{1}{2h}(y_{i+2} - 2y_{i+1} + y_i) + O(h^2)$$

$$= \frac{1}{2h}(-y_{i+2} + 4y_{i+1} - 3y_i) + O(h^2) \tag{4-72}$$

**Example 4-8** Express the second-order derivative of $y$ in terms of forward finite differences with error of order $h^2$.

SOLUTION Rearrange Eq. (4-61) to solve for $h^2D^2$:

$$h^2D^2 = \Delta^2 - h^3D^3 - \tfrac{7}{12}h^4D^4 - \cdots$$

Rearrange Eq. (4-62) to solve for $h^3D^3$:

$$h^3D^3 = \Delta^3 - \tfrac{3}{2}h^4D^4 - \tfrac{5}{4}h^5D^5 - \cdots$$

Combine these two equations to eliminate $h^3D^3$:

$$h^2D^2 = \Delta^2 - (\Delta^3 - \tfrac{3}{2}h^4D^4 - \tfrac{5}{4}h^5D^5 - \cdots) - \tfrac{7}{12}h^4D^4 \cdots$$
$$= \Delta^2 - \Delta^3 + \tfrac{11}{12}h^4D^4 - \cdots$$

Divide through by $h^2$, and apply this operator to the function $y$ at $i$:

$$D^2y_i = \frac{1}{h^2}\Delta^2y_i - \frac{1}{h^2}\Delta^3y_i + \tfrac{11}{12}h^2D^4y_i - \cdots$$

Truncate the series, retaining only the first two terms, and express the operators in terms of their respective definitions:

$$\frac{d^2y_i}{dx^2} = \frac{1}{h^2}(-y_{i+3} + 4y_{i+2} - 5y_{i+1} + 2y_i) + O(h^2) \tag{4-73}$$

The formulas developed in these examples for the first- and second-order derivatives are summarized in Table 4-4, together with those of the third-order derivative.

It should be pointed out that all the finite difference approximations of derivatives obtained in this section and the previous section have coefficients which add up to zero. This is a rule of thumb which applies to all such combinations of finite differences.

From a comparison between Tables 4-2 and 4-4 we conclude that derivatives can be expressed in either backward or forward differences, with formulas that are very similar to each other in the number of terms involved, and in the order of their truncation error. The choice between using forward or backward differences will depend on the geometry of the problem and its boundary conditions. This will be discussed further in Chaps. 5 and 6.

**Table 4-4 Derivatives in terms of forward finite differences**

| Error of order $h$ | |
|---|---|
| $\frac{dy_i}{dx} = \frac{1}{h}(y_{i+1} - y_i) + O(h)$ | (4-70) |
| $\frac{d^2y_i}{dx^2} = \frac{1}{h^2}(y_{i+2} - 2y_{i+1} + y_i) + O(h)$ | (4-71) |
| $\frac{d^3y_i}{dx^3} = \frac{1}{h^3}(y_{i+3} - 3y_{i+2} + 3y_{i+1} - y_i) + O(h)$ | (4-74) |
| **Error of order $h^2$** | |
| $\frac{dy_i}{dx} = \frac{1}{2h}(-y_{i+2} + 4y_{i+1} - 3y_i) + O(h^2)$ | (4-72) |
| $\frac{d^2y_i}{dx^2} = \frac{1}{h^2}(-y_{i+3} + 4y_{i+2} - 5y_{i+1} + 2y_i) + O(h^2)$ | (4-73) |
| $\frac{d^3y_i}{dx^3} = \frac{1}{2h^3}(-3y_{i+4} + 14y_{i+3} - 24y_{i+2} + 18y_{i+1} - 5y_i) + O(h^2)$ | (4-75) |

## 4-5 CENTRAL FINITE DIFFERENCES

As their name implies, central finite differences are *centered* at the pivot position and are evaluated utilizing the values of the function to the right and to the left of the pivot position, but located only $h/2$ distance away from it.

Consider the series of values used in the previous two sections, but with the additional values at the midpoints of the intervals

$$y_{i-2} \quad y_{i-1\frac{1}{2}} \quad y_{i-1} \quad y_{i-1/2} \quad y_i \quad y_{i+1/2} \quad y_{i+1} \quad y_{i+1\frac{1}{2}} \quad y_{i+2}$$

or the equivalent set

$$y(x-2h) \quad y(x-1\tfrac{1}{2}h) \quad y(x-h) \quad y(x-\tfrac{1}{2}h) \quad y(x) \quad y(x+\tfrac{1}{2}h) \quad y(x+h) \quad y(x+1\tfrac{1}{2}h) \quad y(x+2h)$$

The *first central difference* of $y$ at $i$ (or $x$) is defined as

$$\delta y_i = y_{i+1/2} - y_{i-1/2}$$

or

$$\delta y(x) = y(x+\tfrac{1}{2}h) - y(x-\tfrac{1}{2}h) \tag{4-76}$$

The *second central difference* of $y$ at $i$ (or $x$) is defined as

$$\begin{aligned}\delta^2 y_i &= \delta(\delta y_i) = \delta(y_{i+1/2} - y_{i-1/2}) = \delta y_{i+1/2} - \delta y_{i-1/2}\\ &= (y_{i+1} - y_i) - (y_i - y_{i-1})\\ \delta^2 y_i &= y_{i+1} - 2y_i + y_{i-1}\end{aligned}$$

or

$$\delta^2 y(x) = y(x+h) - 2y(x) + y(x-h) \tag{4-77}$$

The *third central difference* of $y$ at $i$ is defined as

$$\begin{aligned}\delta^3 y_i &= \delta(\delta^2 y_i) = \delta(y_{i+1} - 2y_i + y_{i-1})\\ &= \delta y_{i+1} - 2\delta y_i + \delta y_{i-1}\\ &= (y_{i+1\frac{1}{2}} - y_{i+1/2}) - 2(y_{i+1/2} - y_{i-1/2}) + (y_{i-1/2} - y_{i-1\frac{1}{2}})\\ \delta^3 y_i &= y_{i+1\frac{1}{2}} - 3y_{i+1/2} + 3y_{i-1/2} - y_{i-1\frac{1}{2}}\end{aligned} \tag{4-78}$$

Higher-order central differences are similarly derived:

$$\delta^4 y_i = y_{i+2} - 4y_{i+1} + 6y_i - 4y_{i-1} + y_{i-2} \tag{4-79}$$

$$\delta^5 y_i = y_{i+2\frac{1}{2}} - 5y_{i+1\frac{1}{2}} + 10y_{i+1/2} - 10y_{i-1/2} + 5y_{i-1\frac{1}{2}} - y_{i-2\frac{1}{2}} \tag{4-80}$$

Consistent with the other finite differences, the central finite differences also have coefficients which correspond to those of the binomial expansion $(a-b)^n$.

It should be noted that the *odd*-order central differences involve values of the function at the midpoints of the intervals, while the *even*-order central differences involve values at the full intervals. To fully utilize odd- and even-order central differences we need a set of values of the function $y$ which includes twice as many points as that used in either backward or forward differences. This situation is rather uneconomical, especially in the case where these values must be obtained experimentally. To alleviate this difficulty, we make use of the *averager operator* $\mu$, which is defined as

$$\mu = \tfrac{1}{2}[E^{1/2} + E^{-1/2}] \tag{4-81}$$

The averager operator shifts its operand by a half interval to the right of the pivot and by a half interval to the left of the pivot, evaluates it at these two positions, and averages the two values.

Application of the averager on the odd central differences gives the *first averaged central difference* as follows:

$$\begin{aligned}\mu\delta y_i &= \tfrac{1}{2}(E^{1/2}\delta y_i + E^{-1/2}\delta y_i)\\ &= \tfrac{1}{2}(\delta y_{i+1/2} + \delta y_{i-1/2})\\ &= \tfrac{1}{2}[(y_{i+1} - y_i) + (y_i - y_{i-1})]\\ &= \tfrac{1}{2}(y_{i+1} - y_{i-1})\end{aligned} \tag{4-82}$$

The *third averaged central difference* is given by

$$\begin{aligned}\mu\delta^3 y_i &= \tfrac{1}{2}(E^{1/2}\delta^3 y_i + E^{-1/2}\delta^3 y_i)\\ &= \tfrac{1}{2}(\delta^3 y_{i+1/2} + \delta^3 y_{i-1/2})\\ &= \tfrac{1}{2}[(y_{i+2} - 3y_{i+1} + 3y_i - y_{i-1}) + (y_{i+1} - 3y_i + 3y_{i-1} - y_{i-2})]\\ &= \tfrac{1}{2}(y_{i+2} - 2y_{i+1} + 2y_{i-1} - y_{i-2})\end{aligned} \tag{4-83}$$

As expected, the effect of the averager is to remove the midpoint values of the function $y$ from the odd central differences.

It will be shown later in this section that central differences are more accurate than either backward or forward differences when used to evaluate the derivatives of functions.

The relationships between central difference operators and differential operators can now be developed. Equation (4-82), representing the first averaged central difference, is combined with Eqs. (4-17) and (4-22) to yield

$$\begin{aligned}\mu\delta y(x) &= \tfrac{1}{2}[y(x+h) - y(x-h)]\\ &= \tfrac{1}{2}[e^{hD}y(x) - e^{-hD}y(x)]\\ &= \tfrac{1}{2}(e^{hD} - e^{-hD})y(x)\end{aligned} \tag{4-84}$$

which shows that the first averaged central difference operator is given by

$$\mu\delta = \tfrac{1}{2}(e^{hD} - e^{-hD}) = \sinh hD \tag{4-85}$$

Using the infinite series expansions of $e^{hD}$ and $e^{-hD}$, or equivalently the infinite series expansion of the hyperbolic sine,

$$\sinh hD = hD + \frac{(hD)^3}{3!} + \frac{(hD)^5}{5!} + \frac{(hD)^7}{7!} + \cdots \tag{4-86}$$

Eq. (4-85) becomes

$$\mu\delta = hD + \frac{h^3D^3}{6} + \frac{h^5D^5}{120} + \frac{h^7D^7}{5040} + \cdots \tag{4-87}$$

Similarly, using Eq. (4-77) for the second central difference, and combining it with Eqs. (4-17) and (4-22), we obtain

$$\begin{aligned} \delta^2 y(x) &= y(x+h) - 2y(x) + y(x-h) \\ &= e^{hD}y(x) - 2y(x) + e^{-hD}y(x) \\ &= (e^{hD} - 2 + e^{-hD})y(x) \end{aligned} \tag{4-88}$$

which shows that the second central difference operator is equivalent to

$$\delta^2 = e^{hD} - 2 + e^{-hD} = E - 2 + E^{-1} \tag{4-89}$$

Expanding the exponentials in Eq. (4-89) into their infinite series and regrouping, we obtain

$$\delta^2 = h^2D^2 + \frac{h^4D^4}{12} + \frac{h^6D^6}{360} + \frac{h^8D^8}{20{,}160} + \cdots \tag{4-90}$$

The higher-order averaged odd central difference operators are obtained by taking products of Eqs. (4-87) and (4-90). The higher-order even central differences are formulated by taking powers of Eq. (4-90). The third and fourth central operators, thus obtained, are listed below:

$$\mu\delta^3 = h^3D^3 + \frac{h^5D^5}{4} + \frac{h^7D^7}{40} + \cdots \tag{4-91}$$

$$\delta^4 = h^4D^4 + \frac{h^6D^6}{6} + \frac{h^8D^8}{80} + \cdots \tag{4-92}$$

In order to develop the inverse relationships, i.e., equations for the differential operators in terms of the central difference operators, we must first derive an algebraic relationship between $\mu$ and $\delta$†. To do this, we start with Eqs. (4-81) and (4-89). Squaring both sides of Eq. (4-81), we obtain

$$\mu^2 = \tfrac{1}{4}(E + E^{-1} + 2) \tag{4-93}$$

Rearranging Eq. (4-89), we get

$$\delta^2 + 2 = E + E^{-1} \tag{4-94}$$

Combining Eqs. (4-93) and (4-94), and rearranging, we arrive at the desired relationship

$$\mu^2 = \frac{\delta^2}{4} + 1 \tag{4-95}$$

Now taking the inverse of Eq. (4-85),

$$hD = \sinh^{-1} \mu\delta \tag{4-96}$$

The infinite series expansion of the inverse hyperbolic sine is

$$\sinh^{-1} \mu\delta = \mu\delta - \frac{(\mu\delta)^3}{6} + \frac{3(\mu\delta)^5}{40} - \cdots \tag{4-97}$$

therefore, Eq. (4-96) expands to

$$hD = \mu\delta - \frac{\mu^3\delta^3}{6} + \frac{3\mu^5\delta^5}{40} - \cdots \tag{4-98}$$

The even powers of $\mu$ are eliminated from Eq. (4-98) by using Eq. (4-95) to obtain the first differential operator in terms of central difference operators:

$$hD = \mu\left(\delta - \frac{\delta^3}{6} + \frac{\delta^5}{30} - \cdots\right) \tag{4-99}$$

Higher-order differential operators are obtained by raising Eq. (4-99) to the appropriate power and using (4-95) to eliminate the even powers of $\mu$. The second, third, and fourth differential operators obtained this way are shown below:

$$h^2D^2 = \delta^2 - \frac{\delta^4}{12} + \frac{\delta^6}{90} - \cdots \tag{4-100}$$

$$h^3D^3 = \mu\left(\delta^3 - \frac{\delta^5}{4} + \frac{7\delta^7}{120} - \cdots\right) \tag{4-101}$$

$$h^4D^4 = \delta^4 - \frac{\delta^6}{6} + \frac{7\delta^8}{240} - \cdots \tag{4-102}$$

† The steps used in this development are parallel to those used by Salvadori and Baron [4, pp. 83–86], with some modifications made by the author. Permission obtained from the publisher.

The complete set of relationships between central difference operators and differential operators is summarized in Table 4-5. These relationships will be used in the following examples to develop a set of formulas expressing the derivatives in terms of central finite differences. These formulas will have higher accuracy than those developed in the previous two sections using backward and forward finite differences.

**Example 4-9** Express the first-order derivative of $y$ in terms of central finite differences with error of order $h^2$.

Solution Rearrange Eq. (4-87) to solve for $D$:

$$D = \frac{1}{h}\mu\delta - \frac{h^2D^3}{6} - \frac{h^4D^5}{120} - \cdots$$

Apply this operator to the function $y$ at $i$:

$$Dy_i = \frac{1}{h}\mu\delta y_i - \frac{h^2D^3y_i}{6} - \frac{h^4D^5y_i}{120} - \cdots$$

Truncate the series, retaining only the first term:

$$Dy_i = \frac{1}{h}\mu\delta y_i + O(h^2)$$

**Table 4-5 Central finite differences**

| Central difference operators | |
|---|---|
| $\mu\delta = hD + \frac{h^3D^3}{6} + \frac{h^5D^5}{120} + \frac{h^7D^7}{5040} + \cdots$ | (4-87) |
| $\delta^2 = h^2D^2 + \frac{h^4D^4}{12} + \frac{h^6D^6}{360} + \frac{h^8D^8}{20{,}160} + \cdots$ | (4-90) |
| $\mu\delta^3 = h^3D^3 + \frac{h^5D^5}{4} + \frac{h^7D^7}{40} + \cdots$ | (4-91) |
| $\delta^4 = h^4D^4 + \frac{h^6D^6}{6} + \frac{h^8D^8}{80} + \cdots$ | (4-92) |
| **Differential operators** | |
| $hD = \mu\left(\delta - \frac{\delta^3}{6} + \frac{\delta^5}{30} - \cdots\right)$ | (4-99) |
| $h^2D^2 = \delta^2 - \frac{\delta^4}{12} + \frac{\delta^6}{90} - \cdots$ | (4-100) |
| $h^3D^3 = \mu\left(\delta^3 - \frac{\delta^5}{4} + \frac{7\delta^7}{120} - \cdots\right)$ | (4-101) |
| $h^4D^4 = \delta^4 - \frac{\delta^6}{6} + \frac{7\delta^8}{240} - \cdots$ | (4-102) |

Express the differential and averaged central difference operators in terms of their respective definitions:

$$\frac{dy_i}{dx} = \frac{1}{2h}(y_{i+1} - y_{i-1}) + O(h^2) \tag{4-103}$$

Equation (4-103) enables us to evaluate the first-order derivative of $y$ at position $i$ in terms of central finite differences. Comparing this equation with Eq. (4-44) of Example 4-1 and Eq. (4-70) of Example 4-5 reveals that use of central differences increases the accuracy of the formulas, for the same number of terms retained.

**Example 4-10** Express the second-order derivative of $y$ in terms of central finite differences with error of order $h^2$.

Solution Rearrange Eq. (4-90) to solve for $D^2$:

$$D^2 = \frac{1}{h^2}\delta^2 - \frac{h^2D^4}{12} - \frac{h^4D^6}{360} - \cdots$$

Apply this operator to the function $y$ at $i$:

$$D^2y_i = \frac{1}{h^2}\delta^2y_i - \frac{h^2D^4y_i}{12} - \frac{h^4D^6y_i}{360} - \cdots$$

Truncate the series, retaining only the first term:

$$D^2y_i = \frac{1}{h^2}\delta^2y_i + O(h^2)$$

Express the differential and central difference operators in terms of their respective definitions:

$$\frac{d^2y_i}{dx^2} = \frac{1}{h^2}(y_{i+1} - 2y_i + y_{i-1}) + O(h^2) \tag{4-104}$$

**Example 4-11** Express the first-order derivative of $y$ in terms of central finite differences with error of order $h^4$.

Solution Rearrange Eq. (4-87) to solve for $hD$:

$$hD = \mu\delta - \frac{h^3D^3}{6} - \frac{h^5D^5}{120} - \cdots$$

Rearrange Eq. (4-91) to solve for $h^3D^3$:

$$h^3D^3 = \mu\delta^3 - \frac{h^5D^5}{4} - \frac{h^7D^7}{40} - \cdots$$

Combine these two equations to eliminate $h^3D^3$:

$$hD = \mu\delta - \frac{1}{6}\left(\mu\delta^3 - \frac{h^5D^5}{4} - \frac{h^7D^7}{40} - \cdots\right) - \frac{h^5d^5}{120} - \cdots$$

$$= \mu\delta - \tfrac{1}{6}\,\mu\delta^3 + \frac{h^5D^5}{30} + \cdots$$

Divide through by $h$ and apply this operator to the function $y$ at $i$:

$$Dy_i = \frac{1}{h}\,\mu\delta y_i - \frac{1}{6h}\,\mu\delta^3 y_i + \frac{h^4D^5y_i}{30} + \cdots$$

Truncate the series, retaining only the first two terms, and express the operators in terms of their respective definitions:

$$\frac{dy_i}{dx} = \frac{1}{2h}\,(y_{i+1} - y_{i-1}) - \frac{1}{12h}\,(y_{i+2} - 2y_{i+1} + 2y_{i-1} - y_{i-2}) + O(h^4)$$

$$= \frac{1}{12h}\,(-y_{i+2} + 8y_{i+1} - 8y_{i-1} + y_{i-2}) + O(h^4) \tag{4-105}$$

**Example 4-12** Express the second-order derivative of $y$ in terms of central finite differences with error of order $h^4$.

SOLUTION Rearrange Eq. (4-90) to solve for $h^2D^2$:

$$h^2D^2 = \delta^2 - \frac{h^4D^4}{12} - \frac{h^6D^6}{360} - \cdots$$

Rearrange Eq. (4-92) to solve for $h^4D^4$:

$$h^4D^4 = \delta^4 - \frac{h^6D^6}{6} - \frac{h^8D^8}{80} - \cdots$$

Combine these two equations to eliminate $h^4D^4$:

$$h^2D^2 = \delta^2 - \frac{1}{12}\left(\delta^4 - \frac{h^6D^6}{6} - \frac{h^8D^8}{80}\cdots\right) - \frac{h^6D^6}{360}$$

$$= \delta^2 - \tfrac{1}{12}\delta^4 + \frac{h^6D^6}{90} - \cdots$$

Divide through by $h^2$ and apply this operator to the function $y$ at $i$:

$$D^2y_i = \frac{1}{h^2}\,\delta^2 y_i - \frac{1}{12h^2}\,\delta^4 y_i + \frac{h^4D^6y_i}{90} - \cdots$$

Truncate the series, retaining only the first two terms, and express the operators in terms of their respective definitions:

$$\frac{d^2y_i}{dx^2} = \frac{1}{h^2}(y_{i+1} - 2y_i + y_{i-1}) - \frac{1}{12h^2}(y_{i+2} - 4y_{i+1} + 6y_i - 4y_{i-1} + y_{i-2}) + O(h^4)$$

$$= \frac{1}{12h^2}(-y_{i+2} + 16y_{i+1} - 30y_i + 16y_{i-1} - y_{i-2}) + O(h^4) \qquad (4\text{-}106)$$

The formulas derived in Examples 4-9 to 4-12 for the first- and second-order derivatives are summarized in Table 4-6, along with those for the third-order derivatives. Development of formulas with higher accuracy and for the higher-order derivatives are left as exercises for the reader (see Problems).

**Table 4-6 Derivatives in terms of central finite differences**

|  |  |
|---|---|
| Error of order $h^2$ | |
| $\frac{dy_i}{dx} = \frac{1}{2h}(y_{i+1} - y_{i-1}) + O(h^2)$ | (4-103) |
| $\frac{d^2y_i}{dx^2} = \frac{1}{h^2}(y_{i+1} - 2y_i + y_{i-1}) + O(h^2)$ | (4-104) |
| $\frac{d^3y_i}{dx^3} = \frac{1}{2h^3}(y_{i+2} - 2y_{i+1} + 2y_{i-1} - y_{i-2}) + O(h^2)$ | (4-107) |
| Error of order $h^4$ | |
| $\frac{dy_i}{dx} = \frac{1}{12h}(-y_{i+2} + 8y_{i+1} - 8y_{i-1} + y_{i-2}) + O(h^4)$ | (4-105) |
| $\frac{d^2y_i}{dx^2} = \frac{1}{12h^2}(-y_{i+2} + 16y_{i+1} - 30y_i + 16y_{i-1} - y_{i-2}) + O(h^4)$ | (4-106) |
| $\frac{d^3y_i}{dx^3} = \frac{1}{8h^3}(-y_{i+3} + 8y_{i+2} - 13y_{i+1} + 13y_{i-1} - 8y_{i-2} + y_{i-3}) + O(h^4)$ | (4-108) |

## 4-6 DIFFERENCE EQUATIONS AND THEIR SOLUTIONS

The application of forward, backward, or central finite differences in the solution of differential equations transforms these equations to *difference equations* of the form

$$f(y_k, y_{k+1}, \ldots, y_{k+n}) = 0 \qquad (4\text{-}109)$$

In addition, difference equations are obtained from the application of material balances on multistage operations, such as distillation and extraction.

Depending on their origin, difference equations may be linear or nonlinear, homogeneous or nonhomogeneous, with constant or variable coefficients. For the purposes of this book it will be necessary to discuss only the methods of solution of *homogeneous linear difference equations with constant coefficients*.

The *order* of a difference equation is the difference between the highest and the lowest subscript of the dependent variable in the equation, i.e., it is the number of finite steps spanned by the equation. The order of Eq. (4-109) is given by

$$\text{Order} = (k + n) - k = n \tag{4-110}$$

The process of obtaining $y_k$ is called *solving the difference equation*. The methods of obtaining such solutions are analogous to those used in finding analytical solutions of differential equations. As a matter of fact, the theory of difference equations is parallel to the corresponding theory of differential equations [4, 5]. Difference equations resemble ordinary differential equations. For example, Eq. (4-111) is a second-order homogeneous linear ordinary *differential* equation:

$$y'' + 3y' - 4y = 0 \tag{4-111}$$

while Eq. (4-112) is a second-order homogeneous linear *difference* equation:

$$y_{k+2} + 3y_{k+1} - 4y_k = 0 \tag{4-112}$$

The solution to the differential equation (4-111) can be obtained from the methods of differential calculus applied as follows:

1. Replace the derivatives in (4-111) with the differential operators:

$$D^2y + 3Dy - 4y = 0$$

2. Factor out the $y$:

$$(D^2 + 3D - 4)y = 0$$

3. Find the roots of the *characteristic equation*:

$$D^2 + 3D - 4 = 0$$

   These roots are called the *eigenvalues* of the differential equation. In this case they are

$$\lambda_1 = 1 \qquad \text{and} \qquad \lambda_2 = -4$$

4. Construct the solution of the homogeneous differential equation as follows:

$$y = C_1 e^{\lambda_1 x} + C_2 e^{\lambda_2 x}$$
$$= C_1 e^{(1)x} + C_2 e^{(-4)x} \tag{4-113}$$

where $C_1$ and $C_2$ are constants which must be evaluated from the boundary conditions of the differential equation.

Similarly, the solution of the difference equation (4-112) can be obtained by using the shift operator $E$:

$$E^2 y_k + 3Ey_k - 4y_k = 0$$

1. Factor out the $y_k$:

$$(E^2 + 3E - 4)y_k = 0$$

2. Find the roots of the characteristic equation:

$$E^2 + 3E - 4 = 0$$

These roots are

$$\lambda_1 = 1 \qquad \text{and} \qquad \lambda_2 = -4$$

3. Construct the solution of the homogeneous difference equation as follows:†

$$y_k = C_1(\lambda_1)^k + C_2(\lambda_2)^k$$
$$= C_1(1)^k + C_2(-4)^k \tag{4-114}$$

where $C_1$ and $C_2$ are constants which must be evaluated from the boundary conditions of the difference equation.

In the above case, both eigenvalues were real and distinct. When the eigenvalues are real and repeated, the solution for a second-order equation with both roots identical is formed as follows:

$$y_k = (C_1 + C_2 k)\lambda^k \tag{4-115}$$

For an $n$th-order equation, which has $m$ repeated roots ($\lambda_m$) and one distinct root ($\lambda_n$), the general formulation of the solution is obtained by superposition:

† Proofs and derivations of the solutions of difference equations are given in Refs. 5 (p. 186) and 6 (p. 371). Alternatively, the solutions of difference equations can be obtained from the solutions of differential equations by utilizing the relationship between the shift operator and the differential operator [Eq. (4-18)]. This is demonstrated in Prob. 4-7.

$$y_k = (C_1 + C_2k + C_3k^2 + \cdots + C_mk^{m-1})\lambda_m^k + C_n\lambda_n^k \tag{4-116}$$

In the case where the characteristic equation contains two complex roots,

$$\lambda_1 = \alpha + \beta i \qquad \text{and} \qquad \lambda_2 = \alpha - \beta i \tag{4-117}$$

the solution is

$$y_k = C_1(\alpha + \beta i)^k + C_2(\alpha - \beta i)^k \tag{4-118}$$

This solution may be also expressed in terms of trigonometric quantities by utilizing the trigonometric (polar) form of complex numbers:

$$\alpha \pm \beta i = r(\cos\theta \pm i\sin\theta) \tag{4-119}$$

This is obtained by showing the complex number as a vector in the complex plane [7] represented in Fig. 4-2. The *modulus* $r$ of the complex number is obtained from the Pythagorean theorem

$$r = \sqrt{\alpha^2 + \beta^2} \tag{4-120}$$

The values of $\alpha$ and $\beta$ are expressed in terms of the *phase angle* $\theta$:

$$\alpha = r\cos\theta \tag{4-121}$$

$$\beta = r\sin\theta \tag{4-122}$$

and the phase angle is given by

$$\theta = \tan^{-1}\frac{\beta}{\alpha} \tag{4-123}$$

Substituting Eq. (4-119) in Eq. (4-118) and utilizing de Moivre's theorem

$$(\cos\theta \pm i\sin\theta)^k = (\cos k\theta \pm i\sin k\theta) \tag{4-124}$$

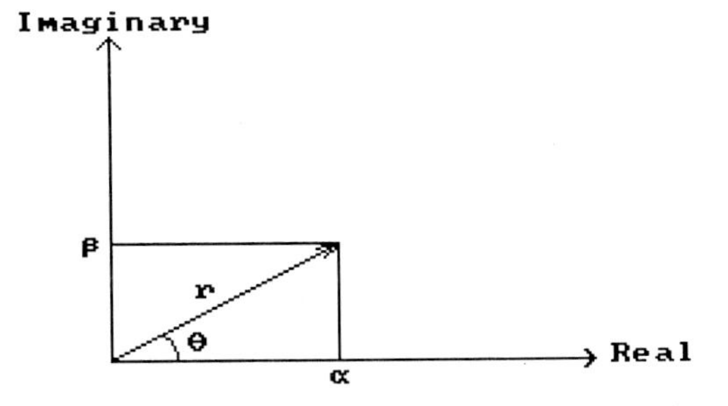


**Figure 4-2** Representation of a complex number in a plane.

we obtain the solution of the difference equation as

$$y_k = r^k[C_1' \cos k\theta + C_2' \sin k\theta] \tag{4-125}$$

where

$$C_1' = C_1 + C_2 \tag{4-126}$$

and

$$C_2' = (C_1 - C_2)i \tag{4-127}$$

It can be concluded from the above discussion that the solution of homogeneous linear difference equations with constant coefficients is of the form

$$y_k = f(k, \lambda) \tag{4-128}$$

where $k$ is the forward-marching counter and $\lambda$ is the vector of eigenvalues of the characteristic equation. The stability and convergence of these solutions depend on the values of the eigenvalues [8]. The following stability cases apply to the solutions of difference equations:

1. The solution is stable, converging without oscillations, when
   *a.* All the eigenvalues are real distinct and have absolute values less than, or equal to, unity.

$$\lambda = \text{real distinct}$$
$$|\lambda| \leq 1.0$$

   *b.* The eigenvalues are real, but repeated, and have absolute values less than unity.

$$\lambda = \text{real repeated}$$
$$|\lambda| < 1.0$$

2. The solution is stable, converging with damped oscillations, when
   *a.* Complex distinct eigenvalues are present and the moduli of the eigenvalues are less than, or equal to, unity.

$$\lambda = \text{complex distinct}$$
$$|r| \leq 1.0$$

   *b.* Complex repeated eigenvalues are present and the moduli of the eigenvalues are less than unity.

$$\lambda = \text{complex repeated}$$
$$|r| < 1.0$$

3. The solution is unstable and nonoscillatory, when
   *a.* All the eigenvalues are real distinct and one or more of these have absolute values greater than unity.

$$\lambda = \text{real distinct}$$
$$|\lambda| > 1.0$$

   *b.* The eigenvalues are real, but repeated, and one or more of these have absolute values equal to, or greater than, unity.

$$\lambda = \text{real repeated}$$
$$|\lambda| \geq 1.0$$

4. The solution is unstable and oscillatory, when
   *a.* Complex distinct eigenvalues are present and the moduli of one or more of these are greater than unity.

$$\lambda = \text{complex distinct}$$
$$|r| > 1.0$$

   *b.* Complex repeated eigenvalues are present and the moduli of one or more of these are equal to, or greater than, unity.

$$\lambda = \text{complex repeated}$$
$$|r| \geq 1.0$$

The numerical solutions of ordinary and partial differential equations are based on the finite difference formulation of these differential equations. Therefore, the stability and convergence considerations of finite difference solutions have important implications on the numerical solutions of differential equations. This topic will be discussed in more detail in Chaps. 5 and 6.

## 4-7 INTERPOLATING POLYNOMIALS

Engineers and scientists often face the task of interpreting and correlating experimental observations, which are usually in the form of discrete data, and are called upon to either integrate or differentiate these data numerically or graphically. This task is facilitated by the use of interpolation/extrapolation formulas. The calculus of finite differences enables us to develop *interpolating polynomials* that can represent experimental data when the actual functionality of these data is not well known. But, even more significantly, these polynomials can be used to approximate functions which are difficult to integrate or differentiate, thus making the task somewhat easier, albeit approximate.

Let us assume that values of a function $f(x)$ are known at a set of $n + 1$ values of the independent variable $x$:

$$
\begin{array}{ll}
x_0 & f(x_0) \\
x_1 & f(x_1) \\
x_2 & f(x_2) \\
x_3 & f(x_3) \\
\vdots & \vdots \\
x_n & f(x_n)
\end{array}
$$

These values are called the *base points* of the function. They are shown graphically in Fig. 4-3*a*.

The general objective in developing interpolating polynomials is to choose a polynomial of the form

$$P_n(x) = a_0 + a_1 x + a_2 x^2 + a_3 x^3 + \cdots + a_n x^n \tag{4-129}$$

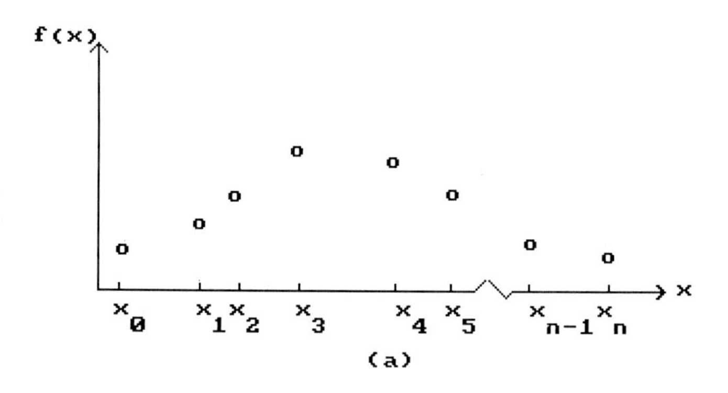


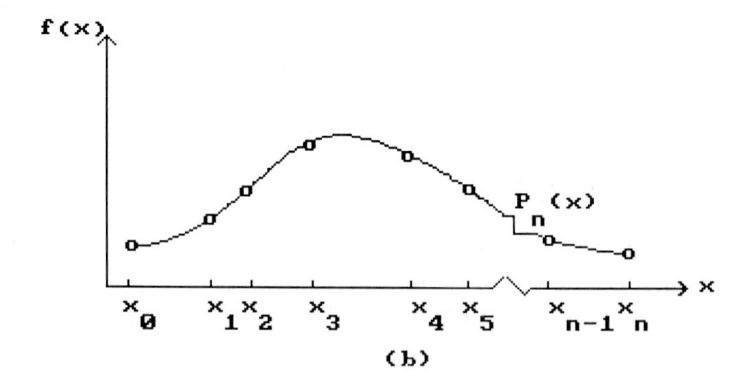


**Figure 4-3** (*a*) Unequally spaced base points of the function $f(x)$. (*b*) Unequally spaced base points with $n$th-degree interpolating polynomial.

so that this equation fits exactly the base points of the function and connects these points with a smooth curve, as shown in Fig. 4-3*b*. This polynomial can then be used to approximate the function at any value of the independent variable $x$ between the base points.

For the given set of $n + 1$ known base points, the polynomial must satisfy the equation

$$P_n(x_i) = f(x_i) \qquad i = 0, 1, 2, \ldots, n \tag{4-130}$$

Substitution of the known values of $(x_i, f(x_i))$ in Eq. (4-129) yields a set of $n + 1$ simultaneous linear algebraic equations whose unknowns are the coefficients $a_0, \ldots, a_n$ of the polynomial equation. The solution of this set of linear algebraic equations may be obtained using the Gauss elimination algorithms discussed in Chap. 3. However, other methods, which are mathematically more elegant and computationally less intensive, have been favored in the development of interpolating polynomials.

In the remainder of this section, we will develop three such interpolation methods: (1) the *Gregory-Newton formulas*, which are based on forward and backward differences of equally spaced points, (2) *Stirling's interpolation formula* based on central differences of equally spaced points, and (3) the *Lagrange polynomials* and the *orthogonal polynomials* for the interpolation of unequally spaced points.

### 4-7.1 Interpolation of Equally Spaced Points

First, we consider a set of known values of the function $f(x)$ at *equally spaced* values of $x$:

$$\begin{array}{ll} x-3h & f(x-3h) \\ x-2h & f(x-2h) \\ x-h & f(x-h) \\ x & f(x) \\ x+h & f(x+h) \\ x+2h & f(x+2h) \\ x+3h & f(x+3h) \end{array}$$

These points are represented graphically in Fig. 4-4*a* and are tabulated in Tables 4-7 and 4-8. The first, second, and third forward differences of these base points are also tabulated in Table 4-7 and the corresponding backward differences in Table 4-8.

The *Gregory-Newton forward interpolation formula* can be derived using the forward finite difference relations derived in Secs. 4-2 and 4-4. Equation (4-17), written for the function $f$,

$$f(x + h) = e^{hD} f(x) \tag{4-131}$$

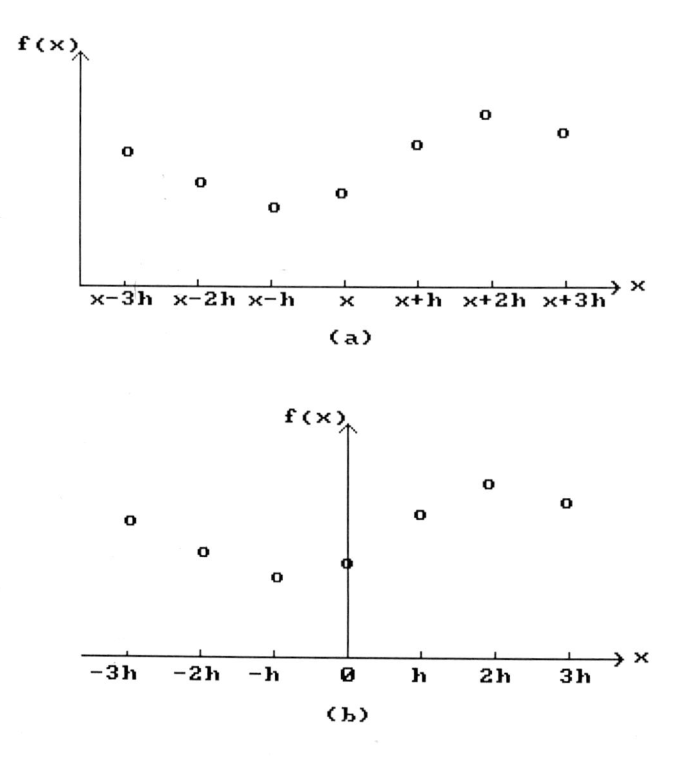


**Figure 4-4** (*a*) Equally spaced base points for interpolating polynomials. (*b*) Equally spaced base points with pivot point shifted to the origin.

relates the value of the function at one interval forward of the pivot point $x$ to the value of the function at the pivot point. Applying this equation for $n$ intervals forward, i.e., replacing $h$ with $nh$, we obtain

$$f(x + nh) = e^{nhD} f(x) \tag{4-132}$$

or equivalently

$$f(x + nh) = (e^{hD})^n f(x) \tag{4-133}$$

We note from Eq. (4-63) that

$$e^{hD} = 1 + \Delta \tag{4-134}$$

Combining Eqs. (4-133) and (4-134) we obtain

$$f(x + nh) = (1 + \Delta)^n f(x) \tag{4-135}$$

The term $(1 + \Delta)^n$ can be expanded using the *binomial series*

**Table 4-7 Forward difference table**

| $i$ | $x_i$ | $f(x_i)$ | $\Delta f(x_i)$ | $\Delta^2 f(x_i)$ | $\Delta^3 f(x_i)$ |
|---|---|---|---|---|---|
| 0 | $x$ | $f(x)$ | $f(x+h)-f(x)$ | $f(x+2h)-2f(x+h)+f(x)$ | $f(x+3h)-3f(x+2h)+3f(x+h)-f(x)$ |
| 1 | $x+h$ | $f(x+h)$ | $f(x+2h)-f(x+h)$ | $f(x+3h)-2f(x+2h)+f(x+h)$ | |
| 2 | $x+2h$ | $f(x+2h)$ | $f(x+3h)-f(x+2h)$ | | |
| 3 | $x+3h$ | $f(x+3h)$ | | | |

**Table 4-8 Backward difference table**

| $i$ | $x_i$ | $f(x_i)$ | $\nabla f(x_i)$ | $\nabla^2 f(x_i)$ | $\nabla^3 f(x_i)$ |
|---|---|---|---|---|---|
| −3 | $x-3h$ | $f(x-3h)$ | | | |
| −2 | $x-2h$ | $f(x-2h)$ | $f(x-2h)-f(x-3h)$ | | |
| −1 | $x-h$ | $f(x-h)$ | $f(x-h)-f(x-2h)$ | $f(x-h)-2f(x-2h)+f(x-3h)$ | |
| 0 | $x$ | $f(x)$ | $f(x)-f(x-h)$ | $f(x)-2f(x-h)+f(x-2h)$ | $f(x)-3f(x-h)+3f(x-2h)-f(x-3h)$ |

$$(1+\Delta)^n = 1 + n\Delta + \frac{n(n-1)}{2!}\Delta^2 + \frac{n(n-1)(n-2)}{3!}\Delta^3 + \frac{n(n-1)(n-2)(n-3)}{4!}\Delta^4 + \cdots \tag{4-136}$$

Therefore, Eq. (4-135) becomes

$$f(x+nh) = f(x) + n\Delta f(x) + \frac{n(n-1)}{2!}\Delta^2 f(x) + \frac{n(n-1)(n-2)}{3!}\Delta^3 f(x) + \frac{n(n-1)(n-2)(n-3)}{4!}\Delta^4 f(x) + \cdots \tag{4-137}$$

When $n$ is a positive integer, the binomial series has $n+1$ terms; therefore, Eq. (4-137) is a polynomial of degree $n$. If $n+1$ base-point values of the function $f$ are known, this polynomial fits all $n+1$ points exactly.

In order to be able to use this polynomial to interpolate the function at a point of interest which lies between the base points, it is convenient to shift the pivot point $x$ to the origin ($x = 0$) as shown in Fig. 4-4$b$. We can now designate the distance of the point of interest from the pivot point as $x$. The value of $n$ is no longer an integer and is replaced by

$$n = \frac{x}{h} \tag{4-138}$$

These substitutions convert Eq. (4-137) to

$$f(x) = f(0) + \frac{x}{h}\Delta f(0) + \frac{x(x-h)}{2!h^2}\Delta^2 f(0) + \frac{x(x-h)(x-2h)}{3!h^3}\Delta^3 f(0) + \frac{x(x-h)(x-2h)(x-3h)}{4!h^4}\Delta^4 f(0) + \cdots \tag{4-139}$$

This is the *Gregory-Newton forward interpolation formula*.

In a similar derivation, using backward differences, the *Gregory-Newton backward interpolation formula* is derived as

$$f(-x) = f(0) - \frac{x}{h}\nabla f(0) + \frac{x(x-h)}{2!h^2}\nabla^2 f(0) - \frac{x(x-h)(x-2h)}{3!h^3}\nabla^3 f(0) + \frac{x(x-h)(x-2h)(x-3h)}{4!h^4}\nabla^4 f(0) - \cdots \tag{4-140}$$

It was stated earlier that the binomial series [Eq. (4-136)] has a finite number of terms, $n+1$, when $n$ is a positive integer. However, in the Gregory-Newton interpolation formulas, $n$ is not usually an integer; therefore, these polynomials have an infinite number of terms. It is known from algebra [9] that if $|\Delta| \le 1$, then the binomial series for $(1+\Delta)^n$ converges to the value of $(1+\Delta)^n$ as the number of terms becomes larger and larger. This implies that the finite differences must be small. This is true for a flat, smooth function, or, alternatively, if the known base points are close together, i.e., if $h$ is small. Of course, the number of terms which can be used in each formula depends on the highest order of finite differences that can be evaluated from the available known data. It is common sense that for evenly spaced data, the accuracy of interpolation is higher for a large number of data points which are closely spaced together.

For a given set of data points, the accuracy of interpolation can be further enhanced by choosing the pivot point as close to the point of interest as possible, so that $x < h$. If this is satisfied, then the series should utilize as many terms as possible, i.e., the number of finite differences in the equation should be maximized. The order of error of the formula applied in each case is equivalent to the order of the finite difference contained in the first truncated term of the series. Examination of Table 4-7 reveals that points at the top of the table have the largest available number of forward differences, while Table 4-8 reveals that points at the bottom of the table have the largest number of backward differences. Therefore, the forward formula should be used for interpolating between points near the top of the table, and the backward formula should be used for interpolation near the bottom of the table [10].

*Stirling's interpolation formula* is based on central differences. Its derivation is similar to that of the Gregory-Newton formulas and can be arrived at by using either the symbolic operator relations or the Taylor series expansion of the function. We will use the latter and expand the function $f(x+nh)$ in a Taylor series around $x$:

$$f(x+nh) = f(x) + \frac{nh}{1!} f'(x) + \frac{n^2h^2}{2!} f''(x) + \frac{n^3h^3}{3!} f'''(x) + \cdots \tag{4-141}$$

We replace the derivatives of $f(x)$ with the differential operators to obtain

$$f(x+nh) = f(x) + \frac{nh}{1!} Df(x) + \frac{n^2h^2}{2!} D^2f(x) + \frac{n^3h^3}{3!} D^3f(x) + \cdots \tag{4-142}$$

The odd-order differential operators in Eq. (4-142) are replaced by averaged central differences and the even-order differential operators by central differences, all taken from Table 4-5 [Eqs. (4-99) to (4-102)]. Substituting these into (4-142) and regrouping of terms yield the formula

$$\begin{aligned} f(x+nh) = f(x) + n\mu\delta f(x) + \frac{n^2}{2!}\delta^2 f(x) + \frac{n(n^2-1)}{3!}\mu\delta^3 f(x) \\ + \frac{n^2(n^2-1)}{4!}\delta^4 f(x) + \frac{n(n^2-1)(n^2-4)}{5!}\mu\delta^5 f(x) \\ + \frac{n^2(n^2-1)(n^2-4)}{6!}\delta^6 f(x) + \cdots \end{aligned} \tag{4-143}$$

Shifting the pivot point to the origin and substituting

$$n = \frac{x}{h} \tag{4-144}$$

we obtain the final form of *Stirling's interpolation formula*

$$f(x) = f(0) + \frac{x}{h}\,\mu\delta f(0) + \frac{x^2}{2!h^2}\,\delta^2 f(0) + \frac{x(x^2-h^2)}{3!h^3}\,\mu\delta^3 f(0) + \frac{x^2(x^2-h^2)}{4!h^4}\,\delta^4 f(0) + \frac{x(x^2-h^2)(x^2-4h^2)}{5!h^5}\,\mu\delta^5 f(0) + \frac{x^2(x^2-h^2)(x^2-4h^2)}{6!h^6}\,\delta^6 f(0) + \cdots \tag{4-145}$$

The general formula for determining the higher-order terms containing odd differences, in the above series, is

$$\left[\frac{x}{(2k-1)!h^{(2k-1)}} \prod_{m=1}^{k-1} (x^2 - m^2h^2)\right] \mu\delta^{(2k-1)} f(0) \tag{4-146}$$

and the formula for terms with even differences is

$$\left[\frac{x^2}{(2k)!h^{(2k)}} \prod_{m=1}^{k-1} (x^2 - m^2h^2)\right] \delta^{2k} f(0) \tag{4-147}$$

where $k = 2, 3, \ldots \infty$.

Other forms of Stirling's interpolation formula exist, which make use of base points spaced at half intervals (i.e., at $h/2$). Our choice of using *averaged* central differences to replace the odd differential operators eliminated the need for having base points located at the midpoints. The central differences for Eq. (4-145) are tabulated in Table 4-9.

### 4-7.2 Interpolation of Unequally Spaced Points

Finally, let us consider a set of unequally spaced base points, such as those shown in Fig. 4-3*a*. Define the interpolating polynomial

$$P_n(x) = \sum_{k=0}^{n} p_k(x) f(x_k) \tag{4-148}$$

which is the sum of the *weighted* values of the function at all $n + 1$ base points. The weights $p_k(x)$ are $n$th-degree polynomial functions corresponding to each base point. Equation (4-148) is actually a linear combination of $n$th-degree polynomials; therefore, $P_n(x)$ is also an $n$th-degree polynomial.

**Table 4-9 Central difference table***

| $i$ | $x_i$ | $f(x_i)$ | $\mu\delta f(x_i)$ | $\delta^2 f(x_i)$ |
|---|---|---|---|---|
| $-3$ | $x-3h$ | $f(x-3h)$ | | |
| $-2$ | $x-2h$ | $f(x-2h)$ | $\frac{1}{2}[f(x-h)-f(x-3h)]$ | $f(x-h)-2f(x-2h)+f(x-3h)$ |
| $-1$ | $x-h$ | $f(x-h)$ | $\frac{1}{2}[f(x)-f(x-2h)]$ | $f(x)-2f(x-h)+f(x-2h)$ |
| $0$ | $x$ | $f(x)$ | $\frac{1}{2}[f(x+h)-f(x-h)]$ | $f(x+h)-2f(x)+f(x-h)$ |
| $1$ | $x+h$ | $f(x+h)$ | $\frac{1}{2}[f(x+2h)-f(x)]$ | $f(x+2h)-2f(x+h)+f(x)$ |
| $2$ | $x+2h$ | $f(x+2h)$ | $\frac{1}{2}[f(x+3h)-f(x+h)]$ | $f(x+3h)-2f(x+2h)+f(x+h)$ |
| $3$ | $x+3h$ | $f(x+3h)$ | | |

| $i$ | $\mu\delta^3 f(x_i)$ | $\delta^4 f(x_i)$ |
|---|---|---|
| $-3$ | | |
| $-2$ | | |
| $-1$ | $\frac{1}{2}[f(x+h)-2f(x)+2f(x-2h)-f(x-3h)]$ | $f(x+h)-4f(x)+6f(x-h)$ $-4f(x-2h)+f(x-3h)$ |
| $0$ | $\frac{1}{2}[f(x+2h)-2f(x+h)+2f(x-h)-f(x-2h)]$ | $f(x+2h)-4f(x+h)+6f(x)$ $-4f(x-h)+f(x-2h)$ |
| $1$ | $\frac{1}{2}[f(x+3h)-2f(x+2h)+2f(x)-f(x-h)]$ | $f(x+3h)-4f(x+2h)+6f(x+h)$ $-4f(x)+f(x-h)$ |
| $2$ | | |
| $3$ | | |

| $i$ | $\mu\delta^5 f(x_i)$ | $\delta^6 f(x_i)$ |
|---|---|---|
| $-3$ | | |
| $-2$ | | |
| $-1$ | | |
| $0$ | $\frac{1}{2}[f(x+3h)-4f(x+2h)+5f(x+h)$ $-5f(x-h)+4f(x-2h)-f(x-3h)]$ | $f(x+3h)-6f(x+2h)+15f(x+h)-20f(x)$ $+15f(x-h)-6f(x-2h)+f(x-3h)$ |
| $1$ | | |
| $2$ | | |
| $3$ | | |

* Read this table from left to right, starting with top section and continuing with middle and bottom sections.

In order for the interpolating polynomial to fit the function exactly at all the base points, each particular weighting polynomial $p_k(x)$ must be chosen so that it has the value of unity when $x = x_k$, and the value of zero at all other base points, i.e.,

$$p_k(x_i) = \begin{cases} 0 & i \neq k \\ 1 & i = k \end{cases} \tag{4-149}$$

The *Lagrange polynomials*, which have the form

$$p_k(x) = C_k \prod_{\substack{i=0 \\ i \neq k}}^{n} (x - x_i) \tag{4-150}$$

satisfy the first part of condition (4-149) since there will be a term $(x_i - x_i)$ in the product series of Eq. (4-150) whenever $x = x_i$. The constant $C_k$ is evaluated to make the Lagrange polynomial satisfy the second part of condition (4-149):

$$C_k = \frac{1}{\prod_{\substack{i=0 \\ i \neq k}}^{n} (x_k - x_i)} \tag{4-151}$$

Combination of Eqs. (4-150) and (4-151) gives the Lagrange polynomials

$$p_k(x) = \prod_{\substack{i=0 \\ i \neq k}}^{n} \left( \frac{x - x_i}{x_k - x_i} \right) \tag{4-152}$$

The interpolating polynomial $P_n(x)$ has a remainder term which can be obtained from Eq. (4-6):

$$R_n(x) = \prod_{i=0}^{n} (x - x_i) \frac{f^{(n+1)}(\xi)}{(n+1)!} \tag{4-153}$$

The accuracy of all interpolating polynomials depends greatly on the spacing between the base points. This is particularly true with unequally spaced points. The Lagrange polynomials, for instance, give satisfactory results between closely spaced points but may fluctuate excessively in the center regions between widely spaced points [10]. The accuracy of interpolating unequally spaced base points is improved by the use of orthogonal polynomials, which are discussed in the next section.

### 4-7.3 Orthogonal Polynomials

*Orthogonal polynomials* are a special category of functions which satisfy the following orthogonality condition with respect to a weighting function $w(x) \geq 0$, on the interval $[a, b]$:

$$\int_a^b w(x) g_n(x) g_m(x)\, dx = \begin{cases} 0 & \text{if } n \neq m \\ c(n) > 0 & \text{if } n = m \end{cases} \tag{4-154}$$

This orthogonality condition can be viewed as the continuous analog of the orthogonality property of two vectors (see Chap. 3)

$$\mathbf{x}^T \mathbf{y} = 0 \tag{3-92}$$

in $n$-dimensional space, where $n$ becomes very large and the elements of the vectors are represented as continuous functions of some independent variable [2].

There are many families of polynomials which obey the orthogonality condition. These are generally known by the name of the mathematician who discovered them: *Legendre*, *Chebyshev*, *Hermite*, and *Laguerre polynomials*† are the most widely used orthogonal polynomials. In this section, we list the Legendre and Chebyshev polynomials. For further discussion the reader is referred to Hamming [11] and Carnahan et al. [2].

The *Legendre polynomials* are orthogonal on the interval $[-1, 1]$ with respect to the weighting function $w(x) = 1$. The orthogonality condition is

$$\int_{-1}^{1} P_n(x) P_m(x)\, dx = \begin{cases} 0 & \text{if } n \neq m \\ \dfrac{2}{2n+1} & \text{if } n = m \end{cases} \tag{4-155}$$

They also satisfy the recurrence relation

$$(n+1)P_{n+1}(x) - (2n+1)xP_n(x) + nP_{n-1}(x) = 0 \tag{4-156}$$

Starting with

$$P_0(x) = 1 \qquad P_1(x) = x$$

the recurrence formula (4-156), or the orthogonality condition (4-155), can be used to generate the Legendre polynomials. These are listed in Table 4-10 and drawn in Fig. 4-5.

The *Chebyshev polynomials* are orthogonal on the interval $[-1, 1]$ with respect to the weighting function $w(x) = 1/\sqrt{1 - x^2}$. Their orthogonality condition is

$$\int_{-1}^{1} \frac{1}{\sqrt{1-x^2}} T_n(x) T_m(x)\, dx = \begin{cases} 0 & \text{if } n \neq m \\ \pi & \text{if } n = m = 0 \\ \dfrac{\pi}{2} & \text{if } n = m > 0 \end{cases} \tag{4-158}$$

and their recurrence relation is

$$T_{n+1} - 2xT_n + T_{n-1} = 0 \tag{4-159}$$

† Adrien-Marie Legendre (1752–1833); Pafnuty Lvovich Chebyshev (1821–1894); Charles Hermite (1822–1901); Edmund Nicolas Laguerre (1834–1886).

## Table 4-10 Orthogonal Polynomials

**Legendre polynomials**

$$
\begin{aligned}
P_0(x) &= 1 \\
P_1(x) &= x \\
P_2(x) &= \frac{3x^2 - 1}{2} \\
P_3(x) &= \frac{5x^3 - 3x}{2} \\
P_4(x) &= \frac{35x^4 - 30x^2 + 3}{8} \\
&\vdots
\end{aligned}
\tag{4-157}
$$

**Chebyshev polynomials**

$$
\begin{aligned}
T_0(x) &= 1 \\
T_1(x) &= x \\
T_2(x) &= 2x^2 - 1 \\
T_3(x) &= 4x^3 - 3x \\
T_4(x) &= 8x^4 - 8x^2 + 1 \\
&\vdots
\end{aligned}
\tag{4-160}
$$

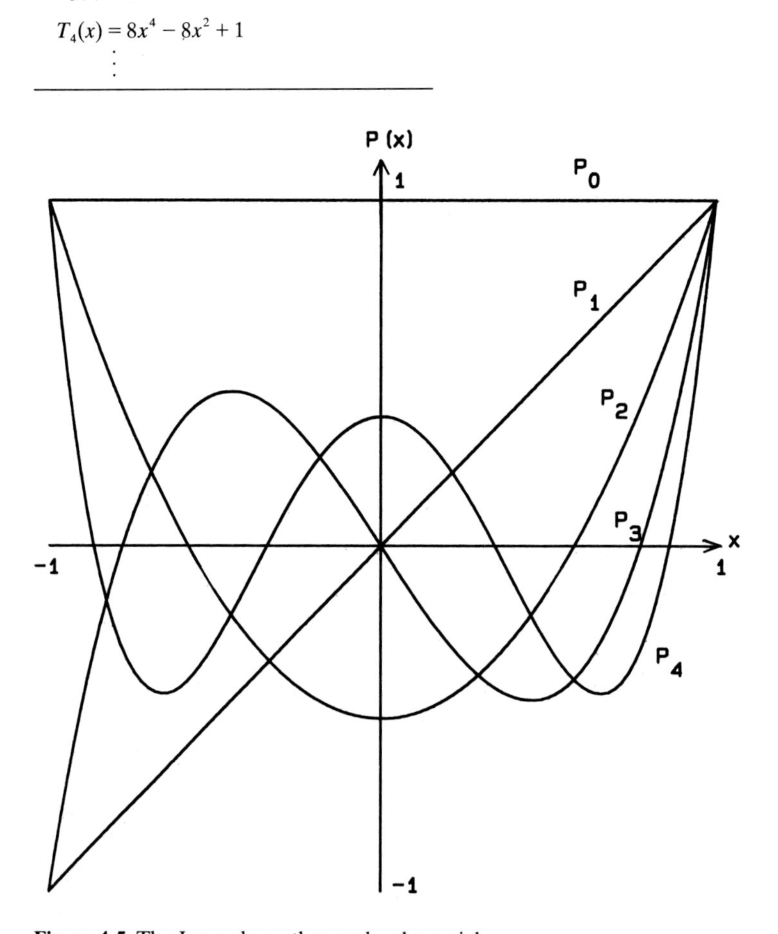


**Figure 4-5** The Legendre orthogonal polynomials.

Starting with

$$T_0(x) = 1 \qquad T_1(x) = x$$

the recurrence formula (4-159), or the orthogonality condition (4-158), can be used to generate the Chebyshev polynomials which are listed in Table 4-10 and drawn in Fig. 4-6.

It should be noticed from Figs. 4-5 and 4-6 that these orthogonal polynomials have their zeros (roots) more closely packed near the ends of the interval of integration. This property can be used to advantage in order to improve the accuracy of interpolation of unequally spaced points. This can be done in the case where the choice of base points is completely free. The interpolation can be performed using the Lagrange interpolation method described in Sec. 4-7.2, but the base points are chosen at the roots of the appropriate orthogonal polynomial. This concept is demonstrated in Chap. 5 in connection with the development of *Gauss quadrature*.

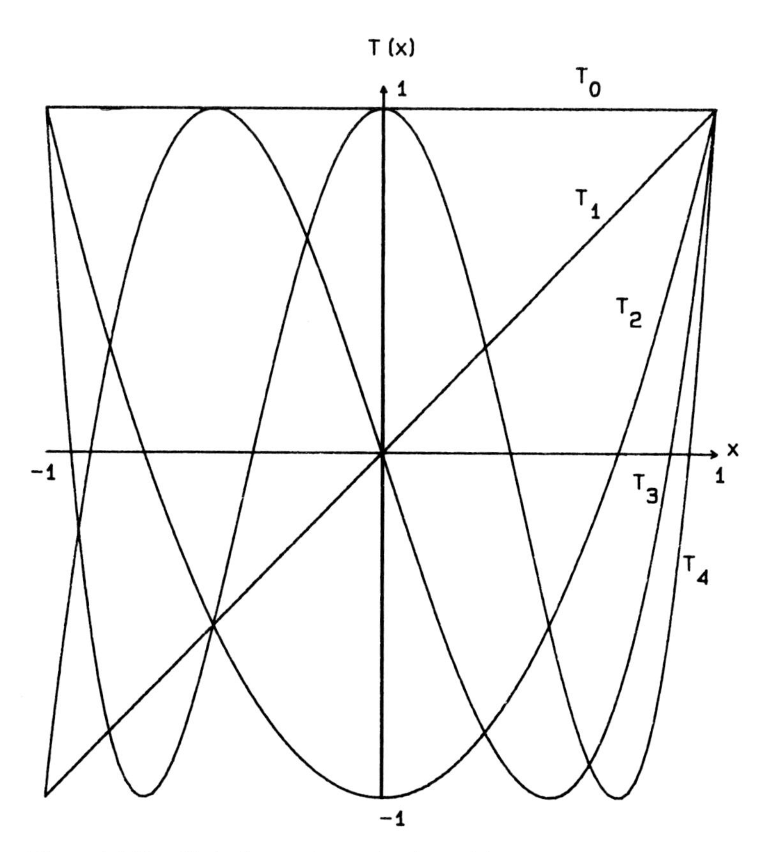


**Figure 4-6** The Chebyshev orthogonal polynomials.

# PROBLEMS

**4-1** Derive the equation which expresses the third-order derivative of $y$ in terms of backward finite differences, with

(*a*) Error of order $h$
(*b*) Error of order $h^2$

**4-2** Repeat Prob. 4-1, using forward finite differences.

**4-3** Derive the equations for the first, second, and third derivatives of $y$ in terms of backward finite differences with error of order $h^3$.

**4-4** Repeat Prob. 4-3, using forward finite differences.

**4-5** Derive the equation which expresses the third-order derivative of $y$ in terms of central finite differences, with

(*a*) Error of order $h^2$
(*b*) Error of order $h^4$

**4-6** Derive the equations for the first, second, and third derivatives of $y$ in terms of central finite differences with error of order $h^6$.

**4-7** Obtain the solution of the difference equation (4-112) directly from the solution of the differential equation (4-111) by utilizing the relationship $E = e^{hD}$.

**4-8** Derive the Gregory-Newton backward interpolation formula.

**4-9** Using the experimental data in Table P4-9

(*a*) Develop the forward difference table.
(*b*) Develop the backward difference table.
(*c*) Apply the Gregory-Newton interpolation formulas to evaluate the function at $x = 10$, 50, 90, 130, 170, and 190.

**4-10** Write a computer program which uses the Gregory-Newton interpolation formulas to evaluate the function $f(x)$ from a set of $n + 1$ equally spaced input values. Write the program in a general fashion so that $n$ can be any positive integer. Make use of both the forward and backward formulas to reduce the error. Provide a plotting routine to show how the program fits the data. Use the experimental data of Prob. 4-9 to verify the program, and evaluate the function at $x = 10$, 30, 50, 70, 90, 110, 130, 150, 170, and 190.

**4-11** Using the experimental data of Prob. 4-9,

(*a*) Develop the central difference table.

**Table P4-9**

| Time, h | Penicillin concentration, units/mL |
|---|---|
| 0 | 0 |
| 20 | 106 |
| 40 | 1600 |
| 60 | 3000 |
| 80 | 5810 |
| 100 | 8600 |
| 120 | 9430 |
| 140 | 10950 |
| 160 | 10280 |
| 180 | 9620 |
| 200 | 9400 |

**Table P4-13**

| $x$ | $f(x)$ |
|---|---|
| 1 | 7 |
| 3 | 3.5 |
| 6 | 3.2 |
| 7 | 3.9 |
| 10 | 8.2 |
| 12 | 9.0 |
| 13 | 9.2 |

(*b*) Apply Stirling's interpolation formula to evaluate the function at $x = 10$, 50, 90, 130, 170, and 190.

**4-12** Write a computer program which uses Stirling's interpolation formula to evaluate the function $f(x)$ from a set of $n + 1$ equally spaced input values. Write the program in a general fashion so that $n$ can be any positive integer. Provide a plotting routine to show how the program fits the data. Use the experimental data of Prob. 4-9 to verify the program and evaluate the function at $x = 10$, 30, 50, 70, 90, 110, 130, 150, 170, and 190.

**4-13** With the set of unequally spaced data points in Table P4-13 use Lagrange polynomials to evaluate the function at $x = 2$, 4, 5, 8, 9, and 11.

**4-14** Write a computer program which uses Lagrange polynomials to evaluate the function $f(x)$ from a set of $n + 1$ unequally spaced input values. Write the program in a general fashion so that $n$ can be any positive integer. Provide a plotting routine to show how the program fits the data. Use the experimental data of Prob. 4-13 to verify the program and evaluate the function at $x = 2$, 4, 5, 8, 9, and 11.

## REFERENCES

1. James, M. L., Smith, G. M., and Wolford, J. C.: *Applied Numerical Methods for Digital Computation with FORTRAN and CSMP*, 2d ed., Harper & Row, Publishers Incorporated, New York, 1977, p. 317.
2. Carnahan, B., Luther, H. A., and Wilkes, J. O.: *Applied Numerical Methods*, John Wiley & Sons, Inc., New York, 1969, p. 9.
3. Lapidus, L.: *Digital Computation for Chemical Engineers*, McGraw-Hill Book Company, New York, 1962, p. 14.
4. Salvadori, M. G., and Baron, M. L.: *Numerical Methods in Engineering*, Prentice-Hall, Inc., Englewood Cliffs, N.J., 1961.
5. Chorlton, F.: *Ordinary Differential and Difference Equations*, D. Van Nostrand Company, Ltd., London, 1965, p. 186.
6. Gel'fond, A. O.: *Calculus of Finite Differences*, English translation of the third Russian edition, Hindustan Publishing Corporation, Delhi, India, 1971, p. 371.
7. Vichnevetsky, R.: *Computer Methods for Partial Differential Equations. Vol. I. Elliptic Equations and the Finite-Element Method*, Prentice-Hall, Inc., Englewood Cliffs, N.J., 1981, p. 36.
8. Johnson, L. W., and Riess, R. D.: *Numerical Analysis*, 2d ed., Addison-Wesley Publishing Company, Inc., Reading, Mass., 1982, p. 417.
9. Kreyszig, E.: *Advanced Engineering Mathematics*, 3d ed., John Wiley & Sons, Inc., New York, 1972, p. 581.
10. Hornbeck, R. W.: *Numerical Methods*, Quantum Publishers, Inc., New York, 1975, p. 39.
11. Hamming, R. W.: *Numerical Method for Scientists and Engineers*, 2d ed., McGraw-Hill Book Company, New York, 1973.

CHAPTER

# FIVE

# NUMERICAL SOLUTION OF ORDINARY DIFFERENTIAL EQUATIONS

## 5-1 INTRODUCTION

Ordinary differential equations arise from the study of the dynamics of physical and chemical systems which have one independent variable. The latter may be either the space variable $x$ or the time variable $t$ depending on the geometry of the system and its boundary conditions.

For example, when a chemical reaction of the type

$$A + B \underset{k_2}{\overset{k_1}{\rightleftharpoons}} C + D \overset{k_3}{\longrightarrow} E \tag{5-1}$$

takes place in a reactor, the material balance can be applied:

$$\text{Input} + \text{generation} = \text{output} + \text{accumulation} \tag{5-2}$$

For a batch reactor, the input and output terms are zero; therefore, the material balance simplifies to

$$\text{Accumulation} = \text{generation} \tag{5-3}$$

Assuming that reaction (5-1) takes place in the liquid phase with negligible changes in volume, Eq. (5-3) written for each component of the reaction will have the form

$$
\begin{aligned}
\frac{dC_A}{dt} &= -k_1 C_A C_B + k_2 C_C C_D \\
\frac{dC_B}{dt} &= -k_1 C_A C_B + k_2 C_C C_D \\
\frac{dC_C}{dt} &= k_1 C_A C_B - k_2 C_C C_D - k_3 C_C^n C_D^m \\
\frac{dC_D}{dt} &= k_1 C_A C_B - k_2 C_C C_D - k_3 C_C^n C_D^m \\
\frac{dC_E}{dt} &= k_3 C_C^n C_D^m
\end{aligned}
\tag{5-4}
$$

where $C_A, C_B, \ldots, C_E$ represent the concentrations of the five chemical components of this reaction. This is a set of simultaneous *first-order nonlinear ordinary differential equations* which describe the dynamic behavior of the chemical reaction. With the methods to be developed in this chapter, these equations, with a set of initial conditions, can be integrated to obtain the time profiles of all the concentrations.

Consider the growth of a microorganism, say a yeast, in a continuous fermentor of the type shown in Fig. 5-1. The volume of the liquid in the fermentor is $V$. The flow rate of nutrients into the fermentor is $F_{in}$ and the flow rate of products out of the fermentor is $F_{out}$. The material balance for the cells is

$$\text{Input} + \text{generation} = \text{output} + \text{accumulation}$$

$$F_{in} X_{in} + r_X V = F_{out} X_{out} + \frac{d(VX)}{dt} \tag{5-5}$$

The material balance for the substrate $S$ is given by

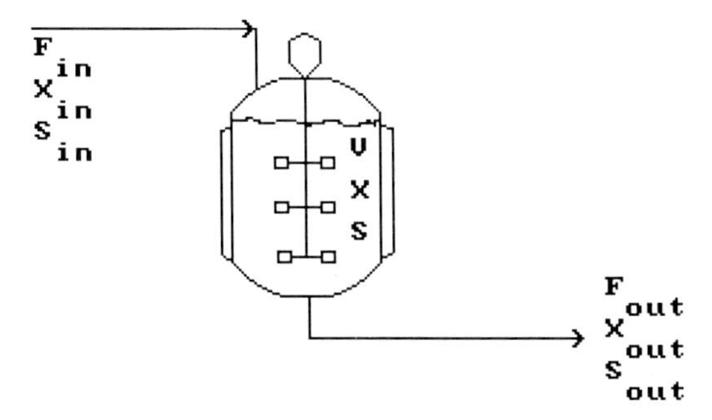


**Figure 5-1** Continuous fermentor.

$$F_{in}S_{in} + r_S V = F_{out}S_{out} + \frac{d(VS)}{dt} \tag{5-6}$$

The *overall* volumetric balance is

$$F_{in} = F_{out} + \frac{dV}{dt} \tag{5-7}$$

If we make the assumption that the fermentor is perfectly mixed, i.e., the concentrations at every point in the fermentor are the same, then

$$\begin{aligned} X &= X_{out} \\ S &= S_{out} \end{aligned} \tag{5-8}$$

and the equations simplify to

$$\frac{d(VX)}{dt} = (F_{in}X_{in} - F_{out}X) + r_X V \tag{5-9}$$

$$\frac{d(VS)}{dt} = (F_{in}S_{in} - F_{out}S) + r_S V \tag{5-10}$$

$$\frac{dV}{dt} = F_{in} - F_{out} \tag{5-11}$$

Further assumptions are made that the flow rates in and out of the fermentor are identical, and that the rates of cell formation and substrate utilization are given by

$$r_X = \frac{\mu_{max}SX}{K + S} \tag{5-12}$$

and

$$r_S = -\frac{1}{Y_S}\frac{\mu_{max}SX}{K + S} \tag{5-13}$$

The set of equations becomes

$$\frac{dX}{dt} = \left(\frac{F_{out}}{V}\right)(X_{in} - X) + \frac{\mu_{max}SX}{K + S} \tag{5-14}$$

$$\frac{dS}{dt} = \left(\frac{F_{out}}{V}\right)(S_{in} - S) - \frac{1}{Y_S}\frac{\mu_{max}SX}{K + S} \tag{5-15}$$

This is a set of simultaneous ordinary differential equations which describe the dynamics of a continuous culture fermentation.

The dynamic behavior of a distillation column may be examined by making material balances around each stage of the column. Figure 5-2 shows a typical stage $n$ with a liquid flow into the stage $L_{n+1}$ and out of the stage $L_n$; and a vapor flow into the stage $V_{n-1}$ and out of the stage $V_n$. The liquid holdup on the stage is designated as $H_n$. There is no generation of material in this process, so the material balance [Eq. (5-2)] becomes

$$\text{Accumulation} = \text{input} - \text{output}$$

$$\frac{dH_n}{dt} = V_{n-1} + L_{n+1} - V_n - L_n \tag{5-16}$$

The liquids and vapors in this operation are multicomponent mixtures of $k$ components. The mole fractions of each component in the liquid and vapor phases are designated by $x_i$ and $y_i$, respectively. Therefore, the material balance for the $i$th component is

$$\frac{d(H_n x_{i,n})}{dt} = V_{n-1} y_{i,n-1} + L_{n+1} x_{i,n+1} - V_n y_{i,n} - L_n x_{i,n} \tag{5-17}$$

The concentrations of liquid and vapor are related by the equilibrium relationship

$$y_{i,n} = f(x_{i,n}) \tag{5-18}$$

If the assumptions of constant molar overflow and negligible delay in vapor flow are made, then $V_{n-1} = V_n$. The delay in liquid flow is

$$\tau \frac{dL_n}{dt} = L_{n-1} - L_n \tag{5-19}$$

where $\tau$ is the hydraulic time constant.

The above equations applied to each stage in a multistage separation process result in a large set of simultaneous ordinary differential equations.

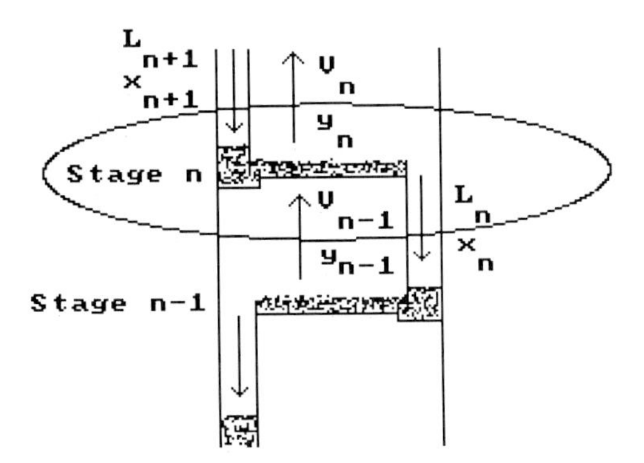


**Figure 5-2** Material balance around stage $n$ of a distillation column.

In all the above examples, the systems were chosen so that the models resulted in sets of *simultaneous first-order ordinary differential equations*. These are the most commonly encountered types of problems in the analysis of multicomponent and/or multistage operations. *Closed-form* solutions for such sets of equations are not usually obtainable. However, *numerical methods* have been thoroughly developed for the solution of sets of simultaneous differential equations. In this chapter we discuss the most useful techniques for the solution of such problems. We first show that higher-order differential equations can be reduced to first order by a series of substitutions.

## 5-2 CLASSIFICATIONS OF ORDINARY DIFFERENTIAL EQUATIONS

Ordinary differential equations are classified according to their *order*, their *linearity*, and their *boundary conditions*.

The order of a differential equation is the order of the highest derivative present in that equation. Examples of first-, second-, and third-order differential equations are given below:

First order:
$$\frac{dy}{dx} + y = kx \tag{5-20}$$

Second order:
$$\frac{d^2y}{dx^2} + y\frac{dy}{dx} = kx \tag{5-21}$$

Third order:
$$\frac{d^3y}{dx^3} + a\frac{d^2y}{dx^2} + b\left(\frac{dy}{dx}\right)^2 = kx \tag{5-22}$$

Ordinary differential equations may be categorized as *linear* and *nonlinear* equations. A differential equation is nonlinear if it contains products of the dependent variable, or of its derivatives, or of both. For example, Eqs. (5-21) and (5-22) are nonlinear because they contain the terms $y(dy/dx)$ and $(dy/dx)^2$, respectively, while Eq. (5-20) is linear. The general form of a linear differential equation of order $n$ may be written as

$$b_0(x)\frac{d^ny}{dx^n} + b_1(x)\frac{d^{n-1}y}{dx^{n-1}} + \cdots + b_{n-1}(x)\frac{dy}{dx} + b_n(x)y = R(x) \tag{5-23}$$

If $R(x) = 0$, the equation is called *homogeneous*. If $R(x) \neq 0$, the equation is *nonhomogeneous*. The coefficients $\{b_i \mid i = 0, 1, \ldots, n\}$ are called *variable coefficients* when they are functions of $x$, and *constant coefficients* when they are scalars. A differential equation is *autonomous* if the independent variable does not appear explicitly in that equation. For example, if Eq. (5-23) is homogeneous with constant coefficients, it is also autonomous.

To obtain a *unique solution* of an $n$th-order differential equation, or of a set of $n$ simultaneous first-order differential equations, it is necessary to specify $n$ values of the dependent variables (or their derivatives) at specific values of the independent variable.

Ordinary differential equations may be classified as *initial-value* problems or *boundary-value* problems. In initial-value problems, the values of the dependent variables and/or their derivatives are *all* known at the initial value of the independent variable†. In boundary-value problems, the dependent variables and/or their derivatives are known at more than one point of the independent variable. If some of the dependent variables (or their derivatives) are specified at the initial value of the independent variable, and the remaining variables (or their derivatives) are specified at the final value of the independent variable, then this is a *two-point boundary-value* problem.

The methods of solution of initial-value problems are developed in Sec. 5-6 and the methods for boundary-value problems are discussed in Sec. 5-7.

## 5-3 TRANSFORMATION TO CANONICAL FORM

Numerical integration of ordinary differential equations is most conveniently performed when the system consists of a set of $n$ simultaneous first-order ordinary differential equations of the form

$$\begin{aligned}\frac{dy_1}{dx} &= f_1(y_1, y_2, \ldots, y_n, x)\\ \frac{dy_2}{dx} &= f_2(y_1, y_2, \ldots, y_n, x)\\ &\vdots\\ \frac{dy_n}{dx} &= f_n(y_1, y_2, \ldots, y_n, x)\end{aligned} \tag{5-24}$$

This is called the *canonical* form of the equations [1]. When the initial conditions are given at a common point $x_0$,

$$\begin{aligned}y_1(x_0) &= y_{1,0}\\ y_2(x_0) &= y_{2,0}\\ &\vdots\\ y_n(x_0) &= y_{n,0}\end{aligned} \tag{5-25}$$

then the system equations (5-24) have solutions of the form

† A problem whose dependent variables and/or their derivatives are *all* known at the final value of the independent variable (rather than the initial value) is identical to the initial-value problem, since only the direction of integration must be reversed. Therefore, hereinafter the term initial-value problem refers to both cases.

$$
\begin{aligned}
y_1 &= F_1(x) \\
y_2 &= F_2(x) \\
&\vdots \\
y_n &= F_n(x)
\end{aligned}
\tag{5-26}
$$

The above problem can be condensed into matrix notation, where the system equations are represented by

$$\frac{d\mathbf{y}}{dx} = \mathbf{f}(\mathbf{y}, x) \tag{5-27}$$

the vector of initial conditions is

$$\mathbf{y}(x_0) = \mathbf{y}_0 \tag{5-28}$$

and the vector of solutions is

$$\mathbf{y} = \mathbf{F}(x) \tag{5-29}$$

Differential equations of higher order, or systems containing equations of mixed order, can be transformed to the canonical form by a series of substitutions. For example, consider the $n$th-order differential equation

$$\frac{d^n z}{dx^n} = G\left(z, \frac{dz}{dx}, \frac{d^2 z}{dx^2}, \ldots, \frac{d^{n-1} z}{dx^{n-1}}, x\right) \tag{5-30}$$

The following transformations

$$
\begin{aligned}
z &= y_1 \\
\frac{dz}{dx} &= \frac{dy_1}{dx} = y_2 \\
\frac{d^2 z}{dx^2} &= \frac{dy_2}{dx} = y_3 \\
&\vdots \\
\frac{d^{n-1} z}{dx^{n-1}} &= \frac{dy_{n-1}}{dx} = y_n \\
\frac{d^n z}{dx^n} &= \frac{dy_n}{dx}
\end{aligned}
\tag{5-31}
$$

when substituted into the $n$th-order equation (5-30), give the equivalent set of $n$ first-order equations of canonical form

$$\frac{dy_1}{dx} = y_2$$

$$\frac{dy_2}{dx} = y_3$$

$$\vdots \tag{5-32}$$

$$\frac{dy_n}{dx} = G(y_1, y_2, y_3, \ldots, y_n, x)$$

If the right-hand side of the differential equations is not a function of the independent variable, i.e.,

$$\frac{d\mathbf{y}}{dx} = \mathbf{f}(\mathbf{y}) \tag{5-33}$$

then the set is *autonomous*. A *nonautonomous* set may be transformed to an autonomous set by an appropriate substitution (see Examples 5-2 and 5-4). If the functions $f(y)$ are linear in terms of $y$, then the equations can be written in matrix form

$$\mathbf{y}' = \mathbf{A}\mathbf{y} \tag{5-34}$$

as in Examples 5-1 and 5-2. Solutions for linear sets of ordinary differential equations are developed in Sec. 5-4. The methods for solution of nonlinear sets are discussed in Secs. 5-6 and 5-7.

A more restricted form of differential equation is

$$\frac{d\mathbf{y}}{dx} = \mathbf{f}(x) \tag{5-35}$$

where the $\mathbf{f}(x)$ are functions of the independent variable only. Solution methods for these equations are developed in Sec. 5-5.

The next four examples demonstrate the technique for converting higher-order linear and nonlinear differential equations to canonical form.

**Example 5-1** Transform the following fourth-order homogeneous linear ordinary differential equation to a set of first-order linear ordinary differential equations:

$$\frac{d^4z}{dt^4} + 5\frac{d^3z}{dt^3} - 2\frac{d^2z}{dt^2} - 6\frac{dz}{dt} + 3z = 0 \tag{5-36}$$

SOLUTION Apply the transformations according to Eqs. (5-31):

$$
\begin{aligned}
z &= y_1 \\
\frac{dz}{dt} &= \frac{dy_1}{dt} = y_2 \\
\frac{d^2z}{dt^2} &= \frac{dy_2}{dt} = y_3 \\
\frac{d^3z}{dt^3} &= \frac{dy_3}{dt} = y_4 \\
\frac{d^4z}{dt^4} &= \frac{dy_4}{dt}
\end{aligned}
\tag{5-37}
$$

Make these substitutions into Eq. (5-36) to obtain the following four equations:

$$
\begin{aligned}
\frac{dy_1}{dt} &= y_2 \\
\frac{dy_2}{dt} &= y_3 \\
\frac{dy_3}{dt} &= y_4 \\
\frac{dy_4}{dt} &= -3y_1 + 6y_2 + 2y_3 - 5y_4
\end{aligned}
\tag{5-38}
$$

This is a set of linear ordinary differential equations which can be represented in matrix form

$$\dot{\mathbf{y}} = \mathbf{A}\mathbf{y} \tag{5-39}$$

where matrix $\mathbf{A}$ is given by

$$\mathbf{A} = \begin{bmatrix} 0 & 1 & 0 & 0 \\ 0 & 0 & 1 & 0 \\ 0 & 0 & 0 & 1 \\ -3 & 6 & 2 & -5 \end{bmatrix} \tag{5-40}$$

The method of obtaining the solution of sets of linear ordinary differential equations is discussed in Sec. 5-4.

**Example 5-2** Transform the fourth-order nonhomogeneous linear ordinary differential equation into its canonical form:

$$\frac{d^4z}{dt^4} + 5\frac{d^3z}{dt^3} - 2\frac{d^2z}{dt^2} - 6\frac{dz}{dt} + 3z = e^{-t} \tag{5-41}$$

SOLUTION The presence of the term $e^{-t}$ on the right-hand side of this equation makes it a nonhomogeneous equation. The left-hand side is identical to that of Eq. (5-36), so that the transformations [Eq. (5-37)] of Example 5-1 are applicable. An additional transformation is needed to replace the $e^{-t}$ term. This transformation is

$$\begin{aligned} y_5 &= e^{-t} \\ \frac{dy_5}{dt} &= -e^{-t} = -y_5 \end{aligned} \tag{5-42}$$

Make the substitutions into Eq. (5-41) to obtain the following set of five linear ordinary differential equations:

$$\begin{aligned} \frac{dy_1}{dt} &= y_2 \\ \frac{dy_2}{dt} &= y_3 \\ \frac{dy_3}{dt} &= y_4 \\ \frac{dy_4}{dt} &= -3y_1 + 6y_2 + 2y_3 - 5y_4 + y_5 \\ \frac{dy_5}{dt} &= -y_5 \end{aligned} \tag{5-43}$$

which also condenses into the matrix form of Eq. (5-39), with the matrix **A** given by

$$\mathbf{A} = \begin{bmatrix} 0 & 1 & 0 & 0 & 0 \\ 0 & 0 & 1 & 0 & 0 \\ 0 & 0 & 0 & 1 & 0 \\ -3 & 6 & 2 & -5 & 1 \\ 0 & 0 & 0 & 0 & -1 \end{bmatrix} \tag{5-44}$$

**Example 5-3** Transform the following third-order nonlinear ordinary differential equation into its canonical form:

$$\frac{d^3z}{dx^3} + z^2 \frac{d^2z}{dx^2} - \left(\frac{dz}{dx}\right)^3 - 2z = 0 \tag{5-45}$$

SOLUTION Apply the following transformations:

$$
\begin{aligned}
z &= y_1 \\
\frac{dz}{dx} &= \frac{dy_1}{dx} = y_2 \\
\frac{d^2 z}{dx^2} &= \frac{dy_2}{dx} = y_3 \\
\frac{d^3 z}{dx^3} &= \frac{dy_3}{dx}
\end{aligned} \tag{5-46}
$$

Make the substitutions into Eq. (5-45) to obtain the set

$$
\begin{aligned}
\frac{dy_1}{dx} &= y_2 \\
\frac{dy_2}{dx} &= y_3 \\
\frac{dy_3}{dx} &= 2y_1 + y_2^3 - y_1^2 y_3
\end{aligned} \tag{5-47}
$$

This is a set of *nonlinear* differential equations which cannot be expressed in matrix form. The methods of solution of nonlinear differential equations are developed in Secs. 5-6 and 5-7.

**Example 5-4** The following third-order linear differential equation has variable coefficients:

$$
\frac{d^3 z}{dt^3} + t^3 \frac{d^2 z}{dt^2} - t^2 \frac{dz}{dt} + 5z = 0 \tag{5-48}
$$

Transform this equation to its canonical form.

SOLUTION Apply the following transformations:

$$
\begin{aligned}
z &= y_1 \\
\frac{dz}{dt} &= \frac{dy_1}{dt} = y_2 \\
\frac{d^2 z}{dt^2} &= \frac{dy_2}{dt} = y_3 \\
\frac{d^3 z}{dt^3} &= \frac{dy_3}{dt} \\
y_4 &= t \\
\frac{dy_4}{dt} &= 1
\end{aligned} \tag{5-49}
$$

Make the substitutions into Eq. (5-48) to obtain the set

$$\begin{aligned}
\frac{dy_1}{dt} &= y_2 \\
\frac{dy_2}{dt} &= y_3 \\
\frac{dy_3}{dt} &= -5y_1 + y_4^2 y_2 - y_4^3 y_3 \\
\frac{dy_4}{dt} &= 1
\end{aligned} \tag{5-50}$$

This is a set of autonomous nonlinear differential equations.

## 5-4 LINEAR ORDINARY DIFFERENTIAL EQUATIONS

The analysis of many *physicochemical* systems yields mathematical models that are sets of *linear* ordinary differential equations with constant coefficients and can be reduced to the form

$$\dot{\mathbf{y}} = \mathbf{A}\mathbf{y} \tag{5-51}$$

with given initial conditions

$$\mathbf{y}(0) = \mathbf{y}_0 \tag{5-52}$$

Such examples abound in chemical engineering. The modeling of the kinetics of monomolecular reactions was demonstrated in Example 3-4. The *unsteady-state* material and energy balances of multiunit processes, without chemical reaction, often yield linear differential equations.

Sets of linear ordinary differential equations with constant coefficients have closed-form solutions which can be readily obtained from the eigenvalues and eigenvectors of the matrix **A**.

In order to develop this solution, let us first consider a single linear differential equation of the type

$$\frac{dy}{dt} = ay \tag{5-53}$$

with the given initial condition

$$y(0) = y_0 \tag{5-54}$$

Equation (5-53) is essentially the scalar form of the matrix set of Eqs. (5-51). The solution of the scalar equation can be obtained by separating the variables and integrating both sides of the equation

$$\int_{y_0}^{y} \frac{dy}{y} = \int_0^t a\,dt$$

$$\ln \frac{y}{y_0} = at$$

$$y = e^{at} y_0 \tag{5-55}$$

In an analogous fashion, the matrix set can be integrated to obtain the solution

$$\mathbf{y} = \mathbf{e}^{\mathbf{A}t} \mathbf{y}_0 \tag{5-56}$$

In this case, **y** and $\mathbf{y}_0$ are *vectors* of the dependent variables and the initial conditions, respectively. The term $\mathbf{e}^{\mathbf{A}t}$ is the matrix exponential function, which can be obtained from Eq. (3-84):

$$\mathbf{e}^{\mathbf{A}t} = \mathbf{I} + \mathbf{A}t + \frac{\mathbf{A}^2 t^2}{2!} + \frac{\mathbf{A}^3 t^3}{3!} + \cdots \tag{3-84}$$

It has been demonstrated [1] that (5-56) is a solution of (5-51) by differentiating it:

$$\begin{aligned}
\frac{d\mathbf{y}}{dt} &= \frac{d}{dt}(\mathbf{e}^{\mathbf{A}t})\mathbf{y}_0 \\
&= \frac{d}{dt}\left(\mathbf{I} + \mathbf{A}t + \frac{\mathbf{A}^2 t^2}{2!} + \frac{\mathbf{A}^3 t^3}{3!} + \cdots\right)\mathbf{y}_0 \\
&= \left(\mathbf{A} + \mathbf{A}^2 t + \frac{\mathbf{A}^3 t^2}{2!} + \cdots\right)\mathbf{y}_0 \\
&= \mathbf{A}\left(\mathbf{I} + \mathbf{A}t + \frac{\mathbf{A}^2 t^2}{2!} + \cdots\right)\mathbf{y}_0 \\
&= \mathbf{A}(\mathbf{e}^{\mathbf{A}t})\mathbf{y}_0 \\
&= \mathbf{A}\mathbf{y}
\end{aligned}$$

The solution of the set of linear ordinary differential equations is very cumbersome to evaluate in the form of Eq. (5-56) because it requires the evaluation of the infinite series of the exponential term $\mathbf{e}^{\mathbf{A}t}$. However, this solution can be modified, by further algebraic manipulation, to express it in terms of the eigenvalues and eigenvectors of the matrix **A**.

In Chap. 3, we showed that a nonsingular matrix **A** of order $n$ has $n$ eigenvectors and $n$ nonzero eigenvalues, whose definitions are given by

$$
\begin{aligned}
\mathbf{A}\mathbf{x}_1 &= \lambda_1 \mathbf{x}_1 \\
\mathbf{A}\mathbf{x}_2 &= \lambda_2 \mathbf{x}_2 \\
&\vdots \\
\mathbf{A}\mathbf{x}_n &= \lambda_n \mathbf{x}_n
\end{aligned} \tag{5-57}
$$

All the above eigenvectors and eigenvalues can be represented in a more compact form as follows:

$$\mathbf{AX} = \mathbf{X\Lambda} \tag{5-58}$$

where the columns of matrix $\mathbf{X}$ are the individual eigenvectors

$$\mathbf{X} = [\mathbf{x}_1, \mathbf{x}_2, \mathbf{x}_3, \ldots, \mathbf{x}_n] \tag{5-59}$$

and $\mathbf{\Lambda}$ is a diagonal matrix with the eigenvalues of $\mathbf{A}$ on its diagonal:

$$\mathbf{\Lambda} = \begin{bmatrix} \lambda_1 & 0 & 0 & \cdots & 0 \\ 0 & \lambda_2 & 0 & \cdots & 0 \\ 0 & 0 & \lambda_3 & \cdots & 0 \\ \cdot & \cdot & \cdot & \cdots & \cdot \\ 0 & 0 & 0 & \cdots & \lambda_n \end{bmatrix} \tag{5-60}$$

If we postmultiply each side of Eq. (5-58) by $\mathbf{X}^{-1}$, we obtain

$$
\begin{aligned}
\mathbf{AXX}^{-1} &= \mathbf{X\Lambda X}^{-1} \\
\mathbf{A} &= \mathbf{X\Lambda X}^{-1}
\end{aligned} \tag{5-61}
$$

Squaring Eq. (5-61),

$$
\begin{aligned}
\mathbf{A}^2 &= [\mathbf{X\Lambda X}^{-1}][\mathbf{X\Lambda X}^{-1}] \\
&= \mathbf{X\Lambda}^2\mathbf{X}^{-1}
\end{aligned} \tag{5-62}
$$

Similarly, raising Eq. (5-61) to any power $n$ we obtain

$$\mathbf{A}^n = \mathbf{X\Lambda}^n\mathbf{X}^{-1} \tag{5-63}$$

Starting with Eq. (3-84) and replacing the matrices $\mathbf{A}, \mathbf{A}^2, \ldots, \mathbf{A}^n$ with their equivalent from Eqs. (5-61) to (5-63), we obtain

$$\mathbf{e}^{\mathbf{A}t} = \mathbf{I} + \mathbf{X\Lambda X}^{-1}t + \mathbf{X\Lambda}^2\mathbf{X}^{-1}\frac{t^2}{2!} + \cdots \tag{5-64}$$

The identity matrix $\mathbf{I}$ can be premultiplied by $\mathbf{X}$ and postmultiplied by $\mathbf{X}^{-1}$ without changing it. Therefore, Eq. (5-64) rearranges to

$$\mathbf{e}^{\mathbf{A}t} = \mathbf{X}\left(\mathbf{I} + \mathbf{\Lambda}t + \frac{\mathbf{\Lambda}^2 t^2}{2!} + \cdots\right)\mathbf{X}^{-1} \tag{5-65}$$

which simplifies to

$$\mathbf{e}^{\mathbf{A}t} = \mathbf{X}\mathbf{e}^{\mathbf{\Lambda}t}\mathbf{X}^{-1} \tag{5-66}$$

where the exponential matrix $\mathbf{e}^{\mathbf{\Lambda}t}$ is defined as

$$\mathbf{e}^{\mathbf{\Lambda}t} = \begin{bmatrix} e^{\lambda_1 t} & 0 & 0 & \cdots & 0 \\ 0 & e^{\lambda_2 t} & 0 & \cdots & 0 \\ 0 & 0 & e^{\lambda_3 t} & \cdots & 0 \\ \cdot & \cdot & \cdot & \cdot & \cdot \\ 0 & 0 & 0 & \cdots & e^{\lambda_n t} \end{bmatrix} \tag{5-67}$$

The solution of the linear differential equations can now be expressed in terms of eigenvalues and eigenvectors by combining Eqs. (5-56) and (5-66):

$$\mathbf{y} = [\mathbf{X}\mathbf{e}^{\mathbf{\Lambda}t}\mathbf{X}^{-1}]\mathbf{y}_0 \tag{5-68}$$

The eigenvalues and eigenvectors of matrix **A** can be calculated using the techniques developed in Chap. 3. This is demonstrated in Example 5-5.

**Example 5-5 Solution of linear differential equations** Utilize the results obtained in Example 3-4 to construct the solution of the set of linear differential equations which describe the monomolecular kinetics of the chemical reaction

$$\mathrm{A} \xrightarrow{k_1} \mathrm{B} \underset{k_3}{\overset{k_2}{\rightleftarrows}} \mathrm{C}$$

The values of the kinetic rate constants are

$$k_1 = 1 \qquad k_2 = 2 \qquad k_3 = 3$$

The initial concentrations of the three components are

$$A_0 = 1 \qquad B_0 = 0 \qquad C_0 = 0$$

SOLUTION The differential equations, which were derived in Example 3-4, are

$$\frac{dA}{dt} = -k_1 A$$

$$\frac{dB}{dt} = k_1 A - k_2 B + k_3 C$$

$$\frac{dC}{dt} = k_2 B - k_3 C$$

In matrix form this set reduces to

$$\dot{\mathbf{y}} = \mathbf{K}\mathbf{y}$$

where $\dot{\mathbf{y}}$ is the vector of derivatives, $\mathbf{y}$ is the vector of concentrations, and $\mathbf{K}$ is the matrix

$$\mathbf{K} = \begin{bmatrix} -1 & 0 & 0 \\ 1 & -2 & 3 \\ 0 & 2 & -3 \end{bmatrix}$$

The solution of a set of linear ordinary differential equations is given by Eq. (5-68):

$$\mathbf{y} = [\mathbf{X}\mathbf{e}^{\Lambda t}\mathbf{X}^{-1}]\mathbf{y}_0 \tag{5-68}$$

The matrix $\mathbf{X}$ consists of the eigenvectors of $\mathbf{K}$, which were determined in Example 3-4:

$$\mathbf{X} = \begin{bmatrix} 0 & 1 & 0 \\ 1 & -0.5 & -1 \\ 0.6667 & -0.5 & 1 \end{bmatrix}$$

The inverse of $\mathbf{X}$ is

$$\mathbf{X}^{-1} = \begin{bmatrix} 0.6 & 0.6 & 0.6 \\ 1 & 0 & 0 \\ 0.1 & -0.4 & 0.6 \end{bmatrix}$$

This was obtained by using the computer program JORDAN.BAS of Example 3-2, as shown in the printout of results given at the end of this example. The eigenvalues of matrix $\mathbf{K}$, which were also determined in Example 3-4, are

$$\lambda_1 = 0 \qquad \lambda_2 = -1 \qquad \lambda_3 = -5$$

Therefore the solution of the differential equations is

$$\mathbf{y} = \begin{bmatrix} 0 & 1 & 0 \\ 1 & -0.5 & -1 \\ 0.6667 & -0.5 & 1 \end{bmatrix} \begin{bmatrix} e^{0t} & 0 & 0 \\ 0 & e^{-t} & 0 \\ 0 & 0 & e^{-5t} \end{bmatrix} \begin{bmatrix} 0.6 & 0.6 & 0.6 \\ 1 & 0 & 0 \\ 0.1 & -0.4 & 0.6 \end{bmatrix} \begin{bmatrix} 1 \\ 0 \\ 0 \end{bmatrix}$$

Multiplying through we obtain

$$y_1 = e^{-t}$$

$$y_2 = 0.6 - 0.5e^{-t} - 0.1e^{-5t}$$

$$y_3 = 0.4 - 0.5e^{-t} + 0.1e^{-5t}$$

Verification of the solution shows that at $t = 0$

$$\mathbf{y} = \begin{bmatrix} 1 \\ 0 \\ 0 \end{bmatrix}$$

and at $t = \infty$

$$\mathbf{y} = \begin{bmatrix} 0 \\ 0.6 \\ 0.4 \end{bmatrix}$$

which is the expected steady-state solution of the above chemical reaction.

RESULTS

```
*******************************************************************
*                                                                 *
*                          EXAMPLE 3-2                            *
*                                                                 *
*               THE GAUSS-JORDAN REDUCTION METHOD                 *
*                                                                 *
*         FOR SIMULTANEOUS LINEAR ALGEBRAIC EQUATIONS             *
*                                                                 *
*                     AND MATRIX INVERSION                        *
*                                                                 *
*                          (JORDAN.BAS)                           *
*                                                                 *
*******************************************************************
YOU MAY USE THIS PROGRAM TO:

            1. SOLVE LINEAR ALGEBRAIC EQUATIONS

            2. FIND THE INVERSE OF A MATRIX

            3. DO BOTH OF THE ABOVE
THE NUMBER OF YOUR SELECTION? 2

NUMBER OF ROWS OF THE MATRIX? 3

ENTER ELEMENTS OF MATRIX

ROW  1

   ELEMENT ( 1 , 1 ) =? 0
   ELEMENT ( 1 , 2 ) =? 1
   ELEMENT ( 1 , 3 ) =? 0

ROW  2

   ELEMENT ( 2 , 1 ) =? 1
   ELEMENT ( 2 , 2 ) =? -0.5
   ELEMENT ( 2 , 3 ) =? -1

ROW  3

   ELEMENT ( 3 , 1 ) =? 0.6667
   ELEMENT ( 3 , 2 ) =? -0.5
   ELEMENT ( 3 , 3 ) =? 1

GIVE THE MINIMUM ALLOWABLE VALUE OF THE PIVOT ELEMENT? 0.00001


*******************************************************************
AUGMENTED MATRIX:

 0              1              0              0

 1             -.5            -1              0

 .6667         -.5             1              0
```

```
*******************************************************************

IS THE AUGMENTED MATRIX CORRECT(Y/N)? Y

DO YOU WANT TO SEE STEP-BY-STEP RESULTS(Y/N)? N

*******************************************************************
INVERSE OF MATRIX:

 .599988         .599988         .599988

 1               0               0

 .099988        -.400012         .599988

*******************************************************************
PRODUCT OF MATRIX AND INVERSE SHOULD BE THE IDENTITY MATRIX:

 1.0000          0.0000          0.0000
-0.0000          1.0000          0.0000
-0.0000         -0.0000          1.0000

*******************************************************************

DO YOU WANT TO REPEAT THE CALCULATIONS
WITH MINOR CHANGES TO THE COEFFICIENTS(Y/N)? N

DO YOU WANT TO RESET ALL THE COEFFICIENTS(Y/N)? N


*********************** END OF PROGRAM ****************************
Ok
```

## 5-5 INTEGRATION FORMULAS

In this section we develop the integration formulas which can be used to integrate the category of differential equations represented by

$$\frac{dy}{dx} = f(x) \tag{5-69}$$

with initial conditions given as

$$y(0) = y_0 \tag{5-70}$$

The methods of solution of this category of differential equations are *always* of the *explicit* type, because the function $f(x)$ can be calculated at any value of $x$, without prior knowledge of the dependent variable $y$. For this reason, these methods are easier to obtain and execute.

Equation (5-69) may be either a scalar equation or a set of equations. However, since the dependent variable does not appear on the right-hand side, the equations are not coupled; therefore, they can be integrated either individually or collectively. The methods of solution developed in this section apply equally well to one, or many, differential equations of the form of (5-69).

In order to perform the integration of (5-69), we rearrange this equation and integrate both sides:

$$\int_{y_0}^{y_n} dy = \int_0^{x_n} f(x)\, dx \tag{5-71}$$

The left side integrates readily to obtain

$$y_n - y_0 = \int_0^{x_n} f(x)\, dx \tag{5-72}$$

If the function $f(x)$ is such that it can be integrated analytically, then numerical methods are not needed for this problem. However, in many cases the function $f(x)$ is very complicated, or the function is only a set of tabulated values of $x$ and $f(x)$, such as experimental data. Under these circumstances, the integral $\int_0^{x_n} f(x)\, dx$ must be evaluated numerically. This operation is known as *numerical quadrature*.

It is known from differential calculus that the integral of a function $f(x)$ is equivalent to the area between the function and the $x$ axis enclosed within the limits of integration, as shown in Fig. 5-3. Any portion of the area that is below

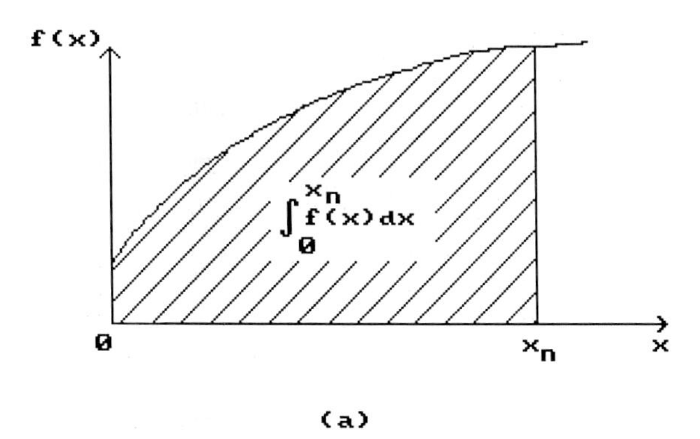


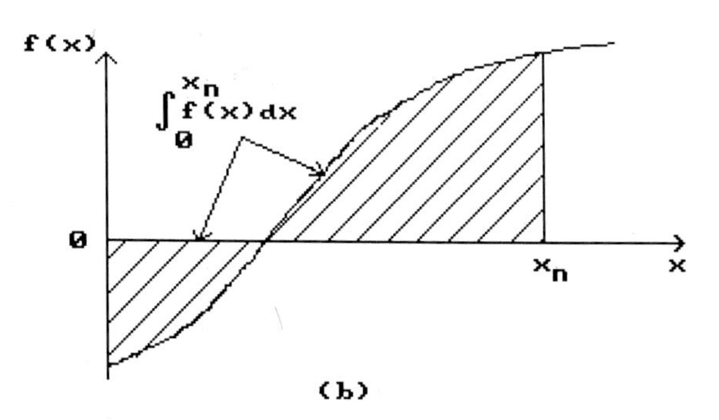


**Figure 5-3** Graphical representation of the integral: (*a*) positive area only, (*b*) positive and negative areas.

the $x$ axis is counted as negative area (Fig. 5-3*b*). Therefore, one way of evaluating the integral $\int_0^{x_n} f(x)\,dx$ is to plot the function graphically and then simply measure the area enclosed by the function. However, this is a very impractical and inaccurate way of evaluating integrals.

A more accurate and systematic way of evaluating integrals is to perform the integration numerically. This is accomplished by first replacing the function $f(x)$ with a polynomial interpolation formula, such as the Gregory-Newton forward interpolation formula (4-139):

$$f(x) = f(0) + \frac{x}{h}\,\Delta f(0) + \frac{x(x-h)}{2!h^2}\,\Delta^2 f(0) + \frac{x(x-h)(x-2h)}{3!h^3}\,\Delta^3 f(0) + \cdots \tag{4-139}$$

Since this interpolation formula fits the function exactly at a finite number of points $(n+1)$, we divide the total interval of integration $[0, x_n]$ into $n$ segments, each of width $h$, as shown in Fig. 5-4. Substitution of the Gregory-Newton formula into Eq. (5-71) gives

$$\int_{y_0}^{y_i} dy = \int_0^{x_i} \left[ f(0) + \frac{x}{h}\,\Delta f(0) + \frac{x(x-h)}{2!h^2}\,\Delta^2 f(0) + \frac{x(x-h)(x-2h)}{3!h^3}\,\Delta^3 f(0) + \cdots \right] dx \tag{5-73}$$

Both sides of this equation can be integrated. The upper limits of integration can be chosen to include an increasing set of segments of integration, each of width $h$. For each case, we retain a number of finite differences in the infinite series equal to the number of segments of integration. This operation yields the well-known *Newton-Cotes closed formulas of integration* [2]. The first three of the Newton-Cotes closed formulas are also known by the names *trapezoidal rule*, *Simpson's $\frac{1}{3}$ rule* and *Simpson's $\frac{3}{8}$ rule*, respectively. These are developed in the next three sections.

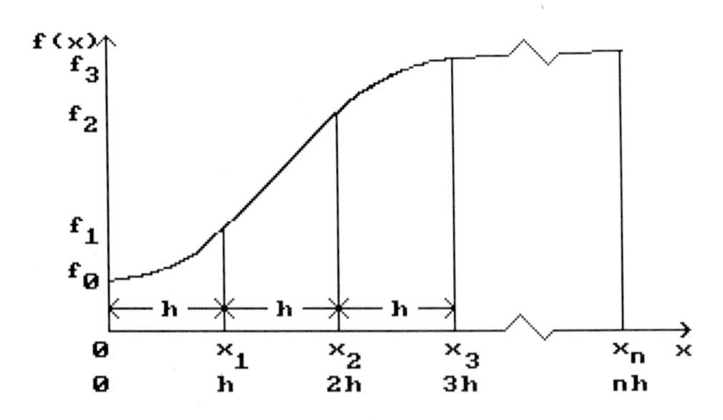


**Figure 5-4** Division of the total interval of integration into $n$ equal segments of width $h$.

### 5-5.1 The Trapezoidal Rule

In developing the first Newton-Cotes closed formula we use one segment of width $h$ and fit the polynomial through two points $(0, f_0)$ and $(h, f_1)$. This is tantamount to fitting a straight line between these points. We retain the first two terms of the Gregory-Newton polynomial (up to, and including, the first forward finite difference), and group together the rest of the terms of the polynomial into the *remainder* term. Thus, the integral equation (5-73) becomes

$$\int_{y_0}^{y_1} dy = \int_0^h \left[ f(0) + \frac{x}{h} \Delta f(0) \right] dx + \int_0^h R_n(x)\, dx \tag{5-74}$$

The first integral on the right-hand side is integrated with respect to $x$ and the first forward difference is replaced with its definition of $\Delta f(0) = f_1 - f_0$, to obtain

$$y_1 - y_0 = \frac{h}{2} [f_0 + f_1] + \int_0^h R_n(x)\, dx \tag{5-75}$$

The remainder term is evaluated as follows:

$$\int_0^h R_n(x)\, dx = \int_0^h \left[ \frac{x(x-h)}{2!h^2} \Delta^2 f(0) + \frac{x(x-h)(x-2h)}{3!h^3} \Delta^3 f(0) + \cdots \right] dx$$

$$= [-\tfrac{1}{12} h\Delta^2 f(0) + \tfrac{1}{24} h\Delta^3 f(0) - \cdots] \tag{5-76}$$

The forward difference operators, $\Delta^2, \Delta^3, \ldots$, are replaced by their equivalent in terms of differential operators [Eqs. (4-61) and (4-62)]

$$\Delta^2 = h^2D^2 + h^3D^3 + \tfrac{7}{12}h^4D^4 + \cdots \tag{4-61}$$

$$\Delta^3 = h^3D^3 + \tfrac{3}{2}h^4D^4 + \tfrac{5}{4}h^5D^5 + \cdots \tag{4-62}$$

so the remainder term becomes

$$\int_0^h R_n(x)\, dx = -\tfrac{1}{12}h^3D^2f(0) - \tfrac{1}{24}h^4D^3f(0) + \cdots \tag{5-77}$$

The infinite series can be represented by one term evaluated at $\xi_1$; therefore,

$$\int_0^h R_n(x)\, dx = -\tfrac{1}{12}h^3D^2f(\xi_1) \tag{5-78}$$

This is a term of order $h^3$ and is abbreviated by $O(h^3)$. Therefore, Eq. (5-75) can be written as

$$y_1 = y_0 + \frac{h}{2}(f_0 + f_1) + O(h^3) \tag{5-79}$$

This equation is known as the *trapezoidal rule*, because the term $(h/2)(f_0 + f_1)$ is essentially the formula for calculating the area of a trapezoid. In this case, the segment of integration is a trapezoid standing on its side. It was mentioned earlier that fitting a polynomial through only two points is equivalent to fitting a straight line through these points. This causes the shape of the integration segment to be a trapezoid, shown as the shaded area in Fig. 5-5. The area between $f(x)$ and the straight line represents the truncation error of the trapezoidal rule. If the function $f(x)$ is actually linear, then the trapezoidal rule calculates the integral *exactly*, because $D^2f(\xi_1) = 0$, which causes the remainder term to vanish.

The trapezoidal rule in the form of Eq. (5-79) gives the integral of only one integration segment of width $h$. To obtain the total integral, Eq. (5-74) must be applied over each of the $n$ segment (with the appropriate limits of integration) to obtain the following series of equations:

$$y_1 = y_0 + \frac{h}{2}(f_0 + f_1) + O(h^3) \tag{5-79}$$

$$y_2 = y_1 + \frac{h}{2}(f_1 + f_2) + O(h^3) \tag{5-80}$$

$$\vdots$$

$$y_n = y_{n-1} + \frac{h}{2}(f_{n-1} + f_n) + O(h^3) \tag{5-81}$$

Addition of all these equations over the total interval gives the *multiple-segment trapezoidal rule*:

$$y_n = y_0 + \frac{h}{2}\left(f_0 + 2\sum_{i=1}^{n-1} f_i + f_n\right) + nO(h^3) \tag{5-82}$$

For simplicity, the error term has been shown as $nO(h^3)$. This is only an

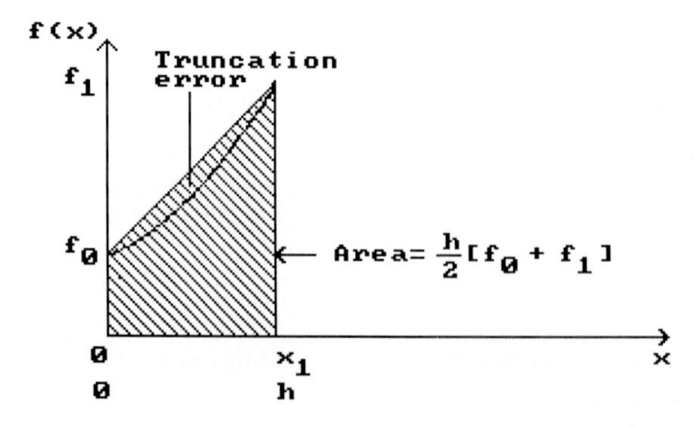


**Figure 5-5** Enlargement of segment showing the application of the trapezoidal rule.

approximation because the remainder term includes the second-order derivative of $f$ evaluated at unknown values of $\xi_i$, each $\xi_i$ being specific for that interval of integration. The absolute value of the error term cannot be calculated, but its relative magnitude can be measured by the order of the term. Since $n$ is inversely proportional to $h$,

$$n = \frac{x_n - 0}{h} \tag{5-83}$$

the error term for the multiple-segment trapezoidal rule becomes

$$nO(h^3) = \frac{x_n}{h} O(h^3) \cong O(h^2) \tag{5-84}$$

That is, the repeated application of the trapezoidal rule over multiple segments has lowered the error term by approximately one order of magnitude. A more rigorous analysis of the truncation error is given in Sec. 5-8.

Thus far, our interpretation of the trapezoidal rule has been based on the concept of calculating the area under the curve $f(x)$. A different approach, one based on the tangential trajectory of the function $y$, is very useful in enhancing our understanding of the numerical integration of differential equations. This approach will be discussed briefly here and will be revisited in Sec. 5-6.

Utilizing the original differential equation (5-59) in the form

$$y' = f(x) \tag{5-85}$$

we can replace the values of the function $f_0$ and $f_1$ in the trapezoidal rule with $y_0'$ and $y_1'$, to obtain

$$y_1 = y_0 + h\left(\frac{y_0' + y_1'}{2}\right) + O(h^3) \tag{5-86}$$

This equation simply states that the value of $y_1$ can be obtained from $y_0$ by taking a step of width $h$ in the average tangential direction of $y$. The latter is calculated from $[(y_0' + y_1')/2]$. This concept is illustrated graphically in Fig. 5-6. The difference between the calculated value of $y_1$ and the true value of $y$ at $x_1$ represents the truncation error of the trapezoidal rule.

## 5-5.2 Simpson's $\frac{1}{3}$ Rule

In the derivation of the second Newton-Cotes closed formula of integration we use two segments of width $h$ (see Fig. 5-7$a$), and fit the polynomial through three points, $(0, f_0)$, $(h, f_1)$, and $(2h, f_2)$. This is equivalent to fitting a parabola through these points. We retain the first three terms of the Gregory-Newton polynomial (up to, and including, the second forward finite difference), and group together the rest of the terms of the polynomial into the

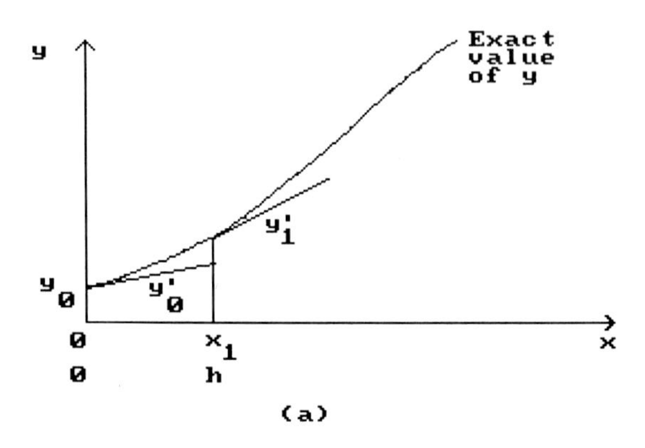


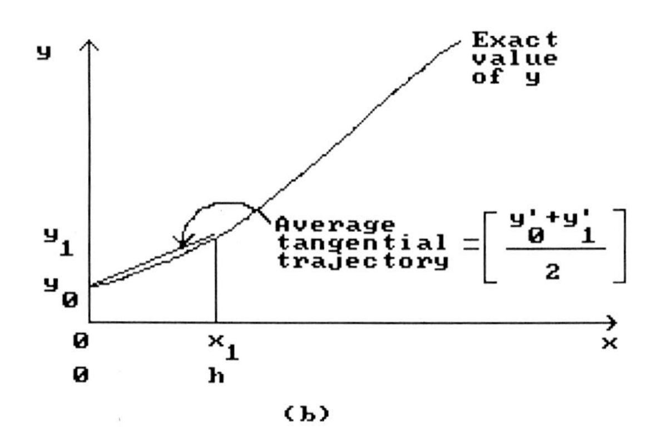


**Figure 5-6** The trapezoidal rule based on the average tangential trajectory of $y$.

remainder term. The integral equation (5-73) becomes

$$\int_{y_0}^{y_2} dy = \int_0^{2h} \left[ f(0) + \frac{x}{h} \Delta f(0) + \frac{x(x-h)}{2!h^2} \Delta^2 f(0) \right] dx + \int_0^{2h} R_n(x)\, dx \qquad (5\text{-}87)$$

Integration of Eq. (5-87) and substitution of the relevant finite difference relations simplify this equation to

$$y_2 = y_0 + \frac{h}{3}(f_0 + 4f_1 + f_2) - \tfrac{1}{90}h^5 D^4 f(\xi_1) \qquad (5\text{-}88)$$

The error term is of order $h^5$ and may be abbreviated by $O(h^5)$. We would have expected to obtain an error term of $O(h^4)$ since three terms were retained in the Gregory-Newton polynomial. However, the term containing $h^4$ in the remainder has a zero coefficient, thus giving this fortuitous result. The final form of the second Newton-Cotes formula, which is better known as *Simpson's* $\frac{1}{3}$ *rule*, is

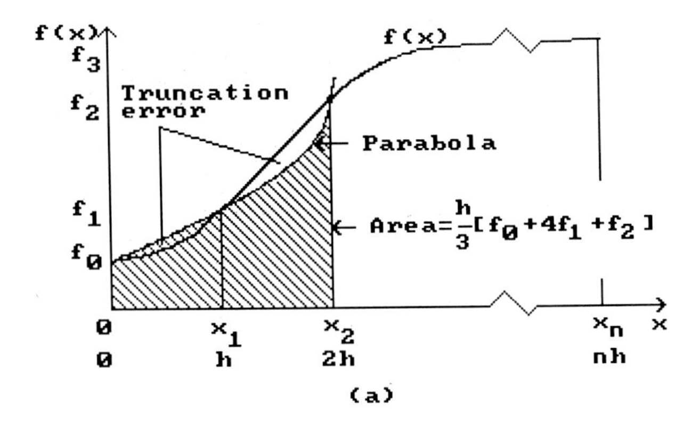


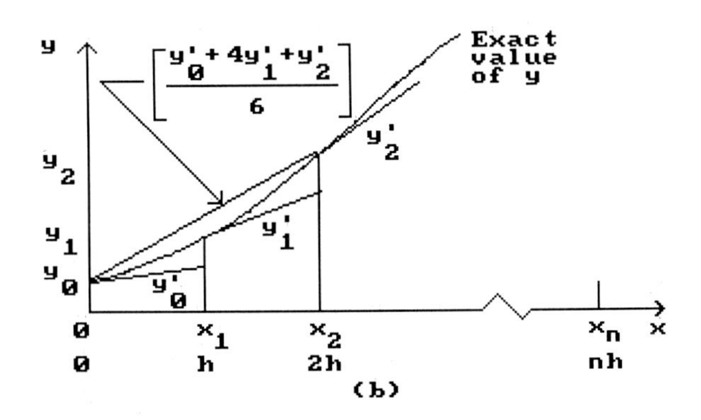


**Figure 5-7** Application of Simpson's $\frac{1}{3}$ rule over two segments of integration: (*a*) based on area, (*b*) based on trajectory.

$$y_2 = y_0 + \frac{h}{3}(f_0 + 4f_1 + f_2) + O(h^5) \tag{5-89}$$

This equation calculates the integral over two segments of integration. Repeated application of Simpson's $\frac{1}{3}$ rule over subsequent pairs of segments, and summation of all the formulas over the total interval, gives the *multiple-segment Simpson's* $\frac{1}{3}$ *rule*:

$$y_n = y_0 + \frac{h}{3}\left(f_0 + 4\sum_{\substack{i=1\\i=\text{odd}}}^{n-1} f_i + 2\sum_{\substack{i=2\\i=\text{even}}}^{n-2} f_i + f_n\right) + O(h^4) \tag{5-90}$$

Since Simpson's $\frac{1}{3}$ rule fits *pairs* of segments, the total interval must be subdivided into an *even* number of segments. The first summation term in Eq. (5-90) sums up the odd-subscripted terms and the second summation adds up to the even-subscripted terms.

The order of error of the multiple-segment Simpson's $\frac{1}{3}$ rule was reduced by one order of magnitude to $O(h^4)$ for the same reason as in Sec. 5-5.1. Simpson's $\frac{1}{3}$ rule is more accurate than the trapezoidal rule but requires additional arithmetic operations.

In a manner analogous to that used for the trapezoidal rule, Simpson's $\frac{1}{3}$ rule can be expressed in terms of the tangential trajectories of $y$:

$$y_2 = y_0 + h\left(\frac{y_0' + 4y_1' + y_2'}{3}\right) + O(h^5) \tag{5-91}$$

The term in parentheses is the *weighted average* of the tangents of $y$ at the three points, whereas the middle point is weighted more heavily (Fig. 5-7$b$).

### 5-5.3 Simpson's $\frac{3}{8}$ Rule

In the derivation of the third Newton-Cotes closed formula of integration we use three segments of width $h$ (see Fig. 5-8) and fit the polynomial through four

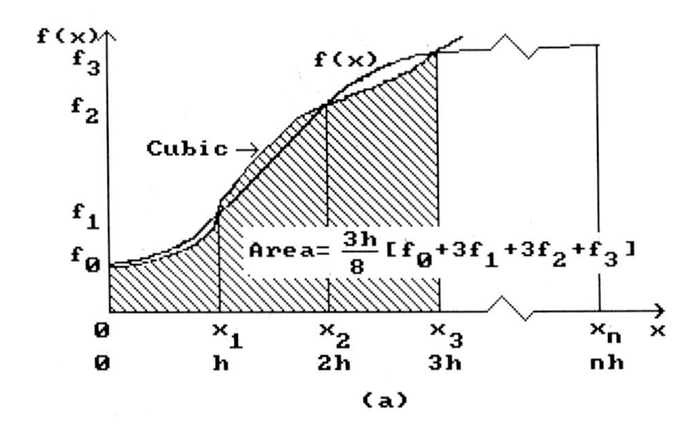


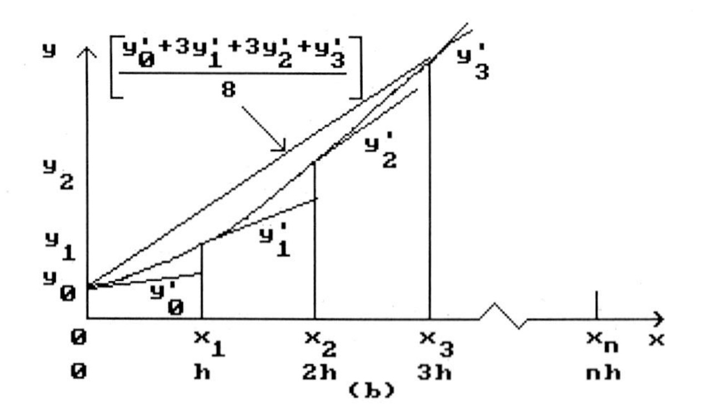


**Figure 5-8** Application of Simpson's $\frac{3}{8}$ rule over three segments of integration: ($a$) based on area, ($b$) based on trajectory.

points, $(0, f_0)$, $(h, f_1)$, $(2h, f_2)$, and $(3h, f_3)$. This, in effect, is equivalent to fitting a cubic equation through the four points. We retain the first four terms of the Gregory-Newton polynomial (up to, and including, the third forward finite difference) and group together the rest of the terms of the polynomial into the remainder term. The integral equation (5-73) becomes

$$\int_{y_0}^{y_3} dy = \int_0^{3h} \left[ f(0) + \frac{x}{h} \Delta f(0) + \frac{x(x-h)}{2!h^2} \Delta^2 f(0) + \frac{x(x-h)(x-2h)}{3!h^3} \Delta^3 f(0) \right] dx + \int_0^{3h} R_n(x)\, dx \tag{5-92}$$

Integration of Eq. (5-92) and substitution of the relevant finite difference relations simplify this equation to

$$y_3 = y_0 + \frac{3h}{8}(f_0 + 3f_1 + 3f_2 + f_3) - \tfrac{3}{80} h^5 D^4 f(\xi_1) \tag{5-93}$$

The error term is of order $h^5$ and may be abbreviated by $O(h^5)$. The final form of this equation, which is better known as *Simpson's $\frac{3}{8}$ rule*, is given by

$$y_3 = y_0 + \frac{3h}{8}(f_0 + 3f_1 + 3f_2 + f_3) + O(h^5) \tag{5-94}$$

The *multiple-segment Simpson's $\frac{3}{8}$ rule* is obtained by the repeated application of Eq. (5-92) over triplets of segments and summation over the total interval of integration:

$$y_n = y_0 + \frac{3h}{8}\left( f_0 + 3 \sum_{\substack{i=1 \\ i \neq 3, 6, 9, \ldots}}^{n-1} f_i + 2 \sum_{i=3, 6, 9, \ldots}^{n-3} f_i + f_n \right) + O(h^4) \tag{5-95}$$

Comparison of the error terms of Simpson's $\frac{1}{3}$ rule and Simpson's $\frac{3}{8}$ rule shows that they are both of the same order, with the latter being only slightly more accurate. For this reason, Simpson's $\frac{1}{3}$ rule is usually preferred because it achieves the same order of accuracy with three points rather than the four points required by the $\frac{3}{8}$ rule [3].

### 5-5.4 Summary of Newton-Cotes Integration

The three Newton-Cotes closed formulas of integration derived in the previous sections are summarized in Table 5-1.

In the derivation of the Newton-Cotes formulas, the function $f(x)$ is approximated by the Gregory-Newton polynomial $P_n(x)$ of degree $n$ with remainder $R_n(x)$. The evaluation of the integral is performed:

$$\int_a^b f(x)\, dx = \int_a^b P_n(x)\, dx + \int_a^b R_n(x)\, dx \tag{5-99}$$

**Table 5-1 Summary of the Newton-Cotes numerical integration (closed) formulas**

| | | |
|---|---|---|
| Trapezoidal rule | $\int_0^{x_1} f(x)\,dx = \frac{h}{2}(f_0 + f_1) - \frac{1}{12}h^3D^2f(\xi)$ | (5-96) |
| Simpson's $\frac{1}{3}$ rule | $\int_0^{x_2} f(x)\,dx = \frac{h}{3}(f_0 + 4f_1 + f_2) - \frac{1}{90}h^5D^4f(\xi)$ | (5-97) |
| Simpson's $\frac{3}{8}$ rule | $\int_0^{x_3} f(x)\,dx = \frac{3h}{8}(f_0 + 3f_1 + 3f_2 + f_3) - \frac{3}{80}h^5D^4f(\xi)$ | (5-98) |
| General quadrature formula | $\int_a^b f(x)\,dx = \sum_{i=0}^{n} w_i f(x_i) + O[h^{n+2}, D^{n+1}f(\xi)]$ | (5-100) |

This results in a formula of the general form

$$\int_a^b f(x)\,dx = \sum_{i=0}^{n} w_i f(x_i) + O[h^{n+2}, D^{n+1}f(\xi)] \tag{5-100}$$

where the $x_i$ are $n+1$ equally spaced base points in the interval $[a, b]$. The weights $w_i$ are determined by fitting the $P_n(x)$ polynomial to the $n+1$ base points. The integral is exact, i.e.,

$$\int_a^b f(x)\,dx = \sum_{i=0}^{n} w_i f(x_i) \tag{5-101}$$

for any function $f(x)$ which is of polynomial form up to degree $n$, because the derivative $D^{n+1}f(\xi)$ is zero for polynomials of degree $\leq n$; thus the error term $O[h^{n+2}, D^{n+1}f(\xi)]$ vanishes.

### 5-5.5 Unequally Spaced Points

In the development of the Newton-Cotes formulas we have assumed that the interval of integration could be divided into segments of equal width. This is usually possible when integrating continuous functions. However, if experimental data are to be integrated, such data may be unequally spaced. If this is the case, the trapezoidal rule could still be used with a variable-width segment. It has been suggested by Chapra and Canale [3] that a combination of the trapezoidal rule with Simpson's rules may be feasible for integrating certain sets of unevenly spaced data points.

If no restrictions are placed on the location of the base points, they may be chosen to be the locations of the roots of certain orthogonal polynomials in order to achieve higher accuracy than the Newton-Cotes formulas for the same number of base points [4]. This concept is used in the *Gauss quadrature* method which is discussed in more detail in the next section.

### 5-5.6 Gauss Quadrature

Gauss quadrature is a powerful method of integration which employs unequally spaced base points. This method uses the Lagrange polynomial to approximate the function and then applies orthogonal polynomials to locate the loci of the base points. In this section we outline the Gauss quadrature method without giving a formal derivation. The interested reader is referred to Carnahan et al. [5] and Vichnevetsky [4] for detailed derivations.

The function $f(x)$ is replaced by the Lagrange polynomial (see Sec. 4-7.2) and its remainder,

$$f(x) = P_n(x) + R_n(x)$$

$$= \sum_{i=0}^{n} L_i(x) f(x_i) + \prod_{i=0}^{n} (x - x_i) \frac{f^{(n+1)}(\xi)}{(n+1)!} \qquad a < \xi < b \tag{5-102}$$

where

$$L_i(x) = \prod_{\substack{j=0 \\ j \neq i}}^{n} \frac{x - x_j}{x_i - x_j} \tag{5-103}$$

The integral $\int_a^b f(x)\,dx$ is evaluated by Eq. (5-99):

$$\int_a^b f(x)\,dx = \int_a^b P_n(x)\,dx + \int_a^b R_n(x)\,dx \tag{5-99}$$

Without loss of generality, the interval $[a, b]$ is changed to $[-1, 1]$. The general transformation equation for converting between $x$ in interval $[a, b]$ and $z$ in interval $[c, d]$ is the following:

$$\frac{x - a}{b - a} = \frac{z - c}{d - c} \tag{5-104}$$

For converting to the interval $[-1, 1]$, this equation becomes

$$z = \frac{2x - (a + b)}{b - a} \tag{5-105}$$

Using Eqs. (5-102), (5-103), and (5-105) in (5-99), the transformed integral is given by

$$\int_{-1}^{1} F(z)\, dz = \sum_{i=0}^{n} w_i F(z_i) + \int_{-1}^{1} R_n(z)\, dz \tag{5-106}$$

where the weights $w_i$ are calculated from

$$w_i = \int_{-1}^{1} L_i(z)\, dz = \int_{-1}^{1} \prod_{\substack{j=0\\ j\neq 1}}^{n} \frac{z - z_j}{z_i - z_j}\, dz \tag{5-107}$$

and the error term is given by

$$\int_{-1}^{1} R_n(z)\, dz = \int_{-1}^{1} \prod_{i=0}^{n} (z - z_i) q_n(z)\, dz \tag{5-108}$$

The $q_n(z)$ and $\Pi_{i=0}^{n}(z - z_i)$ are polynomials of degree $n$ and $n+1$, respectively.

Up to this point, the development of this method is different from that of the Newton-Cotes formulas in only one respect: the use of the Lagrange interpolation formula for unequally spaced points instead of the Gregory-Newton formula. The Gauss quadrature method goes a step beyond this in order to make the error term [Eq. (5-108)] vanish. To do so, the two polynomials in the error term are expanded in terms of *Legendre orthogonal polynomials* (see Sec. 4-7.3). The values of $z_i$ are chosen as the roots of the $(n+1)$st-degree Legendre polynomial. This choice of roots combined with the orthogonality property [Eq. (4-155)] of the Legendre polynomials causes the error term to vanish. Therefore Eq. (5-106) becomes

$$\int_{-1}^{1} F(z)\, dz = \sum_{i=0}^{n} w_i F(z_i) \tag{5-109}$$

Since the vanishing error term was of degree $(n+1)$, Eq. (5-109) yields the integral of the function $F(z)$ *exactly* when $F(z)$ is a polynomial of degree $(2n+1)$ or less [4]. In effect, the judicious choice of the $n+1$ base points at the $n+1$ roots of the Legendre polynomial has increased the accuracy of the integration from $n$ to $2n+1$. As usual, however, the increase in accuracy has been obtained at the cost of having to perform a larger number of arithmetic calculations. Fortunately, the roots $z_i$ of the Legendre polynomials and the values of the weights $w_i$ corresponding to these roots have been calculated for the integration interval $[-1, 1]$. These values are tabulated in standard mathematical tables [6]. Table 5-2 lists the roots and weights of the Gauss-Legendre quadrature for selected values of $n$.

The Gauss quadrature formula developed in this section is known as the

**Table 5-2 Roots of the Legendre polynomials $P_{n+1}(z)$ and the weight factors for the Gauss-Legendre quadrature**

| Roots ($z_i$) | $\int_{-1}^{1} F(z)\,dz = \sum_{i=0}^{n} w_i F(z_i)$ | Weight factors ($w_i$) |
|---|---|---|
| ±0.57735 02691 89626 | Two-point formula ($n+1=2$) | 1.00000 00000 00000 |
| 0.00000 00000 00000 | Three-point formula | 0.88888 88888 88888 |
| ±0.77459 66692 41483 | ($n+1=3$) | 0.55555 55555 55555 |
| ±0.33998 10435 84856 | Four-point formula | 0.65214 51548 62546 |
| ±0.86113 63115 94053 | ($n+1=4$) | 0.34785 48451 37454 |
| 0.00000 00000 00000 | Five-point formula | 0.56888 88888 88889 |
| ±0.53846 93101 05683 | ($n+1=5$) | 0.47862 86704 99366 |
| ±0.90617 98459 38664 | | 0.23692 68850 56189 |
| ±0.23861 91860 83197 | Six-point formula | 0.46791 39345 72691 |
| ±0.66120 93864 66265 | ($n+1=6$) | 0.36076 15730 48139 |
| ±0.93246 95142 03152 | | 0.17132 44923 79170 |
| ±0.14887 43389 81631 | Ten-point formula | 0.29552 42247 14753 |
| ±0.43339 53941 29247 | ($n+1=10$) | 0.26926 67193 09996 |
| ±0.67940 95682 99024 | | 0.21908 63625 15982 |
| ±0.86506 33666 88985 | | 0.14945 13491 50581 |
| ±0.97390 65285 17172 | | 0.06667 13443 08688 |
| 0.00000 00000 00000 | Fifteen-point formula | 0.20257 82419 25561 |
| ±0.20119 40939 97435 | ($n+1=15$) | 0.19843 14853 27111 |
| ±0.39415 13470 77563 | | 0.18616 10001 15562 |
| ±0.57097 21726 08539 | | 0.16626 92058 16994 |
| ±0.72441 77313 60170 | | 0.13957 06779 26154 |
| ±0.84820 65834 10427 | | 0.10715 92204 67172 |
| ±0.93727 33924 00706 | | 0.07036 60474 88108 |
| ±0.98799 25180 20485 | | 0.03075 32419 96117 |

*Source*: Reproduced from Ref. 5, by permission of the publisher.

*Gauss-Legendre quadrature* because of the use of the Legendre polynomials. Other orthogonal polynomials, such as Chebyshev, Laguerre, or Hermite, may be used in a similar manner to develop a variety of Gauss quadrature formulas.

**Example 5-6 Integration formulas: trapezoidal, Simpson's $\frac{1}{3}$, and Simpson's $\frac{3}{8}$ rules** Write a general computer program for integrating differential equations and/or experimental data using the trapezoidal rule, Simpson's $\frac{1}{3}$ rule, or Simpson's $\frac{3}{8}$ rule. Apply this program for the solution of the following problems:

(*a*) A semi-infinite body of liquid with constant density ($\rho$) and viscosity ($\mu$) is bounded on one side by a flat surface (the $xz$-plane). Initially, the fluid and the solid surface are at rest; but at time $t = 0$ the solid surface is set in motion in the positive $x$-direction with a velocity $V$. It is desired to know the velocity of the fluid as a function of $y$ and $t$.

There is no pressure gradient or gravity force in the $x$-direction, and the flow is assumed to be laminar [7]. The partial differential equation which describes the velocity of the fluid is

$$\frac{\partial v_x}{\partial t} = \left(\frac{\mu}{\rho}\right) \frac{\partial^2 v_x}{\partial y^2}$$

with the following initial and boundary conditions:

| | | |
|---|---|---|
| At $t \leq 0$, | $v_x = 0$ | for all $y$ |
| At $y = 0$, | $v_x = V$ | for all $t > 0$ |
| At $y = \infty$, | $v_x = 0$ | for all $t > 0$ |

The analytical solution to this problem, given by Bird, Stewart, and Lightfoot [7], is

$$\frac{v_x}{V} = 1 - \frac{2}{\sqrt{\pi}} \int_0^\eta e^{-\eta^2}\, d\eta \tag{1}$$

where $\eta = y/\sqrt{4\mu t/\rho}$ is a dimensionless variable. The integral term is the well-known *error function*

$$\operatorname{erf}(\eta) = \frac{2}{\sqrt{\pi}} \int_0^\eta e^{-\eta^2}\, d\eta$$

which appears frequently in the solution of partial differential equations. This integral does not possess an analytical solution; therefore, it must be integrated numerically. Incidentally, the right-hand side of Eq. (1) is the *complement of error function*, i.e.,

$$\operatorname{erfc}(\eta) = 1 - \operatorname{erf}(\eta)$$

Using Simpson's $\frac{3}{8}$ rule integrate the error function for $0 \leq \eta \leq 1.0$ and plot the velocity profile $(v_x/V)$ in this range of $\eta$.

(*b*) Two very important quantities in the study of fermentation processes are the carbon dioxide evolution rate and the oxygen uptake rate. These are calculated from experimental analyses of the inlet and exit gases of the fermentor, and the flow rates, temperature, and pressure of these gases. The ratio of carbon dioxide evolution rate to oxygen uptake rate yields the respiratory quotient, which is a good barometer of the metabolic activity of the microorganism. In addition, the above rates can be integrated to obtain the total amounts of carbon dioxide produced and oxygen consumed during the fermentation. These total amounts form the basis of the material balancing technique described in Sec. 3-1. Table E5-6*a* shows a set of rates calculated from the fermentation of *Penicillium chrysogenum*, which produces penicillin antibiotics.

**Table E5-6*a***

| Time of fermentation, h | Carbon dioxide evolution rate, g/h | Oxygen uptake rate, g/h |
|---|---|---|
| 140 | 15.72 | 15.49 |
| 141 | 15.53 | 16.16 |
| 142 | 15.19 | 15.35 |
| 143 | 16.56 | 15.13 |
| 144 | 16.21 | 14.20 |
| 145 | 17.39 | 14.23 |
| 146 | 17.36 | 14.29 |
| 147 | 17.42 | 12.74 |
| 148 | 17.60 | 14.74 |
| 149 | 17.75 | 13.68 |
| 150 | 18.95 | 14.51 |

Using Simpson's $\frac{1}{3}$ rule, calculate the total amounts of carbon dioxide produced and oxygen consumed during this 10-h period of fermentation. Repeat this using the trapezoidal rule and compare the results obtained from the two methods.

METHOD OF SOLUTION

(*a*) In order to evaluate the velocity profile $(v_x/V)$, we integrate the function

$$-\frac{2}{\sqrt{\pi}}\,e^{-\eta^2}$$

in the range $0 \leq \eta \leq 1$ with the initial condition

At $\eta = 0$, $$\frac{v_x}{V} = 1$$

(*b*) In this case, the carbon dioxide evolution rate data and the oxygen uptake rate data are integrated separately. There are 11 data points (10 intervals) for each rate; therefore, we can use either the trapezoidal rule or Simpson's $\frac{1}{3}$ rule for this integration. We first use Simpson's $\frac{1}{3}$ rule and then repeat using the trapezoidal rule, as the problem specifies.

PROGRAM DESCRIPTION This computer program implements the three integration formulas: trapezoidal rule, Simpson's $\frac{1}{3}$ rule, and Simpson's $\frac{3}{8}$ rule. The program is called INTEGR.BAS, and it consists of the sections shown in Table E5-6*b*.

**Table E5-6*b***

| Program section | Line numbers |
|---|---|
| Title | 1–90 |
| Main program | 100–460 |
| Formula options | 110–170 |
| Integration options | 180–230 |
| Output options | 240–290 |
| Subroutine calls | 300–350 |
| Rerun options | 360–450 |
| Subroutines | |
| 1. Input equations | 1000–1490 |
| 2. Input data | 2000–2450 |
| 3. Input integration parameters | 3000–3280 |
| 4. Differential equations | 4000–4999 |
| 5. Integration formulas | 5000–5890 |
| 6. Print table of results | 6000–6100 |
| 7. Plotting options | 7000–7480 |
| 8. Plotting | 8000–8870 |

*Main program* Lines 110–170 offer the options of using the trapezoidal, Simpson's $\frac{1}{3}$, or Simpson's $\frac{3}{8}$ rules.

Lines 180–230 provide the user the choice of either integrating equations or integrating data. When equations are integrated, the program generates the profile of the dependent variable ($y$) between the initial value and the final value of the independent variable ($x$). When data are integrated, the program calculates the total area under the curve between the initial and final data points.

Lines 240–290 provide the options of presenting the output in tables and graphs, or only graphs. These options are not available when integrating data. In the latter case, only the total area is calculated.

Lines 360–450 enable the user to print and/or graph the same results again, or to reset the integration parameters and reintegrate the equations.

*Subroutine 1* This subroutine enables the user to enter the differential equations, and other associated algebraic equations, to be integrated together. A maximum of 980 lines of BASIC statements may be inserted in the program during execution. All statements must obey the rules of BASIC programming. Each line is terminated by the ENTER key. The differential equations must be entered in the form shown below:

$$G(i) = f[x, c(1), c(2), \ldots]$$

For example,

```
G(1) = c(1) * x + c(2) * x^2 - c(3) * SQR(x)
```

where G(i) = the derivatives $dy(i)/dx$
x = the independent variable
y(i) = the dependent variables
f[ ] = the functions to be integrated
c( ) = constants to be entered later

If other algebraic statements are entered along with the differential equations, care must be exercised in the choice of variable names so that no duplication occurs with variables used in the rest of the programs [other than G(i), x, y(i) and c( )]. A cross-reference utility program, if available, may be used to scan the program for variable names. The program automatically numbers the statements, starting at 4010, and stores them on the diskette under a user-specified filename. The user should not specify an extension to the filename (the program itself adds the extension .EQU to identify this file as one containing equations). The file is subsequently merged with the program by the CHAIN MERGE command of line 1480. When line 1480 is executed, the following steps occur in this order:

1. Lines 4010–4995 of the executing program are deleted in order to remove any differential equations left over in subroutine 4 from a previous execution of this program. The first and last lines (4010 and 4995) are comment lines and serve as anchors for the DELETE command. For this reason they must always be present in the program (see note of caution given below).
2. The file FIL$, which contains the newly entered differential equations, is merged into the executing program in subroutine 4 occupying lines 4010–4995.
3. All values of variables are passed on to the merged program by the ALL command.
4. The merged program resumes executing at line 310. (If for any reason, this program is renumbered using the RENUM command, the "310" in line 1480 must be changed *manually* to the new line number corresponding to the old statement 310.)

The file containing the differential equations remains on the diskette and can be reused, until it is erased or overwritten. *Caution* must be exercised when using a file which had been entered earlier. The filename must be specified correctly, otherwise an error will occur in line 1480 because the CHAIN MERGE command fails to locate the file on the diskette. This error will cause execution to stop at line 1480 and, since the DELETE 4010–4995 part of that line has already been executed, lines

4010–4995 have been deleted. The next attempt to run this program will fail again at line 1480 but this time because DELETE 4010–4995 does not find any statements to delete. If this error occurs, simply reload the original program INTEGR.BAS from the disk and rerun it, this time making sure that the filename is specified correctly.

*Subroutine 2* This subroutine is for entering experimental data to be integrated. The user is reminded that the data must be equally spaced, that Simpson's $\frac{1}{3}$ rule requires an even number of intervals (number of points minus one), and that Simpson's $\frac{3}{8}$ rule requires a number of intervals that is divisible by 3. The program performs an automatic check on all the above requirements to ensure that the data are equally spaced and that the appropriate number of data are entered for the particular integration formula. The data are entered in pairs of $x$ and $f(x)$ and are stored on the diskette in a file whose name is specified by the user. The program adds the extension .DAT to the filename. The data stored this way remain on the diskette and can be reused until erased or overwritten.

*Subroutine 3* This subroutine inputs the integration parameters for the differential equations. The program automatically chooses the number of integration steps to be 100, 50, or 33 for the trapezoidal, Simpson's $\frac{1}{3}$, or Simpson's $\frac{3}{8}$ rules, respectively. If necessary, this may be changed by modifying lines 3020–3040 before running the program. The initial and final values of the independent variable are entered by the user during program execution. The increment in the independent variable is automatically calculated. The initial values of the dependent variables, and the constants (if any) in the equations, are entered by the user.

*Subroutine 4* This subroutine allocates space for merging the differential equations that were entered in subroutine 1. During the integration of the differential equations, subroutine 4 is called repeatedly by subroutine 5 in order to evaluate the derivatives. Lines 4010 and 4995 must not be removed for any reason, since these are utilized in the CHAIN MERGE command of line 1480.

*Subroutine 5* This subroutine implements the integration formulas for equations and for data. The trapezoidal rule [Eq. (5-82)] is applied in lines 5040–5170 for equations, and in lines 5620–5690 for data. Simpson's $\frac{1}{3}$ rule [Eq. (5-90)] is implemented in lines 5180–5330 for equations, and 5700–5790 for data. Simpson's $\frac{3}{8}$ rule [Eq. (5-95)] is programmed in lines 5340–5510 for equations, and 5800–5890 for data.

*Subroutine 6* This subroutine prints a table of the integrated profiles of the dependent variables.

*Subroutine 7* This subroutine provides the options for plotting the graphs. When there is more than one dependent variable, they can be plotted separately or on the same graph. It is usually preferable to plot them separately, unless their scales are of the same magnitude. The maxima and minima of the scales can be specified by the user, or can be automatically calculated by the program. The manual choice usually results in better looking graphs.

*Subroutine 8* This is a plotting subroutine, which can draw one or more curves per graph. This is an extended version of the plotting routine used in Examples 2-1 and 2-2. It can automatically choose the scales, expand them to round numbers, and label the axes accordingly.

PROGRAM

```
1 SCREEN 0:WIDTH 80:CLS:KEY OFF
5 PRINT "***********************************************************************"
10 PRINT"*                                                                     *"
15 PRINT"*                          EXAMPLE 5-6                                *"
20 PRINT"*                                                                     *"
25 PRINT"*                     INTEGRATION FORMULAS:                           *"
30 PRINT"*                                                                     *"
35 PRINT"*        TRAPEZOIDAL, SIMPSON'S 1/3, AND SIMPSON'S 3/8 RULES          *"
40 PRINT"*                                                                     *"
45 PRINT"*                          (INTEGR.BAS)                               *"
50 PRINT"*                                                                     *"
90 PRINT"***********************************************************************"
100 '****************************** Main Program ****************************
110 PRINT
120 PRINT:PRINT"                        FORMULA OPTIONS":PRINT
130 PRINT:PRINT"             1. TRAPEZOIDAL RULE"
140 PRINT:PRINT"             2. SIMPSON'S 1/3 RULE"
150 PRINT:PRINT"             3. SIMPSON'S 3/8 RULE"
160 PRINT:PRINT:INPUT"    ENTER YOUR CHOICE(1-3)"; FQ
170 IF FQ<1 OR FQ>3 THEN GOTO 160
180 PRINT:PRINT
190 PRINT:PRINT"                        INTEGRATION OPTIONS":PRINT
200 PRINT:PRINT"             1. INTEGRATE EQUATIONS OR FUNCTIONS"
210 PRINT:PRINT"             2. INTEGRATE DATA"
220 PRINT:PRINT:INPUT"    ENTER YOUR CHOICE(1-2)"; IQ
230 IF IQ<1 OR IQ>2 THEN GOTO 220
240 PRINT:PRINT:IF IQ=2 THEN GOTO 300
250 PRINT:PRINT"                        OUTPUT OPTIONS":PRINT
260 PRINT:PRINT"             1. PRINT TABLE AND DRAW GRAPHS OF RESULTS"
270 PRINT:PRINT"             2. DRAW GRAPHS OF RESULTS ONLY"
280 PRINT:PRINT:INPUT"    ENTER YOUR CHOICE(1-2)"; RQ
290 IF RQ<1 OR RQ>2 THEN GOTO 280
300  ON IQ GOSUB 1000,2000
310 IF IQ=1 THEN GOSUB 3000
320  PRINT:PRINT"INTEGRATING. PLEASE WAIT"
330  GOSUB 5000
340 IF RQ=1 THEN GOSUB 6000
350 GOSUB 7000
360 PRINT:PRINT"                        RERUN OPTIONS":PRINT
370 PRINT:PRINT"             1. SHOW SAME RESULTS AGAIN"
380 PRINT:PRINT"             2. RESET INTEGRATION PARAMETERS AND RECALCULATE"
390 PRINT:PRINT"             3. END PROGRAM"
400 PRINT:PRINT:INPUT"    ENTER YOUR CHOICE(1-3)"; RRQ
410 IF RRQ<1 OR RRQ>3 THEN GOTO 400
420 ON RRQ GOTO 340,430,460
430 PRINT:PRINT:PRINT:PRINT"                          NEW CONDITIONS"
440  GOSUB 3060
450  GOTO 320
460  END
1000 '************* Subroutine 1: Input equations ***********************
1010 '
1020 PRINT CHR$(12)
1030 PRINT:PRINT"                        DIFFERENTIAL EQUATIONS"
1040 PRINT:PRINT"                        **********************"
1050 PRINT:PRINT"THE DIFFERENTIAL EQUATIONS, AND OTHER ASSOCIATED ALGEBRAIC"
1060 PRINT: PRINT"EQUATIONS, MAY BE ENTERED HERE."
```

```
1070 PRINT:PRINT"EACH LINE-STATEMENT MUST OBEY THE RULES OF BASIC PROGRAMMING."
1080 PRINT:PRINT"THE LINES ARE AUTOMATICALLY NUMBERED, STARTING AT 4011."
1090 PRINT:PRINT"THESE STATEMENTS ARE MERGED INTO THE PROGRAM IN SUBROUTINE 4."
1100 PRINT:PRINT"                                    **********************"
1110 PRINT:INPUT "HOW MANY DIFFERENTIAL EQUATIONS IN THE SYSTEM"; N1
1120 DIM Y(N1),Y9(N1),Y1(N1),C(50),G(N1),B1(N1),B2(N1),B3(N1),B4(N1)
1130 PRINT:INPUT "HAVE YOU PREVIOUSLY ENTERED THE EQUATIONS(Y/N)";Q1$
1140 IF Q1$="Y" OR Q1$="y" THEN GOTO 1170
1150 IF Q1$="N" OR Q1$="n" THEN GOTO 1210
1160 GOTO 1130
1170 PRINT:PRINT"THE FOLLOWING EQUATION-FILES EXIST ON YOUR DISK:":PRINT
1180 FILES"*.EQU"
1190 INPUT"GIVE NAME OF FILE TO BE USED (DO NOT ENTER THE EXTENSION): ",FIL$
1200 FIL$=FIL$+".EQU":GOTO 1480
1210 PRINT:INPUT"NAME OF FILE FOR STORING THE DIFFERENTIAL EQUATIONS: ",FIL$
1220 FIL$=FIL$+".EQU"
1230 OPEN FIL$ FOR OUTPUT AS #1
1240 CLS:PRINT:PRINT"                                    **********************"
1250 PRINT:PRINT"ENTER THE DIFFERENTIAL EQUATIONS IN THE FORM SHOWN BELOW:"
1260 PRINT:PRINT"        G(i)= f[x, c(1), c(2),....]"
1270 PRINT:PRINT"                    FOR EXAMPLE"
1280 PRINT:PRINT"        G(1)= c(1)*x + c(2)*x^2 - c(3)*SQR(x)"
1290 PRINT:PRINT"WHERE:  G(i)  ARE THE DERIVATIVES: dy(i)/dx"
1300 PRINT:PRINT"        x     IS THE INDEPENDENT VARIABLE"
1310 PRINT:PRINT"        y(i)  ARE THE DEPENDENT VARIABLES"
1320 PRINT:PRINT"        f[ ]  ARE THE FUNCTIONS TO BE INTEGRATED
1330 PRINT:PRINT"        c( )  ARE CONSTANTS TO BE ENTERED LATER"
1340 PRINT:PRINT"                                 **********************":PRINT
1350 INPUT"GIVE THE EXACT NUMBER OF LINES OF INPUT YOU WILL ENTER(1-980)";LN1
1360 PRINT:PRINT"SEPARATE EACH STATEMENT WITH THE ENTER(RETURN) KEY."
1370 PRINT#1,"4010 'The differential equations are merged in lines 4011-4990
1380 FOR I=1 TO LN1
1390 PRINT:PRINT "         ";4010+I;
1400 LINE INPUT ;EQUATION$
1410 IF EQUATION$="" THEN GOTO 1390
1420 PRINT#1,4010+I; EQUATION$
1430 PRINT
1440 NEXT I
1450 PRINT#1,"4995 'End of space allocated for the differential equations
1460 PRINT CHR$(12)
1470 CLOSE #1
1480 CHAIN MERGE FIL$,310,ALL,DELETE 4010-4995
1490 RETURN
2000 '************* Subroutine 2: Input data *****************************
2010 '
2020 PRINT CHR$(12)
2030 PRINT:PRINT"                              DATA POINTS":PRINT
2040 INPUT "HAVE YOU PREVIOUSLY ENTERED THE DATA POINTS(Y/N)";Q2$
2050 IF Q2$="N" OR Q2$="n" THEN GOTO 2080
2060 IF Q2$="Y" OR Q2$="y" THEN GOTO 2350
2070 GOTO 2040
2080 PRINT:PRINT"THE DATA MUST BE EQUALLY SPACED."
2090 PRINT:PRINT"THE NUMBER OF INTERVALS MUST BE:"
2100 PRINT:PRINT"    TRAPEZOIDAL RULE    : ANY NUMBER"
2110 PRINT:PRINT"    SIMPSON'S 1/3 RULE : EVEN NUMBER"
2120 PRINT:PRINT"    SIMPSON'S 3/8 RULE : NUMBER DIVISIBLE BY 3"
2130 PRINT:PRINT "HOW MANY DATA POINTS:";: INPUT L
2140 IT=(L-1)/FQ - INT((L-1)/FQ)
2150 IF IT=0 THEN GOTO 2190
```

```
2160 PRINT:PRINT"THE NUMBER OF POINTS IS INCONSISTENT WITH THE NUMBER"
2170 PRINT"OF INTERVALS REQUIRED FOR THE METHOD CHOSEN."
2180 GOTO 2130
2190 DIM X(L),FX(L)
2200 PRINT:INPUT"GIVE NAME OF FILE FOR STORING THE DATA ";DFIL$
2210 DFIL$=DFIL$+".DAT"
2220 OPEN DFIL$ FOR OUTPUT AS #2
2230 PRINT #2, L
2240 PRINT:PRINT"ENTER PAIRS OF DATA POINTS:":PRINT
2250 FOR I=0 TO L-1
2260 PRINT "x(";I")=";:INPUT X(I)
2270 PRINT "f( x )=";:INPUT FX(I)
2280 IF I=0 THEN GOTO 2310
2290 DX=X(1)-X(0)
2300 IF X(I)-X(I-1)<>DX THEN PRINT "DATA IS NOT EQUALLY SPACED.":GOTO 2260
2310 PRINT #2, X(I),FX(I) : PRINT
2320 NEXT I
2330 CLOSE #2
2340 RETURN
2350 PRINT:PRINT"THE FOLLOWING DATA-FILES EXIST ON YOUR DISK:":PRINT
2360 FILES"*.DAT"
2370 INPUT"GIVE NAME OF FILE TO BE USED (DO NOT ENTER THE EXTENSION): ",DFIL$
2380 DFIL$=DFIL$+".DAT"
2390 OPEN DFIL$ FOR INPUT AS #2
2400 INPUT #2, L
2410 DIM X(L),FX(L)
2420 PRINT:PRINT"READING THE DATA POINTS:":PRINT
2430 FOR I=0 TO L-1: INPUT #2, X(I),FX(I): PRINT  X(I),FX(I): NEXT I
2440 CLOSE #2
2450 RETURN
3000 '************* Subroutine 3: Input integration parameters ***********
3010 '
3020 IF FQ=1 THEN L=100
3030 IF FQ=2 THEN L=50
3040 IF FQ=3 THEN L=33
3050  DIM A(L,N1)
3060 PRINT:PRINT"                              INTEGRATION PARAMETERS"
3070 PRINT:PRINT"INITIAL VALUE OF INDEPENDENT VARIABLE:";: INPUT X9
3080 PRINT:PRINT"FINAL VALUE OF INDEPENDENT VARIABLE:  ";: INPUT XF
3090 PRINT:PRINT"THE TOTAL INTERVAL OF INTEGRATION IS DIVIDED AS FOLLOWS:"
3100 PRINT:PRINT"    TRAPEZOIDAL RULE   : 100 INTERVALS(100 INTEGRATION STEPS)"
3110 PRINT:PRINT"    SIMPSON'S 1/3 RULE : 100 INTERVALS( 50 INTEGRATION STEPS)"
3120 PRINT:PRINT"    SIMPSON'S 3/8 RULE :  99 INTERVALS( 33 INTEGRATION STEPS)"
3130 D=(XF-X9)/L : X = X9 : A(0,0) = X9
3140 PRINT:PRINT "INCREMENT IN INDEPENDENT VARIABLE IS="; D
3150  FOR K = 1 TO N1
3160 PRINT:PRINT "INITIAL Y(";K;")";: INPUT Y9(K)
3170 Y(K) = Y9(K)
3180 A(0,K) = Y9(K)
3190  NEXT K
3200 PRINT
3210 PRINT"DO YOU WANT TO ENTER CONSTANTS IN THE DIFFERENTIAL EQUATIONS(Y/N)";
3220 INPUT VC$ : IF VC$="N" OR VC$="n" THEN  GOTO 3280
3230 PRINT:INPUT"HOW MANY CONSTANTS";CO
3240 PRINT:PRINT "GIVE VALUES OF CONSTANTS":PRINT
3250  FOR K = 1 TO CO
3260  PRINT "CONSTANT ";K;: INPUT C(K)
3270  NEXT K
3280 RETURN
```

```
4000 '************* Subroutine 4: Differential equations ****************
4010 'The differential equations are merged in lines 4011-4990
4995 'End of space allocated for the differential equations
4999 RETURN
5000 '************* Subroutine 5: Integration formulas ******************
5010 '
5020  IF IQ=2 THEN GOTO 5600
5030  ON FQ GOTO 5040,5180,5340
5040 'Trapezoidal rule
5050 FOR J2=1 TO L
5060 FOR K2=1 TO N1 : Y1(K2)=Y(K2) : NEXT K2
5070 X1=X : GOSUB 4000
5080 FOR K2=1 TO N1 : B1(K2)=G(K2) : NEXT K2
5090 X=X1 + D : GOSUB 4000
5100 FOR K2=1 TO N1 : B2(K2)=G(K2) : NEXT K2
5110 FOR K2=1 TO N1
5120 Y(K2)=Y1(K2) + D*(B1(K2) + B2(K2))/2
5130 A(J2,K2)=Y(K2)
5140 NEXT K2
5150 A(J2,0)=X
5160 NEXT J2
5170 RETURN
5180 'Simpson's 1/3 rule
5190 FOR J2=1 TO L
5200 FOR K2=1 TO N1 : Y1(K2)=Y(K2) : NEXT K2
5210 X1=X : GOSUB 4000
5220 FOR K2=1 TO N1 : B1(K2)=G(K2) : NEXT K2
5230 X=X1 + D/2 : GOSUB 4000
5240 FOR K2=1 TO N1 : B2(K2)=G(K2) : NEXT K2
5250 X=X1 + D : GOSUB 4000
5260 FOR K2=1 TO N1 : B3(K2)=G(K2) : NEXT K2
5270 FOR K2=1 TO N1
5280 Y(K2)=Y1(K2) + D*(B1(K2) + 4*B2(K2) + B3(K2))/6
5290 A(J2,K2)=Y(K2)
5300 NEXT K2
5310 A(J2,0)=X
5320 NEXT J2
5330 RETURN
5340 'Simpson's 3/8 rule
5350 FOR J2=1 TO L
5360 FOR K2=1 TO N1 : Y1(K2)=Y(K2) : NEXT K2
5370 X1=X : GOSUB 4000
5380 FOR K2=1 TO N1 : B1(K2)=G(K2) : NEXT K2
5390 X=X1 + D/3 : GOSUB 4000
5400 FOR K2=1 TO N1 : B2(K2)=G(K2) : NEXT K2
5410 X=X1 + 2*D/3 : GOSUB 4000
5420 FOR K2=1 TO N1 : B3(K2)=G(K2) : NEXT K2
5430 X=X1 + D : GOSUB 4000
5440 FOR K2=1 TO N1 : B4(K2)=G(K2) : NEXT K2
5450 FOR K2=1 TO N1
5460 Y(K2)=Y1(K2) + D*(B1(K2) + 3*B2(K2) + 3*B3(K2) + B4(K2))/8
5470 A(J2,K2)=Y(K2)
5480 NEXT K2
5490 A(J2,0)=X
5500 NEXT J2
5510 RETURN
5600 'Integrating data
5610  ON FQ GOTO 5620,5700,5800
5620 'Trapezoidal rule
```

```
5630 H=X(1)-X(0)
5640 SUM = 0
5650 FOR J2=0 TO L-2
5660 SUM=SUM + H*(FX(J2) + FX(J2+1))/2
5670 NEXT J2
5680 AREA = SUM : PRINT:PRINT "AREA OF CURVE UNDER DATA POINTS = ";AREA
5690 END
5700 'Simpson's 1/3 rule
5710 IT=(L-1)/2-INT((L-1)/2)
5720 IF IT>0 THEN PRINT "NUMBER OF INTERVALS IN SIMPSON'S 1/3 RULE MUST BE EVEN"
:END
5730 H=X(1)-X(0)
5740 SUM = 0
5750 FOR J2=0 TO L-2 STEP 2
5760 SUM=SUM + H*(FX(J2) + 4*FX(J2+1) + FX(J2+2))/3
5770 NEXT J2
5780 AREA = SUM : PRINT:PRINT "AREA OF CURVE UNDER DATA POINTS = ";AREA
5790 END
5800 'Simpson's 3/8 rule
5810 IT=(L-1)/3-INT((L-1)/3)
5820 IF IT>0 THEN PRINT "NUMBER OF INTERVALS IN SIMPSON'S 3/8 RULE MUST BE DIVIS
IBLE BY 3":END
5830 H=X(1)-X(0)
5840 SUM = 0
5850 FOR J2=0 TO L-2 STEP 3
5860 SUM=SUM + 3*H*(FX(J2) + 3*FX(J2+1) + 3*FX(J2+2) + FX(J2+3))/8
5870 NEXT J2
5880 AREA = SUM : PRINT:PRINT "AREA OF CURVE UNDER DATA POINTS = ";AREA
5890 END
6000 '************* Subroutine 6: Print table of results *****************
6010 '
6020 PRINT CHR$(12)
6030 PRINT "             TABLE OF RESULTS"
6040 PRINT:PRINT "INDEPENDENT        DEPENDENT VARIABLES"
6050 FOR J2 = 0 TO L
6060 FOR J1 = 0 TO N1 : PRINT USING "####.####    ";A(J2,J1), : NEXT J1
6070 PRINT
6080 NEXT J2
6090 PRINT:INPUT "PRESS THE ENTER KEY TO CONTINUE",KK$
6100 RETURN
7000 '************* Subroutine 7: Plotting options **********************
7010 '
7020 PRINT CHR$(12):IF N1=1 THEN GOTO 7300
7030 PRINT:PRINT"              PLOTTING OPTIONS"
7040 PRINT:PRINT"    1. PLOT EACH VARIABLE SEPARATELY"
7050 PRINT:PRINT"    2. PLOT ALL VARIABLES ON SAME GRAPH"
7060 PRINT:INPUT"ENTER YOUR CHOICE(1-2)"; G1
7070 IF G1<1 OR G1>2 THEN GOTO 7060
7080 ON G1 GOTO 7090,7300
7090 FOR IP=1 TO N1
7100 NS=IP : NF=IP
7110 YTIT$="Y"+STR$(IP) : XTIT$="X"
7120 PRINT:PRINT
7130 PRINT:PRINT"               SCALING OPTIONS "
7140 PRINT:PRINT"    1. MANUAL CHOICE OF MAXIMA AND MINIMA"
7150 PRINT:PRINT"    2. AUTOMATIC CHOICE OF MAXIMA AND MINIMA"
7160 PRINT:INPUT"ENTER YOUR CHOICE(1-2)"; G2
7170 IF G2<1 OR G2>2 THEN GOTO 7160
7180 IF G2=2 THEN PRINT CHR$(12): GOSUB 8000: GOTO 7250
```

```
7190 PRINT:PRINT"          SCALES FOR VARIABLE ";IP
7200 PRINT:INPUT"MINIMUM X=";MINX :PRINT:INPUT"MAXIMUM X=";MAXX
7210 PRINT:LINE INPUT"SPECIFY X-AXIS TITLE:",XTIT$
7220 PRINT:INPUT"MINIMUM Y=";MINY :PRINT:INPUT"MAXIMUM Y=";MAXY
7230 PRINT:LINE INPUT"SPECIFY Y-AXIS TITLE:",YTIT$
7240 PRINT CHR$(12): GOSUB 8280
7250 FOR PAUSE=1 TO 10000:NEXT PAUSE
7260 LOCATE 25,1:INPUT"PRESS THE ENTER KEY TO SEE NEXT GRAPH",KK$
7270 SCREEN 0:WIDTH 80:PRINT CHR$(12)
7280 NEXT IP
7290 GOTO 7480
7300 NS=1: NF=N1
7310 YTIT$="Y" : XTIT$="X"
7320 PRINT:PRINT
7330 PRINT:PRINT"              SCALING OPTIONS "
7340 PRINT:PRINT"    1. MANUAL CHOICE OF MAXIMA AND MINIMA"
7350 PRINT:PRINT"    2. AUTOMATIC CHOICE OF MAXIMA AND MINIMA"
7360 PRINT:INPUT"ENTER YOUR CHOICE(1-2)"; G2
7370 IF G2<1 OR G2>2 THEN GOTO 7360
7380 IF G2=2 THEN PRINT CHR$(12): GOSUB 8000: GOTO 7450
7390 PRINT:PRINT"          SCALES FOR ALL VARIABLES"
7400 PRINT:INPUT"MINIMUM X=";MINX :PRINT:INPUT"MAXIMUM X=";MAXX
7410 PRINT:LINE INPUT"SPECIFY X-AXIS TITLE:",XTIT$
7420 PRINT:INPUT"MINIMUM Y=";MINY :PRINT:INPUT"MAXIMUM Y=";MAXY
7430 PRINT:LINE INPUT"SPECIFY Y-AXIS TITLE:",YTIT$
7440 GOSUB 8280
7450 FOR PAUSE=1 TO 10000:NEXT PAUSE
7460 LOCATE 25,1:INPUT"PRESS THE ENTER KEY TO CONTINUE",KK$
7470 SCREEN 0:WIDTH 80:PRINT CHR$(12)
7480 RETURN
8000 '************* Subroutine 8: Plotting ******************************
8010 '
8020 '
8030 'Locate maxima and minima for scaling axes
8040 MINX=A(0,0):MAXX=A(0,0):MINY =A(0,NS):MAXY =A(0,NS)
8050 FOR K = 0 TO L
8060 IF A(K ,0) < MINX THEN MINX = A(K ,0)
8070 IF A(K ,0) > MAXX THEN MAXX = A(K ,0)
8080 FOR J = NS TO NF
8090 IF A(K ,J) < MINY THEN MINY = A(K ,J)
8100 IF A(K ,J) > MAXY THEN MAXY = A(K ,J)
8110 NEXT J:NEXT K
8120 IF MAXX=0 GOTO 8160
8130 FACT=INT(LOG(ABS(MAXX))/LOG(10))
8140 TEMP=MAXX/(10^FACT)
8150 MAXX=(INT(TEMP+.999))*(10^FACT)
8160 IF MINX=0 GOTO 8200
8170 FACT=INT(LOG(ABS(MINX))/LOG(10))
8180 TEMP=MINX/(10^FACT)
8190 MINX=(INT(TEMP))*(10^FACT)
8200 IF MAXY=0 GOTO 8240
8210 FACT=INT(LOG(ABS(MAXY))/LOG(10))
8220 TEMP=MAXY/(10^FACT)
8230 MAXY=(INT(TEMP+.999))*(10^FACT)
8240 IF MINY=0 GOTO 8280
8250 FACT=INT(LOG(ABS(MINY))/LOG(10))
8260 TEMP=MINY/(10^FACT)
8270 MINY=(INT(TEMP))*(10^FACT)
8280 CLS:SCREEN 1
```

```
8290 VIEW (56,1)-(296,161)
8300 IF MINY=MAXY THEN MINY=MINY/2: MAXY=2*MAXY
8310 IF MINY=MAXY AND MINY=0 THEN MINY=-1:MAXY=1
8320 WINDOW (MINX,MINY)-(MAXX,MAXY)
8330 'Draw the axes
8340 DX=(MAXX-MINX)/10: DY=(MAXY-MINY)/10
8350 IF MINY/MAXY > 0 THEN  YZ=MINY ELSE YZ=0
8360 X=MINX : PSET(X,YZ)
8370 FOR I=1 TO 10
8380 X=X+DX : YT=YZ + ABS(DY/10)
8390 LINE -(X,YZ) : LINE -(X,YT) : LINE -(X,YZ)
8400 NEXT I
8410 IF MINX/MAXX > 0 THEN  XZ=MINX ELSE XZ=0
8420 Y=MINY
8430 PSET(XZ,Y)
8440 FOR I=1 TO 10
8450 Y=Y+DY : XT=XZ + ABS(DX/10)
8460 LINE -(XZ,Y) : LINE -(XT,Y) : LINE -(XZ,Y)
8470 NEXT I
8480 'Label the axes
8490 DELX=(MAXX-MINX) : DELY=(MAXY-MINY)
8500 FOR I=0 TO 10 STEP 2
8510 LBX=MINX+I*DELX/10 : LBX=INT(LBX*100)/100
8520 LN=LEN(STR$(LBX))
8530 IF I=10 THEN LOCATE 22,40-LN:PRINT LBX:GOTO 8550
8540 LOCATE 22,6+I*3:PRINT LBX
8550 NEXT I
8560 FOR I=0 TO 10
8570 LBY=MINY+I*DELY/10 : LBY=INT(LBY*100)/100
8580 LN=LEN(STR$(LBY))
8590 IF LN>=7 THEN LN=LN-1 : GOTO 8590
8600 LOCATE 21-I*2,7-LN:PRINT LBY
8610 NEXT I
8620 'Center vertical axis title
8630 IF LEN(YTIT$)>16 GOTO 8690
8640 V1=3+(14-LEN(YTIT$))/2
8650 FOR I=1 TO LEN(YTIT$)
8660 LOCATE V1+I,1:PRINT MID$(YTIT$,I,1)
8670 NEXT I
8680 GOTO 8720
8690 FOR I=1 TO 16
8700 LOCATE 3+I,2:PRINT MID$(YTIT$,I,1)
8710 NEXT I
8720 'Center horizontal title
8730 IF LEN(XTIT$)>35 GOTO 8770
8740 LOCATE 23,5+(36-LEN(XTIT$))/2
8750 PRINT XTIT$
8760 GOTO 8780
8770 LOCATE 23,5:PRINT MID$(XTIT$,1,36)
8780 'Draw the function
8790 FOR J=NS TO NF
8800 X=A(0,0):Y=A(0,J)
8810 PSET(X,Y)
8820 FOR I=1 TO L
8830 X=A(I,0) :Y=A(I,J)
8840 LINE -(X,Y)
8850 NEXT I
8860 NEXT J
8870 RETURN
```

## Results

```
*******************************************************************
*                                                                 *
*                          EXAMPLE 5-6                            *
*                                                                 *
*                    INTEGRATION FORMULAS:                        *
*                                                                 *
*       TRAPEZOIDAL, SIMPSON'S 1/3, AND SIMPSON'S 3/8 RULES       *
*                                                                 *
*                         (INTEGR.BAS)                            *
*                                                                 *
*******************************************************************
                        FORMULA OPTIONS

             1. TRAPEZOIDAL RULE

             2. SIMPSON'S 1/3 RULE

             3. SIMPSON'S 3/8 RULE

    ENTER YOUR CHOICE(1-3)? 3

                        INTEGRATION OPTIONS

             1. INTEGRATE EQUATIONS OR FUNCTIONS

             2. INTEGRATE DATA

    ENTER YOUR CHOICE(1-2)? 1

                        OUTPUT OPTIONS

             1. PRINT TABLE AND DRAW GRAPHS OF RESULTS

             2. DRAW GRAPHS OF RESULTS ONLY

    ENTER YOUR CHOICE(1-2)? 1

                        DIFFERENTIAL EQUATIONS
                        **********************

THE DIFFERENTIAL EQUATIONS, AND OTHER ASSOCIATED ALGEBRAIC

EQUATIONS, MAY BE ENTERED HERE.

EACH LINE-STATEMENT MUST OBEY THE RULES OF BASIC PROGRAMMING.

THE LINES ARE AUTOMATICALLY NUMBERED, STARTING AT 4011.

THESE STATEMENTS ARE MERGED INTO THE PROGRAM IN SUBROUTINE 4.
```

```
                        **********************

HOW MANY DIFFERENTIAL EQUATIONS IN THE SYSTEM? 1

HAVE YOU PREVIOUSLY ENTERED THE EQUATIONS(Y/N)? N

NAME OF FILE FOR STORING THE DIFFERENTIAL EQUATIONS: ERRORFUN

                        **********************

ENTER THE DIFFERENTIAL EQUATIONS IN THE FORM SHOWN BELOW:

        G(i)= f[x, c(1), c(2),....]

                    FOR EXAMPLE

        G(1)= c(1)*x + c(2)*x^2 - c(3)*SQR(x)

WHERE:  G(i)  ARE THE DERIVATIVES: dy(i)/dx

        x     IS THE INDEPENDENT VARIABLE

        y(i)  ARE THE DEPENDENT VARIABLES

        f[ ]  ARE THE FUNCTIONS TO BE INTEGRATED

        c( )  ARE CONSTANTS TO BE ENTERED LATER

                        **********************

GIVE THE EXACT NUMBER OF LINES OF INPUT YOU WILL ENTER(1-980)? 1

SEPARATE EACH STATEMENT WITH THE ENTER(RETURN) KEY.

        4011 G(1)=-(2/SQR(3.1416))*EXP(-X^2)


                        INTEGRATION PARAMETERS

INITIAL VALUE OF INDEPENDENT VARIABLE:? 0

FINAL VALUE OF INDEPENDENT VARIABLE:  ? 1

THE TOTAL INTERVAL OF INTEGRATION IS DIVIDED AS FOLLOWS:

    TRAPEZOIDAL RULE   : 100 INTERVALS(100 INTEGRATION STEPS)

    SIMPSON'S 1/3 RULE : 100 INTERVALS( 50 INTEGRATION STEPS)

    SIMPSON'S 3/8 RULE :  99 INTERVALS( 33 INTEGRATION STEPS)

INCREMENT IN INDEPENDENT VARIABLE IS= 3.030303E-02

INITIAL Y( 1 )? 1

DO YOU WANT TO ENTER CONSTANTS IN THE DIFFERENTIAL EQUATIONS(Y/N)? N

INTEGRATING. PLEASE WAIT
```

```
          TABLE OF RESULTS

INDEPENDENT        DEPENDENT VARIABLES
   0.0000         1.0000
   0.0303         0.9658
   0.0606         0.9317
   0.0909         0.8977
   0.1212         0.8639
   0.1515         0.8303
   0.1818         0.7971
   0.2121         0.7642
   0.2424         0.7317
   0.2727         0.6997
   0.3030         0.6683
   0.3333         0.6374
   0.3636         0.6071
   0.3939         0.5774
   0.4242         0.5485
   0.4545         0.5203
   0.4848         0.4929
   0.5152         0.4663
   0.5455         0.4405
   0.5758         0.4155
   0.6061         0.3914
   0.6364         0.3681
   0.6667         0.3458
   0.6970         0.3243
   0.7273         0.3037
   0.7576         0.2840
   0.7879         0.2652
   0.8182         0.2472
   0.8485         0.2302
   0.8788         0.2139
   0.9091         0.1986
   0.9394         0.1840
   0.9697         0.1703
   1.0000         0.1573

PRESS THE ENTER KEY TO CONTINUE

            SCALING OPTIONS

   1. MANUAL CHOICE OF MAXIMA AND MINIMA

   2. AUTOMATIC CHOICE OF MAXIMA AND MINIMA

ENTER YOUR CHOICE(1-2)? 1
         SCALES FOR ALL VARIABLES

MINIMUM X=? 0

MAXIMUM X=? 1

SPECIFY X-AXIS TITLE:DIMENSIONLESS VARIABLE

MINIMUM Y=? 0

MAXIMUM Y=? 1

SPECIFY Y-AXIS TITLE:vx/V
```

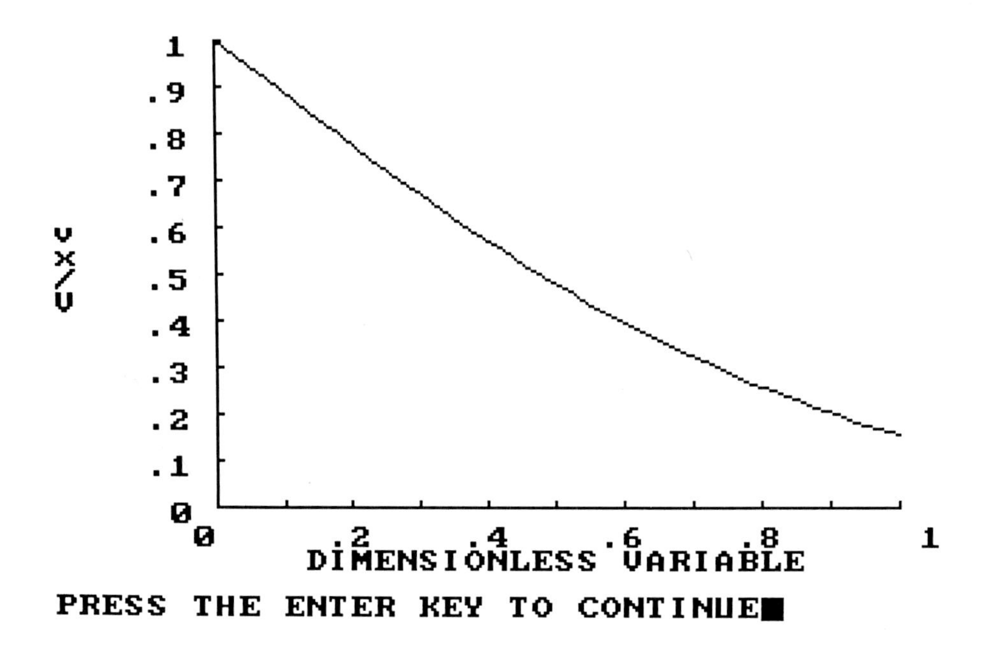


```
                    RERUN OPTIONS

          1. SHOW SAME RESULTS AGAIN

          2. RESET INTEGRATION PARAMETERS AND RECALCULATE

          3. END PROGRAM

   ENTER YOUR CHOICE(1-3)? 3
Ok
*******************************************************************
*                                                                 *
*                         EXAMPLE 5-6                             *
*                                                                 *
*                    INTEGRATION FORMULAS:                        *
*                                                                 *
*      TRAPEZOIDAL, SIMPSON'S 1/3, AND SIMPSON'S 3/8 RULES        *
*                                                                 *
*                        (INTEGR.BAS)                             *
*                                                                 *
*******************************************************************

                    FORMULA OPTIONS

          1. TRAPEZOIDAL RULE

          2. SIMPSON'S 1/3 RULE

          3. SIMPSON'S 3/8 RULE

   ENTER YOUR CHOICE(1-3)? 2
```

```
                        INTEGRATION OPTIONS

         1. INTEGRATE EQUATIONS OR FUNCTIONS

         2. INTEGRATE DATA

ENTER YOUR CHOICE(1-2)? 2

                            DATA POINTS

HAVE YOU PREVIOUSLY ENTERED THE DATA POINTS(Y/N)? N

THE DATA MUST BE EQUALLY SPACED.

THE NUMBER OF INTERVALS MUST BE:
    TRAPEZOIDAL RULE    : ANY NUMBER

    SIMPSON'S 1/3 RULE : EVEN NUMBER

    SIMPSON'S 3/8 RULE : NUMBER DIVISIBLE BY 3
HOW MANY DATA POINTS:? 11

GIVE NAME OF FILE FOR STORING THE DATA ? CO2RATE

ENTER PAIRS OF DATA POINTS:
x( 0 )=? 140
f( x )=? 15.72

x( 1 )=? 141
f( x )=? 15.53

x( 2 )=? 142
f( x )=? 15.19

x( 3 )=? 143
f( x )=? 16.56

x( 4 )=? 144
f( x )=? 16.21

x( 5 )=? 145
f( x )=? 17.39

x( 6 )=? 146
f( x )=? 17.36

x( 7 )=? 147
f( x )=? 17.42

x( 8 )=? 148
f( x )=? 17.60

x( 9 )=? 149
f( x )=? 17.75

x( 10 )=? 150
f( x )=? 18.95

INTEGRATING. PLEASE WAIT

AREA OF CURVE UNDER DATA POINTS =  168.6634
Ok
```

```
*****************************************************************
*                                                               *
*                         EXAMPLE 5-6                           *
*                                                               *
*                    INTEGRATION FORMULAS:                      *
*                                                               *
*     TRAPEZOIDAL, SIMPSON'S 1/3, AND SIMPSON'S 3/8 RULES       *
*                                                               *
*                         (INTEGR.BAS)                          *
*                                                               *
*****************************************************************


                       FORMULA OPTIONS

          1. TRAPEZOIDAL RULE

          2. SIMPSON'S 1/3 RULE

          3. SIMPSON'S 3/8 RULE

    ENTER YOUR CHOICE(1-3)? 2


                       INTEGRATION OPTIONS

          1. INTEGRATE EQUATIONS OR FUNCTIONS

          2. INTEGRATE DATA

    ENTER YOUR CHOICE(1-2)? 2


                          DATA POINTS

    HAVE YOU PREVIOUSLY ENTERED THE DATA POINTS(Y/N)? N

    THE DATA MUST BE EQUALLY SPACED.

    THE NUMBER OF INTERVALS MUST BE:

        TRAPEZOIDAL RULE    : ANY NUMBER

        SIMPSON'S 1/3 RULE : EVEN NUMBER

        SIMPSON'S 3/8 RULE : NUMBER DIVISIBLE BY 3

    HOW MANY DATA POINTS:? 11

    GIVE NAME OF FILE FOR STORING THE DATA ? O2RATE
```

```
ENTER PAIRS OF DATA POINTS:

x( 0 )=? 140
f( x )=? 15.49

x( 1 )=? 141
f( x )=? 16.16

x( 2 )=? 142
f( x )=? 15.35

x( 3 )=? 143
f( x )=? 15.13

x( 4 )=? 144
f( x )=? 14.20

x( 5 )=? 145
f( x )=? 14.23

x( 6 )=? 146
f( x )=? 14.29

x( 7 )=? 147
f( x )=? 12.74

x( 8 )=? 148
f( x )=? 14.74

x( 9 )=? 149
f( x )=? 13.68

x( 10 )=? 150
f( x )=? 14.51

INTEGRATING. PLEASE WAIT

AREA OF CURVE UNDER DATA POINTS =  144.9733
Ok
******************************************************************
*                                                                *
*                          EXAMPLE 5-6                           *
*                                                                *
*                     INTEGRATION FORMULAS:                      *
*                                                                *
*    TRAPEZOIDAL, SIMPSON'S 1/3, AND SIMPSON'S 3/8 RULES         *
*                                                                *
*                         (INTEGR.BAS)                           *
*                                                                *
******************************************************************

                        FORMULA OPTIONS

            1. TRAPEZOIDAL RULE

            2. SIMPSON'S 1/3 RULE

            3. SIMPSON'S 3/8 RULE
```

```
ENTER YOUR CHOICE(1-3)? 1

                    INTEGRATION OPTIONS

        1. INTEGRATE EQUATIONS OR FUNCTIONS

        2. INTEGRATE DATA

ENTER YOUR CHOICE(1-2)? 2

                        DATA POINTS

HAVE YOU PREVIOUSLY ENTERED THE DATA POINTS(Y/N)? Y

THE FOLLOWING DATA-FILES EXIST ON YOUR DISK:

B:\
                 O2RATE  .DAT      CO2RATE .DAT
 200704 Bytes free

GIVE NAME OF FILE TO BE USED (DO NOT ENTER THE EXTENSION): CO2RATE

READING THE DATA POINTS:

 140           15.72
 141           15.53
 142           15.19
 143           16.56
 144           16.21
 145           17.39
 146           17.36
 147           17.42
 148           17.6
 149           17.75
 150           18.95

INTEGRATING. PLEASE WAIT

AREA OF CURVE UNDER DATA POINTS =  168.345
Ok
```

```
******************************************************************
*                                                                *
*                          EXAMPLE 5-6                           *
*                                                                *
*                     INTEGRATION FORMULAS:                      *
*                                                                *
*       TRAPEZOIDAL, SIMPSON'S 1/3, AND SIMPSON'S 3/8 RULES      *
*                                                                *
*                         (INTEGR.BAS)                           *
*                                                                *
******************************************************************
```

```
                    FORMULA OPTIONS

        1. TRAPEZOIDAL RULE

        2. SIMPSON'S 1/3 RULE

        3. SIMPSON'S 3/8 RULE

ENTER YOUR CHOICE(1-3)? 1



                    INTEGRATION OPTIONS

        1. INTEGRATE EQUATIONS OR FUNCTIONS

        2. INTEGRATE DATA

 ENTER YOUR CHOICE(1-2)? 2



                    DATA POINTS

HAVE YOU PREVIOUSLY ENTERED THE DATA POINTS(Y/N)? Y

THE FOLLOWING DATA-FILES EXIST ON YOUR DISK:

B:\
                 O2RATE   .DAT      CO2RATE .DAT
 200704 Bytes free

GIVE NAME OF FILE TO BE USED (DO NOT ENTER THE EXTENSION): O2RATE

READING THE DATA POINTS:

 140           15.49
 141           16.16
 142           15.35
 143           15.13
 144           14.2
 145           14.23
 146           14.29
 147           12.74
 148           14.74
 149           13.68
 150           14.51

 INTEGRATING. PLEASE WAIT

 AREA OF CURVE UNDER DATA POINTS =  145.52
 Ok
```

DISCUSSION OF RESULTS

*Part a* The results obtained by the application of Simpson's $\frac{3}{8}$ rule are shown in the table of results and are plotted on the accompanying graph. This curve is a representation of the fluid velocity profiles in space and time since the dimensionless variable $\eta$ is a function of both dimensions. We can verify the solution at the boundaries of space and time:

1. The results show that at $\eta = 0$, the velocity of the fluid is identical to that of the moving plate (that is, $v_x/V = 1$). The variable $\eta$ attains a value of zero at only two situations in the space-time domain:
   *a*. In the fluid next to the moving plate (at $y = 0$ and at all $t$)
   *b*. After an infinite amount of time has elapsed (at $t = \infty$ and at all $y$)
   Situation *a* is consistent with the boundary conditions given in the statement of the problem. Situation *b* is an expected result, since given a long enough time, all the fluid will be moving along at the same velocity as the plate.
2. The results also show that

$$\lim_{\eta \to \infty} \frac{v_x}{V} = 0$$

   The variable $\eta$ becomes infinity under the following circumstances:
   *a*. In the fluid far away from the plate (at $y = \infty$ and all $t$)
   *b*. At the initial time (at $t = 0$ and all $y$)
   Both these situations are specified as boundary conditions of the problem.

*Part b* The integration of the experimental data, utilizing both Simpson's $\frac{1}{3}$ and the trapezoidal rule, yield the total amounts of carbon dioxide and oxygen shown in Table E5-6*c*.

**Table E5-6*c***

| | Simpson's $\frac{1}{3}$ | Trapezoidal |
|---|---|---|
| Total $CO_2$, g | 168.66 | 168.34 |
| Total $O_2$, g | 144.97 | 145.52 |

## 5-6 NONLINEAR ORDINARY DIFFERENTIAL EQUATIONS—INITIAL-VALUE PROBLEMS

In this section we develop numerical solutions for a set of ordinary differential equations in their canonical form:

$$\frac{d\mathbf{y}}{dx} = \mathbf{f}(x, \mathbf{y}) \tag{5-27}$$

with the vector of initial conditions given by

$$\mathbf{y}(x_0) = \mathbf{y}_0 \tag{5-28}$$

In order to be able to illustrate these methods graphically, we treat **y** as a single variable rather than as a vector of variables. The formulas developed for the solution of a single differential equation are readily expandable to those for a set of differential equations which must be solved *simultaneously*. This concept is demonstrated in Sec. 5-6.3.

We begin the development of these methods by first rearranging Eq. (5-27) and integrating both sides between the limits of $x_i \le x \le x_{i+1}$ and $y_i \le y \le y_{i+1}$.

$$\int_{y_i}^{y_{i+1}} dy = \int_{x_i}^{x_{i+1}} f(x, y)\, dx \tag{5-110}$$

The left side integrates readily to obtain

$$y_{i+1} - y_i = \int_{x_i}^{x_{i+1}} f(x, y)\, dx \tag{5-111}$$

In Sec. 5-5 we developed the integration formulas by first replacing the function $f(x)$ with an interpolating polynomial, and then evaluating the integral $f(x)\, dx$ between the appropriate limits. A similar technique could be followed here; however, we choose to take the alternative approach of working with the left side of Eq. (5-111) and using finite differences for its approximation. It should be emphasized that either technique yields the same set of formulas; however, we consider the latter approach to be more elegant because it works directly with the tangential trajectories of the dependent variable $y$ rather than with the areas under the function $f(x, y)$.

### 5-6.1 The Euler Methods

One of the earliest techniques developed for the solution of ordinary differential equations is the *Euler† method*. This is simply obtained by recognizing that the left side of Eq. (5-111) is the first forward finite difference of $y$ at position $i$

$$y_{i+1} - y_i = \Delta y_i \tag{4-50}$$

which, when rearranged, gives a "forward marching" formula for evaluating $y$:

$$y_{i+1} = y_i + \Delta y_i \tag{5-112}$$

The forward difference term $\Delta y_i$ is obtained from Eq. (4-57) applied to $y$ at position $i$:

† Leonhard Euler (1707–1783).

$$\Delta y_i = hDy_i + \frac{h^2D^2y_i}{2} + \frac{h^3D^3y_i}{6} + \cdots \tag{5-113}$$

In the Euler method, the above series is truncated after the first term to obtain

$$\Delta y_i = hDy_i + O(h^2) \tag{5-114}$$

The combination of Eqs. (5-112) and (5-114) gives the *explicit Euler formula* for integrating differential equations

$$y_{i+1} = y_i + hDy_i + O(h^2) \tag{5-115}$$

The derivative $Dy_i$ is replaced by its equivalent $y'_i$ or $f(x_i, y_i)$† to give the more commonly used form of the explicit Euler method

$$y_{i+1} = y_i + hf(x_i, y_i) + O(h^2) \tag{5-116}$$

This equation simply states that the next value of $y$ is obtained from the previous value by moving a step of width $h$ in the tangential direction of $y$. This is demonstrated graphically in Fig. 5-9*a*. This Euler formula is rather inaccurate because it has a truncation error of only $O(h^2)$. If $h$ is large the trajectory of $y$ can quickly deviate from its true value, as shown in Fig. 5-9*b*.

The accuracy of the Euler method can be improved by utilizing a combination of forward and backward differences. Note that the first forward difference of $y$ at $i$ is equal to the first backward difference of $y$ at $i+1$

$$\Delta y_i = y_{i+1} - y_i = \nabla y_{i+1} \tag{5-117}$$

Therefore, the forward marching formula in terms of backward differences is

$$y_{i+1} = y_i + \nabla y_{i+1} \tag{5-118}$$

The backward difference term $\nabla y_{i+1}$ is obtained from Eq. (4-31) applied to $y$ at position $i+1$

$$\nabla y_{i+1} = hDy_{i+1} - \frac{h^2D^2y_{i+1}}{2} + \frac{h^3D^3y_{i+1}}{6} - \cdots \tag{5-119}$$

Combining Eqs. (5-118) and (5-119),

$$y_{i+1} = y_i + hf(x_{i+1}, y_{i+1}) + O(h^2) \tag{5-120}$$

† From here on the terms $y'_i$ and $f(x_i, y_i)$ will be used interchangeably. The reader should remember that these are equal to each other through the differential equation (5-27).

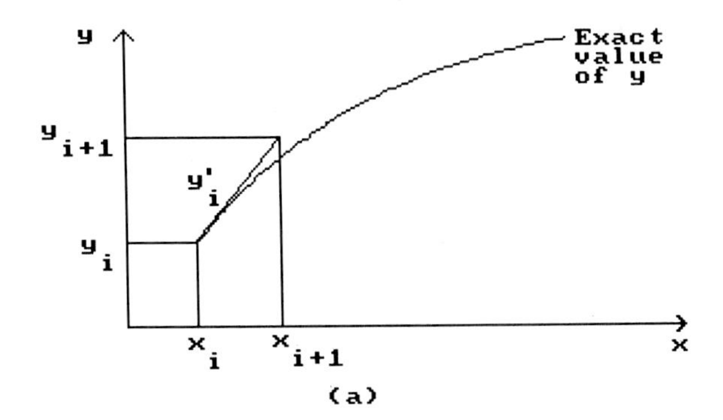


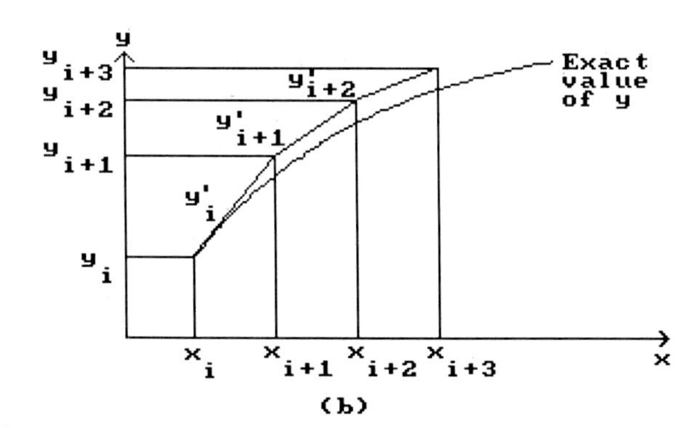


**Figure 5-9** The explicit Euler method of integration: (*a*) single step, (*b*) several steps.

This is called the *implicit Euler formula* (or backward Euler) because it involves the calculation of the function $f$ (gradient of $y$) at an unknown value of $y_{i+1}$. Equation (5-120) can be viewed as taking a step forward from position $i$ to $i + 1$ in a gradient direction which must be evaluated at $i + 1$.

Implicit equations cannot be solved individually but must be set up as sets of simultaneous algebraic equations. When these sets are linear, the problem can be solved by the application of the Gauss elimination methods developed in Chap. 3. If the set consists of nonlinear equations, the problem is much more difficult and must be solved using Newton's method for simultaneous nonlinear algebraic equations developed in Chap. 2.

In the case of the Euler methods, the problem can be simplified by first applying the explicit method to *predict* a value of $y_{i+1}$,

$$(y_{i+1})_{\text{Pr}} = y_i + hf(x_i, y_i) + O(h^2) \tag{5-121}$$

and then using this predicted value in the implicit method to get a *corrected* value:

$$(y_{i+1})_{\text{Cor}} = y_i + hf(x_{i+1}, (y_{i+1})_{\text{Pr}}) + O(h^2) \tag{5-122}$$

This combination of steps is known as the *Euler predictor-corrector* (or *modified Euler*) method, whose application is demonstrated graphically in Fig. 5-10.

The explicit, as well as the implicit, forms of the Euler methods have error of order $(h^2)$. However, when used in combination with each other, as predictor-corrector, their accuracy is enhanced, yielding an error of order $(h^3)$. This conclusion can be reached by adding Eqs. (5-112) and (5-118)

$$y_{i+1} = y_i + \tfrac{1}{2}(\Delta y_i + \nabla y_{i+1}) \tag{5-123}$$

and utilizing (5-113) and (5-119) to obtain

$$y_{i+1} = y_i + \frac{h}{2}\left[f(x_i, y_i) + f(x_{i+1}, y_{i+1})\right] + O(h^3) \tag{5-124}$$

The terms of order $(h^2)$ cancel out because they have opposite sign, thus giving a formula of higher accuracy. Equation (5-124) is essentially the same as the

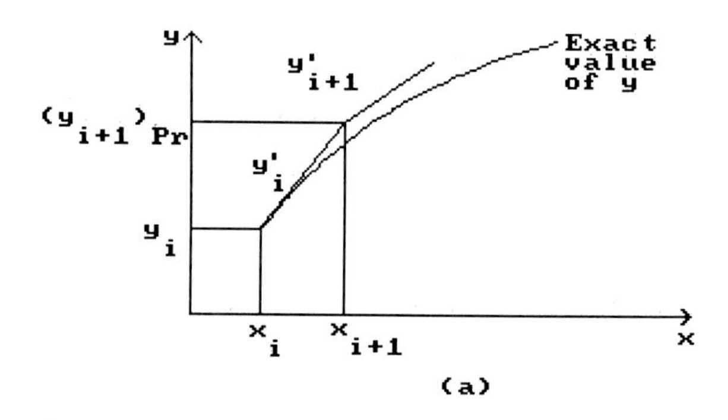


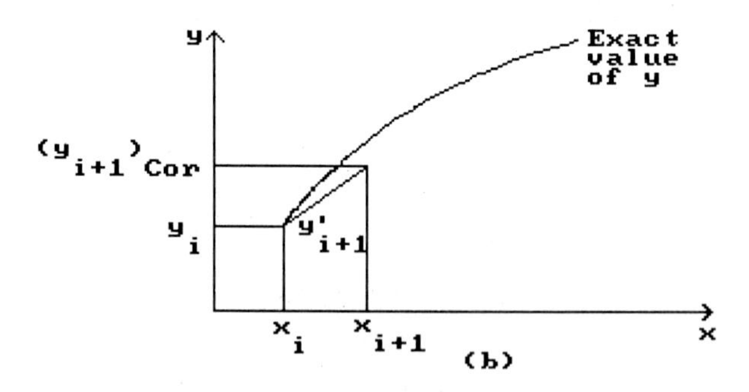


**Figure 5-10** The Euler predictor-corrector method: (*a*) value of $y_{i+1}$ is predicted and $y'_{i+1}$ is calculated, (*b*) value of $y_{i+1}$ is corrected.

trapezoidal rule [Eq. (5-79)], with the only difference being in the way the function is evaluated at $(x_{i+1}, y_{i+1})$.

It has been shown [8] that the Euler implicit formula is more stable than the explicit one. The stability of these methods will be discussed in Sec. 5-9.

It can be seen by writing Eq. (5-124) in the form

$$y_{i+1} = y_i + \tfrac{1}{2}hf(x_i, y_i) + \tfrac{1}{2}hf(x_{i+1}, y_{i+1}) + O(h^3) \tag{5-125}$$

that the modified Euler method uses the weighted trajectories of the function $y$ evaluated at two positions which are located one full step of width $h$ apart and weighted equally. In this form, Eq. (5-125) is also known as the Crank-Nicolson method.

Equation (5-125) can be written in a more general form as

$$y_{i+1} = y_i + w_1k_1 + w_2k_2 \tag{5-126}$$

where, in this case,

$$k_1 = hf(x_i, y_i) \tag{5-127}$$

$$k_2 = hf(x_i + c_2h, y_i + a_{21}k_1) \tag{5-128}$$

The choice of the weighting factors $w_1$ and $w_2$ and the positions $i$ and $i+1$ at which to evaluate the trajectories is dictated by the accuracy required of the integration formula, i.e., by the number of terms retained in the infinite series expansion.

This concept forms the basis for a whole series of integration formulas, with increasingly higher accuracies, for ordinary differential equations. These are discussed in the following section.

### 5-6.2 The Runge-Kutta Methods

The most widely used methods of integration for ordinary differential equations are the series of methods called Runge-Kutta second, third, and fourth order, plus a number of other techniques which are variations on the Runge-Kutta theme. These methods are based on the concept of weighted trajectories formulated at the end of Sec. 5-6.1. In a more general fashion, the forward marching integration formula for the differential equation (5-27) is given by the recurrence equation

$$y_{i+1} = y_i + w_1k_1 + w_2k_2 + w_3k_3 + \cdots + w_mk_m \tag{5-129}$$

where each of the trajectories $k_i$ are evaluated by

$$
\begin{aligned}
k_1 &= hf(x_i, y_i) \\
k_2 &= hf(x_i + c_2 h, y_i + a_{21} k_1) \\
k_3 &= hf(x_i + c_3 h, y_i + a_{31} k_1 + a_{32} k_2) \\
&\vdots \\
k_m &= hf(x_i + c_m h, y_i + a_{m1} k_1 + a_{m2} k_2 + \cdots + a_{m,m-1} k_{m-1})
\end{aligned} \tag{5-130}
$$

These equations can be written in a compact form as

$$y_{i+1} = y_i + \sum_{j=1}^{m} w_j k_j \tag{5-131}$$

$$k_j = hf\left(x_i + c_j h, y_i + \sum_{l=1}^{j-1} a_{jl} k_l\right) \tag{5-132}$$

where $c_1 = 0$ and $a_{1j} = 0$. The value of $m$, which determines the complexity and accuracy of the method, is set when $m + 1$ terms are retained in the infinite series expansion of $y_{i+1}$

$$y_{i+1} = y_i + hy'_1 + \frac{h^2 y''_i}{2!} + \frac{h^3 y'''_i}{3!} + \cdots \tag{5-133}$$

or

$$y_{i+1} = y_i + hDy_i + \frac{h^2 D^2 y_i}{2!} + \frac{h^3 D^3 y_i}{3!} + \cdots \tag{5-134}$$

The procedure for deriving the Runge-Kutta methods can be divided into five steps which are demonstrated below in the derivation of the *second-order Runge-Kutta* formulas.

*Step 1* Choose the value of $m$, which fixes the accuracy of the formula to be obtained. For second-order Runge-Kutta $m = 2$. Truncate the series (5-134) after the $m + 1$ term:

$$y_{i+1} = y_i + hDy_i + \frac{h^2 D^2 y_i}{2!} + O(h^3) \tag{5-135}$$

*Step 2* Replace each derivative of $y$ in (5-135) by its equivalent in $f$, remembering that $f$ is a function of both $x$ and $y(x)$:

$$Dy_i = f_i \tag{5-136}$$

$$
\begin{aligned}
D^2 y_i &= \frac{df}{dx} = \left(\frac{\partial f}{\partial x}\frac{dx}{dx} + \frac{\partial f}{\partial y}\frac{dy}{dx}\right)_i \\
&= (f_x + f f_y)_i
\end{aligned} \tag{5-137}
$$

Combine Eqs. (5-135) to (5-137) and regroup the terms:

$$y_{i+1} = y_i + hf_i + \frac{h^2}{2} f_{x_i} + \frac{h^2}{2} f_i f_{y_i} + O(h^3) \tag{5-138}$$

*Step 3* Write Eq. (5-131) with $m$ terms in the summation:

$$y_{i+1} = y_i + w_1 k_1 + w_2 k_2 \tag{5-139}$$

where

$$k_1 = hf(x_i, y_i) \tag{5-140}$$

$$k_2 = hf(x_i + c_2 h, y_i + a_{21} k_1) \tag{5-141}$$

*Step 4* Expand the $f$ functions in Taylor series:

$$f(x_i + c_2 h, y_i + a_{21} k_1) = f_i + c_2 h f_{x_i} + a_{21} h f_{y_i} f_i + O(h^2) \tag{5-142}$$

Combine Eqs. (5-139) to (5-142) and regroup the terms:

$$y_{i+1} = y_i + (w_1 + w_2) h f_i + (w_2 c_2) h^2 f_{x_i} + (w_2 a_{21}) h^2 f_i f_{y_i} + O(h^3) \tag{5-143}$$

*Step 5* In order for Eqs. (5-138) and (5-143) to be identical the coefficients of the corresponding terms must be equal to one another. This results in a set of simultaneous nonlinear algebraic equations in the unknown constants $w_j$, $c_j$, and $a_{jl}$. For this second-order Runge-Kutta method there are three equations and four unknowns:

$$\begin{aligned} w_1 + w_2 &= 1.0 \\ w_2 c_2 &= 0.5 \\ w_2 a_{21} &= 0.5 \end{aligned} \tag{5-144}$$

It turns out that there are always more unknowns than equations. The degrees of freedom allow us to choose some of the parameters. For second-order Runge-Kutta, there is one degree of freedom. For third- and fourth-order Runge-Kutta, there are two degrees of freedom. For fifth-order Runge-Kutta, there are at least five degrees of freedom. This freedom of choice of parameters gives rise to a very large number of different forms of the Runge-Kutta formulas. It is usually desirable to choose first the values of the $c_j$ constants, thus fixing the positions along the independent variable where the functions

$$f\left(x_i + c_j h, y_i + \sum_{l=1}^{j-1} a_{jl} k_l\right)$$

are to be evaluated. An important consideration in choosing the free parameters is to minimize the *roundoff error* of the calculation. Discussion of the effect of the roundoff error will be given in Sec. 5-9.

For the second-order Runge-Kutta method, which we are currently deriving, let us choose $c_2 = 1$. The rest of the parameters are evaluated from Eqs. (5-144):

$$w_1 = w_2 = \tfrac{1}{2} \qquad a_{21} = 1 \tag{5-145}$$

With this set of parameters the second-order Runge-Kutta formula is

$$\left.\begin{aligned} y_{i+1} &= y_i + \tfrac{1}{2}(k_1 + k_2) \\ k_1 &= hf(x_i, y_i) \\ k_2 &= hf(x_i + h, y_i + k_1) \end{aligned}\right\} \qquad O(h^3) \tag{5-146}$$

This is identical to the Euler predictor-corrector of Eq. (5-124).

A different version of the second-order Runge-Kutta is obtained by choosing to evaluate the function at the midpoints (that is, $c_2 = \frac{1}{2}$). This yields the formula

$$\left.\begin{aligned} y_{i+1} &= y_i + k_2 \\ k_1 &= hf(x_i, y_i) \\ k_2 &= hf(x_i + \tfrac{1}{2}h, y_i + \tfrac{1}{2}k_1) \end{aligned}\right\} \qquad O(h^3) \tag{5-147}$$

Higher-order Runge-Kutta formulas are derived in an analogous manner. Several of these are listed in Table 5-3. The fourth-order Runge-Kutta, which has an error of $O(h^5)$, is probably the most widely used numerical integration method for ordinary differential equations.

### 5-6.3 Simultaneous Differential Equations

It was mentioned at the beginning of Sec. 5-6 that the methods of solution of a single differential equation are readily adaptable for solving sets of simultaneous differential equations. To illustrate this, we use the set of $n$ simultaneous ordinary differential equations

$$\begin{aligned} \frac{dy_1}{dx} &= f_1(x, y_1, y_2, \ldots, y_n) \\ \frac{dy_2}{dx} &= f_2(x, y_1, y_2, \ldots, y_n) \\ &\vdots \\ \frac{dy_n}{dx} &= f_n(x, y_1, y_2, \ldots, y_n) \end{aligned} \tag{5-151}$$

**Table 5-3 Summary of the Runge-Kutta integration formulas**

Second order

$$\left.\begin{aligned} y_{i+1} &= y_i + \tfrac{1}{2}(k_1 + k_2) \\ k_1 &= hf(x_i, y_i) \\ k_2 &= hf(x_i + h, y_i + k_1) \end{aligned}\right\} \quad O(h^3) \tag{5-146}$$

Third order

$$\left.\begin{aligned} y_{i+1} &= y_i + \tfrac{1}{6}(k_1 + 4k_2 + k_3) \\ k_1 &= hf(x_i, y_i) \\ k_2 &= hf\left(x_i + \frac{h}{2}, y_i + \frac{k_1}{2}\right) \\ k_3 &= hf(x_i + h, y_i + 2k_2 - k_1) \end{aligned}\right\} \quad O(h^4) \tag{5-148}$$

Fourth order

$$\left.\begin{aligned} y_{i+1} &= y_i + \tfrac{1}{6}(k_1 + 2k_2 + 2k_3 + k_4) \\ k_1 &= hf(x_i, y_i) \\ k_2 &= hf\left(x_i + \frac{h}{2}, y_i + \frac{k_1}{2}\right) \\ k_3 &= hf\left(x_i + \frac{h}{2}, y_i + \frac{k_2}{2}\right) \\ k_4 &= hf(x_i + h, y_i + k_3) \end{aligned}\right\} \quad O(h^5) \tag{5-149}$$

Fourth-order Runge-Kutta-Gill

$$\left.\begin{aligned} y_{i+1} &= y_i + \tfrac{1}{6}(k_1 + 2bk_2 + 2dk_3 + k_4) \\ k_1 &= hf(x_i, y_i) \\ k_2 &= hf\left(x_i + \frac{h}{2}, y_i + \frac{k_1}{2}\right) \\ k_3 &= hf\left(x_i + \frac{h}{2}, y_i + ak_1 + bk_2\right) \\ k_4 &= hf(x_i + h, y_i + ck_2 + dk_3) \\ a &= \frac{\sqrt{2} - 1}{2} \qquad b = \frac{2 - \sqrt{2}}{2} \\ c &= -\frac{\sqrt{2}}{2} \qquad d = 1 + \frac{\sqrt{2}}{2} \end{aligned}\right\} \quad O(h^5) \tag{5-150}$$

and expand the fourth-order Runge-Kutta formulas to

$$
\begin{aligned}
y_{i+1,j} &= y_{ij} + \tfrac{1}{6}(k_{1j} + 2k_{2j} + 2k_{3j} + k_{4j}) && j = 1, 2, \ldots, n \\
k_{1j} &= hf_j(x_i, y_{i1}, y_{i2}, \ldots, y_{in}) && j = 1, 2, \ldots, n \\
k_{2j} &= hf_j\left(x_i + \frac{h}{2}, y_{i1} + \frac{k_{11}}{2}, y_{i2} + \frac{k_{12}}{2}, \ldots, y_{in} + \frac{k_{1n}}{2}\right) && j = 1, 2, \ldots, n \\
k_{3j} &= hf_j\left(x_i + \frac{h}{2}, y_{i1} + \frac{k_{21}}{2}, y_{i2} + \frac{k_{22}}{2}, \ldots, y_{in} + \frac{k_{2n}}{2}\right) && j = 1, 2, \ldots, n \\
k_{4j} &= hf_j(x_i + h, y_{i1} + k_{31}, y_{i2} + k_{32}, \ldots, y_{in} + k_{3n}) && j = 1, 2, \ldots, n
\end{aligned}
\tag{5-152}
$$

This method is easily programmable using nested loops, as will be demonstrated in Example 5-7.

The Runge-Kutta family of integration techniques, developed above, are called *single-step* methods. The value of $y_{i+1}$ is obtained from $y_i$ and the trajectories of $y$ within the single step from $(x_i, y_i)$ to $(x_{i+1}, y_{i+1})$. This procedure marches forward, taking single steps of width $h$, over the entire interval of integration. These methods are very suitable for solving initial-value problems because they are *self-starting* from a given initial point of integration.

Other categories of integration techniques, called the *multiple-step* methods, have been developed. These compute the value of $y_{i+1}$ utilizing several previously known, or calculated, values of $y$ ($y_i$, $y_{i-1}$, $y_{i-2}$, etc.) as the base points. For this reason the multiple-step methods are *nonself-starting*. For the solution of initial-value problems, where only $y$ is known, the multiple-step methods must be "primed" by first utilizing a self-starting procedure to obtain the requisite number of base points. The multiple-step methods, known by the names of Adams, Milne, and others, will not be covered here. A complete discussion of these techniques is given by Lapidus and Seinfeld [2].

**Example 5-7 Fourth-order Runge-Kutta and Euler predictor-corrector methods for integrating simultaneous ordinary differential equations** Write a general computer program for integrating simultaneous ordinary differential equations using the fourth-order Runge-Kutta and Euler predictor-corrector methods. Apply this program for the solution of differential equations which simulate the dynamics of interacting populations, as described in the next section.

POPULATION DYNAMICS The best-known mathematical representation of population dynamics between interacting species is the Lotka-Volterra model [9]. For the case of two competing species, these equations take the general form

$$\frac{dN_1}{N_1 dt} = f_1(N_1, N_2) \qquad \frac{dN_2}{N_2\, dt} = f_2(N_1, N_2)$$

where $N_1$ is the population density of species 1 and $N_2$ is the population density of species 2. The functions $f_1$ and $f_2$ describe the specific growth rates of the two populations. Under certain assumptions, these functions can be expressed in terms of $N_1$, $N_2$, and a set of constants whose values depend on natural birth and death rates and on the interactions between the two species. Numerous examples of such interactions can be cited from ecological and microbiological studies. The predator-prey problem, which has been studied extensively, presents a very interesting ecological example of population dynamics. On the other hand, the interaction between bacteria and phages in a fermentor is a well-known nemesis to industrial microbiologists.

Let us now consider in detail the classical predator-prey problem, i.e., the interaction between two wild-life species, the prey, which is a herbivore, and the predator, a carnivore. These two animals coinhabit a region where the prey have an abundant supply of natural vegetation for food, and the predators depend on the prey for their entire supply of food. This is a simplification of the real ecological system where more than two species coexist, and where predators usually feed on a variety of prey. The Lotka-Volterra equations have also been formulated for such complex systems; however, for the sake of this problem, our ecological system will contain only two interacting species. An excellent example of such an ecological system is Isle Royale National Park, a 210-square mile archipelago in Lake Superior. The park comprises a single large island and many small islands which extend off the main island. "Moose arrived on Isle Royale around 1900, probably swimming in from Canada. By 1930 their unchecked numbers approached 3000, ravaging vegetation. In 1949, across an ice bridge from Ontario, came a predator—the wolf. Since 1958 the longest study of its kind [10, 11] still seeks to define the complete cycle in the ebb and flow of predator and prey populations, with wolves fluctuating from 14 to 50 and moose from 500 to 1400" [12].

In order to formulate the predator-prey problem, we make the following assumptions:

(*a*) In the absence of the predator the prey has a natural birth rate $b$ and a natural death rate $d$. Since an abundant supply of food is available, the birth rate is higher than the death rate; therefore, the net specific growth rate $\alpha$ is positive; i.e.,

$$\frac{dN_1}{N_1\, dt} = b - d = \alpha$$

(*b*) In the presence of the predator the prey is consumed at a rate proportional to the number of predators present $\beta N_2$, as shown below:

$$\frac{dN_1}{N_1\,dt} = \alpha - \beta N_2 \tag{1}$$

(*c*) In the absence of the prey the predator has a negative specific growth rate $(-\gamma)$ since the inevitable consequence of such a situation is the starvation of the predator:

$$\frac{dN_2}{N_2\,dt} = -\gamma$$

(*d*) In the presence of the prey the predator has an ample supply of food which enables it to survive and reproduce at a rate proportional to the abundance of the prey $\delta N_1$. Under these circumstances, the specific growth rate of the predator is

$$\frac{dN_2}{N_2\,dt} = -\gamma + \delta N_1 \tag{2}$$

Equations (1) and (2) constitute the Lotka-Volterra model for the one-predator–one-prey problem. Rearranging these two equations to put them in the canonical form,

$$\frac{dN_1}{dt} = \alpha N_1 - \beta N_1 N_2 \tag{1}$$

$$\frac{dN_2}{dt} = -\gamma N_2 + \delta N_1 N_2 \tag{2}$$

This is a set of simultaneous first-order nonlinear ordinary differential equations. The solution of these equations first requires the determination of the constants $\alpha$, $\beta$, $\gamma$, and $\delta$, and the specification of boundary conditions. The latter could be either initial or final conditions. In population dynamics, it is more customary to specify initial population densities, since actual numerical values of the population densities may be known at some point in time, which can be called the initial starting time. However, it is not inconceivable that one may want to specify final values of the population densities to be accomplished as targets in a well-managed ecological system. In this problem, we will specify the initial population densities of the prey and predator to be

$$N_1(t_0) = N_1^0 \qquad \text{and} \qquad N_2(t_0) = N_2^0 \tag{3}$$

Equations (1) to (3) constitute the complete mathematical formulation of the predator-prey problem based on assumptions *a* to *d*. Different assumptions would yield another set of differential equations. In addition, the choice of constants and initial conditions influence the solution of the differential equations and generate a diverse set of qualitative behavior patterns for the two populations. Depending on the form of the differential equations and the values of the constants chosen, the solution patterns may vary from stable damped oscillations, where the species reach their respective stable symbiotic population densities, to highly unstable situations in which one of the species is driven to extinction while the other explodes to extreme population density.

The literature on the solution of the Lotka-Volterra problems is voluminous. Several references on this topic are given at the end of this chapter. A closed-form analytical solution of this system of nonlinear ordinary differential equations is not possible. The equations must be integrated numerically using any of the numerical integration methods, such as the fourth-order Runge-Kutta method, described earlier in this chapter. However, before numerical integration is attempted the stability of these equations must be examined thoroughly. In a recent treatise on this subject, J. Vandermeer [13] examined the stability of the solutions of these equations around equilibrium points. These points are located by setting the derivatives in Eqs. (1) and (2) to zero,

$$\frac{dN_1}{dt} = \alpha N_1 - \beta N_1 N_2 = 0$$

$$\frac{dN_2}{dt} = -\gamma N_2 + \delta N_1 N_2 = 0$$

and rearranging these equations to obtain the values of $N_1$ and $N_2$ at the equilibrium point in terms of the constants

$$N_1^* = \frac{\gamma}{\delta} \qquad N_2^* = \frac{\alpha}{\beta}$$

where * denotes the equilibrium values of the population densities. Vandermeer states that: "Sometimes only one point $(N_1^*, N_2^*)$ will satisfy the equilibrium equations. At other times, multiple points will satisfy the equilibrium equations. . . . The neighborhood stability analysis is undertaken in the neighborhood of a single equilibrium point." The stability is determined by examining the eigenvalues of the Jacobian matrix evaluated at equilibrium,

$$\mathbf{J} = \begin{bmatrix} \left(\dfrac{\partial f_1}{\partial N_1}\right)^* & \left(\dfrac{\partial f_1}{\partial N_2}\right)^* \\ \left(\dfrac{\partial f_2}{\partial N_1}\right)^* & \left(\dfrac{\partial f_2}{\partial N_2}\right)^* \end{bmatrix}$$

where $f_1$ and $f_2$ are the right-hand sides of Eqs. (1) and (2), respectively.

The eigenvalues of the Jacobian matrix can be obtained by the solution of the following equation (as described in Chap. 3):

$$|\mathbf{J} - \lambda \mathbf{I}| = 0$$

For the problem of two differential equations, there are two eigenvalues which can possibly have both real and imaginary parts. These eigenvalues take the general form

$$\lambda_1 = a_1 + b_1 i$$

$$\lambda_2 = a_2 + b_2 i$$

where $i = \sqrt{-1}$. The values of the real parts ($a_1$, $a_2$) and imaginary parts ($b_1$, $b_2$) determine the nature of the stability (or instability) in the neighborhood of the equilibrium point. These possibilities are summarized in Table E5-7*a*.

NUMERICAL SOLUTIONS Many combinations of values of constants and initial conditions exist that would generate solutions to Eqs. (1) and (2). In order to obtain a realistic solution to these equations, we utilize the data of Allen [10] and Peterson [11] on the moose-wolf populations of Isle Royale National Park (see Prob. 5-9). From these data, which cover the period 1960–1983, we estimate the average values of the moose and wolf populations and use these as equilibrium values:

$$N_1^* = \frac{\gamma}{\delta} = 800 \qquad \text{and} \qquad N_2^* = \frac{\alpha}{\beta} = 27$$

In addition, we estimate the period of oscillation to be 25 years. This was based on the moose data; the wolf data show a shorter period. For this reason we predict that the predator equation may not be a good representation of the data. Lotka has shown that the period of oscillation around the equilibrium point is approximated by

**Table E5-7*a***

| $a_1, a_2$ | $b_1, b_2$ | Stability analysis |
|---|---|---|
| Negative | Zero | Stable, nonoscillatory |
| Positive | Zero | Unstable, nonoscillatory |
| One positive<br>One negative | Zero | Metastable, saddle point |
| Negative | Nonzero | Stable, oscillatory |
| Positive | Nonzero | Unstable, oscillatory |
| Zero | Nonzero | Neutrally stable, oscillatory |

$$\tau = \frac{2\pi}{\sqrt{\alpha\gamma}}$$

These three equations have four unknowns. By assuming the value of $\alpha$ to be 0.3 (this is an estimate of the net specific growth rate of the prey in the absence of the predator), the complete set of constants is

$$\alpha = 0.3$$

$$\beta = 0.01111$$

$$\gamma = 0.2106$$

$$\delta = 0.0002632$$

The initial conditions are taken from Allen [10] for 1960, the earliest date for which complete data are available. These are

$$N_1(1960) = 700 \qquad \text{and} \qquad N_2(1960) = 22$$

We will integrate the predator-prey equations for the period 1960–1985 using the above constants and initial conditions and will compare the simulation with the actual data.

The Jacobian matrix around the equilibrium point is

$$\mathbf{J} = \begin{bmatrix} \alpha - \beta N_2^* & -\beta N_1^* \\ \delta N_2^* & -\gamma + \delta N_1^* \end{bmatrix} = \begin{bmatrix} 0 & -8.888 \\ 0.007106 & 0 \end{bmatrix}$$

The eigenvalues of the Jacobian matrix are

$$\lambda_{1,2} = \pm 0.2513i$$

The real parts are zero and the imaginary parts are nonzero; therefore, this case yields neutrally stable oscillatory solutions, as will be shown in the discussion of results.

It should be observed that whenever the Lotka-Volterra problem has the form of Eqs. (1) and (2), the real parts of the eigenvalues of the Jacobian matrix are zero. This implies that the solution always has neutrally stable oscillatory behavior. This is explainable by the fact that assumptions *a* to *d* did not include the crowding effect each population may have on its own fertility or mortality. For example, Eq. (1) can be rewritten with the additional term $\epsilon N_1^2$:

$$\frac{dN_1}{dt} = \alpha N_1 - \beta N_1 N_2 - \epsilon N_1^2$$

The new term introduces a negative density-dependency of the specific growth rate of the prey on its own population. This term can be viewed as either a contribution to the death rate or a reduction of the birth rate caused by overcrowding of the species.

In the original formulation of the problem [Eqs. (1) and (2)], the two species, if left to themselves, without third-party interference and barring natural catastrophes, will continue their symbiosis in a stable but oscillatory pattern. We confirm this conclusion by integrating the predator-prey equations for the period 1960–2060.

PROGRAM DESCRIPTION This computer program implements the fourth-order Runge-Kutta and the Euler predictor-corrector methods for integrating simultaneous ordinary differential equations. The program is called ODE.BAS, and it consists of the sections listed in Table E5-7*b*.

The structure of this program is parallel to that of INTEGR.BAS in Example 5-6, with the following modifications:

1. ODE.BAS does not provide the option of integrating experimental data.
2. The differential equations in subroutine 1 are of the form

$$G(i) = f[x, y(1), y(2), \ldots, y(n), c(1), c(2), \ldots]$$

**Table E5-7*b***

| Program section | Line numbers |
|---|---|
| Title | 1–90 |
| Main program | 100–460 |
| Formula options | 110–170 |
| Results options | 240–290 |
| Subroutine calls | 300–350 |
| Rerun options | 360–450 |
| Subroutines | |
| 1. Input equations | 1000–1490 |
| 3. Input integration parameters | 3000–3220 |
| 4. Differential equations | 4000–4999 |
| 5. Integration methods | 5000–5400 |
| 6. Print table of results | 6000–6100 |
| 7. Plotting options | Same as in INTEGR.BAS of Example 5-6 |
| 8. Plotting | Same as in INTEGR.BAS of Example 5-6 |

3. Users can specify their own choice of number of integration steps in subroutine 3.
4. Subroutine 5 contains the fourth-order Runge-Kutta method [Eqs. (5-152)] and the Euler predictor-corrector method [Eqs. (5-121) and (5-122)] for simultaneous differential equations.
5. Subroutine 6 uses different formatting characters in line 6060 for printing the results.

Subroutines 7 and 8 are identical to those of INTEGR.BAS; therefore they are not listed here.

PROGRAM

```
1 SCREEN 0:WIDTH 80:CLS:KEY OFF
5 PRINT "**********************************************************************"
10 PRINT"*                                                                    *"
15 PRINT"*                            EXAMPLE 5-7                             *"
20 PRINT"*                                                                    *"
25 PRINT"*   4th ORDER RUNGE-KUTTA AND EULER PREDICTOR-CORRECTOR METHODS      *"
30 PRINT"*                                                                    *"
35 PRINT"*   FOR INTEGRATING SIMULTANEOUS ORDINARY DIFFERENTIAL EQUATIONS     *"
40 PRINT"*                                                                    *"
45 PRINT"*                             (ODE.BAS)                              *"
50 PRINT"*                                                                    *"
90 PRINT"**********************************************************************"
100 '***************************** Main Program ***************************
110 PRINT
120 PRINT:PRINT"                        FORMULA OPTIONS":PRINT
130 PRINT:PRINT"                 1. RUNGE-KUTTA 4th ORDER"
140 PRINT:PRINT"                 2. EULER PREDICTOR-CORRECTOR"
160 PRINT:PRINT:INPUT"    ENTER YOUR CHOICE(1-2)"; FQ
170 IF FQ<1 OR FQ>2 THEN GOTO 160
240 PRINT:PRINT
250 PRINT:PRINT"                        OUTPUT OPTIONS":PRINT
260 PRINT:PRINT"                 1. PRINT TABLE AND DRAW GRAPHS OF RESULTS"
270 PRINT:PRINT"                 2. DRAW GRAPHS OF RESULTS ONLY"
280 PRINT:PRINT:INPUT"    ENTER YOUR CHOICE(1-2)"; RQ
290 IF RQ<1 OR RQ>2 THEN GOTO 280
300 GOSUB 1000
310 GOSUB 3000
320  PRINT:PRINT"INTEGRATING. PLEASE WAIT"
330  GOSUB 5000
340 IF RQ=1 THEN GOSUB 6000
350 GOSUB 7000
360 PRINT:PRINT"                        RERUN OPTIONS":PRINT
370 PRINT:PRINT"                 1. SHOW SAME RESULTS AGAIN"
380 PRINT:PRINT"                 2. RESET INTEGRATION PARAMETERS AND RECALCULATE"
390 PRINT:PRINT"                 3. END PROGRAM"
400 PRINT:PRINT:INPUT"    ENTER YOUR CHOICE(1-3)"; RRQ
410 IF RRQ<1 OR RRQ>3 THEN GOTO 400
420 ON RRQ GOTO 340,430,460
430 PRINT:PRINT:PRINT:PRINT"                        NEW CONDITIONS"
440  GOSUB 3040
450  GOTO 320
460  END
1000 '************* Subroutine 1: Input equations ***********************
1010 '
1020 PRINT CHR$(12)
1030 PRINT:PRINT"                        DIFFERENTIAL EQUATIONS"
1040 PRINT:PRINT"                        **********************"
1050 PRINT:PRINT"THE DIFFERENTIAL EQUATIONS, AND OTHER ASSOCIATED ALGEBRAIC"
1060 PRINT: PRINT"EQUATIONS, MAY BE ENTERED HERE."
1070 PRINT:PRINT"EACH LINE-STATEMENT MUST OBEY THE RULES OF BASIC PROGRAMMING."
1080 PRINT:PRINT"THE LINES ARE AUTOMATICALLY NUMBERED, STARTING AT 4011."
1090 PRINT:PRINT"THESE STATEMENTS ARE MERGED INTO THE PROGRAM IN SUBROUTINE 4."
1100 PRINT:PRINT"                        **********************"
1110 PRINT:INPUT "HOW MANY DIFFERENTIAL EQUATIONS IN THE SYSTEM"; N1
1120 DIM Y(N1),Y9(N1),Y1(N1),C(50),G(N1),B1(N1),B2(N1),B3(N1),B4(N1)
1130 PRINT:INPUT "HAVE YOU PREVIOUSLY ENTERED THE EQUATIONS(Y/N)";Q1$
1140 IF Q1$="Y" OR Q1$="y" THEN GOTO 1170
1150 IF Q1$="N" OR Q1$="n" THEN GOTO 1210
```

```
1160 GOTO 1130
1170 PRINT:PRINT"THE FOLLOWING EQUATION-FILES EXIST ON YOUR DISK:":PRINT
1180 FILES"*.EQU"
1190 INPUT"GIVE NAME OF FILE TO BE USED (DO NOT ENTER THE EXTENSION): ",FIL$
1200 FIL$=FIL$+".EQU":GOTO 1480
1210 PRINT:INPUT"NAME OF FILE FOR STORING THE DIFFERENTIAL EQUATIONS: ",FIL$
1220 FIL$=FIL$+".EQU"
1230 OPEN FIL$ FOR OUTPUT AS #1
1240 CLS:PRINT:PRINT"                                  **********************"
1250 PRINT:PRINT"ENTER THE DIFFERENTIAL EQUATIONS IN THE FORM SHOWN BELOW:"
1260 PRINT:PRINT"     G(i)= f[x, y(1), y(2),....,y(n),c(1), c(2),....]"
1270 PRINT:PRINT"                  FOR EXAMPLE"
1280 PRINT:PRINT"     G(1)= c(1)*x + c(2)*y(1) - c(3)*y(1)*y(2)"
1285 PRINT:PRINT"     G(2)= c(4)*y(1)^2 - c(5)*y(2)/y(1)"
1290 PRINT:PRINT"WHERE:  G(i)  ARE THE DERIVATIVES: dy(i)/dx"
1300 PRINT:PRINT"        x     IS THE INDEPENDENT VARIABLE"
1310 PRINT:PRINT"        y(i)  ARE THE DEPENDENT VARIABLES"
1320 PRINT:PRINT"        f[ ]  ARE THE FUNCTIONS TO BE INTEGRATED
1330 PRINT:PRINT"        c( )  ARE CONSTANTS TO BE ENTERED LATER"
1340 PRINT:PRINT"                               *********************":PRINT
1350 INPUT"GIVE THE EXACT NUMBER OF LINES OF INPUT YOU WILL ENTER(1-980)";LN1
1360 PRINT:PRINT"SEPARATE EACH STATEMENT WITH THE ENTER(RETURN) KEY."
1370 PRINT#1,"4010 'The differential equations are merged in lines 4011-4990
1380 FOR I=1 TO LN1
1390 PRINT:PRINT "       ";4010+I;
1400 LINE INPUT ;EQUATION$
1410 IF EQUATION$="" THEN GOTO 1390
1420 PRINT#1,4010+I; EQUATION$
1430 PRINT
1440 NEXT I
1450 PRINT#1,"4995 'End of space allocated for the differential equations
1460 PRINT CHR$(12)
1470 CLOSE #1
1480 CHAIN MERGE FIL$,310,ALL,DELETE 4010-4995
1490 RETURN
3000 '************* Subroutine 3: Input integration parameters ***********
3010 PRINT:PRINT "                        INTEGRATION PARAMETERS"
3020 PRINT:PRINT "NUMBER OF INTEGRATION STEPS:";: INPUT L
3030  DIM A(L,N1)
3040 PRINT:PRINT "INITIAL VALUE OF INDEPENDENT VARIABLE:";: INPUT X9
3050 PRINT:PRINT "FINAL VALUE OF INDEPENDENT VARIABLE:  ";: INPUT XF
3100 D=(XF-X9)/L : X = X9 : A(0,0) = X9
3110 PRINT:PRINT "INCREMENT IN INDEPENDENT VARIABLE IS="; D
3120  FOR K = 1 TO N1
3130 PRINT:PRINT "INITIAL Y(";K;")";: INPUT Y9(K)
3140 Y(K) = Y9(K)
3150 A(0,K) = Y9(K)
3160  NEXT K
3165 PRINT
3170 PRINT"DO YOU WANT TO ENTER CONSTANTS IN THE DIFFERENTIAL EQUATIONS(Y/N)";
3175 INPUT VC$ : IF VC$="N" OR VC$="n" THEN  GOTO 3220
3177 PRINT:INPUT"HOW MANY CONSTANTS";CO
3180 PRINT:PRINT "GIVE VALUES OF CONSTANTS":PRINT
3190  FOR K = 1 TO CO
3200  PRINT "CONSTANT ";K;: INPUT C(K)
3210  NEXT K
3220 RETURN
4000 '************* Subroutine 4: Differential equations ****************
4010 'The differential equations are merged in lines 4011-4990
4995 'End of space allocated for the differential equations
4999 RETURN
```

```
5000 '************* Subroutine 5: Integration methods *******************
5010 '
5020 ON FQ GOTO 5030, 5260
5030 'Runge-Kutta 4th  order method
5040  FOR J2 = 1 TO L
5050  FOR K2 = 1 TO N1 : Y1(K2) = Y(K2) : NEXT K2
5060 X1 = X : GOSUB 4000
5070  FOR K3 = 1 TO N1
5080 B1(K3) = G(K3):Y(K3) = Y1(K3) + D * G(K3) / 2
5090  NEXT K3
5100 X = X1 + D / 2 : GOSUB 4000
5110  FOR K4 = 1 TO N1
5120 B2(K4) = G(K4):Y(K4) = Y1(K4) + D * G(K4) / 2
5130  NEXT K4
5140  GOSUB 4000
5150  FOR K5 = 1 TO N1
5160 B3(K5) = G(K5):Y(K5) = Y1(K5) + D * G(K5)
5170  NEXT K5
5180 X = X1 + D : GOSUB 4000
5190  FOR K6 = 1 TO N1
5200 Y(K6) = Y1(K6) + D * (B1(K6) + G(K6) + 2 * (B2(K6) + B3(K6))) / 6
5210 A(J2,K6) = Y(K6)
5220  NEXT K6
5230 A(J2,0) = X
5240  NEXT J2
5250  RETURN
5260 'Euler predictor-corrector method
5270  FOR J2 = 1 TO L
5280  FOR K2 = 1 TO N1 : Y1(K2) = Y(K2) : NEXT K2
5290 X1 = X : GOSUB 4000
5300  FOR K3 = 1 TO N1
5310 Y(K3) = Y1(K3) + D * G(K3)
5320  NEXT K3
5330 X = X1 + D : GOSUB 4000
5340  FOR K6 = 1 TO N1
5350 Y(K6) = Y1(K6) + D * G(K6)
5360 A(J2,K6) = Y(K6)
5370  NEXT K6
5380 A(J2,0) = X
5390  NEXT J2
5400  RETURN
6000 '************* Subroutine 6: Print table of results ****************
6010 '
6020 PRINT CHR$(12)
6030 PRINT "            TABLE OF RESULTS"
6040 PRINT:PRINT "INDEPENDENT       DEPENDENT VARIABLES"
6050 FOR J2 = 0 TO L
6060 FOR J1 = 0 TO N1 : PRINT USING "####.##    ";A(J2,J1), : NEXT J1
6070 PRINT
6080 NEXT J2
6090 PRINT:INPUT "PRESS THE ENTER KEY TO CONTINUE",KK$
6100 RETURN
7000 '************* Subroutine 7: Plotting options *********************

                    (Same as in Example 5-6)

8000 '************* Subroutine 8: Plotting ******************************

                    (Same as in Example 5-6)
```

## Results

```
*****************************************************************
*                                                               *
*                         EXAMPLE 5-7                           *
*                                                               *
*   4th ORDER RUNGE-KUTTA AND EULER PREDICTOR-CORRECTOR METHODS *
*                                                               *
*   FOR INTEGRATING SIMULTANEOUS ORDINARY DIFFERENTIAL EQUATIONS *
*                                                               *
*                          (ODE.BAS)                            *
*                                                               *
*****************************************************************

                          FORMULA OPTIONS

            1. RUNGE-KUTTA 4th ORDER

            2. EULER PREDICTOR-CORRECTOR

    ENTER YOUR CHOICE(1-2)? 1

                          OUTPUT OPTIONS

            1. PRINT TABLE AND DRAW GRAPHS OF RESULTS

            2. DRAW GRAPHS OF RESULTS ONLY

    ENTER YOUR CHOICE(1-2)? 1

                          DIFFERENTIAL EQUATIONS
                          **********************

THE DIFFERENTIAL EQUATIONS, AND OTHER ASSOCIATED ALGEBRAIC

EQUATIONS, MAY BE ENTERED HERE.

EACH LINE-STATEMENT MUST OBEY THE RULES OF BASIC PROGRAMMING.

THE LINES ARE AUTOMATICALLY NUMBERED, STARTING AT 4011.

THESE STATEMENTS ARE MERGED INTO THE PROGRAM IN SUBROUTINE 4.
                          **********************

HOW MANY DIFFERENTIAL EQUATIONS IN THE SYSTEM? 2

HAVE YOU PREVIOUSLY ENTERED THE EQUATIONS(Y/N)? N

NAME OF FILE FOR STORING THE DIFFERENTIAL EQUATIONS: PRPR
```

```
                        **********************

ENTER THE DIFFERENTIAL EQUATIONS IN THE FORM SHOWN BELOW:

     G(i)= f[x, y(1), y(2),....,y(n),c(1), c(2),....]

                  FOR EXAMPLE

     G(1)= c(1)*x + c(2)*y(1) - c(3)*y(1)*y(2)

     G(2)= c(4)*y(1)^2 - c(5)*y(2)/y(1)

WHERE:  G(i)  ARE THE DERIVATIVES: dy(i)/dx

        x     IS THE INDEPENDENT VARIABLE

        y(i)  ARE THE DEPENDENT VARIABLES

        f[ ]  ARE THE FUNCTIONS TO BE INTEGRATED

        c( )  ARE CONSTANTS TO BE ENTERED LATER

                        **********************

GIVE THE EXACT NUMBER OF LINES OF INPUT YOU WILL ENTER(1-980)? 2

SEPARATE EACH STATEMENT WITH THE ENTER(RETURN) KEY.

        4011 G(1)=C(1)*Y(1) - C(2)*Y(1)*Y(2)

        4012 G(2)=-C(3)*Y(2)+ C(4)*Y(1)*Y(2)

                        INTEGRATION PARAMETERS

NUMBER OF INTEGRATION STEPS:? 100

INITIAL VALUE OF INDEPENDENT VARIABLE:? 1960

FINAL VALUE OF INDEPENDENT VARIABLE:  ? 1985

INCREMENT IN INDEPENDENT VARIABLE IS= .25

INITIAL Y( 1 )? 700

INITIAL Y( 2 )? 22

DO YOU WANT TO ENTER CONSTANTS IN THE DIFFERENTIAL EQUATIONS(Y/N)? Y

HOW MANY CONSTANTS? 4

GIVE VALUES OF CONSTANTS

CONSTANT  1 ? 0.3
CONSTANT  2 ? 0.01111
CONSTANT  3 ? 0.2106
CONSTANT  4 ? 0.0002632

INTEGRATING. PLEASE WAIT
```

TABLE OF RESULTS

| INDEPENDENT | DEPENDENT VARIABLES | |
|---|---|---|
| 1960.00 | 700.00 | 22.00 |
| 1960.25 | 709.93 | 21.86 |
| 1960.50 | 720.27 | 21.74 |
| 1960.75 | 730.98 | 21.63 |
| 1961.00 | 742.06 | 21.54 |
| 1961.25 | 753.48 | 21.47 |
| 1961.50 | 765.21 | 21.41 |
| 1961.75 | 777.23 | 21.37 |
| 1962.00 | 789.51 | 21.35 |
| 1962.25 | 802.02 | 21.34 |
| 1962.50 | 814.72 | 21.35 |
| 1962.75 | 827.58 | 21.38 |
| 1963.00 | 840.55 | 21.43 |
| 1963.25 | 853.58 | 21.50 |
| 1963.50 | 866.64 | 21.58 |
| 1963.75 | 879.66 | 21.69 |
| 1964.00 | 892.60 | 21.81 |
| 1964.25 | 905.40 | 21.95 |
| 1964.50 | 917.99 | 22.11 |
| 1964.75 | 930.32 | 22.29 |
| 1965.00 | 942.31 | 22.49 |
| 1965.25 | 953.89 | 22.71 |
| 1965.50 | 965.01 | 22.95 |
| 1965.75 | 975.57 | 23.21 |
| 1966.00 | 985.52 | 23.49 |
| 1966.25 | 994.78 | 23.79 |
| 1966.50 | 1003.28 | 24.10 |
| 1966.75 | 1010.94 | 24.43 |
| 1967.00 | 1017.71 | 24.78 |
| 1967.25 | 1023.51 | 25.14 |
| 1967.50 | 1028.29 | 25.51 |
| 1967.75 | 1032.00 | 25.90 |
| 1968.00 | 1034.58 | 26.30 |
| 1968.25 | 1036.00 | 26.71 |
| 1968.50 | 1036.23 | 27.13 |
| 1968.75 | 1035.25 | 27.56 |
| 1969.00 | 1033.04 | 27.98 |
| 1969.25 | 1029.61 | 28.41 |
| 1969.50 | 1024.98 | 28.84 |
| 1969.75 | 1019.15 | 29.27 |
| 1970.00 | 1012.18 | 29.68 |
| 1970.25 | 1004.09 | 30.09 |
| 1970.50 | 994.96 | 30.49 |
| 1970.75 | 984.83 | 30.87 |
| 1971.00 | 973.80 | 31.24 |
| 1971.25 | 961.94 | 31.59 |
| 1971.50 | 949.33 | 31.91 |
| 1971.75 | 936.08 | 32.21 |
| 1972.00 | 922.27 | 32.49 |
| 1972.25 | 908.01 | 32.73 |
| 1972.50 | 893.39 | 32.95 |
| 1972.75 | 878.52 | 33.14 |
| 1973.00 | 863.48 | 33.29 |
| 1973.25 | 848.37 | 33.42 |
| 1973.50 | 833.28 | 33.51 |
| 1973.75 | 818.30 | 33.56 |

```
1974.00      803.49      33.59
1974.25      788.94      33.58
1974.50      774.70      33.54
1974.75      760.84      33.47
1975.00      747.40      33.36
1975.25      734.44      33.23
1975.50      721.99      33.08
1975.75      710.09      32.89
1976.00      698.76      32.69
1976.25      688.03      32.46
1976.50      677.92      32.21
1976.75      668.43      31.94
1977.00      659.58      31.66
1977.25      651.38      31.36
1977.50      643.83      31.04
1977.75      636.93      30.72
1978.00      630.68      30.38
1978.25      625.09      30.04
1978.50      620.13      29.69
1978.75      615.82      29.34
1979.00      612.14      28.98
1979.25      609.09      28.62
1979.50      606.66      28.26
1979.75      604.85      27.90
1980.00      603.64      27.54
1980.25      603.03      27.19
1980.50      603.01      26.84
1980.75      603.57      26.49
1981.00      604.71      26.15
1981.25      606.42      25.82
1981.50      608.68      25.50
1981.75      611.51      25.18
1982.00      614.87      24.87
1982.25      618.78      24.57
1982.50      623.22      24.29
1982.75      628.19      24.01
1983.00      633.67      23.74
1983.25      639.66      23.49
1983.50      646.15      23.25
1983.75      653.14      23.02
1984.00      660.61      22.80
1984.25      668.56      22.60
1984.50      676.96      22.41
1984.75      685.82      22.24
1985.00      695.12      22.08

PRESS THE ENTER KEY TO CONTINUE



           PLOTTING OPTIONS

   1. PLOT EACH VARIABLE SEPARATELY

   2. PLOT ALL VARIABLES ON SAME GRAPH

ENTER YOUR CHOICE(1-2)? 1
```

```
              SCALING OPTIONS

      1. MANUAL CHOICE OF MAXIMA AND MINIMA

      2. AUTOMATIC CHOICE OF MAXIMA AND MINIMA

ENTER YOUR CHOICE(1-2)? 1

           SCALES FOR VARIABLE  1

MINIMUM X=? 1960

MAXIMUM X=? 1985

SPECIFY X-AXIS TITLE:YEAR

MINIMUM Y=? 0

MAXIMUM Y=? 1500

SPECIFY Y-AXIS TITLE:PREY


              SCALING OPTIONS

      1. MANUAL CHOICE OF MAXIMA AND MINIMA

      2. AUTOMATIC CHOICE OF MAXIMA AND MINIMA

ENTER YOUR CHOICE(1-2)? 1

           SCALES FOR VARIABLE  2

MINIMUM X=? 1960

MAXIMUM X=? 1985

SPECIFY X-AXIS TITLE:YEAR

MINIMUM Y=? 0

MAXIMUM Y=? 50

SPECIFY Y-AXIS TITLE:PREDATOR
```

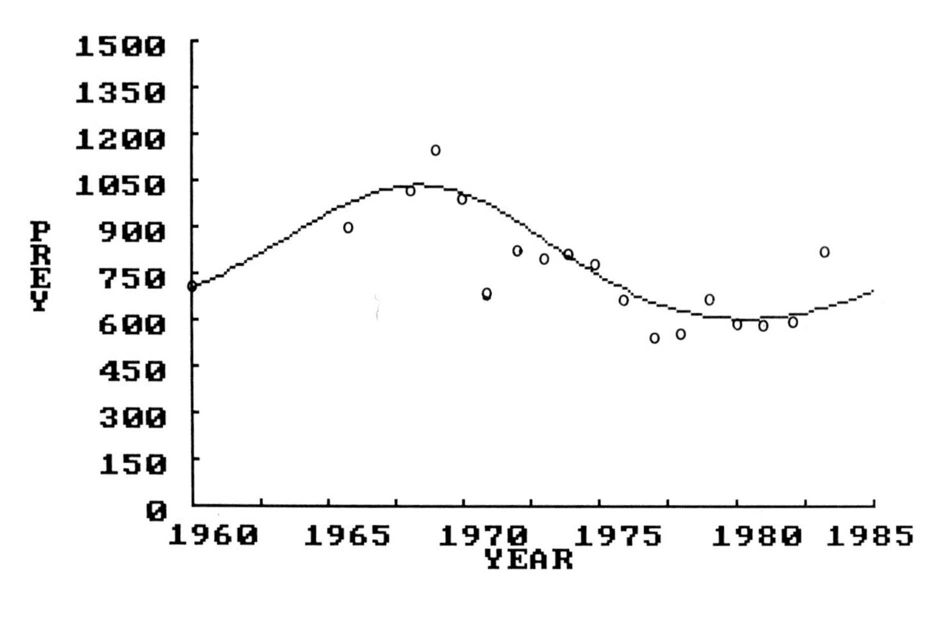


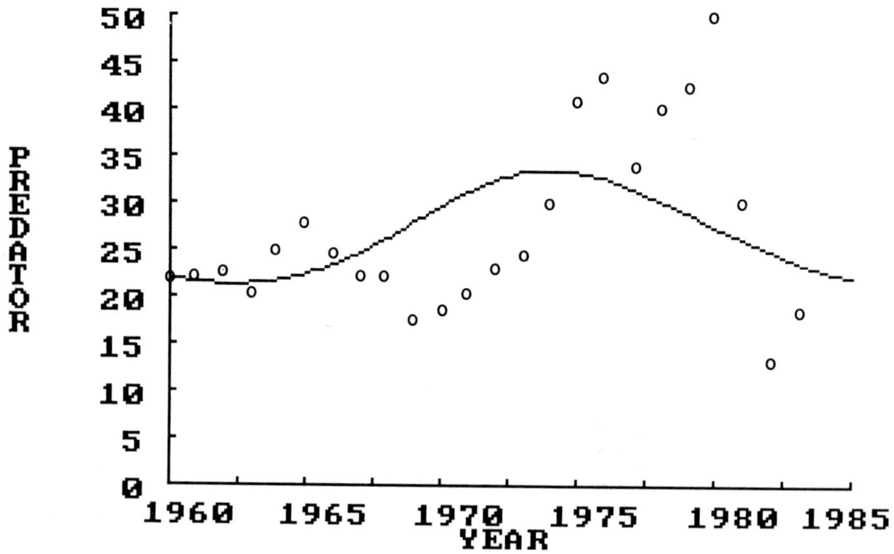


```
                    RERUN OPTIONS

        1. SHOW SAME RESULTS AGAIN

        2. RESET INTEGRATION PARAMETERS AND RECALCULATE

        3. END PROGRAM
ENTER YOUR CHOICE(1-3)? 2
```

```
                    NEW CONDITIONS

INITIAL VALUE OF INDEPENDENT VARIABLE:? 1960

FINAL VALUE OF INDEPENDENT VARIABLE:  ? 2060

INCREMENT IN INDEPENDENT VARIABLE IS= 1

INITIAL Y( 1 )? 700

INITIAL Y( 2 )? 22

DO YOU WANT TO ENTER CONSTANTS IN THE DIFFERENTIAL EQUATIONS(Y/N)? Y

HOW MANY CONSTANTS? 4

GIVE VALUES OF CONSTANTS

CONSTANT  1 ? 0.3
CONSTANT  2 ? 0.01111
CONSTANT  3 ? 0.2106
CONSTANT  4 ? 0.0002632

INTEGRATING. PLEASE WAIT


          TABLE OF RESULTS

INDEPENDENT        DEPENDENT VARIABLES
1960.00      700.00       22.00
1961.00      742.06       21.54
1962.00      789.51       21.35
1963.00      840.54       21.43
1964.00      892.60       21.81
1965.00      942.30       22.49
1966.00      985.51       23.49
1967.00     1017.70       24.78
1968.00     1034.57       26.30
1969.00     1033.03       27.98
1970.00     1012.17       29.68
1971.00      973.81       31.24
1972.00      922.28       32.49
1973.00      863.50       33.29
1974.00      803.52       33.59
1975.00      747.43       33.36
1976.00      698.79       32.69
1977.00      659.61       31.66
1978.00      630.71       30.38
1979.00      612.16       28.98
1980.00      603.65       27.55
1981.00      604.71       26.16
1982.00      614.87       24.87
1983.00      633.66       23.74
1984.00      660.59       22.80
1985.00      695.10       22.08
1986.00      736.37       21.59
1987.00      783.22       21.36
1988.00      833.92       21.40
1989.00      886.01       21.74
```

```
1990.00     936.22     22.39
1991.00     980.51     23.35
1992.00    1014.36     24.60
1993.00    1033.38     26.10
1994.00    1034.30     27.77
1995.00    1015.86     29.47
1996.00     979.53     31.06
1997.00     929.38     32.35
1998.00     871.18     33.22
1999.00     811.04     33.58
2000.00     754.23     33.42
2001.00     704.50     32.80
2002.00     664.05     31.80
2003.00     633.82     30.56
2004.00     613.96     29.16
2005.00     604.20     27.73
2006.00     604.07     26.33
2007.00     613.09     25.03
2008.00     630.81     23.88
2009.00     656.73     22.91
2010.00     690.31     22.16
2011.00     730.77     21.64
2012.00     776.99     21.37
2013.00     827.31     21.38
2014.00     879.39     21.68
2015.00     930.05     22.29
2016.00     975.34     23.21
2017.00    1010.77     24.42
2018.00    1031.90     25.90
2019.00    1035.25     27.55
2020.00    1019.26     29.26
2021.00     985.03     30.87
2022.00     936.35     32.21
2023.00     878.83     33.13
2024.00     818.62     33.56
2025.00     761.15     33.47
2026.00     710.37     32.90
2027.00     668.65     31.95
2028.00     637.10     30.73
2029.00     615.93     29.35
2030.00     604.90     27.91
2031.00     603.57     26.50
2032.00     611.46     25.19
2033.00     628.09     24.01
2034.00     652.99     23.02
2035.00     685.63     22.24
2036.00     725.26     21.69
2037.00     770.83     21.39
2038.00     820.74     21.37
2039.00     872.75     21.63
2040.00     923.80     22.20
2041.00     970.01     23.07
2042.00    1006.95     24.25
2043.00    1030.13     25.70
2044.00    1035.88     27.33
2045.00    1022.35     29.04
2046.00     990.30     30.67
2047.00     943.17     32.06
2048.00     886.44     33.04
2049.00     826.25     33.53
2050.00     768.17     33.51
```

```
2051.00     716.38      32.99
2052.00     673.43      32.09
2053.00     640.55      30.89
2054.00     618.07      29.53
2055.00     605.77      28.09
2056.00     603.23      26.68
2057.00     609.96      25.35
2058.00     625.50      24.15
2059.00     649.38      23.14
2060.00     681.06      22.33
PRESS THE ENTER KEY TO CONTINUE

           PLOTTING OPTIONS

    1. PLOT EACH VARIABLE SEPARATELY

    2. PLOT ALL VARIABLES ON SAME GRAPH

ENTER YOUR CHOICE(1-2)? 1

           SCALING OPTIONS

    1. MANUAL CHOICE OF MAXIMA AND MINIMA

    2. AUTOMATIC CHOICE OF MAXIMA AND MINIMA

ENTER YOUR CHOICE(1-2)? 1

         SCALES FOR VARIABLE  1

MINIMUM X=? 1960

MAXIMUM X=? 2060

SPECIFY X-AXIS TITLE:YEAR

MINIMUM Y=? 0

MAXIMUM Y=? 1500

SPECIFY Y-AXIS TITLE:PREY

           SCALING OPTIONS

    1. MANUAL CHOICE OF MAXIMA AND MINIMA

    2. AUTOMATIC CHOICE OF MAXIMA AND MINIMA

ENTER YOUR CHOICE(1-2)? 1

         SCALES FOR VARIABLE  2

MINIMUM X=? 1960

MAXIMUM X=? 2060

SPECIFY X-AXIS TITLE:YEAR

MINIMUM Y=? 0

MAXIMUM Y=? 50

SPECIFY Y-AXIS TITLE:PREDATOR
```

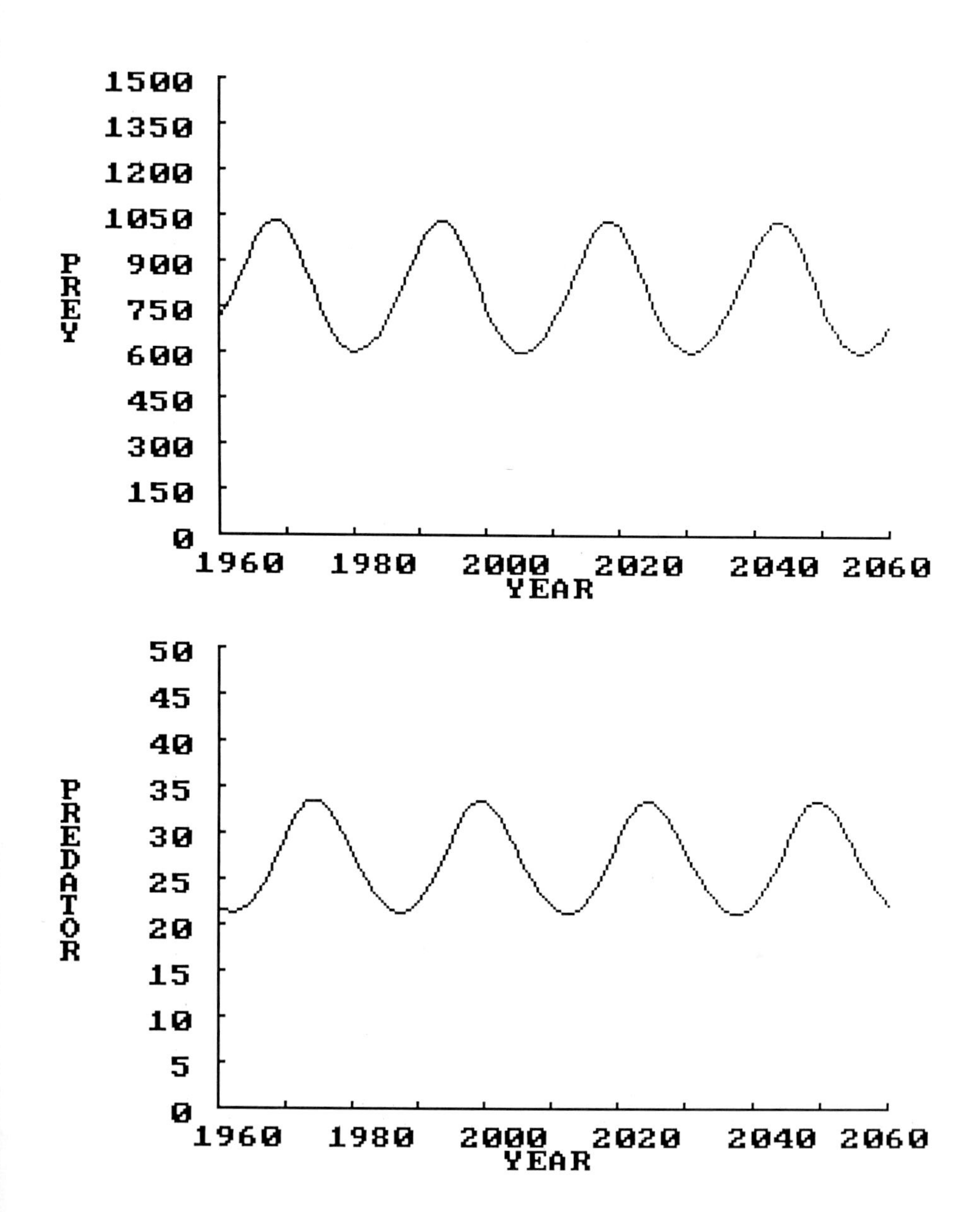


```
                    RERUN OPTIONS

        1. SHOW SAME RESULTS AGAIN

        2. RESET INTEGRATION PARAMETERS AND RECALCULATE

        3. END PROGRAM

    ENTER YOUR CHOICE(1-3)? 3
Ok
```

DISCUSSION OF RESULTS

*Period 1960–1985* These results show a complete cycle of interaction between the prey and the predator. The data of Allen [10] and Peterson [11] have been superimposed (manually) on the model-generated profiles of the two populations. As predicted earlier, the model is not a good representation of the predator data; however, a rather remarkable agreement is obtained for the prey population. This is partially explainable by the fact that the model does not account for overcrowding effects, for the availability of supplementary food sources for the predator, and for extreme variations in weather conditions which influence the fertility and mortality rates for both species.

From careful consideration of the predator data we may conclude that this population is oscillating in an unstable pattern. Our earlier stability analysis would indicate that equations which have eigenvalues with positive real parts and nonzero imaginary parts would exhibit such behavior more closely. It is left as a problem for the student to examine other forms of the Lotka-Volterra equations which may be better representations of the real predator-prey data (see Prob. 5-9).

*Period 1960–2060* Under the assumptions made in this example for the Lotka-Volterra predator-prey ecological system, the two species will continue to interact in an oscillatory but stable pattern forever.

## 5-7 NONLINEAR ORDINARY DIFFERENTIAL EQUATIONS—BOUNDARY-VALUE PROBLEMS

Ordinary differential equations with boundary conditions specified at two or more points of the independent variable are classified as boundary-value problems. There are many chemical engineering applications which result in ordinary differential equations of the boundary-value type. To mention only a few examples:

1. Diffusion with chemical reaction in the study of chemical catalysis or enzyme catalysis
2. Heat and mass transfer in boundary-layer problems
3. Application of rigorous optimization methods, such as Pontryagin's maximum principle or the calculus of variations
4. Discretization of nonlinear elliptic partial differential equations [14]

The diversity of problems of the boundary-value type have generated a variety of methods for their solution. The system equations in these problems could be linear or nonlinear, and the boundary conditions could be linear or nonlinear, separated or mixed, two-point or multipoint. Comprehensive discussions of the solutions of boundary-value problems are given by Kubíček and Hlaváček [14] and by Aziz [15]. In this chapter we have chosen to discuss algorithms which are applicable to the solution of nonlinear (as well as linear) boundary-value problems. These are the *Newton* method, the *finite difference* method, and the *collocation* methods. The last two methods will be discussed again in Chap. 6 in connection with the solution of partial differential equations of the boundary-value type.

The canonical form of a two-point boundary-value problem with linear boundary conditions is

$$\frac{dy_i}{dt} = f_i(t, y_1, y_2, \ldots, y_n) \qquad t_0 \le t \le t_f;\ i = 1, 2, \ldots, n \tag{5-153}$$

where the boundary conditions are split between the initial point $t_0$ and the final point $t_f$. The first $r$ equations have initial conditions specified and the last $n - r$ equations have final conditions given:

$$y_i(t_0) = y_{i,0} \qquad i = 1, 2, \ldots, r \tag{5-154}$$

$$y_j(t_f) = y_{j,f} \qquad j = r + 1, \ldots, n \tag{5-155}$$

A second-order two-point boundary-value problem may be expressed in the form

$$\frac{d^2y}{dx^2} = f\left(x, y, \frac{dy}{dx}\right) \qquad x_0 \le x \le x_f \tag{5-156}$$

subject to the boundary conditions

$$a_0 y(x_0) + b_0 y'(x_0) = \gamma_0 \tag{5-157}$$

$$a_f y(x_f) + b_f y'(x_f) = \gamma_f \tag{5-158}$$

where the subscript 0 designates conditions at the left boundary (initial) and the subscript $f$ identifies conditions at the right boundary (final).

This problem can be transformed to the canonical form (5-153) by the appropriate substitutions described in Sec. 5-3.

### 5.7-1 The Newton Method

The Newton method converts the boundary-value problem to an initial-value one to take advantage of the powerful algorithms available for the integration of initial-value problems (see Sec. 5-6). The algorithm of the Newton method is outlined in the following five steps:

1. The unspecified initial conditions of the system differential equations are guessed.

2. A set of *variational equations* are developed which show the sensitivities of the dependent variables with respect to the guessed initial values.
3. The system and variational equations are integrated forward as a set of simultaneous initial-value differential equations.
4. The guessed initial conditions are corrected using the variations (sensitivities) calculated in step 3.
5. Steps 2 through 4 are repeated with the corrected initial conditions, until the specified terminal values are achieved within a small convergence criterion.

The above general algorithm forms the basis for a family of methods called *shooting* methods [16]. These may vary in their choice of initial or final conditions and in the integration of the equations in one direction or two directions. Newton's technique, which is the most widely known of the shooting methods, can be applied successfully to boundary-value problems of any complexity as long as the resulting initial-value problem is stable and a set of good guesses for the unspecified conditions can be made [14].

We first develop the Newton method for a set of two differential equations

$$\frac{dy_1}{dt} = f_1(t, y_1, y_2) \tag{5-159}$$

$$\frac{dy_2}{dt} = f_2(t, y_1, y_2) \tag{5-160}$$

with split boundary conditions

$$y_1(t_0) = y_{1,0} \tag{5-161}$$

$$y_2(t_f) = y_{2,f} \tag{5-162}$$

We guess the initial condition for $y_2$

$$y_2(t_0) = \gamma \tag{5-163}$$

If the system equations are integrated forward, the two trajectories may look like those in Fig. 5-11. Since the value of $y_2(t_0)$ was only a guess, the trajectory of $y_2$ misses its target at $t_f$; that is, it does not satisfy the boundary condition of (5-162). For the given guess of $\gamma$, the calculated value of $y_2$ at $t_f$ is designated as $y_2(t_f, \gamma)$. The desirable objective is to find the value of $\gamma$ which forces $y_2(t_f, \gamma)$ to satisfy the specified boundary condition, i.e.,

$$y_2(t_f, \gamma) = y_{2,f} \tag{5-164}$$

Rearrange Eq. (5-164) to

$$\phi(\gamma) = y_2(t_f, \gamma) - y_{2,f} = 0 \tag{5-165}$$

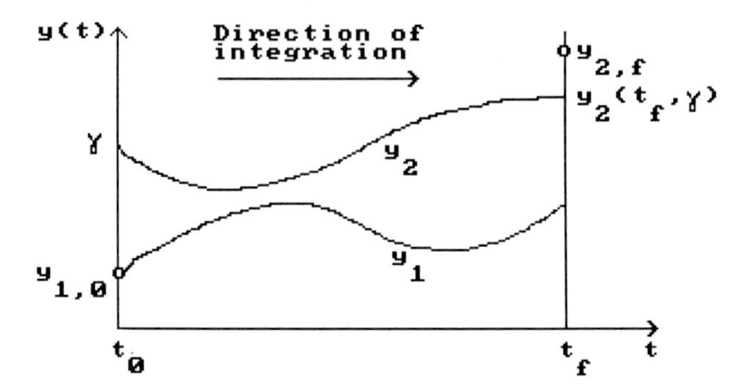


**Figure 5-11** Forward integration using a guessed initial condition $\gamma$. The ○ designates the known boundary points.

The function $\phi(\gamma)$ can be expanded in a Taylor series around $\gamma$:

$$\phi(\gamma + \Delta\gamma) = \phi(\gamma) + \frac{\partial \phi}{\partial \gamma}\Delta\gamma + O[(\Delta\gamma)^2] \tag{5-166}$$

In order for the system to converge, i.e., for the trajectory of $y_2$ to hit the specified boundary value at $t_f$,

$$\lim_{\Delta\gamma \to 0} \phi(\gamma + \Delta\gamma) = 0 \tag{5-167}$$

Therefore, Eq. (5-166) becomes

$$0 = \phi(\gamma) + \frac{\partial \phi}{\partial \gamma}\Delta\gamma + O[(\Delta\gamma)^2] \tag{5-168}$$

Truncation and rearrangement gives†

$$\Delta\gamma = \frac{-\phi(\gamma)}{\left[\dfrac{\partial \phi}{\partial \gamma}\right]} \tag{5-169}$$

Using the definition of $\phi(\gamma)$ [Eq (5-165)], taking its partial derivative, and combining with (5-169), we obtain

$$\Delta\gamma = \frac{-[y_2(t_f, \gamma) - y_{2,f}]}{\left[\dfrac{\partial y_2(t_f, \gamma)}{\partial \gamma}\right]} \tag{5-170}$$

† The reader should be able to recognize this equation as a form of the Newton-Raphson equation of Chap. 2.

For reasons which will become evident when we discuss this method for $n$ differential equations, we express Eq. (5-170) as

$$\Delta\gamma = [\mathbf{J}(t_f, \gamma)]^{-1}\delta y \tag{5-171}$$

where $\mathbf{J}(t_f, \gamma)$ is the *Jacobian matrix* (in this case it is only a $1 \times 1$ matrix) evaluated at $t = t_f$:

$$\mathbf{J}(t_f, \gamma) = \left[\frac{\partial y_2(t_f, \gamma)}{\partial \gamma}\right] \tag{5-172}$$

and $\delta y$ is the difference between the *specified* final boundary value $y_{2,f}$ and the *calculated* final value $y_2(t_f, \gamma)$ obtained from using the guessed $\gamma$:

$$\delta y = -[y_2(t_f, \gamma) - y_{2,f}] \tag{5-173}$$

The value of $\Delta\gamma$ is the correction to be applied to the guessed $\gamma$ to obtain a new guess:

$$(\gamma)_{\text{new}} = (\gamma)_{\text{old}} + \Delta\gamma \tag{5-174}$$

The one-dimensional Jacobian contains $\partial y_2(t_f, \gamma)/\partial\gamma$, which is the sensitivity of the endpoint of the trajectory of $y_2$ with respect to the choice of the guessed initial condition $\gamma$. The sensitivity is calculated from the variational equations. The latter are derived by taking the partial derivatives of the system equations with respect to $\gamma$:

$$\frac{\partial}{\partial \gamma}\left(\frac{dy_1}{dt}\right) = \frac{\partial f_1}{\partial y_1}\frac{\partial y_1}{\partial \gamma} + \frac{\partial f_1}{\partial y_2}\frac{\partial y_2}{\partial \gamma} \tag{5-175}$$

$$\frac{\partial}{\partial \gamma}\left(\frac{dy_2}{dt}\right) = \frac{\partial f_2}{\partial y_1}\frac{\partial y_1}{\partial \gamma} + \frac{\partial f_2}{\partial y_2}\frac{\partial y_2}{\partial \gamma} \tag{5-176}$$

It is assumed that the functions $f_i$ are continuously differentiable with respect to $y_j$ and are continuous with respect to $t$ in a sufficiently large region [14]. The left side of Eqs. (5-175) and (5-176) can be rearranged to give

$$\frac{d}{dt}\left(\frac{\partial y_1}{\partial \gamma}\right) = \frac{\partial f_1}{\partial y_1}\left(\frac{\partial y_1}{\partial \gamma}\right) + \frac{\partial f_1}{\partial y_2}\left(\frac{\partial y_2}{\partial \gamma}\right) \tag{5-177}$$

$$\frac{d}{dt}\left(\frac{\partial y_2}{\partial \gamma}\right) = \frac{\partial f_2}{\partial y_1}\left(\frac{\partial y_1}{\partial \gamma}\right) + \frac{\partial f_2}{\partial y_2}\left(\frac{\partial y_2}{\partial \gamma}\right) \tag{5-178}$$

Denoting the sensitivities as

$$p_1 = \frac{\partial y_1}{\partial \gamma} \qquad p_2 = \frac{\partial y_2}{\partial \gamma} \tag{5-179}$$

Eqs. (5-177) and (5-178) become

$$\frac{dp_1}{dt} = \frac{\partial f_1}{\partial y_1} p_1 + \frac{\partial f_1}{\partial y_2} p_2 \tag{5-180}$$

$$\frac{dp_2}{dt} = \frac{\partial f_2}{\partial y_1} p_1 + \frac{\partial f_2}{\partial y_2} p_2 \tag{5-181}$$

This is the set of *variational equations* which must be solved simultaneously with the system equations. The initial conditions for $p_1$ and $p_2$ are obtained from the definition of these variables applied at $t = t_0$:

$$p_1(t_0) = \frac{\partial y_1(t_0)}{\partial \gamma} = 0 \tag{5-182}$$

$$p_2(t_0) = \frac{\partial y_2(t_0)}{\partial \gamma} = 1 \tag{5-183}$$

We restate the five-step algorithm of the Newton method as follows:

1. The missing initial condition of the system equations is guessed by Eq. (5-163).
2. The variational equations are defined by Eqs. (5-180) and (5-181) with initial conditions (5-182) and (5-183).
3. The system equations (5-159) and (5-160) and the variational equations (5-180) and (5-181) are integrated forward simultaneously.
4. The correction $\Delta\gamma$ to be applied to $\gamma$ is calculated from Eq. (5-171), which utilizes the integrated results of step 3. The new value of $\gamma$ is obtained from Eq. (5-174).
5. Steps 2 and 3 are repeated, each time with a corrected value of $\gamma$, until $|\delta y| \le e$.

In order to avoid divergence it may sometimes be necessary to take a fractional correction step [17], i.e.,

$$(\gamma)_{\text{new}} = (\gamma)_{\text{old}} + \alpha\, \Delta\gamma \tag{5-185}$$

where $0 < \alpha \le 1$.

This algorithm can now be generalized to apply to a set of $n$ simultaneous system equations

$$\frac{dy_i}{dt} = f_i(t, y_1, y_2, \ldots, y_n) \qquad i = 1, 2, \ldots, n \tag{5-186}$$

whose boundary conditions are split between the initial point and the final point. The first $r$ equations have *initial* conditions specified, and the last $n - r$

equations have *final* conditions given:

$$y_i(t_0) = y_{i,0} \qquad i = 1, 2, \ldots, r \tag{5-187}$$

$$y_j(t_f) = y_{j,f} \qquad j = r+1, \ldots, n \tag{5-188}$$

In order to apply Newton's procedure to integrate the system equations forward, the missing $n - r$ initial conditions are guessed as follows:

$$y_j(t_0) = \gamma_j \qquad j = r+1, \ldots, n \tag{5-189}$$

The variational equations are developed by taking the partial derivative of each of the system equations with respect to each of the initial guesses and rearranging to obtain

$$\frac{d(\partial y_i/\partial \gamma_j)}{dt} = \sum_{k=1}^{n} \frac{\partial f_i}{\partial y_k}\left(\frac{\partial y_k}{\partial \gamma_j}\right) \qquad \begin{cases} i = 1, 2, \ldots, n \\ j = r+1, \ldots, n \end{cases} \tag{5-190}$$

By denoting the sensitivities as

$$p_{ij} = \frac{\partial y_i}{\partial \gamma_j} \tag{5-191}$$

the variational equations simplify to

$$\frac{dp_{ij}}{dt} = \sum_{k=1}^{n} \frac{\partial f_i}{\partial y_k} p_{kj} \qquad \begin{cases} i = 1, 2, \ldots, n \\ j = r+1, \ldots, n \end{cases} \tag{5-192}$$

This is a set of $n \times (n - r)$ ordinary differential equations whose initial conditions are given by

$$p_{ij}(t_0) = \begin{cases} 1 & \text{when } i = j \\ 0 & \text{elsewhere} \end{cases} \qquad \{ j = r+1, \ldots, n \qquad \{ i = 1, 2, \ldots, n \tag{5-193}$$

The system equations (5-186) with the given initial conditions (5-187) and the guessed initial conditions (5-189), together with the variational equations (5-192) and their initial conditions (5-193), are integrated simultaneously in the forward direction. At the right-hand boundary ($t_f$), the *Jacobian matrix* is determined from the values of the sensitivities at $t_f$:

$$\mathbf{J}(t_f, \gamma) = \begin{bmatrix} \left.\dfrac{\partial y_{r+1}}{\partial \gamma_{r+1}}\right|_{t_f} & \left.\dfrac{\partial y_{r+1}}{\partial \gamma_{r+2}}\right|_{t_f} & \cdots & \left.\dfrac{\partial y_{r+1}}{\partial \gamma_n}\right|_{t_f} \\ \cdots & \cdots & \cdots & \cdots \\ \left.\dfrac{\partial y_n}{\partial \gamma_{r+1}}\right|_{t_f} & \cdots & \cdots & \left.\dfrac{\partial y_n}{\partial \gamma_n}\right|_{t_f} \end{bmatrix} \tag{5-194}$$

The correction of the guessed initial values is implemented by the equation

$$(\boldsymbol{\gamma})_{\text{new}} = (\boldsymbol{\gamma})_{\text{old}} + [\mathbf{J}(t_f, \boldsymbol{\gamma})]^{-1}\, \boldsymbol{\delta y} \tag{5-195}$$

where the vector $\boldsymbol{\delta y}$ is the difference between the specified final boundary values and the calculated final values using the guessed initial conditions

$$\boldsymbol{\delta y} = -\begin{bmatrix} y_{r+1}(t_f, \boldsymbol{\gamma}) & -y_{r+1,f} \\ \cdots & \cdots \\ y_n(t_f, \boldsymbol{\gamma}) & -y_{n,f} \end{bmatrix} \tag{5-196}$$

The Newton method described here is similar to the *method of adjoints* described by Kubíček and Hlaváček [14], the only difference being that in the latter method the system equations and variational equations are integrated in opposite directions. The application of the Newton method is demonstrated in Example 5-8.

### 5-7.2 The Finite Difference Method

The *finite-difference* method replaces the derivatives in the differential equations with finite difference approximations at each point in the interval of integration, thus converting the differential equations to a large set of simultaneous nonlinear algebraic equations. To demonstrate this method, we use, as before, the set of two differential equations

$$\frac{dy_1}{dt} = f_1(t, y_1, y_2) \tag{5-159}$$

$$\frac{dy_2}{dt} = f_2(t, y_1, y_2) \tag{5-160}$$

with split boundary conditions

$$y_1(t_0) = y_{1,0} \tag{5-161}$$

$$y_2(t_f) = y_{2,f} \tag{5-162}$$

Next, we express the derivatives of $y$ in terms of forward finite differences using Eq. (4-70):

$$\frac{dy_{1,i}}{dt} = \frac{1}{h}(y_{1,i+1} - y_{1,i}) + O(h) \tag{5-197}$$

$$\frac{dy_{2,i}}{dt} = \frac{1}{h}(y_{2,i+1} - y_{2,i}) + O(h) \tag{5-198}$$

For higher accuracy, we could have used Eq. (4-72), which has error of order

$(h^2)$, instead of Eq. (4-70). In either case the steps of obtaining the solution to the boundary-value problem are identical.

Combining Eqs. (5-197) and (5-198) with (5-159) and (5-160) we obtain

$$y_{1,i+1} - y_{1,i} = hf_1(t, y_{1,i}, y_{2,i}) \tag{5-199}$$

$$y_{2,i+1} - y_{2,i} = hf_2(t, y_{1,i}, y_{2,i}) \tag{5-200}$$

We divide the interval of integration into $n$ segments of equal length and write Eqs. (5-199) and (5-200) for $i = 0, 1, 2, \ldots, n-1$. These form a set of $2n$ simultaneous nonlinear algebraic equations in $2n + 2$ variables. The two boundary conditions provide values for two of these variables

$$y_1(t_0) = y_{1,0} \tag{5-161}$$

$$y_2(t_f) = y_{2,f} = y_{2,n} \tag{5-162}$$

Therefore, the system of $2n$ equations in $2n$ unknowns can be solved using Newton's method for simultaneous nonlinear algebraic equations, described in Chap. 2. It should be emphasized, however, that the problem of solving a large set of nonlinear algebraic equations is not a trivial task. It requires, first, a good initial guess of all the values of $y_{ji}$, and it involves the evaluation of the $2n \times 2n$ Jacobian matrix. Kubíček and Hlaváček [14] state that computational experience with the finite difference technique has shown that for a practical engineering problem, this method is more difficult to apply than the shooting method. They recommend that the finite difference method be used only for problems which are too unstable to integrate by the shooting methods. On the other hand, if the differential equations are *linear*, the resulting set of simultaneous algebraic equations will also be linear. In such a case, the solution can be obtained by a straightforward application of the Gauss elimination procedure.

### 5-7.3 Collocation Methods

These methods are based on the concept of interpolation of unequally spaced points, that is, choosing a function, usually a polynomial, that approximates the solution of a differential equation in the range of integration, $a \le x \le b$, and determining the coefficients of that function from a set of base points [18].

Let us consider the second-order two-point boundary-value problem in the form

$$y'' = f(x, y, y') \qquad 0 \le x \le 1 \tag{5-201}$$

with the boundary conditions

$$y(0) = 0 \qquad \text{and} \qquad y(1) = 1 \tag{5-202}$$

Suppose that there is a polynomial

$$P_n(x) = c_0 + c_1 x + c_2 x^2 + \cdots + c_n x^n \tag{5-203}$$

which is a good approximation to the solution $y(x)$ of Eq. (5-201). We call this the *trial function* and we set

$$y(x) \cong P_n(x) \tag{5-204}$$

We take the derivatives of both sides of (5-204) and substitute in (5-201):

$$P_n''(x) \cong f(x, P_n(x), P_n'(x)) \tag{5-205}$$

We then form the *residual*

$$R(x) = P_n''(x) - f(x, P_n(x), P_n'(x)) \tag{5-206}$$

The objective is to determine the coefficients $\{c_i \mid i = 0, 1, \ldots, n\}$ of the polynomial $P_n(x)$ so as to make the residual as small as possible over the range of integration of the differential equation. This is accomplished by making the following integral vanish:

$$\int_0^1 W_k R(x)\, dx = 0 \tag{5-207}$$

where $W_k$ are weighting functions to be chosen. This technique is called the *method of weighted residuals*.

The *collocation method* chooses the weighting functions to be the *Dirac delta* (*unit impulse*) function

$$W_k = \delta(x - x_k) \tag{5-208}$$

which has the property that

$$\int_0^1 a(x)\delta(x - x_k)\, dx = a(x_k)$$

Therefore, the integral (5-207) becomes

$$\int_0^1 W_k R(x)\, dx = R(x_k) = 0 \tag{5-209}$$

Combining Eqs. (5-206) and (5-209), we have

$$P_n''(x_k) - f(x_k, P_n(x_k), P_n'(x_k)) = 0 \tag{5-210}$$

This implies that at a given number of *collocation points*, $\{x_k \mid k = 0, 1, \ldots, n\}$, the coefficients of the polynomial (5-203) are chosen so that Eq. (5-210) is satisfied, i.e., the polynomial is an *exact* solution of the differential equation at those collocation points. The larger the number of collocation points, the closer the trial function would resemble the true solution $y(x)$ of the differential equation.

Equation (5-210) contains the $n+1$ yet-to-be-determined coefficients $\{c_i \mid i = 0, 1, \ldots, n\}$ of the polynomial. These can be calculated by choosing $n+1$ collocation points. Since it is necessary to satisfy the boundary conditions of the problem, two collocation points are already fixed in this case of the second-order boundary-value problem. At $x = 0$,

$$y(0) = 0 = c_0 \tag{5-211}$$

and at $x = 1$,

$$y(1) = 1 = c_0 + c_1 + \cdots + c_n = \sum_{i=1}^{n} c_i \tag{5-212}$$

Therefore, we have the freedom to choose the remaining $n-1$ collocation points $\{x_k \mid k = 1, 2, \ldots, n-1\}$ and then write Eq. (5-210) for each point:

$$\begin{aligned} P_n''(x_1) - f(x_1, P_n(x_1), P_n'(x_1)) &= 0 \\ &\vdots \\ P_n''(x_{n-1}) - f(x_{n-1}, P_n(x_{n-1}), P_n'(x_{n-1})) &= 0 \end{aligned} \tag{5-213}$$

Equations (5-211) to (5-213) constitute a complete set of $n+1$ simultaneous nonlinear algebraic equations in $n+1$ unknowns. The solution of this problem requires the application of Newton's method (Chap. 2) for simultaneous nonlinear equations.

If the collocation points are chosen at equidistant intervals within the interval of integration, then the collocation method is equivalent to polynomial interpolation of equally spaced points and to the finite difference method. This is not at all surprising since the development of interpolating polynomials and finite differences were all based on expanding the functions in Taylor series (see Chap. 4). It is not necessary, however, to choose the collocation points at equidistant intervals. In fact, it is more advantageous to locate the collocation points at the roots of the appropriate orthogonal polynomials, as the following discussion shows.

The *orthogonal collocation method*,† which is an extension of the method

† The discussion of the orthogonal collocation method presented here is based on the work by Finlayson [8, 19], who has published extensively on the application of this method to the solution of problems in chemical engineering.

just described, provides a mechanism for automatically picking the collocation points by making use of orthogonal polynomials. This method chooses the trial function $y(x)$ to be the linear combination

$$y(x) = \sum_{i=1}^{N+2} a_i P_{i-1}(x) \tag{5-214}$$

of a series of orthogonal polynomials $P_m(x)$:

$$\begin{aligned}
P_0(x) &= c_{0,0} \\
P_1(x) &= c_{1,0} + c_{1,1}x \\
P_2(x) &= c_{2,0} + c_{2,1}x + c_{2,2}x^2 \\
&\vdots \\
P_m(x) &= c_{m,0} + c_{m,1}x + c_{m,2}x^2 + \cdots + c_{m,m}x^m:
\end{aligned} \tag{5-215}$$

This set of polynomials can be written in a condensed form:

$$P_m(x) = \sum_{j=0}^{m} c_{mj}x^j \qquad m = 0, 1, \ldots, N+1 \tag{5-216}$$

The coefficients $c_{mj}$ are chosen so that the polynomials obey the orthogonality condition defined in Sec. 4-7.3:

$$\int_a^b w(x)P_k(x)P_m(x)\,dx = 0 \qquad k = 0, 1, 2, \ldots, m-1 \tag{5-217}$$

When $P_m(x)$ is chosen to be the *Legendre* set of orthogonal polynomials [Eqs. (4-157)], the weight $w(x)$ is unity. The standard interval of integration for Legendre polynomials is $[-1, 1]$. The transformation equation (5-104) is used to transform the Legendre polynomials to the interval $[0, 1]$, which applies to our problem at hand. The resulting set of transformed Legendre polynomials is

$$\begin{aligned}
P_0 &= 1 \\
P_1 &= -1 + 2x \\
P_2 &= 1 - 6x + 6x^2 \\
P_3 &= -1 + 12x - 30x^2 + 20x^3 \\
&\vdots
\end{aligned} \tag{5-218}$$

The roots of these polynomials, which are listed in Table 5-4, are simply obtained by applying the transformation formula to the roots of the regular Legendre polynomials from Table 5-2.

**Table 5-4 Roots of the transformed Legendre polynomials**

| $N$ | $x_j$ |
|---|---|
| 1 | 0.500000000 |
| 2 | 0.2113248654 |
| | 0.7886751346 |
| 3 | 0.1127016654 |
| | 0.5000000000 |
| | 0.8872983346 |

The second-order two-point boundary-value problem given by Eqs. (5-201) and (5-202) has $N+2$ collocation points, including the two known boundary values. The locations of the $N$ internal collocation points are determined from the roots of the $P_N(x)=0$ polynomial. The coefficients $a_i$ in Eq. (5-214) must be determined so that the boundary conditions are satisfied. Equation (5-214) can be written for the $N+2$ points as

$$y(x_j) = \sum_{i=1}^{N+2} d_i x_j^{i-1} \tag{5-219}$$

where the terms of the polynomials have been regrouped.

The derivatives of $y$ are taken as

$$\frac{dy(x_j)}{dx} = \sum_{i=1}^{N+2} d_i(i-1)x_j^{i-2} \tag{5-220}$$

$$\frac{d^2y(x_j)}{dx^2} = \sum_{i=1}^{N+2} d_i(i-1)(i-2)x_j^{i-3} \tag{5-221}$$

These equations can be written in matrix form†

$$\frac{\mathbf{dy}}{\mathbf{dx}} = \mathbf{CQ}^{-1}\mathbf{y} = \mathbf{Ay} \tag{5-222}$$

$$\frac{\mathbf{d^2y}}{\mathbf{dx}^2} = \mathbf{DQ}^{-1}\mathbf{y} = \mathbf{By} \tag{5-223}$$

where

$$\mathbf{Q}_{ji} = x_j^{i-1} \qquad \mathbf{C}_{ji} = (i-1)x_j^{i-2} \qquad \mathbf{D}_{ji} = (i-1)(i-2)x_j^{i-3} \tag{5-224}$$

† See Finlayson [8, p. 77].

The **A** and **B** matrices for the transformed Legendre polynomials (5-218) are given in Table 5-5.

**Table 5-5 Matrices of the transformed Legendre polynomials**

| $N$ | **A** | **B** |
|---|---|---|
| 1 | $\begin{bmatrix} -3 & 4 & -1 \\ -1 & 0 & 1 \\ 1 & -4 & 3 \end{bmatrix}$ | $\begin{bmatrix} 4 & -8 & 4 \\ 4 & -8 & 4 \\ 4 & -8 & 4 \end{bmatrix}$ |
| 2 | $\begin{bmatrix} -7 & 8.196 & -2.196 & 1 \\ -2.732 & 1.732 & 1.732 & -0.7321 \\ 0.7321 & -1.732 & -1.732 & 2.732 \\ -1 & 2.196 & -8.196 & 7 \end{bmatrix}$ | $\begin{bmatrix} 24 & -37.18 & 25.18 & -12 \\ 16.39 & -24 & 12 & -4.392 \\ -4.392 & 12 & -24 & 16.39 \\ -12 & 25.18 & -37.18 & 24 \end{bmatrix}$ |

*Source*: Reproduced from Ref. 8, by permission of the publisher.

The two-point boundary-value problem of Eq. (5-201) can now be expressed in terms of the orthogonal collocation method as

$$\mathbf{By} = \mathbf{f}(x, y, \mathbf{A}y) \tag{5-225}$$

or

$$\sum_{j}^{N+2} = 1\, \mathbf{B}_{ij} y_j = f\left(x_j,\ y_j,\ \sum_{j=1}^{N+2} \mathbf{A}_{ij} y_j\right) \tag{5-226}$$

with the boundary conditions

$$y_1 = 0 \qquad \text{and} \qquad y_{N+2} = 1 \tag{5-227}$$

Equations (5-226) and (5-227) constitute a set of $N+2$ simultaneous nonlinear algebraic equations whose solution can be obtained using Newton's method for nonlinear algebraic equations.

The orthogonal collocation method is more accurate than either the finite difference method or the collocation method. The choice of collocation points at the roots of the orthogonal polynomials reduces the error considerably.

**Example 5-8 Boundary-value problems: The Newton method** The equations which describe the state of the system in a batch penicillin fermentation are [20]

Cell mass production:

$$\frac{dy_1}{dt} = b_1 y_1 - \frac{b_1}{b_2} y_1^2 \qquad y_1(0) = 0.03 \tag{1}$$

Penicillin synthesis:

$$\frac{dy_2}{dt} = b_3 y_1 \qquad y_2(0) = 0.0 \tag{2}$$

where $y_1$ = concentration of cell mass (dimensionless)
$y_2$ = concentration of penicillin (dimensionless)
$t$ = time (dimensionless), $0 \le t \le 1$

The parameters $b_i$ are functions of temperature, $\theta$:

$$b_1 = w_1\left[\frac{1.0 - w_2(\theta - w_3)^2}{1.0 - w_2(25 - w_3)^2}\right]$$

$$b_2 = w_4\left[\frac{1.0 - w_2(\theta - w_3)^2}{1.0 - w_2(25 - w_3)^2}\right] \tag{3}$$

$$b_3 = w_5\left[\frac{1.0 - w_2(\theta - w_6)^2}{1.0 - w_2(25 - w_6)^2}\right]$$

$$b_i \ge 0$$

where $w_1$ = 13.1 (value of $b_1$ at 25°C obtained from fitting the model to experimental data)
$w_2 = 0.005$
$w_3$ = 30°C
$w_4$ = 0.94 (value of $b_2$ at 25°C)
$w_5$ = 1.71 (value of $b_3$ at 25°C)
$w_6$ = 20°C
$\theta$ = temperature, °C

These parameter-temperature functions are inverted parabolas which reach their peak at 30°C, for $b_1$ and $b_2$, and at 20°C, for $b_3$. The values of the parameters decrease by a factor of 2 over a 10° change in temperature on either side of the peak. The inequality, $b_i \ge 0$, restricts the values of the parameters to the positive regime. These functions have shapes typical of those encountered in microbial or enzyme-catalyzed reactions.

The *maximum principle of Pontryagin* (see Ref. 21) has been applied to the above model to determine the optimal temperature profile, which maximizes the concentration of penicillin at the final time of the fermentation, $t_f = 1$. The maximum principle algorithm when applied to the state equations, (1) and (2), yields the following additional equations:

The *adjoint equations*:

$$\frac{dy_3}{dt} = -y_3 b_1 + 2y_3 \frac{b_1}{b_2} y_1 - y_4 b_3 \qquad y_3(1) = 0 \tag{4}$$

$$\frac{dy_4}{dt} = 0 \qquad y_4(1) = 1.0 \tag{5}$$

The *Hamiltonian*:

$$H = y_3\left(b_1 y_1 - \frac{b_1}{b_2} y_1^2\right) + y_4(b_3 y_1)$$

The *necessary condition* for maximum:

$$\frac{\partial H}{\partial \theta} = 0 \tag{6}$$

Equations (1) to (6) form a two-point boundary-value problem. Apply Newton's method to obtain the solution of this problem and show the profiles of the state variables, the adjoint variables, and the optimal temperature.

METHOD OF SOLUTION The fundamental numerical problem of optimal control theory is the solution of the two-point boundary-value problem, which invariably arises from the application of the maximum principle to determine optimal control profiles. The state and adjoint equations, coupled together through the necessary condition for optimality, constitute a set of simultaneous differential equations which are often unstable. This difficulty is further complicated, in certain problems, when the necessary condition is not solvable explicitly for the control variable $\theta$. Several numerical methods have been developed for the solution of this class of problems.

We begin the solution of this example by first considering the boundary conditions: The initial values, $y_1(0)$ and $y_2(0)$, of the two state variables are known, and the final values, $y_3(1)$ and $y_4(1)$, of the two adjoint variables are fixed. Therefore, two initial conditions for the adjoint variables must be guessed. We designate these guesses, in accordance with Eq. (5-189), as follows:

$$y_3(0) = \gamma_3 \qquad \text{and} \qquad y_4(0) = \gamma_4 \tag{7}$$

We derive the variational equations by applying formula (5-190). We take the derivative of the first system differential equation (1) with respect to $\gamma_3$, and then with respect to $\gamma_4$. We repeat this procedure on the second (2), third (4), and fourth (5) system equations, to obtain the following eight variational equations:

$$
\begin{aligned}
&\frac{d}{dt}\left(\frac{\partial y_1}{\partial \gamma_3}\right) = b_1\left(\frac{\partial y_1}{\partial \gamma_3}\right) - 2\,\frac{b_1}{b_2}\,y_1\left(\frac{\partial y_1}{\partial \gamma_3}\right) \\
&\frac{d}{dt}\left(\frac{\partial y_1}{\partial \gamma_4}\right) = b_1\left(\frac{\partial y_1}{\partial \gamma_4}\right) - 2\,\frac{b_1}{b_2}\,y_1\left(\frac{\partial y_1}{\partial \gamma_4}\right) \\
&\frac{d}{dt}\left(\frac{\partial y_2}{\partial \gamma_3}\right) = b_3\left(\frac{\partial y_1}{\partial \gamma_3}\right) \\
&\frac{d}{dt}\left(\frac{\partial y_2}{\partial \gamma_4}\right) = b_3\left(\frac{\partial y_1}{\partial \gamma_4}\right) \\
&\frac{d}{dt}\left(\frac{\partial y_3}{\partial \gamma_3}\right) = -b_1\left(\frac{\partial y_3}{\partial \gamma_3}\right) + 2\,\frac{b_1}{b_2}\,y_3\left(\frac{\partial y_1}{\partial \gamma_3}\right) + 2\,\frac{b_1}{b_2}\,y_1\left(\frac{\partial y_3}{\partial \gamma_3}\right) - b_3\left(\frac{\partial y_4}{\partial \gamma_3}\right) \\
&\frac{d}{dt}\left(\frac{\partial y_3}{\partial \gamma_4}\right) = -b_1\left(\frac{\partial y_3}{\partial \gamma_4}\right) + 2\,\frac{b_1}{b_2}\,y_3\left(\frac{\partial y_1}{\partial \gamma_4}\right) + 2\,\frac{b_1}{b_2}\,y_1\left(\frac{\partial y_3}{\partial \gamma_4}\right) - b_3\left(\frac{\partial y_4}{\partial \gamma_4}\right) \\
&\frac{d}{dt}\left(\frac{\partial y_4}{\partial \gamma_3}\right) = 0 \\
&\frac{d}{dt}\left(\frac{\partial y_4}{\partial \gamma_4}\right) = 0
\end{aligned}
\tag{8}
$$

The initial conditions of the variational equations are determined by applying formula (5-193):

$$
\begin{aligned}
&\left(\frac{\partial y_1}{\partial \gamma_3}\right)_{t=0} = 0 \qquad \left(\frac{\partial y_1}{\partial \gamma_4}\right)_{t=0} = 0 \\
&\left(\frac{\partial y_2}{\partial \gamma_3}\right)_{t=0} = 0 \qquad \left(\frac{\partial y_2}{\partial \gamma_4}\right)_{t=0} = 0 \\
&\left(\frac{\partial y_3}{\partial \gamma_3}\right)_{t=0} = 1 \qquad \left(\frac{\partial y_3}{\partial \gamma_4}\right)_{t=0} = 0 \\
&\left(\frac{\partial y_4}{\partial \gamma_3}\right)_{t=0} = 0 \qquad \left(\frac{\partial y_4}{\partial \gamma_4}\right)_{t=0} = 1
\end{aligned}
\tag{9}
$$

Finally, we must express the necessary condition [Eq. (6)] in terms of the system variables:

$$
\frac{\partial H}{\partial \theta} = y_3\left[y_1\left(\frac{\partial b_1}{\partial \theta}\right) - y_1^2\,\frac{\partial (b_1/b_2)}{\partial \theta}\right] + y_4 y_1\left(\frac{\partial b_3}{\partial \theta}\right) = 0
$$

Differentiating the parameter-temperature functions with respect to the temperature $\theta$, and substituting in the above, we obtain

$$\theta = \frac{w_3 w_1 y_3 + w_6 w_5 y_4}{w_1 w_3 + w_5 y_4} \tag{10}$$

This is an explicit solution of the control variable in terms of the adjoint variables. The problem for this example was intentionally chosen so that such a solution could be obtained. In the case where this is not possible, the necessary condition must be solved by gradient search techniques.

The complete set of equations for the solution of this two-point boundary-value problem consists of

1. The four system equations with their known boundary values [(1), (2), (4), (5)]
2. The eight variational equations (8) with their initial conditions (9)
3. The guessed initial conditions for two system variables (7)
4. The parameter-temperature functions (3)
5. The explicit formulation (10) of the necessary condition
6. Equation (5-194) for construction of the Jacobian matrix
7. Equation (5-196) for calculation of the $\boldsymbol{\delta y}$ vector
8. Equations (5-171) and (5-185) for correcting the guessed initial conditions

PROGRAM DESCRIPTION This computer program implements the Newton method for two-point boundary-value problems. The program is called BOUNDARY.BAS, and it consists of the sections listed in Table E5-8.

*Main program* Lines 110–160 provide the options of printing final results only or step-by-step results. We recommend using the second option when a new problem, whose solution is not well known, is being solved.

Lines 170–180 call the subroutines for entering the equations and the integration parameters.

Line 210 is the beginning of the Newton iteration. An upper limit of 30 iterations is used. This can be easily changed to suit the particular needs of the problem.

Lines 270–390 construct the Jacobian matrix utilizing Eq. (5-194) and calculate the $\boldsymbol{\delta y}$ vector from Eq. (5-196).

Lines 400–490 perform the inversion of the Jacobian matrix using subroutine 2.

Lines 500–600 calculate the correction vector $\boldsymbol{\Delta \gamma}$ using Eq. (5-171).

Lines 610–670 check for convergence of the boundary condtions. The convergence test is

$$\left| \frac{\delta y_i}{y_i(t_f)} \right| \leq \epsilon \qquad \text{for } y_i(t_f) \neq 0$$

or

$$|\delta y_i| \leq \epsilon \qquad \text{for } y_i(t_f) = 0$$

**Table E5-8**

| Program section | Line numbers |
|---|---|
| Title | 1–90 |
| Main program | 100–960 |
| Output options | 110–160 |
| Subroutine calls | 170–260 |
| Construction of Jacobian matrix | 270–390 |
| Inversion of Jacobian matrix | 400–490 |
| Calculation of correction vector | 500–600 |
| Check for convergence | 610–670 |
| Correction of initial conditions | 680–750 |
| Nonconvergence options | 760–830 |
| Call to printing and plotting subroutines | 840–855 |
| Rerun options | 860–950 |
| Subroutines | |
| 1. Input equations | 1000–1590 |
| 2. Matrix inversion | 2000–2450 |
| 3. Input integration parameters | 3000–3490 |
| 4. Differential equations | 4000–4999 |
| 5. Integration methods | Same as in Example 5-7 |
| 6. Print table of results | 6000–6090 |
| 7. Plotting options<br>8. Plotting | Same as in Example 5-6 |
| 9. Calculate control variable | 9000–9140 |

The value $\epsilon$ is specified by the user in the input phase of the program (subroutine 3).

Lines 680–750 correct the guessed initial conditions using Eq. (5-185):

$$\boldsymbol{\gamma}_{\text{new}} = \boldsymbol{\gamma}_{\text{old}} + \alpha \, \Delta\boldsymbol{\gamma} \tag{5-185}$$

The fractional correction step ($\alpha = \frac{1}{2}$) is used to reduce the possibility of divergence and to eliminate oscillations around the solution.

Lines 760–830 are executed only if the Newton method does not converge in 30 iterations. The options given are to see the nonconverged results, to reset the parameters, or to end the program.

Lines 840–850 call the subroutines which print the results and plot the system variables.

Line 855 calls subroutine 9 which calculates and plots the optimal profile of the control variable. (See discussion of subroutine 9 for note of caution.)

Lines 860–950 provide the options for rerunning or ending the program.

*Subroutine 1* This subroutine, which is similar to the corresponding one used in Example 5-7, enables the user to enter the system and variational equations. The differential equations must be in the form

$$G(i) = f[x, y(1), y(2), \ldots, y(n)]$$

The vector **y** includes all the dependent variables—system and variational. The differential equations must be numbered in the following order:

| | |
|---|---|
| First: | System equations with known initial conditions |
| Second: | System equations with unknown initial conditions |
| Last: | Variational equations in the exact order as shown in this example |

The differential equations and the associated algebraic equations may be entered during the execution of the program. However, this example requires a large number of lines of input, 25 lines to be exact. Input errors during execution of the program are not correctable once the line has been terminated with the ENTER key. For this reason, we found it more convenient to prepare subroutine 4 before executing BOUNDARY.BAS, and to store it as an ASCII file named PEN.EQU. This is done using the command

```
SAVE "PEN.EQU", A
```

This step is equivalent to what the program does when the equations are entered during execution time. The file PEN.EQU can then be merged with the program during execution (lines 1250, 1310, and 1580).

*Subroutine 2* This part of the program uses the Gauss-Jordan reduction method to invert the Jacobian matrix. Checks are included to ensure that the matrix is not singular, and that the inverse satisfies the requirement

$$\mathbf{A}^{-1}\mathbf{A} = \mathbf{I}$$

where **I** is the identity matrix.

*Subroutine 3* The following integration parameters are entered during execution of this subroutine: number of integration steps, initial and final values of the independent variable, known initial and final conditions of the dependent variables, guessed initial conditions, and the convergence criterion for the boundary conditions. The program generates the initial conditions of the variational equations using formula (5-193). For this reason it is important that the equations be entered in the correct order as specified above.

*Subroutine 4* This subroutine contains all the differential and algebraic equations needed for the solution of the problem. As was mentioned earlier, these equations can be entered during execution of the program or they can be prepared a priori and saved as an ASCII file.

*Subroutine 5* This subroutine is identical to the corresponding one used in Example 5-7. It contains the Runge-Kutta and Euler predictor-corrector integration methods. Only the former method is used in this example.

*Subroutine 6* This subroutine prints the table of results. It is similar to that used in Example 5-7, except that only every $ST$th point is printed out (line 6045) in order to limit the length of the printout.

*Subroutines 7 and 8* These subroutines are identical to those used in Example 5-6.

*Subroutine 9* This subroutine calculates the final profile of the control variable (after convergence of the Newton method). It stores this profile in the last column of matrix **A** and calls subroutine 7 to plot this variable. Consequently, this subroutine is specific for this example only and should be either modified or removed from the program when a different problem is being solved.

Program

```
1 SCREEN 0:WIDTH 80:CLS:KEY OFF
5 PRINT "********************************************************************"
10 PRINT"*                                                                  *"
15 PRINT"*                           EXAMPLE 5-8                            *"
20 PRINT"*                                                                  *"
25 PRINT"*             BOUNDARY-VALUE PROBLEMS: THE NEWTON METHOD           *"
30 PRINT"*                                                                  *"
35 PRINT"*                          (BOUNDARY.BAS)                          *"
40 PRINT"*                                                                  *"
90 PRINT"*******************************************************************"
100 '*************************** Main Program ***************************
110 PRINT:PRINT
120 PRINT:PRINT"                            OUTPUT OPTIONS":PRINT
130 PRINT:PRINT"                1. PRINT FINAL RESULTS ONLY"
140 PRINT:PRINT"                2. PRINT STEP-BY-STEP RESULTS"
150 PRINT:PRINT:INPUT"ENTER YOUR CHOICE(1-2)"; RQ
160 IF RQ<1 OR RQ>2 THEN GOTO 150
170 GOSUB 1000
180 GOSUB 3000
190 PRINT CHR$(12)
200 FQ=1:EPS=.001:N=NFC
210 FOR ITER=1 TO 30
220 PRINT:PRINT"INTEGRATING. PLEASE WAIT"
230  GOSUB 5000
240 IF RQ=1 THEN PRINT "ITERATION=";ITER
250 IF RQ=2 THEN PRINT CHR$(12);"ITERATION=";ITER
260 IF RQ=2 THEN GOSUB 6000
270 'Construct the Jacobian matrix and the DELTAY vector
280 K=NSE+NIC*NFC
290 IF RQ=2 THEN PRINT:PRINT"JACOBIAN MATRIX:":PRINT
300 FOR I=1 TO NFC
310 FOR J=1 TO NFC
320 K=K+1
330 JAC(I,J)=A(L,K)
340 AA(I,J)=JAC(I,J) : BB(I,J)=JAC(I,J)
350 IF RQ=2 THEN PRINT JAC(I,J),
360 NEXT J
370 IF RQ=2 THEN PRINT
380 DELTAY(I)=-(A(L,NIC+I)-YF(NIC+I))
390 NEXT I
400 'Invert the Jacobian
410 GOSUB 2000
420 IF RQ=2 THEN PRINT:PRINT"INVERSE OF JACOBIAN MATRIX:":PRINT
430 FOR I=1 TO NFC
440 FOR J=1 TO NFC
450 JACINV(I,J)=AA(I,NFC+J)
460 IF RQ=2 THEN PRINT JACINV(I,J),
470 NEXT J
480 IF RQ=2 THEN PRINT
490 NEXT I
500 'Calculate the correction vector DELGAMMA
510 IF RQ=2 THEN PRINT
520 FOR I=1 TO NFC
530 SUM=0
540 FOR J=1 TO NFC
550 SUM=SUM + JACINV(I,J)*DELTAY(J)
```

```
560 NEXT J
570 DELGAMMA(I)=SUM
580 IF RQ=2 THEN PRINT"DELTAY(";I;")=";DELTAY(I)
590 IF RQ=2 THEN PRINT"DELGAMMA(";I;")=";DELGAMMA(I)
600 NEXT I
610 'Check for convergence
620 COUNT=0
630 FOR I=1 TO NFC
640 IF YF(NIC+I)=0 THEN CCC=DELTAY(I) ELSE CCC=DELTAY(I)/YF(NIC+I)
650 IF ABS(CCC)<=CONV THEN COUNT=COUNT+1
660 IF COUNT=NFC THEN PRINT "*********** CONVERGED ***********":GOTO 840
670 NEXT I
680 'Correct the guessed initial conditions(use 1/2 step)
690  FOR I = 1 TO NFC
700 Y9(NIC+I)=Y9(NIC+I)+DELGAMMA(I)/2
710 NEXT I
720 'Reset all initial conditions
730 A(0,0)=X9 : X=X9
740 FOR I=1 TO N1 : Y(I)=Y9(I) : A(0,I)=Y9(I) : NEXT I
750 NEXT ITER
760 PRINT CHR$(12);"    ITERATION LIMIT EXCEEDED. RESULTS MAY NOT BE CORRECT."
770 PRINT:PRINT"                         OPTIONS IN CASE OF NONCONVERGENCE":PRINT
780 PRINT:PRINT"               1. SEE THE NONCONVERGED RESULTS "
790 PRINT:PRINT"               2. RESET INTEGRATION PARAMETERS AND RECALCULATE"
800 PRINT:PRINT"               3. END PROGRAM"
810 PRINT:PRINT:INPUT"    ENTER YOUR CHOICE(1-3)"; NRRQ
820 IF NRRQ<1 OR NRRQ>3 THEN GOTO 810
830 ON NRRQ GOTO 840,930,960
840 PRINT CHR$(12);"ITERATION=";ITER :GOSUB 6000
850 SWAP N1,NSE: GOSUB 7000: SWAP N1,NSE
855 GOSUB 9000
860 PRINT:PRINT"                              RERUN OPTIONS":PRINT
870 PRINT:PRINT"               1. SHOW SAME RESULTS AGAIN"
880 PRINT:PRINT"               2. RESET INTEGRATION PARAMETERS AND RECALCULATE"
890 PRINT:PRINT"               3. END PROGRAM"
900 PRINT:PRINT:INPUT"    ENTER YOUR CHOICE(1-3)"; RRQ
910 IF RRQ<1 OR RRQ>3 THEN GOTO 900
920 ON RRQ GOTO 840,930,960
930 PRINT:PRINT:PRINT:PRINT"                             NEW CONDITIONS"
940  GOSUB 3040 'Enter Subroutine 3 after the dimension statement
950  GOTO 210
960  END
1000 '************* Subroutine 1: Input equations ***********************
1010 '
1020 PRINT CHR$(12)
1030 PRINT:PRINT"                         DIFFERENTIAL EQUATIONS"
1040 PRINT:PRINT"                         **********************"
1050 PRINT:PRINT"THE SYSTEM AND VARIATIONAL DIFFERENTIAL EQUATIONS, AND OTHER "
1060 PRINT: PRINT"ASSOCIATED ALGEBRAIC EQUATIONS, MAY BE ENTERED HERE."
1070 PRINT:PRINT"EACH LINE-STATEMENT MUST OBEY THE RULES OF BASIC PROGRAMMING."
1080 PRINT:PRINT"THE LINES ARE AUTOMATICALLY NUMBERED, STARTING AT 4011."
1090 PRINT:PRINT"THESE STATEMENTS ARE MERGED INTO THE PROGRAM IN SUBROUTINE 4."
1100 PRINT:PRINT"                         **********************"
1110 PRINT:INPUT "HOW MANY SYSTEM DIFFERENTIAL EQUATIONS "; NSE
1120 PRINT:INPUT "   HOW MANY OF THESE HAVE KNOWN INITIAL CONDITIONS "; NIC
1130 PRINT:INPUT "   HOW MANY OF THESE HAVE KNOWN FINAL CONDITIONS "; NFC
1140 IF NFC=NSE-NIC THEN GOTO 1170
1150 PRINT:PRINT"****************** ERROR ********************"
1160 PRINT"TOTAL NUMBER OF BOUNDARY CONDITIONS MUST BE =";NSE:GOTO 1110
```

```
1170 PRINT:INPUT "HOW MANY VARIATIONAL DIFFERENTIAL EQUATIONS "; NVE
1180 IF NVE=NSE*NFC THEN GOTO 1210
1190 PRINT:PRINT"****************** ERROR *********************"
1200 PRINT"THE NUMBER OF VARIATIONAL EQUATIONS MUST BE =";NSE*NFC:GOTO 1170
1210 N1=NSE+NVE
1220 DIM Y(N1),Y9(N1),YF(N1),Y1(N1),C(50),G(N1),B1(N1),B2(N1),B3(N1),B4(N1)
1230 DIM JAC(NFC,NFC),JACINV(NFC,NFC),DELTAY(NFC),DELGAMMA(NFC)
1240 DIM AA(NFC,2*NFC),BB(NFC,NFC),CC(NFC,NFC)
1250 PRINT:INPUT "HAVE YOU PREVIOUSLY ENTERED THE EQUATIONS(Y/N)";Q1$
1260 IF Q1$="Y" OR Q1$="y" THEN GOTO 1290
1270 IF Q1$="N" OR Q1$="n" THEN GOTO 1330
1280 GOTO 1250
1290 PRINT:PRINT"THE FOLLOWING EQUATION-FILES EXIST ON YOUR DISK:":PRINT
1300 FILES"*.EQU"
1310 INPUT"GIVE NAME OF FILE TO BE USED (DO NOT ENTER THE EXTENSION): ",FIL$
1320 FIL$=FIL$+".EQU":GOTO 1570
1330 PRINT:INPUT"NAME OF FILE FOR STORING THE DIFFERENTIAL EQUATIONS: ",FIL$
1340 FIL$=FIL$+".EQU"
1350 OPEN FIL$ FOR OUTPUT AS #1
1360 CLS:PRINT:PRINT"                    *********************"
1370 PRINT:PRINT"ENTER THE DIFFERENTIAL EQUATIONS IN THE FORM SHOWN BELOW:"
1380 PRINT:PRINT"     G(i)= f[x, y(1), y(2),....,y(n)]"
1390 PRINT:PRINT"THE EQUATIONS MUST BE NUMBERED IN THE FOLLOWING ORDER:"
1400 PRINT:PRINT"     First: System equations with known initial conditions"
1410 PRINT:PRINT"     Second:System equations with unknown initial conditions"
1420 PRINT:PRINT"     Last:  Variational equations in the exact order as shown"
1430 PRINT"           in this example"
1440 PRINT:PRINT"                  *********************":PRINT
1450 INPUT"GIVE THE EXACT NUMBER OF LINES OF INPUT YOU WILL ENTER(1-980)";LN1
1460 PRINT:PRINT"SEPARATE EACH STATEMENT WITH THE ENTER(RETURN) KEY."
1470 PRINT#1,"4010 'The differential equations are merged in lines 4011-4990
1480 FOR I=1 TO LN1
1490 PRINT:PRINT "       ";4010+I;
1500 LINE INPUT ;EQUATION$
1510 IF EQUATION$="" THEN GOTO 1490
1520 PRINT#1,4010+I; EQUATION$
1530 PRINT
1540 NEXT I
1550 PRINT#1,"4995 'End of space allocated for the differential equations
1560 CLOSE #1
1570 PRINT CHR$(12)
1580 CHAIN MERGE FIL$,180,ALL,DELETE 4010-4995
1590 RETURN
2000 '************* Subroutine 2: Matrix inversion **********************
2010 'Augment with the identity matrix
2020 FOR I = 1 TO N
2030 FOR J = N+1 TO 2*N
2040 AA(I,J) = 0
2050 NEXT J
2060 AA(I,I+N) = 1
2070 NEXT I
2080 'Beginning of the Gauss-Jordan reduction procedure
2090 FOR K = 1 TO N
2100 'Apply partial pivoting strategy
2110 MAXPIVOT = ABS(AA(K,K)):NPIVOT=K
2120 FOR I = K TO N
2130 IF MAXPIVOT > = ABS(AA(I,K)) GOTO 2150
2140 MAXPIVOT=ABS(AA(I,K)) : NPIVOT=I
2150 NEXT I
```

```
2160 IF MAXPIVOT > = EPS GOTO 2180
2170 PRINT"CAUTION: PIVOT ELEMENT SMALLER THAN";EPS;". MATRIX MAY BE SINGULAR."
2180 IF NPIVOT = K GOTO 2220
2190 FOR J=K TO 2*N
2200 SWAP  AA(NPIVOT,J),AA(K,J)
2210 NEXT J
2220 DD=AA(K,K)
2230 FOR J = 2*N TO K STEP -1
2240 AA(K,J) = AA(K,J)/DD
2250 NEXT J
2260 FOR I = 1 TO N
2270 IF I=K GOTO 2320
2280 MULT=AA(I,K)
2290 FOR J = 2*N TO K STEP -1
2300 AA(I,J) = AA(I,J) - MULT* AA(K,J)
2310 NEXT J
2320 NEXT I:NEXT K
2330 'Check the product of matrix and inverse
2340 FOR I = 1 TO N
2350 FOR J = 1 TO N
2360 CC(I,J)=0
2370 FOR K = 1 TO N
2380 CC(I,J)=CC(I,J) + BB(I,K)*AA(K,N+J)
2390 NEXT K
2400 IF I=J AND ABS(CC(I,J)-1)<EPS THEN GOTO 2430
2410 IF I<>J AND ABS(CC(I,J))<EPS THEN GOTO 2430
2420 PRINT "CAUTION: INVERSE MAY BE INCORRECT."
2430 NEXT J
2440 NEXT I
2450 RETURN
3000 '************* Subroutine 3: Input integration parameters ***********
3010 PRINT:PRINT "                              INTEGRATION PARAMETERS"
3020 PRINT:PRINT "NUMBER OF INTEGRATION STEPS:";: INPUT L
3030  DIM A(L,N1)
3040 PRINT:PRINT "INITIAL VALUE OF INDEPENDENT VARIABLE:";: INPUT X9
3050 PRINT:PRINT "FINAL VALUE OF INDEPENDENT VARIABLE:  ";: INPUT XF
3060 D=(XF-X9)/L : X = X9 : A(0,0) = X9
3070 PRINT:PRINT "INCREMENT IN INDEPENDENT VARIABLE IS="; D
3080 PRINT:PRINT"ENTER KNOWN INITIAL CONDITIONS:"
3090  FOR K = 1 TO NIC
3100 PRINT:PRINT "INITIAL Y(";K;")";: INPUT Y9(K)
3110 Y(K) = Y9(K)
3120 A(0,K) = Y9(K)
3130  NEXT K
3140 PRINT:PRINT"ENTER KNOWN FINAL CONDITIONS:"
3150  FOR K = NIC+1 TO NSE
3160 PRINT:PRINT "FINAL Y(";K;")";: INPUT YF(K)
3170  NEXT K
3180 PRINT:PRINT"ENTER GUESSED INITIAL CONDITIONS:"
3190  FOR K = NIC+1 TO NSE
3200 PRINT:PRINT "INITIAL Y(";K;")";: INPUT Y9(K)
3210 Y(K) = Y9(K)
3220 A(0,K) = Y9(K)
3230  NEXT K
3240 PRINT
3250 INPUT"GIVE THE CONVERGENCE CRITERION FOR THE BOUNDARY CONDITIONS";CONV
3260 PRINT
3270 'Set the initial conditions for the variational equations
3280 PRINT"INITIAL CONDITIONS FOR VARIATIONAL EQUATIONS ARE SET AS FOLLOWS:"
```

```
3290 K=NSE
3300 FOR I=1 TO NSE
3310 FOR J=NIC+1 TO NSE
3320 K=K+1
3330 Y9(K)=0
3340 IF I=J THEN Y9(K)=1
3350 Y(K)=Y9(K)  : A(0,K)=Y9(K)
3360 PRINT:PRINT "INITIAL Y(";K;")";Y9(K)
3370 NEXT J
3380 NEXT I
3390 PRINT
3400 PRINT"PLEASE CHECK THE ABOVE TO MAKE SURE THAT THEY CORRESPOND TO YOUR"
3410 PRINT"VARIATIONAL EQUATIONS.":PRINT
3420 PRINT"DO YOU WANT TO ENTER CONSTANTS IN THE DIFFERENTIAL EQUATIONS(Y/N)";
3430 INPUT VC$ : IF VC$="N" OR VC$="n" THEN  GOTO 3490
3440 PRINT:INPUT"HOW MANY CONSTANTS";CO
3450 PRINT:PRINT "GIVE VALUES OF CONSTANTS":PRINT
3460  FOR K = 1 TO CO
3470  PRINT "CONSTANT ";K;: INPUT C(K)
3480  NEXT K
3490 RETURN
4000 '************* Subroutine 4: Differential equations *****************
4010 'The differential equations are merged in lines 4011-4990
4011 W1=13.1
4012 W2=.005
4013 W3=30
4014 W4=.94
4015 W5=1.71
4016 W6=20
4017 THETA=(W3*W1*Y(3)+W6*W5*Y(4))/(W1*Y(3)+W5*Y(4))
4018 BB1=W1*(1-W2*(THETA-W3)^2)/(1-W2*(25-W3)^2)
4019 IF BB1<0 THEN BB1=0
4020 BB2=W4*(1-W2*(THETA-W3)^2)/(1-W2*(25-W3)^2)
4021 EE=W1/W4
4022 BB3=W5*(1-W2*(THETA-W6)^2)/(1-W2*(25-W6)^2)
4023 IF BB3<0 THEN BB3=0
4024 G(1)=BB1*Y(1)  - EE*Y(1)^2
4025 G(2)=BB3*Y(1)
4026 G(3)=-Y(3)*BB1 + 2*Y(3)*EE*Y(1)  - Y(4)*BB3
4027 G(4)=0
4028 G(5)=BB1*Y(5)  - 2*EE*Y(1)*Y(5)
4029 G(6)=BB1*Y(6)  - 2*EE*Y(1)*Y(6)
4030 G(7)=BB3*Y(5)
4031 G(8)=BB3*Y(6)
4032 G(9)=2*Y(3)*EE*Y(5)  - BB1*Y(9)  + 2*EE*Y(1)*Y(9)  - BB3*Y(11)
4033 G(10)=2*Y(3)*EE*Y(6)-BB1*Y(10)  + 2*EE*Y(1)*Y(10)  - BB3*Y(12)
4034 G(11)=0
4035 G(12)=0
4995 'End of space allocated for the differential equations
4999 RETURN

5000 '************* Subroutine 5: Integration methods *******************

                    (Same as in Example 5-7)


6000 '************* Subroutine 6: Print table of results *****************
6010 '
6020 PRINT
```

```
6030 PRINT "           TABLE OF RESULTS"
6040 PRINT:PRINT "INDEPENDENT        DEPENDENT VARIABLES"
6045 ST=INT(L/10) 'Prints every ST(th) point
6050 FOR J2 = 0 TO L STEP ST
6060 FOR J1 = 0 TO N1 : PRINT A(J2,J1), : NEXT J1
6070 PRINT
6080 NEXT J2
6090 ' PRINT:INPUT "PRESS THE ENTER KEY TO CONTINUE",KK$
6100 RETURN

7000 '************* Subroutine 7: Plotting options ***********************

                    (Same as in Example 5-6)

8000 '************* Subroutine 8: Plotting *******************************

                    (Same as in Example 5-6)

9000 '************* Subroutine 9: Calculate control variable *************
9010 '
9020 'This subroutine applies to Example 5-8 only. It should be removed
9030 'from the program, if a different problem is being solved.
9040 '
9050 'It calculates the control variable, places it in the last column of
9060 'matrix A, and plots it using Subroutines 7 and 8
9070 '
9080 FOR I=0 TO L
9090 A(I,N1)=(W3*W1*A(I,3)+W6*W5*A(I,4))/(W1*A(I,3)+W5*A(I,4))
9100 NEXT I
9110 NS=N1: NF=N1
9120 PRINT CHR$(12);"PLOTTING OPTIONS FOR THE CONTROL VARIABLE"
9130 GOSUB 7310
9140 RETURN
```

## RESULTS

```
*****************************************************************
*                                                               *
*                          EXAMPLE 5-8                          *
*                                                               *
*          BOUNDARY-VALUE PROBLEMS: THE NEWTON METHOD           *
*                                                               *
*                         (BOUNDARY.BAS)                        *
*                                                               *
*****************************************************************


                         OUTPUT OPTIONS

            1. PRINT FINAL RESULTS ONLY

            2. PRINT STEP-BY-STEP RESULTS

ENTER YOUR CHOICE(1-2)? 1

                     DIFFERENTIAL EQUATIONS

                     **********************

THE SYSTEM AND VARIATIONAL DIFFERENTIAL EQUATIONS, AND OTHER

ASSOCIATED ALGEBRAIC EQUATIONS, MAY BE ENTERED HERE.

EACH LINE-STATEMENT MUST OBEY THE RULES OF BASIC PROGRAMMING.

THE LINES ARE AUTOMATICALLY NUMBERED, STARTING AT 4011.

THESE STATEMENTS ARE MERGED INTO THE PROGRAM IN SUBROUTINE 4.

                     **********************

HOW MANY SYSTEM DIFFERENTIAL EQUATIONS ? 4

   HOW MANY OF THESE HAVE KNOWN INITIAL CONDITIONS ? 2

   HOW MANY OF THESE HAVE KNOWN FINAL CONDITIONS ? 2

HOW MANY VARIATIONAL DIFFERENTIAL EQUATIONS ? 8

HAVE YOU PREVIOUSLY ENTERED THE EQUATIONS(Y/N)? Y

THE FOLLOWING EQUATION-FILES EXIST ON YOUR DISK:

C:\BOOK
PEN     .EQU      ERRORFUN.EQU      PRPR    .EQU
 1239040 Bytes free

GIVE NAME OF FILE TO BE USED (DO NOT ENTER THE EXTENSION): PEN
```

```
                    INTEGRATION PARAMETERS

NUMBER OF INTEGRATION STEPS:? 100

INITIAL VALUE OF INDEPENDENT VARIABLE:? 0

FINAL VALUE OF INDEPENDENT VARIABLE:  ? 1

INCREMENT IN INDEPENDENT VARIABLE IS= .01


ENTER KNOWN INITIAL CONDITIONS:

INITIAL Y( 1 )? 0.03

INITIAL Y( 2 )? 0.0


ENTER KNOWN FINAL CONDITIONS:

FINAL Y( 3 )? 0.0

FINAL Y( 4 )? 1.0


ENTER GUESSED INITIAL CONDITIONS:

INITIAL Y( 3 )? 5.0

INITIAL Y( 4 )? 1.0

GIVE THE CONVERGENCE CRITERION FOR THE BOUNDARY CONDITIONS? 0.01

INITIAL CONDITIONS FOR VARIATIONAL EQUATIONS ARE SET AS FOLLOWS:

INITIAL Y( 5 ) 0

INITIAL Y( 6 ) 0

INITIAL Y( 7 ) 0

INITIAL Y( 8 ) 0

INITIAL Y( 9 ) 1

INITIAL Y( 10 ) 0

INITIAL Y( 11 ) 0

INITIAL Y( 12 ) 1

PLEASE CHECK THE ABOVE TO MAKE SURE THAT THEY CORRESPOND TO YOUR
VARIATIONAL EQUATIONS.

DO YOU WANT TO ENTER CONSTANTS IN THE DIFFERENTIAL EQUATIONS(Y/N)? N
```

```
INTEGRATING. PLEASE WAIT
ITERATION= 1

INTEGRATING. PLEASE WAIT
ITERATION= 2

INTEGRATING. PLEASE WAIT
ITERATION= 3

INTEGRATING. PLEASE WAIT
ITERATION= 4

INTEGRATING. PLEASE WAIT
ITERATION= 5

INTEGRATING. PLEASE WAIT
ITERATION= 6

INTEGRATING. PLEASE WAIT
ITERATION= 7

INTEGRATING. PLEASE WAIT
ITERATION= 8

INTEGRATING. PLEASE WAIT
ITERATION= 9

INTEGRATING. PLEASE WAIT
ITERATION= 10

INTEGRATING. PLEASE WAIT
ITERATION= 11

INTEGRATING. PLEASE WAIT
ITERATION= 12

INTEGRATING. PLEASE WAIT
ITERATION= 13

INTEGRATING. PLEASE WAIT
ITERATION= 14

INTEGRATING. PLEASE WAIT
ITERATION= 15

INTEGRATING. PLEASE WAIT
ITERATION= 16
*********** CONVERGED ************
```

```
ITERATION= 16

         TABLE OF RESULTS

INDEPENDENT       DEPENDENT VARIABLES
 0             .03          0            3.612108      1
 0             0            0            0             1
 0             0            1
9.999999E-02                .1217076     7.49188E-03   .9120284
 1             0            0            0             0
 .2703253     -6.441563E-02              0             1

 .2            .3790081     3.948357E-02               .3067431
 1             0            0            0             0
 .1203554     -.1279936     0            1
 .3            .7074292     .1263725     .1703922      1
 0             0            0            0             .13885
-.3311489      0            1
 .3999999      .8752784     .2612774     .1395406      1
 0             0            0            0             .3450434
-1.106793      0            1
 .4999998      .9245143     .4153208     .1325024      1
 0             0            0            0             1.153655
-4.034623      0            1
 .5999998      .936302      .5745071     .1306722      1
 0             0            0            0             4.173964
-14.94613      0            1
 .6999996      .9385049     .7349584     .1292121      1
 0             0            0            0             15.38569
-55.44557      0            1
 .7999996      .9367136     .8959228     .1242631      1
 0             0            0            0             56.97019
-205.6582      0            1
 .8999994      .9250812     1.057977     .1027429      1
 0             0            0            0             211.5638
-764.0884      0            1
 .9999994      .8345025     1.223038     7.532826E-03
 1             0            0            0             0
 818.6711     -2957.12      0            1
```

```
          PLOTTING OPTIONS

    1. PLOT EACH VARIABLE SEPARATELY

    2. PLOT ALL VARIABLES ON SAME GRAPH

ENTER YOUR CHOICE(1-2)? 1


           SCALING OPTIONS

    1. MANUAL CHOICE OF MAXIMA AND MINIMA

    2. AUTOMATIC CHOICE OF MAXIMA AND MINIMA

ENTER YOUR CHOICE(1-2)? 1
```

```
          SCALES FOR VARIABLE  1

MINIMUM X=? 0

MAXIMUM X=? 1

SPECIFY X-AXIS TITLE:TIME

MINIMUM Y=? 0

MAXIMUM Y=? 2

SPECIFY Y-AXIS TITLE:CELL
```

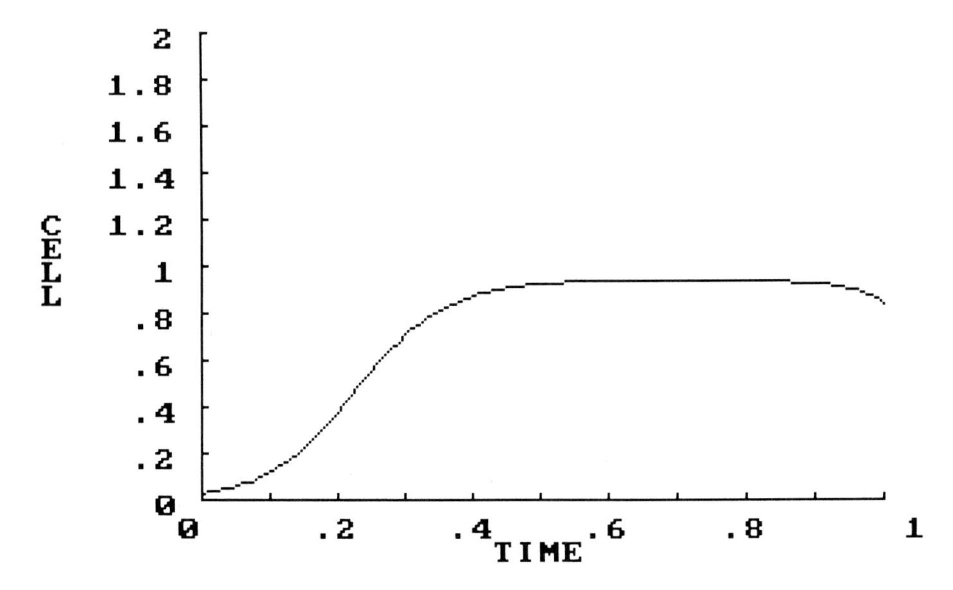


```
             SCALING OPTIONS
    1. MANUAL CHOICE OF MAXIMA AND MINIMA

    2. AUTOMATIC CHOICE OF MAXIMA AND MINIMA

ENTER YOUR CHOICE(1-2)? 1

          SCALES FOR VARIABLE  2

MINIMUM X=? 0

MAXIMUM X=? 1

SPECIFY X-AXIS TITLE:TIME

MINIMUM Y=? 0

MAXIMUM Y=? 2

SPECIFY Y-AXIS TITLE:PENICILLIN
```

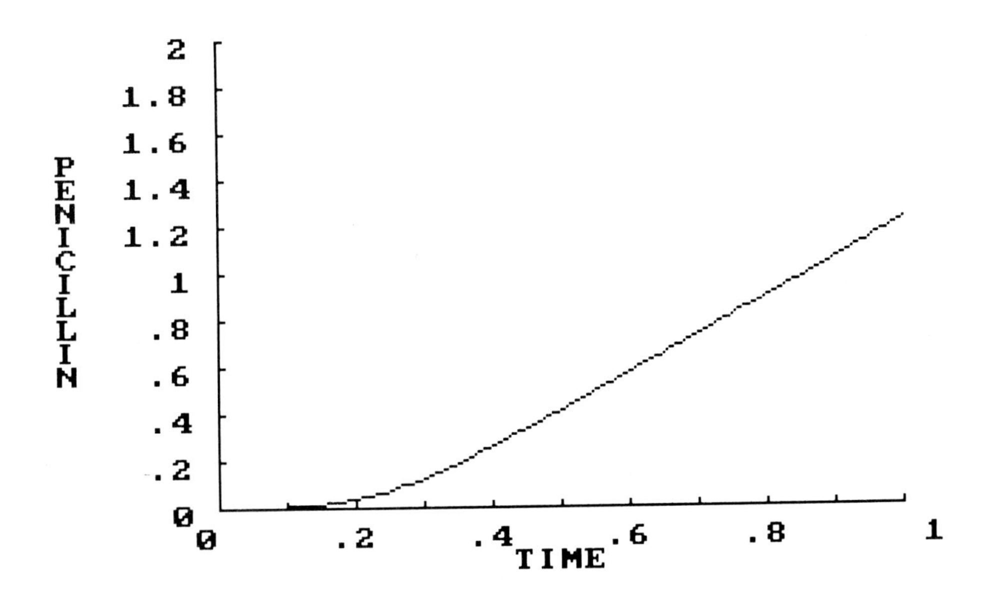


```
            SCALING OPTIONS

    1. MANUAL CHOICE OF MAXIMA AND MINIMA

    2. AUTOMATIC CHOICE OF MAXIMA AND MINIMA

ENTER YOUR CHOICE(1-2)? 1
          SCALES FOR VARIABLE  3

MINIMUM X=? 0

MAXIMUM X=? 1

SPECIFY X-AXIS TITLE:TIME

MINIMUM Y=? 0

MAXIMUM Y=? 4

SPECIFY Y-AXIS TITLE:1st ADJOINT


            SCALING OPTIONS

    1. MANUAL CHOICE OF MAXIMA AND MINIMA

    2. AUTOMATIC CHOICE OF MAXIMA AND MINIMA

ENTER YOUR CHOICE(1-2)? 1
```

```
          SCALES FOR VARIABLE  4

MINIMUM X=? 0

MAXIMUM X=? 1

SPECIFY X-AXIS TITLE:TIME

MINIMUM Y=? 0

MAXIMUM Y=? 2

SPECIFY Y-AXIS TITLE:2nd ADJOINT
```

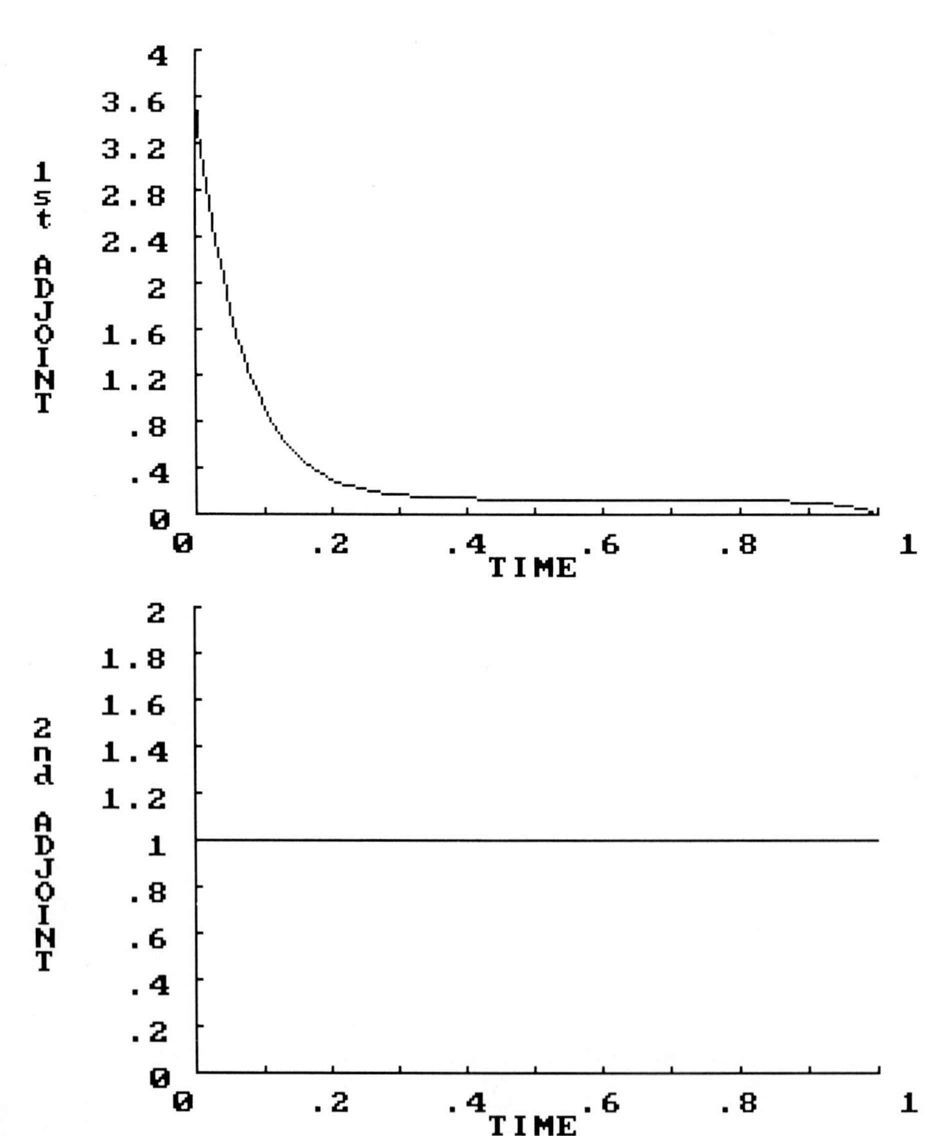

```
PLOTTING OPTIONS FOR THE CONTROL VARIABLE

            SCALING OPTIONS

    1. MANUAL CHOICE OF MAXIMA AND MINIMA

    2. AUTOMATIC CHOICE OF MAXIMA AND MINIMA

ENTER YOUR CHOICE(1-2)? 1

          SCALES FOR ALL VARIABLES

MINIMUM X=? 0

MAXIMUM X=? 1

SPECIFY X-AXIS TITLE:TIME

MINIMUM Y=? 20

MAXIMUM Y=? 30

SPECIFY Y-AXIS TITLE:TEMPERATURE
```

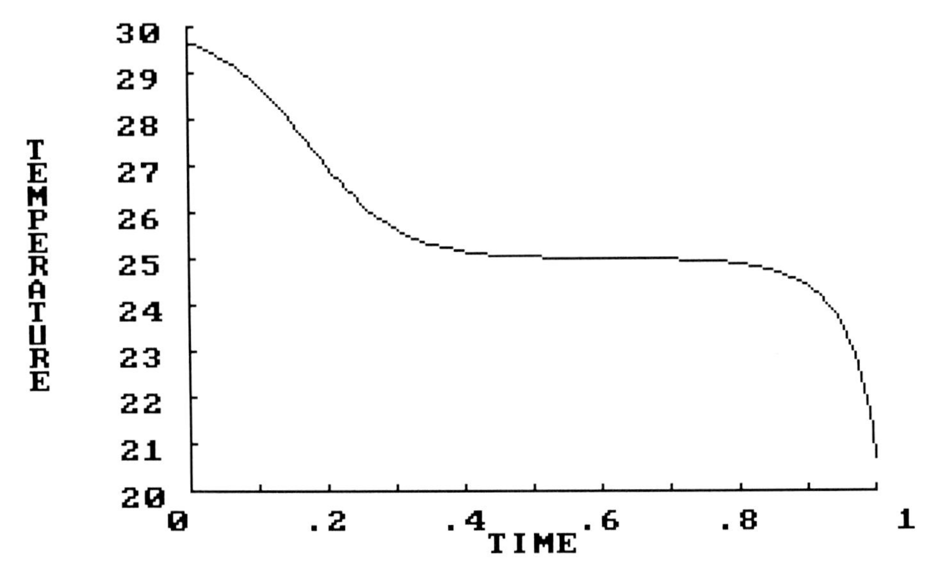


```
                    RERUN OPTIONS

          1. SHOW SAME RESULTS AGAIN

          2. RESET INTEGRATION PARAMETERS AND RECALCULATE

          3. END PROGRAM

   ENTER YOUR CHOICE(1-3)? 3
Ok
```

DISCUSSION OF RESULTS The choice of values for the missing initial conditions for $y_3$ and $y_4$ is an important factor in the convergence of the Newton method. The value of $\gamma_3 = 5.0$ was chosen as the guessed initial condition for $y_3$ after some trial and error. The value of $\gamma_4 = 1.0$ was an obvious choice since the right-hand side of Eq. (5) is zero. The Newton method converged to the correct solution in 16 iterations. The converged value of $\gamma_3$ is 3.6121.

The table of results lists the values of the independent and dependent variables (system and variational) for the period of integration $0 \le t \le 1$. The accompanying graphs show the profiles of the four system variables and the optimal control variable (temperature). For this particular formulation of the penicillin fermentation, the maximum principle indicates that the optimal temperature profile varies from 30 to 20°C in the pattern shown on the last graph.

## 5-8 ERROR PROPAGATION, STABILITY, AND CONVERGENCE

Topics of paramount importance in the numerical integration of differential equations are the *error propagation*, *stability*, and *convergence* of these solutions. Two types of stability considerations enter in the solution of ordinary differential equations: *inherent stability* (or instability) and *numerical stability* (or instability). Inherent stability is determined by the mathematical formulation of the problem and is dependent on the eigenvalues of the Jacobian matrix of the differential equations (see Example 5-7). On the other hand, numerical stability is a function of the error propagation in the numerical integration method. The behavior of error propagation depends on the values of the characteristic roots of the difference equations which yield the numerical solution. In this section, we concern ourselves with numerical stability considerations as they apply to the numerical integration of ordinary differential equations.

There are three types of errors which are present in the application of numerical integration methods. These are the *truncation error*, the *roundoff error*, and the *propagation error*. The truncation error is a function of the number of terms which are retained in the approximation of the solution from the infinite series expansion. The truncation error may be reduced by retaining a larger number of terms in the series or by reducing the step size of integration $h$. The plethora of available numerical methods of integration of ordinary differential equations provides a choice of increasingly higher accuracy (lower truncation error), at an escalating cost in the number of arithmetic operations to be performed, and with the concomitant accumulation of roundoff errors.

In Chap. 1 we mentioned that computers carry numbers using a finite number of significant figures. A roundoff error is introduced in the calculation when the computer rounds up or down (or just chops) the number to $n$ significant figures. Roundoff errors may be reduced significantly by the use of double precision. However, even a very small roundoff error may affect the accuracy of the solution, especially in numerical integration methods which march forward (or backward) for hundreds or thousands of steps, each step being performed using rounded numbers.

The truncation and roundoff errors in numerical integration accumulate and propagate, creating the propagation error, which, in some cases, may grow in exponential or oscillatory pattern, thus invalidating the correct solution.

Figure 5-12 illustrates the propagation of error in the Euler integration method. Starting with a known initial condition $y_0$, the method calculates the value $y_i$, which contains the truncation error for this step and a small roundoff error introduced by the computer. The error has been magnified in order to illustrate it more clearly. The next step starts with $y_1$ as the initial point and calculates $y_2$. But since $y_1$ already contains truncation and roundoff errors, the value obtained for $y_2$ contains these errors propagated, in addition to the new truncation and roundoff errors from the second step. The same process occurs in subsequent steps.

Error propagation in numerical integration methods is a complex operation which depends on several factors. Roundoff error, which contributes to propagation error, is entirely determined by the accuracy of the computer being used. The truncation error is fixed by the choice of method being applied, by the step size of integration, and by the values of the derivatives of the functions being integrated. For these reasons, it is necessary to examine error propagation and stability of each method individually and in connection with the differential equations to be integrated. Some techniques work well with one class of differential equations but fail with others.

In the sections that follow, we examine systematically the error propagation and stability of several numerical integration methods and suggest ways of reducing these errors by the appropriate choice of step size and integration algorithm.

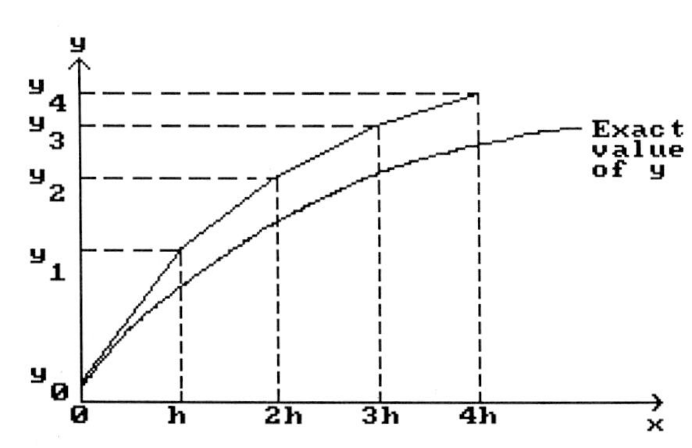


**Figure 5-12** Error propagation of the Euler method.

### 5-8.1 Stability and Error Propagation of Euler Methods

Let us consider the initial-value differential equation in the linear form

$$\frac{dy}{dx} = \lambda y \tag{5-228}$$

where the initial condition is given as

$$y(x_0) = y_0 \tag{5-229}$$

We assume that $\lambda$ is real and $y_0$ is finite. The analytical solution of this differential equation is

$$y(x) = y_0 e^{\lambda x} \tag{5-230}$$

This solution is *inherently stable* for $\lambda < 0$. Under these conditions

$$\lim_{x \to \infty} y(x) = 0 \tag{5-231}$$

Next, we examine the stability of the numerical solution of this problem obtained from using the explicit Euler method. Momentarily we ignore the truncation and roundoff errors. Applying Eq. (5-115), we obtain the recurrence equation

$$y_{n+1} = y_n + h\lambda y_n \tag{5-232}$$

which rearranges to the following *first-order homogeneous difference equation*:

$$y_{n+1} - (1 + h\lambda)y_n = 0 \tag{5-233}$$

Using the methods described in Sec. 4-6, we obtain the characteristic equation

$$E - (1 + h\lambda) = 0 \tag{5-234}$$

whose root is

$$\mu_1 = (1 + h\lambda) \tag{5-235}$$

From this, we obtain the solution of the difference equation (5-233) as

$$y_n = C(1 + h\lambda)^n \tag{5-236}$$

The constant $C$ is calculated from the initial condition, at $x = x_0$,

$$n = 0 \qquad y_n = y_0 = C \tag{5-237}$$

Therefore, the final form of the solution is

$$y_n = y_0(1 + h\lambda)^n \tag{5-238}$$

The differential equation is an initial-value problem; therefore, $n$ can increase without bound. Since the solution $y_n$ is a function of $(1 + h\lambda)^n$, its behavior is determined by the value of $1 + h\lambda$. A numerical solution is said to be *absolutely stable* if

$$\lim_{n\to\infty} y_n = 0 \tag{5-239}$$

The solution of the differential equation (5-228) using the explicit Euler method is absolutely stable if

$$|1 + h\lambda| \leq 1 \tag{5-240}$$

Since $1 + h\lambda$ is the root of the characteristic equation (5-234), an alternative definition of absolute stability is

$$|\mu_i| \leq 1 \qquad i = 1, 2, \ldots, k \tag{5-241}$$

where more than one root exists in the multistep numerical methods.

Returning to the problem at hand, the inequality (5-240) is rearranged to

$$-2 \leq h\lambda \leq 0 \tag{5-242}$$

This inequality sets the limits of the integration step size for a stable solution as follows: Since $h$ is positive, then $\lambda < 0$ and

$$h \leq \frac{2}{|\lambda|} \tag{5-243}$$

Inequality (5-243) is a finite *general stability boundary*, and for this reason, the explicit Euler method is called *conditionally stable*. Any method with an infinite general stability boundary can be called *unconditionally stable* [2].

At the outset of our discussion, we assumed that $\lambda$ was real in order to simplify the derivation. This assumption is not necessary: $\lambda$ can be a complex number. In the earlier discussion of the stability of difference equations (Sec. 4-6) we mentioned that "a solution is stable converging with damped oscillations when complex roots are present and the moduli of the roots are less than or equal to unity."

$$|r| \leq 1 \tag{5-244}$$

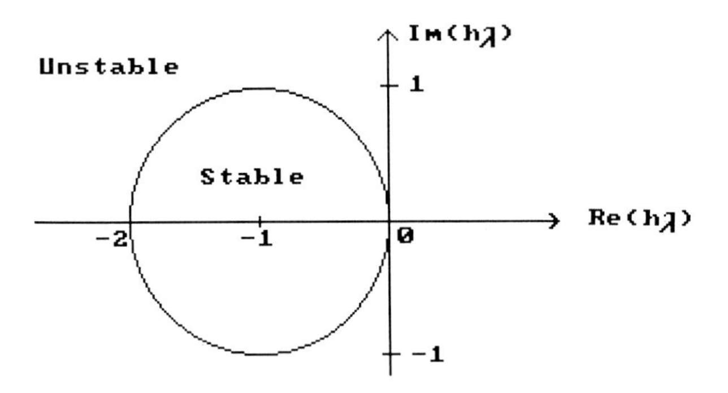


**Figure 5-13** Stability region in the complex plane for the Euler method.

The two inequalities (5-242) and (5-244) describe the circle with a radius of unity on the complex plane shown in Fig. 5-13. The set of values of $h\lambda$ inside the circle yields stable numerical solutions of Eq. (5-228) using the Euler integration method.

We now return to the consideration of the truncation and roundoff errors of the Euler method and develop a difference equation which describes the propagation of the error in the numerical solution. We work with the nonlinear form of the initial-value problem

$$\frac{dy}{dx} = f(x, y) \tag{5-245}$$

where the initial condition is given by

$$y(x_0) = y_0 \tag{5-246}$$

We define the accumulated error of the numerical solution at step $n + 1$ as

$$\epsilon_{n+1} = y_{n+1} - y(x_{n+1}) \tag{5-247}$$

where $y(x_{n+1})$ is the *exact* value of $y$, and $y_{n+1}$ is the *calculated value of* $y$ at $x_{n+1}$. We then write the exact solution $y(x_{n+1})$ as a Taylor series expansion, showing as many terms as needed for the Euler method:

$$y(x_{n+1}) = y(x_n) + hf(x_n, y(x_n)) + T_{E,\,n+1} \tag{5-248}$$

where $T_{E,\,n+1}$ is the local truncation error for step $n + 1$. We also write the calculated value $y_{n+1}$ obtained from the explicit Euler formula

$$y_{n+1} = y_n + hf(x_n, y_n) + R_{E,n+1} \tag{5-249}$$

where $R_{E,n+1}$ is the roundoff error introduced by the computer in step $n+1$.

Combining Eqs. (5-247) to (5-249) we have

$$\epsilon_{n+1} = y_n - y(x_n) + h[f(x_n, y_n) - f(x_n, y(x_n))] - T_{E,n+1} + R_{E,n+1} \tag{5-250}$$

which simplifies to

$$\epsilon_{n+1} = \epsilon_n + h[f(x_n, y_n) - f(x_n, y(x_n))] - T_{E,n+1} + R_{E,n+1} \tag{5-251}$$

The mean-value theorem

$$f(x_n, y_n) - f(x_n, y(x_n)) = \left.\frac{\partial f}{\partial y}\right|_{\alpha, x_n} [y_n - y(x_n)] \qquad y_n < \alpha < y(x_n) \tag{5-252}$$

can be used to further modify the error equation (5-251) to

$$\epsilon_{n+1} - \left[1 + h \left.\frac{\partial f}{\partial y}\right|_{\alpha, x_n}\right]\epsilon_n = -T_{E,n+1} + R_{E,n+1} \tag{5-253}$$

This is a *first-order nonhomogeneous difference equation with varying coefficients* which can be solved only by iteration. However, by making the following simplifying assumptions,

$$\begin{aligned} T_{E,n+1} &= T_E = \text{constant} \\ R_{E,n+1} &= R_E = \text{constant} \\ \left.\frac{\partial f}{\partial y}\right|_{\alpha, x_n} &= \lambda = \text{constant}\dagger \end{aligned} \tag{5-254}$$

Eq. (5-253) simplifies to

$$\epsilon_{n+1} - (1 + h\lambda)\epsilon_n = -T_E + R_E \tag{5-255}$$

whose solution is given by the sum of the homogeneous and particular solutions [22, p. 34]

$$\epsilon_n = C_1(1 + h\lambda)^n + \frac{-T_E + R_E}{1 - (1 + h\lambda)} \tag{5-256}$$

† Under this assumption, Eq. (5-245) becomes identical to Eq. (5-228).

Comparison of Eqs. (5-233) and (5-255) reveals that the characteristic equations for the solution $y_n$ and the error $\epsilon_n$ are identical. The truncation and roundoff error terms in (5-255) introduce the particular solution. The constant $C_1$ is calculated by assuming that the initial condition of the differential equation has no error, that is, $\epsilon_0 = 0$. The final form of the equation which describes the behavior of the propagation error is

$$\epsilon_n = \frac{-T_E + R_E}{h\lambda} [(1 + h\lambda)^n - 1] \tag{5-257}$$

A great deal of insight can be gained by examining Eq. (5-257) thoroughly. As expected, the value of $1 + h\lambda$ is the determining factor in the behavior of the propagation error. Consider first the case of a fixed finite step size $h$, with the number of integration steps increasing to a very large $n$. The limit on the error as $n \to \infty$ is

$$\lim_{n\to\infty} |\epsilon_n| = \frac{-T_E + R_E}{h\lambda} \qquad \text{for } |1 + h\lambda| < 1 \tag{5-258}$$

$$\lim_{n\to\infty} |\epsilon_n| = \infty \qquad \text{for } |1 + h\lambda| > 1 \tag{5-259}$$

In the first case, $\lambda < 0$, $0 < h < 2/|\lambda|$, the error is bounded and the numerical solution is stable. The numerical solution differs from the exact solution by only the finite quantity $(-T_E + R_E)/h\lambda$, which is a function of the truncation error, the roundoff error, the step size, and the eigenvalue of the differential equation.

In the second case, $\lambda > 0$, $h > 0$, the error is unbounded and the numerical solution is unstable. In this case, however, the exact solution is *inherently unstable*. For this reason we introduce the concept of *relative error* defined as

$$\text{Relative error} = \frac{\epsilon_n}{y_n} \tag{5-260}$$

Utilizing Eqs. (5-238) and (5-257) we obtain the relative error as

$$\frac{\epsilon_n}{y_n} = \frac{-T_E + R_E}{y_0 h\lambda} \left[1 - \frac{1}{(1 + h\lambda)^n}\right] \tag{5-261}$$

The relative error is bounded for $\lambda > 0$ and unbounded for $\lambda < 0$. So we conclude that for inherently stable differential equations, the absolute propagation error is the pertinent criterion for numerical stability, while for inherently unstable differential equations, the relative propagation error must be investigated.

Let us now consider a fixed interval of integration, $0 \le x \le \alpha$, so that

$$h = \frac{\alpha}{n} \tag{5-262}$$

and we increase the number of integration steps to a very large $n$. This, of course, causes $h \to 0$. A numerical method is said to be *convergent* if

$$\lim_{h \to 0} |\epsilon_n| = 0 \tag{5-263}$$

In the absence of roundoff error, the Euler method, and most other integration methods, are convergent because

$$\lim_{h \to 0} T_E = 0 \tag{5-264}$$

and

$$\lim_{h \to 0} |\epsilon_n| = 0 \tag{5-265}$$

However, roundoff error is *never* absent in numerical calculations. As $h \to 0$ the roundoff error is *the* crucial factor in the propagation of error:

$$\lim_{h \to 0} |\epsilon_n| = R_E \lim_{h \to 0} \frac{(1 + h\lambda)^n - 1}{h\lambda} \tag{5-266}$$

Application of L'Hospital's rule shows that the roundoff error propagates unbounded as the number of integration steps becomes very large:

$$\lim_{h \to 0} \epsilon_n = R_E[\infty] \tag{5-267}$$

This is the "catch 22" of numerical methods: A smaller step size of integration reduces the truncation error but requires a larger number of steps, thereby increasing the roundoff error.

A similar analysis of the *implicit Euler method* (backward Euler) results in the following two equations, for the solution

$$y_{n+1} = \frac{y_0}{(1 - h\lambda)^n} \tag{5-268}$$

and the propagation error

$$\epsilon_{n+1} = \frac{-T_E + R_E}{h\lambda}(1 - \lambda h)\left[\frac{1}{(1 - h\lambda)^n} - 1\right] \tag{5-269}$$

For $\lambda < 0$ and $0 < h < \infty$, the solution is stable,

$$\lim_{n \to \infty} y_n = 0 \tag{5-270}$$

and the error is bounded:

$$\lim_{n \to \infty} \epsilon_n = -\frac{-T_E + R_E}{h\lambda}(1 - \lambda h) \tag{5-271}$$

No limitation is placed on the step size; therefore, the implicit Euler method is *unconditionally stable* for $\lambda < 0$. On the other hand, when $\lambda > 0$ the following inequality must be true for a stable solution:

$$|1 - h\lambda| \leq 1 \tag{5-272}$$

This imposes the limit on the step size:

$$-2 \le -h\lambda \le 0 \tag{5-273}$$

It can be concluded that the implicit Euler method has a wider range of stability than the explicit Euler method (see Table 5-6).

## 5-8.2 Stability and Error Propagation of Runge-Kutta Methods

Using methods parallel to those of the previous section, the recurrence equations and the corresponding roots for the Runge-Kutta methods can be derived [2]. For the differential equation (5-228), these are

Second-order Runge-Kutta:

$$y_{n+1} = (1 + h\lambda + \tfrac{1}{2}h^2\lambda^2)y_n \tag{5-274}$$

$$\mu_1 = 1 + h\lambda + \tfrac{1}{2}h^2\lambda^2 \tag{5-275}$$

Third-order Runge-Kutta:

$$y_{n+1} = (1 + h\lambda + \tfrac{1}{2}h^2\lambda^2 + \tfrac{1}{6}h^3\lambda^3)y_n \tag{5-276}$$

$$\mu_1 = 1 + h\lambda + \tfrac{1}{2}h^2\lambda^2 + \tfrac{1}{6}h^3\lambda^3 \tag{5-277}$$

Fourth-order Runge-Kutta:

$$y_{n+1} = (1 + h\lambda + \tfrac{1}{2}h^2\lambda^2 + \tfrac{1}{6}h^3\lambda^3 + \tfrac{1}{24}h^4\lambda^4)y_n \tag{5-278}$$

$$\mu_1 = 1 + h\lambda + \tfrac{1}{2}h^2\lambda^2 + \tfrac{1}{6}h^3\lambda^3 + \tfrac{1}{24}h^4\lambda^4 \tag{5-279}$$

The condition for absolute stability

$$|\mu_i| \le 1 \qquad i = 1, 2, \ldots, k \tag{5-241}$$

applies to all the above methods. The absolute real stability boundaries for these methods are listed in Table 5-6, and the regions of stability in the complex plane are shown on Fig. 5-14.

**Table 5-6 Real stability boundaries**

| Method | Boundary |
|---|---|
| Explicit Euler | $-2 \le h\lambda \le 0$ |
| Implicit Euler | $\begin{cases} 0 < h < \infty & \text{for } \lambda < 0 \\ -2 \le -h\lambda \le 0 & \text{for } \lambda > 0 \end{cases}$ |
| Second-order Runge-Kutta | $-2 \le h\lambda \le 0$ |
| Third-order Runge-Kutta | $-2.5 \le h\lambda \le 0$ |
| Fourth-order Runge-Kutta | $-2.785 \le h\lambda \le 0$ |

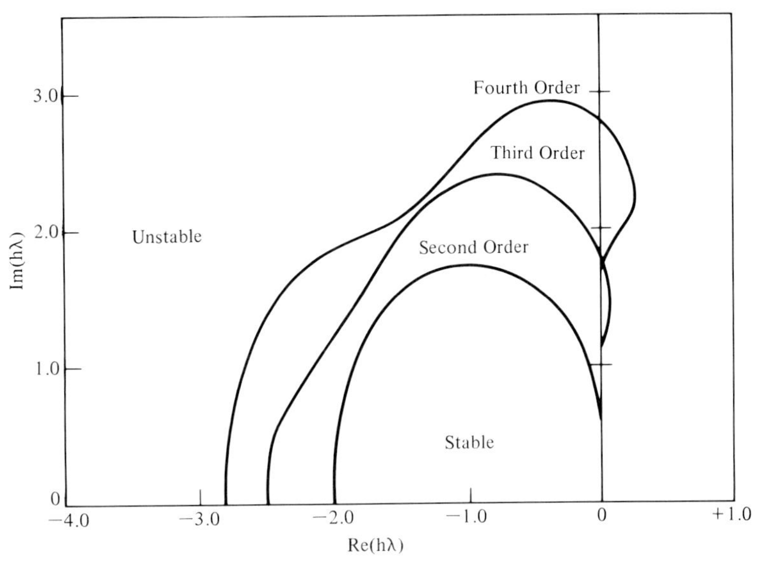


**Figure 5-14** Stability regions in the complex plane for Runge-Kutta methods of order 2, 3, and 4. *(Adapted from L. Lapidus and J.H. Seinfeld, Ref. 2, by permission of the publisher.)*

### 5-8.3 Step-Size Control

The discussion of stability analysis in the previous sections made the simplifying assumption that the value of $\lambda$ remains constant throughout the integration. This is true for linear equations such as (5-228); however, for the nonlinear equation (5-245), the value of $\lambda$ may vary considerably over the interval of integration. The step size of integration must be chosen using the maximum possible value of $\lambda$, thus resulting in the minimum step size. This, of course, will guarantee stability at the expense of computation time. For problems in which computation time becomes excessive, it is possible to develop strategies for automatically adjusting the step size at each step of the integration [23]. The recent improvements in the computational speed of microcomputers, as well as supercomputers, have reduced the need for automatic step-size control for most engineering problems.

## 5-9 STIFF DIFFERENTIAL EQUATIONS

In Sec. 5-8 we showed that the stability of the numerical solution of differential equations depends on the value of $h\lambda$ and that $\lambda$ together with the stability boundary of the method determine the step size of integration. In the case of the linear differential equation

$$\frac{dy}{dx} = \lambda y \tag{5-228}$$

$\lambda$ is the eigenvalue of that equation, and it remains a constant throughout the integration. The nonlinear differential equation

$$\frac{dy}{dx} = f(x, y) \tag{5-245}$$

can be linearized at each step using the mean-value theorem (5-252), so that $\lambda$ can be obtained from the partial derivative of the function with respect to $y$:

$$\lambda = \left. \frac{\partial f}{\partial y} \right|_{\alpha, x_n} \tag{5-280}$$

The value of $\lambda$ is no longer a constant but varies in magnitude at each step of the integration.

This analysis can be extended to a set of simultaneous nonlinear differential equations:

$$\begin{aligned} \frac{dy_1}{dx} &= f_1(x, y_1, y_2, \ldots, y_n) \\ \frac{dy_2}{dx} &= f_2(x, y_1, y_2, \ldots, y_n) \\ &\vdots \\ \frac{dy_n}{dx} &= f_n(x, y_1, y_2, \ldots, y_n) \end{aligned} \tag{5-281}$$

Linearization of the set produces the Jacobian matrix

$$\mathbf{J} = \begin{bmatrix} \dfrac{\partial f_1}{\partial y_1} & \cdots & \dfrac{\partial f_1}{\partial y_n} \\ \cdots & \cdots & \cdots \\ \dfrac{\partial f_n}{\partial y_1} & \cdots & \dfrac{\partial f_n}{\partial y_n} \end{bmatrix} \tag{5-282}$$

The eigenvalues $\{\lambda_i \mid i = 1, 2, \ldots, n\}$ of the Jacobian matrix are the determining factors in the stability analysis of the numerical solution. The step size of integration is determined by the stability boundary of the method and the maximum eigenvalue.

When the eigenvalues of the Jacobian matrix of the differential equations are all of the same order of magnitude, no unusual problems arise in the integration of the set. However, when the maximum eigenvalue is several orders of magnitude larger than the minimum eigenvalue, the equations are said to be *stiff*. The *stiffness ratio* (SR) of such a set is defined as

$$\mathrm{SR} = \frac{\max\limits_{1 \le k \le n} |\mathrm{Re}\,(\lambda_k)|}{\min\limits_{1 \le k \le n} |\mathrm{Re}\,(\lambda_k)|} \tag{5-283}$$

The step size of integration is determined by the largest eigenvalue and the final time of integration is usually fixed by the smallest eigenvalue; therefore, integration of differential equations using explicit methods may be time-intensive. Finlayson [8] recommends using implicit methods for integrating stiff differential equations in order to reduce computation time.

# PROBLEMS

**5-1** Derive the second-order Runge-Kutta method of Eq. (5-147) using central differences.

**5-2** The solution of the following second-order linear ordinary differential equation

$$\frac{d^2x}{dt^2} - 3\frac{dx}{dt} - 10x = 0$$

is to be determined using numerical techniques. The initial conditions for this equation are, at $t = 0$,

$$x\Big|_0 = 3 \qquad \text{and} \qquad \frac{dx}{dt}\Big|_0 = 15$$

(*a*) Transform the above differential equation into a set of first-order linear differential equations with appropriate initial conditions.

(*b*) Find the solution using eigenvalues and eigenvectors, and evaluate the variables in the range $0 \leq t \leq 1.0$.

(*c*) Use the fourth-order Runge-Kutta method to verify the results of part *b*.

**5-3** A radioactive material (A) decomposes according to the series reaction:

$$\mathrm{A} \xrightarrow{k_1} \mathrm{B} \xrightarrow{k_2} \mathrm{C}$$

where $k_1$ and $k_2$ are the rate constants and B and C are the intermediate and final products, respectively. The rate equations are

$$\frac{dC_A}{dt} = -k_1 C_A$$

$$\frac{dC_B}{dt} = k_1 C_A - k_2 C_B$$

$$\frac{dC_C}{dt} = k_2 C_B$$

where $C_A$, $C_B$, and $C_C$ are the concentrations of materials A, B, and C, respectively. The values of the rate constants are

$$k_1 = 3\ \mathrm{s}^{-1} \qquad k_2 = 1\ \mathrm{s}^{-1}$$

Initial conditions are

$$C_A(0) = 1\ \mathrm{lb \cdot mol/ft^3}$$

$$C_B(0) = 0$$

$$C_C(0) = 0$$

(*a*) Use the eigenvalue-eigenvector method to determine the concentrations $C_A$, $C_B$, and $C_C$ as a function of time $t$.

(*b*) At time $t = 1$ s and $t = 10$ s, what are the concentrations of A, B, and C?

(*c*) Sketch the concentration profiles for A, B, and C.

**5-4** (*a*) Integrate the following differential equations:

$$\frac{dC_A}{dt} = -4C_A + C_B \qquad C_A(0) = 100.0$$

$$\frac{dC_B}{dt} = 4C_A - 4C_B \qquad C_B(0) = 0.0$$

for the time period $0 \le t \le 5$, using (1) the Euler predictor-corrector method, (2) the fourth-order Runge-Kutta method.

(*b*) Which method would give a solution closer to the analytical solution?

(*c*) Why do these methods give different results?

**5-5** In the study of fermentation kinetics, the logistic law

$$\frac{dy_1}{dt} = k_1 y_1\left(1 - \frac{y_1}{k_2}\right)$$

has been used frequently to describe the dynamics of cell growth. This equation is a modification of the logarithmic law

$$\frac{dy_1}{dt} = k_1 y_1$$

The term $(1 - y_1/k_2)$ in the logistic law accounts for cessation of growth due to a limiting nutrient.

The logistic law has been used successfully in modeling the growth of *penicillium chrysogenum*, a penicillin-producing organism [20]. In addition, the rate of production of penicillin has been mathematically quantified by the equation

$$\frac{dy_2}{dt} = k_3 y_1 - k_4 y_2$$

Penicillin $(y_2)$ is produced at a rate proportional to the concentration of the cell $(y_1)$ and is degraded by hydrolysis, which is proportional to the concentration of the penicillin itself.

(*a*) Discuss other possible interpretations of the logistic law.

(*b*) Show that $k_2$ is equivalent to the maximum cell concentration which can be reached under given conditions.

(*c*) Apply the fourth-order Runge-Kutta integration method to find the numerical solution of the cell and penicillin equations. Use the following constants and initial conditions:

$$k_1 = 0.03120 \qquad k_3 = 3.374$$

$$k_2 = 47.70 \qquad k_4 = 0.01268$$

at $t = 0$, $y_1(0) = 5.0$, and $y_2(0) = 0.0$; the range of $t$ is $0 \le t \le 212$ h.

**5-6** The conversion of glucose to gluconic acid is a simple oxidation of the aldehyde group of the sugar to a carboxyl group. This transformation can be achieved by a microorganism in a fermentation process. The enzyme glucose oxidase, present in the microorganism, converts glucose to gluconolactone. In turn, the gluconolactone hydrolyzes to form the gluconic acid. The overall mechanism of the fermentation process which performs this transformation can be described as follows:

Cell growth:

$$\text{Glucose} + \text{cells} \rightarrow \text{cells}$$

Glucose oxidation:

$$\text{Glucose} + O_2 \xrightarrow{\text{Glucose oxidase}} \text{gluconolactone} + H_2O_2$$

Gluconolactone hydrolysis:

$$\text{Gluconolactone} + H_2O \rightarrow \text{gluconic acid}$$

Peroxide decomposition:

$$H_2O_2 \xrightarrow{\text{Catalase}} H_2O + \tfrac{1}{2}O_2$$

A mathematical model of the fermentation of the bacterium *Pseudomonas ovalis*, which produces gluconic acid, has been developed by Rai and Constantinides [24]. This model, which describes the dynamics of the logarithmic growth phase, can be summarized as follows:

Rate of cell growth:

$$\frac{dy_1}{dt} = b_1 y_1 \left(1.0 - \frac{y_1}{b_2}\right)$$

Rate of gluconolactone formation:

$$\frac{dy_2}{dt} = \frac{b_3 y_1 y_4}{b_4 + y_4} - 0.9082 b_5 y_2$$

Rate of gluconic acid formation:

$$\frac{dy_3}{dt} = b_5 y_2$$

Rate of glucose consumption:

$$\frac{dy_4}{dt} = -1.011\left(\frac{b_3 y_1 y_4}{b_4 + y_4}\right)$$

where $y_1$ = concentration of cell
$y_2$ = concentration of gluconolactone
$y_3$ = concentration of gluconic acid
$y_4$ = concentration of glucose
$b_1$ to $b_5$ = parameters of the system which are functions of temperature and pH

At the operating conditions of 30°C and pH 6.6, the values of the five parameters were determined from experimental data to be

$$b_1 = 0.949 \qquad b_4 = 37.51$$
$$b_2 = 3.439 \qquad b_5 = 1.169$$
$$b_3 = 18.72$$

At these conditions, develop the time profiles of all variables, $y_1$ to $y_4$, for the period $0 \le t \le 9$ h.

The initial conditions at the start of this period are

$$y_1(0) = 0.5\ \text{U.O.D./mL} \qquad y_3(0) = 0.0\ \text{mg/mL}$$

$$y_2(0) = 0.0\ \text{mg/mL} \qquad y_4(0) = 50.0\ \text{mg/mL}$$

**5-7** A gaseous feedstock containing 40% A, 40% B, and 20% inerts will be processed in a reactor where the following chemical reaction takes place:

$$\text{A} + 2\text{B} \rightarrow \text{C}$$

The reaction rate is

$$r = kC_\text{A}C_\text{B}^2$$

where $k = 0.01\ \text{s}^{-1}(\text{g}\cdot\text{mol/L})^{-2}$ at 500°C
$C_A$ = concentration of A, g · mol/L
$C_\text{B}$ = concentration of B, g · mol/L

Choose a basis of 100 g · mol of feed and assume that all gases behave as ideal gases. Calculate the following:

(*a*) The time needed to produce a product containing 11.8% B in a batch reactor operated at 500°C and at constant pressure of 10 atm.

(*b*) The time needed to produce a product containing 11.8% B in a batch reactor operating at constant volume. The temperature of the reactor is 500°C and the initial pressure is 10 atm.

**5-8** The steady-state simulation of continuous contact countercurrent processes involving simultaneous heat and mass transfer may be described as a nonlinear boundary-value problem [14]. For instance, for a continuous adiabatic gas absorption contactor unit, the model can be written in the following form:

$$\frac{dY_A}{dt} = N\left[\frac{x_A J_A}{P} \exp\left(-\frac{A_A}{T_L}\right) - \frac{Y_A}{1 + Y_A + Y_B}\right]$$

$$\frac{dY_B}{dt} = GN\left[\frac{(1 - x_A)J_B}{P} \exp\left(-\frac{A_B}{T_L}\right) - \frac{Y_B}{1 + Y_A + Y_B}\right]$$

$$\frac{dT_G}{dt} = HN(T_L - T_G)$$

$$\frac{dT_L}{dt} = \frac{\phi}{RC_L}\frac{dY_A}{dt} + \frac{r_{B0} + \mu}{RC_L}\frac{dY_B}{dt} + \frac{C_G}{C_L}\frac{dT_G}{dt}$$

$$\frac{dR}{dt} = \frac{dY_A}{dt} + \frac{dY_B}{dt}$$

$$\frac{dx_A}{dt} = \frac{1}{R}\frac{dY_A}{dt} - \frac{x_A}{R}\frac{dR}{dt}$$

Thermodynamic and physical property data for the system ammonia-air-water are

$$J_A = 1.36 \times 10^{11}\ \text{N/m}^2 \qquad \phi = 1.08 \times 10^5\ \text{J/kmol}$$

$$A_A = 4.212 \times 10^3\ \text{K} \qquad r_{B0} = 1.36 \times 10^5\ \text{J/kmol}$$

$$J_B = 6.23 \times 10^{10}\ \text{N/m}^2 \qquad \mu = 0.0\ \text{J/mol}$$

$$A_B = 5.003 \times 10^3\ \text{K} \qquad C_L = 232\ \text{J/kmol}$$

$$G = 1.41 \qquad C_G = 93\ \text{J/kmol}$$

$$H = 1.11 \qquad P = 10^5\ \text{N/m}^2$$

$$N = 10$$

The inlet stream conditions are

$$Y_A(0) = 0.05 \qquad T_L(1) = 293$$

$$Y_B(0) = 0.0 \qquad R(1) = 1.0$$

$$T_G(0) = 298 \qquad x_A(1) = 0.0$$

Calculate the profiles of all dependent variables using the Newton method.

**5-9** The data of Table P5-9 are estimates of the populations of moose and wolves on Isle Royale National Park for the period 1960–1983 [Refs. 10 and 11]. Based on the discussion of Example 5-7, extend the Lotka-Volterra predator-prey equations to simulate more accurately the dynamics of these two populations. Introduce the effect of overcrowding, account for at least one additional source of food for the predator, or attempt to quantify other influences you believe are important in the life cycle of these two species.

**Table P5-9**

| Year | Moose Population | Wolf population |
|---|---|---|
| 1960 | 700 | 22 |
| 1961 | — | 22 |
| 1962 | — | 23 |
| 1963 | — | 20 |
| 1964 | — | 25 |
| 1965 | — | 28 |
| 1966 | 881 | 24 |
| 1967 | — | 22 |
| 1968 | 1000 | 22 |
| 1969 | 1150 | 17 |
| 1970 | 966 | 18 |
| 1971 | 674 | 20 |
| 1972 | 836 | 23 |
| 1973 | 802 | 24 |
| 1974 | 815 | 30 |
| 1975 | 778 | 41 |
| 1976 | 641 | 43 |
| 1977 | 507 | 33 |
| 1978 | 543 | 40 |
| 1979 | 675 | 42 |
| 1980 | 577 | 50 |
| 1981 | 570 | 30 |
| 1982 | 590 | 13 |
| 1983 | 811 | 23 |

**Table P5-10**

| Time, s | $c(t)$, mg/L |
|---|---|
| 0 | 1 |
| 1 | 2 |
| 2 | 4 |
| 3 | 7 |
| 4 | 6 |
| 5 | 5 |
| 6 | 2 |
| 7 | 1 |
| 8 | 0 |

**5-10** In studying the mixing characteristics of chemical reactors, a sharp pulse of a nonreacting tracer material is injected into the reactor at time $t = 0$. The concentration of material in the effluent from the reactor is measured as a function of time $c(t)$. The *exit age distribution* for the reactor is defined as

$$E(t) = \frac{c(t)}{\int_0^\infty c(t)\,dt}$$

and the *cumulative age distribution function* is defined as

$$F(t) = \int_0^t E(t)\,dt$$

The *mean residence time* of the reactor is calculated from

$$\theta = \frac{V}{q} = \int_0^\infty tE(t)\,dt$$

where $V$ is the volume of the reactor and $q$ is the flow rate.

The exit concentration data shown in Table P5-10 were obtained from a tracer experiment studying the mixing characteristics of a continuous flow reactor.

Calculate the exit age distribution, the cumulative age distribution, and the mean residence time of this reactor.

# REFERENCES


1. Himmelblau, D. M., and Bischoff, K. B.: *Process Analysis and Simulation. Deterministic Systems*, John Wiley & Sons, Inc., New York, 1968, p. 330.
2. Lapidus, L., and Seinfeld, J. H.: *Numerical Solution of Ordinary Differential Equations*, Academic Press, Inc., New York, 1971, p. 9.
3. Chapra, S. C., and Canale, R. P.: *Numerical Methods for Engineers with Personal Computer Applications*, McGraw-Hill Book Company, New York, 1985, p. 402.
4. Vichnevetsky, R.: *Computer Methods for Partial Differential Equations*, vol. I, Prentice-Hall, Inc., Englewood Cliffs, N.J., 1981.

5. Carnahan, B., Luther, H. A., and Wilkes, J. O.: *Applied Numerical Methods*, John Wiley & Sons, Inc., New York, 1969.
6. *Tables of Functions and Zeros of Functions*, National Bureau of Standards Applied Math. Series 37, Washington, D.C., 1954.
7. Bird, R. B., Stewart, W. E., and Lightfoot, E. N.: *Transport Phenomena*, John Wiley & Sons, Inc., New York, 1960, p. 124.
8. Finlayson, B. A.: *Nonlinear Analysis in Chemical Engineering*, McGraw-Hill Book Company, New York, 1980, p. 37.
9. Lotka, A. J.: *Elements of Mathematical Biology*, Dover Publications, Inc., New York, 1956.
10. Allen, D. L.: *Wolves of Minong*, Houghton Mifflin Company, Boston, 1973.
11. Peterson, R. O., Page, R. E., and Dodge, K. M.: "Wolves, Moose and the Allometry of Population Cycles," *Science*, vol. 224, 1984, p. 1350.
12. Elliot, J. L.: "Isle Royale: A North Woods Park Primeval," *National Geographic*, vol. 167, April 1985, p. 534.
13. Vandermeer, J.: *Elementary Mathematical Ecology*, John Wiley & Sons, Inc., New York, 1981.
14. Kubíček, M., and Hlaváček, V.: *Numerical Solution of Nonlinear Boundary Value Problems with Applications*, Prentice-Hall, Inc., Englewood Cliffs, N.J., 1983.
15. Aziz, A. K. (ed.): *Numerical Solutions of Boundary Value Problems for Ordinary Differential Equations*, Academic Press, Inc., New York, 1975.
16. Keller, H. B.: *Numerical Solution of Two Point Boundary Value Problems*, SIAM Reg. Conf. Ser. Appl. Math., No. 24, Philadelphia, Pa., 1976.
17. Bryson, Jr., A. E., and Ho, Y-C.: *Applied Optimal Control*, Blaidsdell Publishing Company, Waltham, Mass., 1969.
18. Johnson, L. W., and Riess, R. D.: *Numerical Analysis*, 2d ed., Addison-Wesley Publishing Company, Reading, Mass., 1982, p. 437.
19. Finlayson, B. A.: "Packed Bed Reactor Analysis by Orthogonal Collocation," *Chem. Eng. Sci.*, vol. 26, 1971, p. 1081.
20. Constantinides, A., Spencer, J. L., and Gaden, Jr., E. L.: "Optimization of Batch Fermentation Processes, I. Development of Mathematical Models for Batch Penicillin Fermentations," *Biotech. Bioeng.*, vol. 12, 1970, p. 803.
21. Constantinides, A., Spencer, J. L., and Gaden, Jr., E. L.: "Optimization of Batch Fermentation Processes, II. Optimum Temperature Profiles for Batch Penicillin Fermentations," *Biotech. Bioeng.*, vol. 12, 1970, p. 1081.
22. Lapidus, L.: *Digital Computation for Chemical Engineers*, McGraw-Hill Book Company, New York, 1962, p. 94.
23. Gear, C. W.: "The Automatic Integration of Ordinary Differential Equations," *Commun. Assoc. Comput. Mach.*, vol. 14, 1971, p. 176.
24. Rai, V. R., and Constantinides, A.: "Mathematical Modeling and Optimization of the Gluconic Acid Fermentation," AICHE Symp. Ser., vol. 69, no. 132, 1973, p. 114.

CHAPTER

# SIX

# NUMERICAL SOLUTION OF PARTIAL DIFFERENTIAL EQUATIONS

## 6-1 INTRODUCTION

The laws of conservation of mass, momentum, and energy form the basis of the field of transport phenomena. These laws applied to the flow of fluids result in the *equations of change* which describe the change of velocity, temperature, and concentration with respect to time and position in the system [1]. The dynamics of such systems, which have more than one independent variable, are modeled by *partial differential equations*.

For example, the mass balance

$$\begin{pmatrix}\text{Rate of mass}\\ \text{accumulation}\end{pmatrix} = \begin{pmatrix}\text{rate of}\\ \text{mass in}\end{pmatrix} - \begin{pmatrix}\text{rate of}\\ \text{mass out}\end{pmatrix} \tag{6-1}$$

applied to a stationary volume element $\Delta x\, \Delta y\, \Delta z$ through which pure fluid is flowing (Fig. 6-1) results in the *equation of continuity*†

$$\frac{\partial \rho}{\partial t} = -\left(\frac{\partial}{\partial x}\rho v_x + \frac{\partial}{\partial y}\rho v_y + \frac{\partial}{\partial z}\rho v_z\right) \tag{6-2}$$

where $\rho$ is the density of the fluid, and $v_x$, $v_y$, and $v_z$ are the velocity components in the three rectangular coordinates.

† For detailed derivation of these equations see Bird, Stewart, and Lightfoot [1].

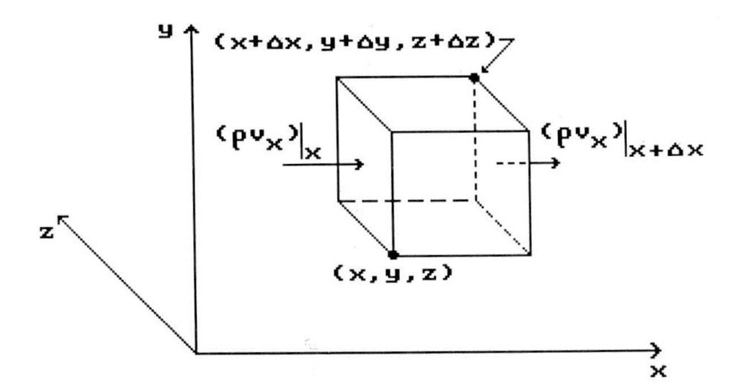


**Figure 6-1** Volume element $\Delta x\, \Delta y\, \Delta z$ for fluid flow.

The application of a momentum balance

$$\begin{pmatrix}\text{Rate of}\\ \text{momentum}\\ \text{accumulation}\end{pmatrix} = \begin{pmatrix}\text{rate of}\\ \text{momentum}\\ \text{in}\end{pmatrix} - \begin{pmatrix}\text{rate of}\\ \text{momentum}\\ \text{out}\end{pmatrix} + \begin{pmatrix}\text{sum of forces}\\ \text{acting on}\\ \text{system}\end{pmatrix} \tag{6-3}$$

on the volume element $\Delta x\, \Delta y\, \Delta z$, for isothermal flow of fluid, yields the *equations of motion* in the three directions

$$\frac{\partial}{\partial t}\rho v_j = -\left(\frac{\partial}{\partial x}\rho v_x v_j + \frac{\partial}{\partial y}\rho v_y v_j + \frac{\partial}{\partial z}\rho v_z v_j\right) - \left(\frac{\partial}{\partial x}\tau_{xj} + \frac{\partial}{\partial y}\tau_{yj} + \frac{\partial}{\partial z}\tau_{zj}\right) - \frac{\partial p}{\partial j} + \rho g_j \qquad j = x,\, y\,,\ \text{or}\ z \tag{6-4}$$

where $\tau_{ij}$ are the components of the shear-stress tensor, $p$ is pressure, and $g_j$ are the components of the gravitational acceleration.

The application of the following energy balance

$$\begin{pmatrix}\text{Rate of}\\ \text{accumulation}\\ \text{of energy}\end{pmatrix} = \begin{pmatrix}\text{rate of}\\ \text{energy in}\\ \text{by convection}\end{pmatrix} - \begin{pmatrix}\text{rate of}\\ \text{energy out}\\ \text{by convection}\end{pmatrix} + \begin{pmatrix}\text{net rate of}\\ \text{heat addition}\\ \text{by conduction}\end{pmatrix} - \begin{pmatrix}\text{net rate of}\\ \text{work done by}\\ \text{system on}\\ \text{surroundings}\end{pmatrix} \tag{6-5}$$

on the volume element $\Delta x\, \Delta y\, \Delta z$, for nonisothermal flow of fluid, results in the *equation of energy*

$$\rho C_v\left(\frac{\partial T}{\partial t} + v_x \frac{\partial T}{\partial x} + v_y \frac{\partial T}{\partial y} + v_z \frac{\partial T}{\partial z}\right) = -\left(\frac{\partial q_x}{\partial x} + \frac{\partial q_y}{\partial y} + \frac{\partial q_z}{\partial z}\right)$$
$$- T\left(\frac{\partial p}{\partial T}\right)_\rho \left(\frac{\partial v_x}{\partial x} + \frac{\partial v_y}{\partial y} + \frac{\partial v_z}{\partial z}\right) - \left(\tau_{xx} \frac{\partial v_x}{\partial x} + \tau_{yy} \frac{\partial v_y}{\partial y} + \tau_{zz} \frac{\partial v_z}{\partial z}\right)$$
$$- \left[\tau_{xy}\left(\frac{\partial v_x}{\partial y} + \frac{\partial v_y}{\partial x}\right) + \tau_{xz}\left(\frac{\partial v_x}{\partial z} + \frac{\partial v_z}{\partial x}\right) + \tau_{yz}\left(\frac{\partial v_y}{\partial z} + \frac{\partial v_z}{\partial y}\right)\right] \tag{6-6}$$

where $T$ is the temperature, $C_v$ is the heat capacity at constant volume, and $q_i$ are the components of the energy flux given by *Fourier's law of heat conduction*:

$$q_x = -k \frac{\partial T}{\partial x}$$
$$q_y = -k \frac{\partial T}{\partial y} \tag{6-7}$$
$$q_z = -k \frac{\partial T}{\partial z}$$

where $k$ is the thermal conductivity.

For heat conduction in solids, where the velocity terms are zero, Eq. (6-6) simplifies considerably. When combined with (6-7) it gives the well-known three-dimensional unsteady-state heat conduction equation

$$\rho C_p \frac{\partial T}{\partial t} = k\left(\frac{\partial^2 T}{\partial x^2} + \frac{\partial^2 T}{\partial y^2} + \frac{\partial^2 T}{\partial z^2}\right) \tag{6-8}$$

where $C_p$, the heat capacity at constant pressure, replaces $C_v$, and $k$ has been assumed to be constant within the solid.

The equation of continuity for component A in a binary mixture (components A and B) of constant fluid density $\rho$ and constant diffusivity $D_{AB}$ is

$$\frac{\partial c_A}{\partial t} + \left(v_x \frac{\partial c_A}{\partial x} + v_y \frac{\partial c_A}{\partial y} + v_z \frac{\partial c_A}{\partial z}\right) = D_{AB}\left(\frac{\partial^2 c_A}{\partial x^2} + \frac{\partial^2 c_A}{\partial y^2} + \frac{\partial^2 c_A}{\partial z^2}\right) + R_A \tag{6-9}$$

where $c_A$ = molar concentration of A and $R_A$ = molar rate of production of component A. This equation reduces to *Fick's second law of diffusion* when $R_A = 0$ and $v_x = v_y = v_z = 0$:

$$\frac{\partial c_A}{\partial t} = D_{AB}\left(\frac{\partial^2 c_A}{\partial x^2} + \frac{\partial^2 c_A}{\partial y^2} + \frac{\partial^2 c_A}{\partial z^2}\right) \tag{6-10}$$

Equation (6-10) is the three-dimensional unsteady-state diffusion equation, which has the same form as the respective heat conduction equation (6-8).

The most commonly encountered partial differential equations in chemical engineering are of first and second order. Our discussion in this chapter focuses on these two categories. In the next two sections we attempt to classify these equations and their boundary conditions, and in the remainder of the chapter we develop the numerical methods, using finite difference and finite element analysis, for the numerical solution of first- and second-order partial differential equations.

## 6-2 CLASSIFICATION OF PARTIAL DIFFERENTIAL EQUATIONS

Partial differential equations are classified according to their *order*, *linearity*, and *boundary conditions*.

The order of a partial differential equation is determined by the highest-order partial derivative present in that equation. Examples of first-, second-, and third-order partial differential equations are given below:

First order:
$$\frac{\partial u}{\partial x} - \alpha \frac{\partial u}{\partial y} = 0 \tag{6-11}$$

Second order:
$$\frac{\partial^2 u}{\partial x^2} + u \frac{\partial u}{\partial y} = 0 \tag{6-12}$$

Third order:
$$\left(\frac{\partial^3 u}{\partial x^3}\right)^2 + \frac{\partial^2 u}{\partial x\, \partial y} + \frac{\partial u}{\partial y} = 0 \tag{6-13}$$

Partial differential equations are categorized into *linear*, *quasilinear*, and *nonlinear* equations. Consider, for example, the following second-order equation:

$$a(\cdot) \frac{\partial^2 u}{\partial y^2} + 2b(\cdot) \frac{\partial^2 u}{\partial x\, \partial y} + c(\cdot) \frac{\partial^2 u}{\partial x^2} + d(\cdot) = 0 \tag{6-14}$$

If the coefficients are constants, or functions of the independent variables only $[(\cdot) \equiv (x, y)]$, then Eq. (6-14) is linear. If the coefficients are functions of the dependent variable and/or any of its derivatives of lower order than that of the differential equation $[(\cdot) \equiv (x, y, u, \partial u/\partial x, \partial u/\partial y)]$ then the equation is quasilinear. Finally, if the coefficients are functions of derivatives of the same order as that of the equation $[(\cdot) \equiv (x, y, u, \partial^2 u/\partial x^2, \partial^2 u/\partial y^2, \partial^2 u/\partial x\, \partial y)]$, then the equation is nonlinear [2]. In accordance with these definitions, Eq. (6-11) is linear, (6-12) is quasilinear, and (6-13) is nonlinear.

Linear second-order partial differential equations in two independent variables are further classified into three *canonical* forms: *elliptic*, *parabolic*, and *hyperbolic*. The general form of this class of equations is

$$a \frac{\partial^2 u}{\partial x^2} + 2b \frac{\partial^2 u}{\partial x\, \partial y} + c \frac{\partial^2 u}{\partial y^2} + d \frac{\partial u}{\partial x} + e \frac{\partial u}{\partial y} + fu + g = 0 \tag{6-15}$$

where the coefficients are either constants or functions of the independent variables only. The three canonical forms are determined by the following criterion:

$$b^2 - ac < 0 \qquad \text{elliptic} \tag{6-16a}$$

$$b^2 - ac = 0 \qquad \text{parabolic} \tag{6-16b}$$

$$b^2 - ac > 0 \qquad \text{hyperbolic} \tag{6-16c}$$

If $g = 0$, then (6-15) is a *homogeneous* differential equation.

The classic examples of second-order partial differential equations which conform to the three canonical forms are

Laplace's equation (elliptic):

$$\frac{\partial^2 u}{\partial x^2} + \frac{\partial^2 u}{\partial y^2} = 0 \tag{6-17}$$

Heat conduction or diffusion equation (parabolic):

$$\alpha \frac{\partial^2 u}{\partial x^2} = \frac{\partial u}{\partial t} \tag{6-18}$$

Wave equation (hyperbolic):

$$a^2 \frac{\partial^2 u}{\partial x^2} = \frac{\partial^2 u}{\partial t^2} \tag{6-19}$$

A similar classification for second-order partial differential equations with *three* independent variables is given by Tychonov and Samarski [3]. This classification includes elliptic, parabolic, hyperbolic, and *ultrahyperbolic*. The majority of partial differential equations in engineering and physics are of second order with two, three, or four independent variables. Most of these equations have canonical forms; however, the names elliptic, parabolic, and hyperbolic have been also applied to equations which are not of second order but which possess similar properties [4].

The method of solution of partial differential equations depends on their canonical form, as will be demonstrated in the rest of this chapter. Since the coefficients of these equations can be functions of the independent variables it is possible that an equation may shift from one canonical form to another over the range of integration of $(x, y)$.

## 6-3 INITIAL AND BOUNDARY CONDITIONS

The initial and boundary conditions associated with the partial differential equations must be specified in order to obtain unique numerical solutions to these equations. In general, boundary conditions for partial differential equations are divided into three categories. These are demonstrated below, using the one-dimensional unsteady-state heat conduction equation

$$\alpha \frac{\partial^2 T}{\partial x^2} = \frac{\partial T}{\partial t} \tag{6-20}$$

This is identical to Eq. (6-18). It is derived from (6-8) by assuming that the temperature gradients in the $y$ and $z$ dimensions are zero. Equation (6-20) essentially describes the change in temperature within a solid slab (e.g., the wall of a furnace), where heat transfer takes place in the $x$-direction (see Fig. 6-2).

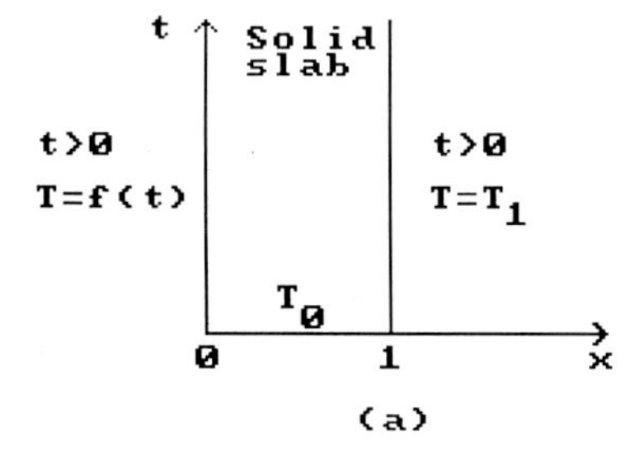


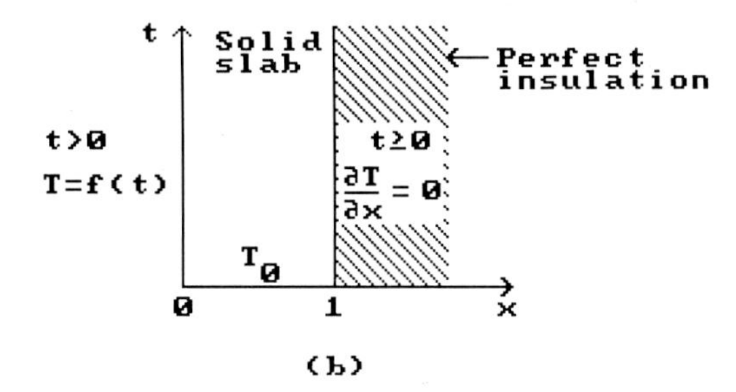


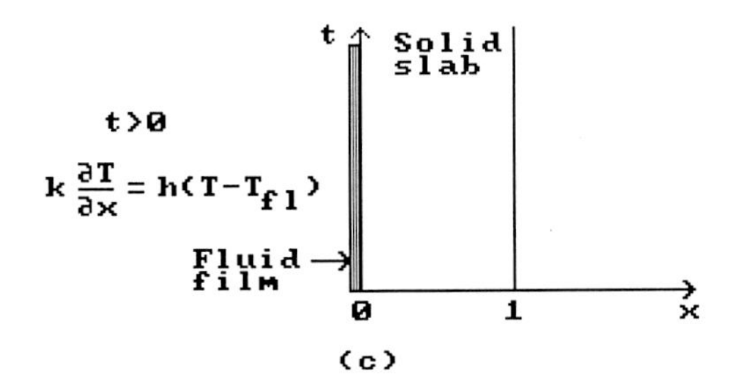


**Figure 6-2** Examples of initial and boundary conditions for the heat conduction problem: (*a*) Dirichlet conditions, (*b*) Cauchy conditions (Dirichlet and Neumann), (*c*) Robbins condition.

Following are the three categories of conditions.

**Dirichlet conditions (first kind)** The values of the dependent variable are given at fixed values of the independent variable. Examples of Dirichlet conditions for the heat conduction equation are

$$T = f(x) \qquad \text{at } t = 0 \text{ and } 0 \le x \le 1$$

or

$$T = T_0 \qquad \text{at } t = 0 \text{ and } 0 \le x \le 1$$

These are alternative initial conditions which specify that the initial temperature inside the slab (wall) is a function of position $f(x)$ or a constant $T_0$ (Fig. 6-2*a*).

Boundary conditions of the first kind are expressed as

$$T = f(t) \qquad \text{at } x = 0 \text{ and } t > 0$$

and

$$T = T_1 \qquad \text{at } x = 1 \text{ and } t > 0$$

These boundary conditions specify the value of the dependent variable at the left boundary as a function of time $f(t)$ (this may be the condition inside a furnace which is maintained at a preprogrammed temperature profile) and at the right boundary as a constant $T_1$ (e.g., the room temperature at the outside of the furnace) (Fig. 6-2*a*).

**Neumann conditions (second kind)** The derivative of the dependent variable is given as a constant or as a function of the independent variable. For example,

$$\frac{\partial T}{\partial x} = 0 \qquad \text{at } x = 1 \text{ and } t \ge 0$$

This condition specifies that the temperature gradient at the right boundary is zero. In the heat conduction problem this can be theoretically accomplished by attaching perfect insulation at the right boundary (Fig. 6-2*b*).

**Cauchy conditions** A problem which combines both Dirichlet and Neumann conditions is said to have Cauchy conditions (Fig. 6-2*b*).

**Robbins conditions (third kind)** The derivative of the dependent variable is given as a function of the dependent variable. For the heat conduction problem the heat flux at the solid-fluid interface may be related to the difference between the temperature at the interface and that in the fluid, i.e.,

$$k \frac{\partial T}{\partial x} = h(T - T_{\text{fluid}}) \qquad \text{at } x = 0 \text{ and } t \ge 0$$

where $h$ is the heat transfer coefficient of the fluid (Fig. 6-2*c*).

On the basis of their initial and boundary conditions, partial differential equations may be further classified into *initial-value* or *boundary-value* problems. In the first case, at least one of the independent variables has an *open region*. In the unsteady-state heat conduction problem the time variable has the range $0 \leq t \leq \infty$, where no condition has been specified at $t = \infty$; therefore, this is an initial-value problem. When the region is *closed* for all independent variables and conditions are specified at all boundaries, then the problem is of the boundary-value type. An example of this is the three-dimensional steady-state heat conduction problem described by the equation

$$\frac{\partial^2 T}{\partial x^2} + \frac{\partial^2 T}{\partial y^2} + \frac{\partial^2 T}{\partial z^2} = 0 \tag{6-21}$$

with the boundary conditions given at all boundaries:

$$\left.\begin{matrix} T(0, y, z) \\ T(1, y, z) \\ T(x, 0, z) \\ T(x, 1, z) \\ T(x, y, 0) \\ T(x, y, 1) \end{matrix}\right\} = \text{specified} \tag{6-22}$$

## 6-4 SOLUTION OF PARTIAL DIFFERENTIAL EQUATIONS USING FINITE DIFFERENCES

### 6-4.1 Finite Difference Approximations

In Chap. 4 we developed the methods of finite differences and demonstrated that ordinary derivatives can be approximated, with any degree of desired accuracy, by replacing the differential operators with finite difference operators.

In this section we apply similar procedures in expressing partial derivatives in terms of finite differences. Since partial differential equations involve more than one independent variable we first establish two-dimensional and three-dimensional grids, in two and three independent variables, respectively, as shown in Fig. 6-3.

The notation $(i, j)$ is used to designate the pivot point for the two-dimensional space and $(i, j, k)$ for the three-dimensional space, where $i$, $j$, and $k$ are the counters in the $x$, $y$, and $z$ directions, respectively. For unsteady-state problems, in which time is one of the independent variables, one of the above counters may be used to designate the time dimension. On the other hand, if all four independent variables are present in the equation, a fourth subscript is added. In order to keep the notation as simple as possible, we add subscripts only when needed.

The distances between grid points are designated as $\Delta x = h$, $\Delta y = k$, and $\Delta z = l$. Again, when time is one of the independent variables, it replaces one

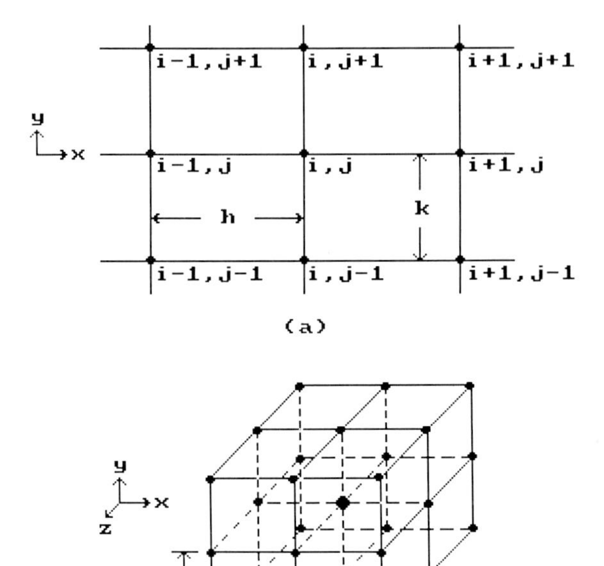


**Figure 6-3** Finite difference grids: (*a*) two-dimensional grid, (*b*) three-dimensional grid.

of the above, or else an additional term is introduced. The use of the letter $k$, both as a counter in the $z$-direction and as an incremental distance in the $y$-direction, may be confusing when these happen to appear in the same equation. However, the reader should remember that when $k$ is used as a counter it appears as a subscript, whereas when it is used as an incremental distance it is a variable.

We now express first, second, and mixed partial derivatives in terms of finite differences. We show the development of these approximations using central differences, and in addition we summarize in tabular form the formulas obtained from using forward and backward differences.

The partial derivative of $u$ with respect to $x$ implies that $y$ and $z$ are held constant; therefore,

$$\left.\frac{\partial u}{\partial x}\right|_{i,j,k} \equiv \left.\frac{du}{dx}\right|_{i,j,k} \tag{6-23}$$

Using Eq. (4-103), which is the approximation of the first-order derivative in terms of central differences, and converting it to the three-dimensional space, we obtain

$$\left.\frac{\partial u}{\partial x}\right|_{i,j,k} = \frac{1}{2h}\left(u_{i+1,j,k} - u_{i-1,j,k}\right) + O(h^2) \tag{6-24}$$

Similarly, the first-order partial derivatives in the $y$- and $z$-directions are given by

$$\left.\frac{\partial u}{\partial y}\right|_{i,j,k} = \frac{1}{2k}(u_{i,j+1,k} - u_{i,j-1,k}) + O(k^2) \tag{6-25}$$

$$\left.\frac{\partial u}{\partial z}\right|_{i,j,k} = \frac{1}{2l}(u_{i,j,k+1} - u_{i,j,k-1}) + O(l^2) \tag{6-26}$$

In an analogous manner, the second-order partial derivatives are expressed in terms of central differences by using Eq. (4-104):

$$\left.\frac{\partial^2 u}{\partial x^2}\right|_{i,j,k} = \frac{1}{h^2}(u_{i+1,j,k} - 2u_{i,j,k} + u_{i-1,j,k}) + O(h^2) \tag{6-27}$$

$$\left.\frac{\partial^2 u}{\partial y^2}\right|_{i,j,k} = \frac{1}{k^2}(u_{i,j+1,k} - 2u_{i,j,k} + u_{i,j-1,k}) + O(k^2) \tag{6-28}$$

$$\left.\frac{\partial^2 u}{\partial z^2}\right|_{i,j,k} = \frac{1}{l^2}(u_{i,j,k+1} - 2u_{i,j,k} + u_{i,j,k-1}) + O(l^2) \tag{6-29}$$

Finally, the mixed partial derivative is developed as follows:

$$\left.\frac{\partial^2 u}{\partial y\, \partial x}\right|_{i,j,k} = \frac{\partial}{\partial y}\left[\left.\frac{\partial u}{\partial x}\right|_{i,j,k}\right]$$

This is equivalent to applying $\partial u/\partial x$ at points $(i, j+1, k)$ and $(i, j-1, k)$, so

$$\begin{aligned}\left.\frac{\partial^2 u}{\partial y\, \partial x}\right|_{i,j,k} &= \frac{1}{2k}\left[\frac{1}{2h}(u_{i+1,j+1,k} - u_{i-1,j+1,k}) \right. \\ &\qquad \left. - \frac{1}{2h}(u_{i+1,j-1,k} - u_{i-1,j-1,k})\right] + O(h^2 + k^2) \\ &= \frac{1}{4hk}(u_{i+1,j+1,k} - u_{i-1,j+1,k} - u_{i+1,j-1,k} + u_{i-1,j-1,k}) + O(h^2 + k^2)\end{aligned} \tag{6-30}$$

The above central difference approximations of partial derivatives are summarized in Table 6-1. The corresponding approximations obtained from using forward and backward differences are shown in Tables 6-2 and 6-3, respectively. Equivalent sets of formulas, which are more accurate than the above, may be developed by using finite difference approximations which have higher accuracies [such as Eqs. (4-105) and (4-106) for central differences, (4-72) and (4-73) for forward differences, and (4-46) and (4-47) for backward

**Table 6-1 Finite difference approximations of partial derivatives using central differences**

| Derivative | Central difference | Error | |
|---|---|---|---|
| $\left.\frac{\partial u}{\partial x}\right\rvert_{i,j,k}$ | $\frac{1}{2h}(u_{i+1,j,k} - u_{i-1,j,k})$ | $O(h^2)$ | (6-24) |
| $\left.\frac{\partial u}{\partial y}\right\rvert_{i,j,k}$ | $\frac{1}{2k}(u_{i,j+1,k} - u_{i,j-1,k})$ | $O(k^2)$ | (6-25) |
| $\left.\frac{\partial u}{\partial z}\right\rvert_{i,j,k}$ | $\frac{1}{2l}(u_{i,j,k+1} - u_{i,j,k-1})$ | $O(l^2)$ | (6-26) |
| $\left.\frac{\partial^2 u}{\partial x^2}\right\rvert_{i,j,k}$ | $\frac{1}{h^2}(u_{i+1,j,k} - 2u_{i,j,k} + u_{i-1,j,k})$ | $O(h^2)$ | (6-27) |
| $\left.\frac{\partial^2 u}{\partial y^2}\right\rvert_{i,j,k}$ | $\frac{1}{k^2}(u_{i,j+1,k} - 2u_{i,j,k} + u_{i,j-1,k})$ | $O(k^2)$ | (6-28) |
| $\left.\frac{\partial^2 u}{\partial z^2}\right\rvert_{i,j,k}$ | $\frac{1}{l^2}(u_{i,j,k+1} - 2u_{i,j,k} + u_{i,j,k-1})$ | $O(l^2)$ | (6-29) |
| $\left.\frac{\partial^2 u}{\partial y\,\partial x}\right\rvert_{i,j,k}$ | $\frac{1}{4hk}(u_{i+1,j+1,k} - u_{i-1,j+1,k} - u_{i+1,j-1,k} + u_{i-1,j-1,k})$ | $O(h^2 + k^2)$ | (6-30) |

**Table 6-2 Finite difference approximations of partial derivatives using forward differences**

| Derivative | Forward difference | Error | |
|---|---|---|---|
| $\left.\frac{\partial u}{\partial x}\right\rvert_{i,j,k}$ | $\frac{1}{h}(u_{i+1,j,k} - u_{i,j,k})$ | $O(h)$ | (6-31) |
| $\left.\frac{\partial u}{\partial y}\right\rvert_{i,j,k}$ | $\frac{1}{k}(u_{i,j+1,k} - u_{i,j,k})$ | $O(k)$ | (6-32) |
| $\left.\frac{\partial u}{\partial z}\right\rvert_{i,j,k}$ | $\frac{1}{l}(u_{i,j,k+1} - u_{i,j,k})$ | $O(l)$ | (6-33) |
| $\left.\frac{\partial^2 u}{\partial x^2}\right\rvert_{i,j,k}$ | $\frac{1}{h^2}(u_{i+2,j,k} - 2u_{i+1,j,k} + u_{i,j,k})$ | $O(h)$ | (6-34) |
| $\left.\frac{\partial^2 u}{\partial y^2}\right\rvert_{i,j,k}$ | $\frac{1}{k^2}(u_{i,j+2,k} - 2u_{i,j+1,k} + u_{i,j,k})$ | $O(k)$ | (6-35) |
| $\left.\frac{\partial^2 u}{\partial z^2}\right\rvert_{i,j,k}$ | $\frac{1}{l^2}(u_{i,j,k+2} - 2u_{i,j,k+1} + u_{i,j,k})$ | $O(l)$ | (6-36) |
| $\left.\frac{\partial^2 u}{\partial y\,\partial x}\right\rvert_{i,j,k}$ | $\frac{1}{hk}(u_{i+1,j+1,k} - u_{i,j+1,k} - u_{i+1,j,k} + u_{i,j,k})$ | $O(h + k)$ | (6-37) |

**Table 6-3 Finite difference approximations of partial derivatives using backward differences**

| Derivative | Backward difference | Error | |
|---|---|---|---|
| $\left.\dfrac{\partial u}{\partial x}\right\|_{i,j,k}$ | $\dfrac{1}{h}(u_{i,j,k} - u_{i-1,j,k})$ | $O(h)$ | (6-38) |
| $\left.\dfrac{\partial u}{\partial y}\right\|_{i,j,k}$ | $\dfrac{1}{k}(u_{i,j,k} - u_{i,j-1,k})$ | $O(k)$ | (6-39) |
| $\left.\dfrac{\partial u}{\partial z}\right\|_{i,j,k}$ | $\dfrac{1}{l}(u_{i,j,k} - u_{i,j,k-1})$ | $O(l)$ | (6-40) |
| $\left.\dfrac{\partial^2 u}{\partial x^2}\right\|_{i,j,k}$ | $\dfrac{1}{h^2}(u_{i,j,k,} - 2u_{i-1,j,k} + u_{i-2,j,k})$ | $O(h)$ | (6-41) |
| $\left.\dfrac{\partial^2 u}{\partial y^2}\right\|_{i,j,k}$ | $\dfrac{1}{k^2}(u_{i,j,k} - 2u_{i,j-1,k} + u_{i,j-2,k})$ | $O(k)$ | (6-42) |
| $\left.\dfrac{\partial^2 u}{\partial z^2}\right\|_{i,j,k}$ | $\dfrac{1}{l^2}(u_{i,j,k} - 2u_{i,j,k-1} + u_{i,j,k-2})$ | $O(l)$ | (6-43) |
| $\left.\dfrac{\partial^2 u}{\partial y\,\partial x}\right\|_{i,j,k}$ | $\dfrac{1}{hk}(u_{i,j,k} - u_{i,j-1,k} - u_{i-1,j,k} + u_{i-1,j-1,k})$ | $O(h+k)$ | (6-44) |

differences]. However, the more accurate formulas are not commonly used because they involve a larger number of terms and require more extensive computation times.

The use of finite difference approximations is demonstrated in the following sections of this chapter in setting up the numerical solutions of elliptic, parabolic, and hyperbolic partial differential equations.

## 6-4.2 Solution of Elliptic Partial Differential Equations

Elliptic differential equations are often encountered in steady-state heat conduction and diffusion operations. For example, in three-dimensional steady-state heat conduction in solids, Eq. (6-8) becomes

$$\frac{\partial^2 T}{\partial x^2} + \frac{\partial^2 T}{\partial y^2} + \frac{\partial^2 T}{\partial z^2} = 0 \tag{6-45}$$

Similarly, Fick's second law of diffusion [Eq. (6-10)] simplifies to

$$\frac{\partial^2 c_A}{\partial x^2} + \frac{\partial^2 c_A}{\partial y^2} + \frac{\partial^2 c_A}{\partial z^2} = 0 \tag{6-46}$$

when steady state is assumed.

We begin our discussion of numerical solutions of elliptic differential equations by first examining the two-dimensional problem in its general form (Laplace's equation):

$$\frac{\partial^2 u}{\partial x^2} + \frac{\partial^2 u}{\partial y^2} = 0 \tag{6-17}$$

We replace each second-order partial derivative by its approximation in central differences [Eqs. (6-27) and (6-28)] to obtain

$$\frac{1}{h^2}(u_{i+1,\,j} - 2u_{i,\,j} + u_{i-1,\,j}) + \frac{1}{k^2}(u_{i,\,j+1} - 2u_{i,\,j} + u_{i,\,j-1}) = 0 \tag{6-47}$$

which rearranges to

$$-2\left(\frac{1}{h^2} + \frac{1}{k^2}\right)u_{i,\,j} + \left(\frac{1}{h^2}\right)u_{i+1,\,j} + \left(\frac{1}{h^2}\right)u_{i-1,\,j} + \left(\frac{1}{k^2}\right)u_{i,\,j+1} + \left(\frac{1}{k^2}\right)u_{i,\,j-1} = 0 \tag{6-48}$$

This is a linear algebraic equation involving the value of the dependent variable at five adjacent grid points.

A rectangular-shaped object which is divided into $m$ segments in the $x$-direction and $n$ segments in the $y$-direction has $(m+1) \times (n+1)$ *total* grid points and $(m-1) \times (n-1)$ *internal* grid points. Equation (6-48), written for each of the internal points, constitutes a set of $(m-1) \times (n-1)$ simultaneous linear algebraic equations in $(m+1) \times (n+1) - 4$ unknowns (the four corner points do not appear in these equations). The boundary conditions provide the additional information for the solution of the problem. If the boundary conditions are of the Dirichlet type the values of the dependent variable are known at all the external grid points. On the other hand, if the boundary conditions at any of the external surfaces are of the Neumann or Robbins type, which specify partial derivatives at the boundaries, these conditions must also be replaced by finite difference approximations.

We demonstrate this by specifying a Neumann condition at the left boundary, i.e.,

$$\frac{\partial u}{\partial x} = \alpha \qquad \text{at } x = 0 \text{ and all } y \tag{6-49}$$

where $\alpha$ is a constant. Replacing the partial derivative in (6-49) with a central difference approximation we obtain

$$\frac{1}{2h}(u_{i+1,\,j} - u_{i-1,\,j}) = \alpha \tag{6-50}$$

This is valid only at $x = 0$ where $i = 0$; therefore, (6-50) becomes

$$u_{-1,\,j} = u_{1,\,j} - 2h\alpha \tag{6-51}$$

The points $(-1, j)$ are located outside the object; therefore, $u_{-1,j}$ have fictitious values. Their calculation, however, is necessary for the evaluation of the Neumann boundary condition.† Equation (6-51), written for all $y$ $(j = 0, 1, \ldots, n)$, provides $n+1$ additional equations but at the same time introduces $n+1$ additional variables. To counter this, Eq. (6-48) is also written for the $n+1$ points along this boundary (at $x=0$), thus providing the necessary number of independent equations for the solution of the problem.

Since Eq. (6-48) and the appropriate boundary conditions constitute a set of linear algebraic equations, the Gauss methods for the solution of such equations may be used. Equation (6-48) is actually a *predominantly diagonal* system; therefore, the Gauss–Seidel method (Sec. 3-7) is especially suitable for the solution of this problem. Rearranging Eq. (6-48) to solve for $u_{i,j}$

$$u_{i,j} = \frac{\dfrac{1}{h^2}(u_{i+1,j} + u_{i-1,j}) + \dfrac{1}{k^2}(u_{i,j+1} + u_{i,j-1})}{2\left(\dfrac{1}{h^2} + \dfrac{1}{k^2}\right)} \tag{6-52}$$

which can be used in the iterative Gauss–Seidel substitution method. An initial estimate of all $u_{i,j}$ is needed, but this can be easily obtained from averaging the Dirichlet boundary conditions.

The Gauss-Seidel method is guaranteed to converge for a predominantly diagonal system of equations. However, its convergence may be quite slow in the solution of elliptic differential equations. The *overrelaxation* method can be used to accelerate the rate of the convergence [5]. This technique applies the following weighting algorithm in evaluating the new values of $u_{i,j}$ at each iteration of the Gauss-Seidel method:

$$(u_{i,j})_{\text{new}} = w[u_{i,j} \text{ from Eq. (6-52)}] + (1-w)(u_{i,j} \text{ from previous iteration}) \tag{6-53}$$

The *relaxation parameter* $w$ can be assigned values from the following ranges:

$$0 < w < 1 \qquad \text{for underrelaxation}$$

$$1 < w \leq 2 \qquad \text{for overrelaxation}$$

When $w = 1$, this method is exactly the same as the unmodified Gauss-Seidel. Methods for estimating the *optimal* $w$ are given by Lapidus and Pinder [2, p. 410], who also show that the overrelaxation method is five to one hundred times faster (depending on step size and convergence criterion) than the Gauss-Seidel method.

† Replacing the $\partial u/\partial x$ in (6-49) with a forward difference does not require the use of the fictitious points; however, this reduces the accuracy of the calculation by one order.

In the case where an equidistant grid can be used, i.e., when $h = k$, Eq. (6-52) simplifies to

$$u_{i,j} = \frac{u_{i+1,j} + u_{i-1,j} + u_{i,j+1} + u_{i,j-1}}{4} \tag{6-54}$$

which simply shows that the value of the dependent variable at the pivotal point $(i, j)$ in the Laplace equation is the arithmetic average of the values at the grid points to the right and left of and above and below the pivot point. This is demonstrated by the computational molecule of Fig. 6-4, which is sometimes referred to as a "5-point star."

The three-dimensional elliptic partial differential equation

$$\frac{\partial^2 u}{\partial x^2} + \frac{\partial^2 u}{\partial y^2} + \frac{\partial^2 u}{\partial z^2} = 0 \tag{6-55}$$

can be similarly converted to algebraic equations using finite difference approximations in three-dimensional space. Applying Eqs. (6-27) to (6-29) from Table 6-1 to replace the three partial derivatives of (6-55), we obtain

$$\frac{1}{h^2}(u_{i+1,j,k} - 2u_{i,j,k} + u_{i-1,j,k}) + \frac{1}{k^2}(u_{i,j+1,k} - 2u_{i,j,k} + u_{i,j-1,k}) + \frac{1}{l^2}(u_{i,j,k+1} - 2u_{i,j,k} + u_{i,j,k-1}) = 0 \tag{6-56}$$

For the equidistant grid $(h = k = l)$, the above equation reduces to

$$u_{i,j,k} = \frac{u_{i+1,j,k} + u_{i-1,j,k} + u_{i,j+1,k} + u_{i,j-1,k} + u_{i,j,k+1} + u_{i,j,k-1}}{6} \tag{6-57}$$

In parallel with the two-dimensional case, the value of the dependent variable at the pivot point $(i, j, k)$ is the arithmetic average of the values at the grid points adjacent to the pivot point. The computational molecule for the three-dimensional elliptic equation is shown in Fig. 6-5.

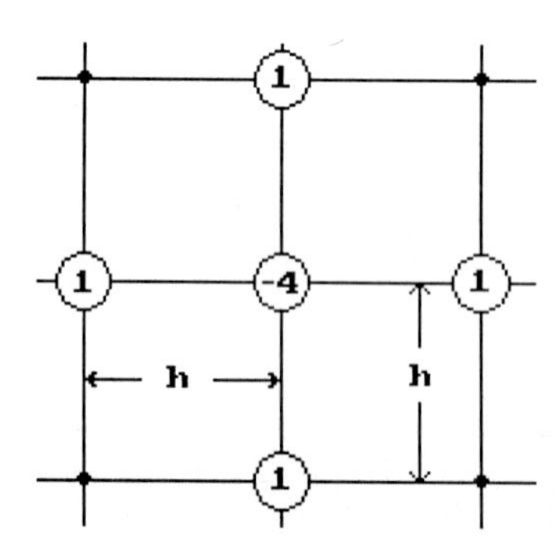


**Figure 6-4** Computational molecule for the Laplace equation using equidistant grid. The number in each circle is the coefficient of that point in the difference equation.

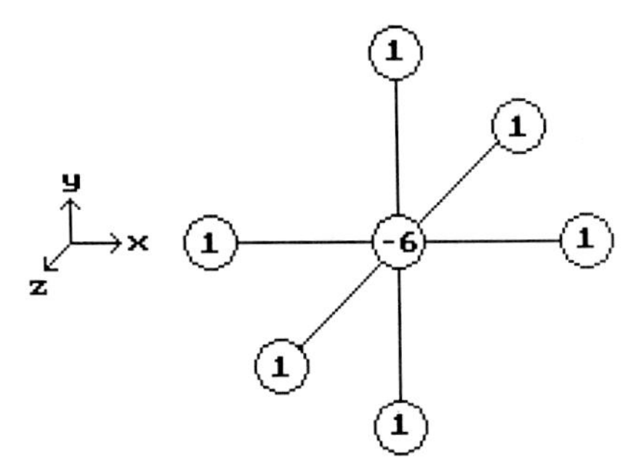


**Figure 6-5** Computational molecule for the three-dimensional elliptic differential equation using equidistant grid.

The *nonhomogeneous* form of the Laplace equation is the *Poisson* equation

$$\frac{\partial^2 u}{\partial x^2} + \frac{\partial^2 u}{\partial y^2} = f(x, y) \tag{6-58}$$

which also belongs to the class of elliptic partial differential equations. A form of the Poisson equation

$$\frac{\partial^2 T}{\partial x^2} + \frac{\partial^2 T}{\partial y^2} = -\frac{Q'(x, y)}{k} \tag{6-59}$$

is used to describe heat conduction in a two-dimensional solid plate with an internal heat source. $Q'(x, y)$ is the heat generated per unit volume per time and $k$ is the thermal conductivity of the material. The finite difference formulation of the Poisson equation is

$$u_{i,j} = \frac{\frac{1}{h^2}(u_{i+1,j} + u_{i-1,j}) + \frac{1}{k^2}(u_{i,j+1} + u_{i,j-1})}{2\left(\frac{1}{h^2} + \frac{1}{k^2}\right)} - \frac{f_{i,j}}{2\left(\frac{1}{h^2} + \frac{1}{k^2}\right)} \tag{6-60}$$

The numerical solution of the Laplace and Poisson elliptic partial differential equations is demonstrated in Example 6-1.

**Example 6-1 Elliptic partial differential equations** Write a computer program to determine the numerical solution of Laplace's equation

$$\frac{\partial^2 u}{\partial x^2} + \frac{\partial^2 u}{\partial y^2} = 0 \tag{6-17}$$

and Poisson's equation

$$\frac{\partial^2 u}{\partial x^2} + \frac{\partial^2 u}{\partial y^2} = f \tag{6-58}$$

for a rectangular object of variable width and height. The object could have Dirichlet, Neumann, or Cauchy boundary conditions. The value of $f$ in Poisson's equation should be assumed to be a constant. Use this program to find the solution of the following problems:

(*a*) A thin square metal plate of dimensions 1 ft $\times$ 1 ft is subjected to four heat sources which maintain the temperatures on its four edges as follows:

$$u(0, y) = 500°\text{F}$$

$$u(1, y) = 200°\text{F}$$

$$u(x, 0) = 1000°\text{F}$$

$$u(x, 1) = 100°\text{F}$$

The flat sides of the plate are insulated so that no heat is transferred through these sides. Calculate the temperature profiles within the plate.

(*b*) Perfect insulation is installed on two edges (right and bottom) of the plate of part *a*. The other two edges are exposed to heat sources. The set of Dirichlet and Neumann boundary conditions is

$$u(0, y) = 500°\text{F}$$

$$\left.\frac{\partial u}{\partial x}\right|_{1,\, y} = 0$$

$$u(x, 0) = 1000°\text{F}$$

$$\left.\frac{\partial u}{\partial y}\right|_{x,\, 1} = 0$$

Calculate the temperature profiles within the plate and compare these with the results of part *a*.

(*c*) The thin metal plate of part *a* is made of an alloy which has a melting point of 1500°F and a thermal conductivity of 10 Btu/(h · ft · °F). The plate is subjected to an electric current which creates a uniform heat source within the plate. The amount of heat generated is $Q' = 200{,}000$ Btu/(h · ft$^3$). The edges of the plate are in contact with heat sinks which maintain the temperature at 100°F on all four edges. Examine the temperature profiles within the plate to ascertain whether the alloy will begin to melt under these conditions.

(*d*) Determine the optimum value of the overrelaxation parameter for the conditions used in part *a*.

METHOD OF SOLUTION The Gauss-Seidel method with overrelaxation is used in this example. This method requires initial guesses as starting values for the dependent variable. These are calculated as follows:

1. *For internal grid points*: The weighted average of the Dirichlet conditions is used as the initial starting value.
2. *For boundaries with Dirichlet condition*: These are set to the value specified by the boundary condition and do not change during the calculation.
3. *For boundary with Neumann condition*: A forward (or backward) difference approximation is applied across the plate using the Dirichlet condition on the opposite boundary. However, in the case where the opposite boundary also has a Neumann condition, then the points on these two boundaries are given the same initial guesses as internal points (i.e., the weighted average of the Dirichlet conditions).
4. *For corner point with two Dirichlet conditions*: The value is set to the average of the two Dirichlet conditions and does not change during the calculation.
5. *For corner point with one Dirichlet and one Neumann condition*: This is set to the value of the one Dirichlet condition and does not change during the calculation.
6. *For corner point with two Neumann conditions*: This is given the same initial guess as an internal point (i.e., the weighted average of the Dirichlet conditions).

The value of the dependent variable at the internal grid points is recalculated using the recursive equation (6-60). When the Laplace equation is being solved, $f_{ij}$ is set to zero, which makes (6-60) identical to (6-52). For the Poisson equation, the value of $f_{ij}$ is constant throughout the plate.

At each boundary, if the condition is of the Dirichlet type, the values of $u$ remain unchanged. However, if the condition is of the Neumann type, a modified equation (6-60) is used to recalculate $u_{i,j}$. The modification applies Eq. (6-51)—or its equivalent for the appropriate boundary—in order to replace the fictitious value of $u$ in (6-60).

The overrelaxation method, Eq. (6-53), is used to establish the new value of $u_{i,j}$ at each iteration. Convergence is accomplished when the newly calculated values on *all* grid points are within a convergence criterion $\varepsilon$ of the corresponding values in the previous iteration.

PROGRAM DESCRIPTION This computer program, which is called ELLIPTIC.BAS, consists of the sections listed in Table E6-1*a*.

The program follows closely the method of solution described above, so no further discussion is necessary.

**Table E6-1*a***

| Program section | Line numbers |
|---|---|
| Title | 1–90 |
| Main program | 100–380 |
| Equation options | 110–150 |
| Output options | 160–200 |
| Parameter input | 210–250 |
| Subroutine calls | 260–280 |
| Rerun options | 290–370 |
| Subroutines | |
| 1. Boundary conditions | 1000–1790 |
| 2. Finite difference equations | 2000–2420 |
| 3. Print results | 3000–3070 |

PROGRAM

```
1 SCREEN 0:WIDTH 80:CLS:KEY OFF
5 PRINT "*********************************************************************"
10 PRINT"*                                                                    *"
15 PRINT"*                           EXAMPLE 6-1                              *"
20 PRINT"*                                                                    *"
25 PRINT"*              ELLIPTIC PARTIAL DIFFERENTIAL EQUATIONS               *"
30 PRINT"*                                                                    *"
35 PRINT"*                          (ELLIPTIC.BAS)                            *"
40 PRINT"*                                                                    *"
90 PRINT"*********************************************************************"
100 '**************************** Main Program ***************************
110 PRINT:PRINT"                        TYPE OF EQUATION"
120 PRINT:PRINT"            1. LAPLACE'S EQUATION"
130 PRINT:PRINT"            2. POISSON'S EQUATION"
140 PRINT:INPUT"ENTER YOUR CHOICE(1-2)"; RQ1
150 IF RQ1<1 OR RQ1>2 THEN GOTO 140
160 PRINT:PRINT"                        OUTPUT OPTIONS"
170 PRINT:PRINT"            1. PRINT FINAL RESULTS ONLY"
180 PRINT:PRINT"            2. PRINT STEP-BY-STEP RESULTS"
190 PRINT:INPUT"ENTER YOUR CHOICE(1-2)"; RQ2
200 IF RQ2<1 OR RQ2>2 THEN GOTO 190
210 PRINT:PRINT"                        PARAMETERS"
220 PRINT:INPUT"CONVERGENCE CRITERION (NEW VALUE - PREVIOUS VALUE)=";EPS
230 PRINT:INPUT"OVERRELAXATION PARAMETER=";OMEGA
240 IF RQ1=1 THEN GOTO 260
250 PRINT:INPUT"POISSON CONSTANT(UNITS/AREA)=";PSSN
260 PRINT CHR$(12):GOSUB 1000
270 GOSUB 2000
280 TIT$="CONVERGED RESULTS":PRINT CHR$(12):GOSUB 3000
290 PRINT:PRINT"                        RERUN OPTIONS"
300 PRINT:PRINT"            1. SHOW SAME RESULTS AGAIN"
310 PRINT:PRINT"            2. RESET INTEGRATION PARAMETERS AND RECALCULATE"
320 PRINT:PRINT"            3. END PROGRAM"
330 PRINT:PRINT:INPUT"ENTER YOUR CHOICE(1-3)"; RQ3
340 IF RQ3<1 OR RQ3>3 THEN GOTO 330
350 ON RQ3 GOTO 360,370,380
360 GOSUB 3000:GOTO 290
370 PRINT CHR$(12):RUN 100
380  END
1000 '************* Subroutine 1: Boundary conditions *******************
1010 '
1020 SCREEN 1
1030 A$(1)=" LEFT ":A$(2)="RIGHT ":A$(3)=" TOP  ":A$(4)="BOTTOM"
1040 PSET (100,10):LINE -(100,75):LINE -(200,75):LINE -(200,10):LINE -(100,10)
1050 LINE (97,22)-(100,25):LINE -(103,22)
1060 LINE (112,7)-(115,10):LINE -(112,13)
1070 LOCATE 1,9:PRINT "(0,0)"
1080 LOCATE 6,7:PRINT "HEIGHT": LOCATE 1,17:PRINT "WIDTH"
1090 LOCATE 13,1:PRINT"SPECIFY DIMENSIONS:"
1100 PRINT:INPUT"     WIDTH  (X-DIRECTION) =";WDTH
1110 PRINT:INPUT"     HEIGHT (Y-DIRECTION) =";HEIGHT
1120 PRINT:PRINT"NUMBER OF DIVISIONS:"
1130 PRINT:INPUT"     IN X-DIRECTION =";NX
1140 PRINT:INPUT"     IN Y-DIRECTION =";NY
1150 DIM U(NX,NY)
1160 FOR I=1 TO 4
1170 CLS
1180 PSET (100,10):LINE -(100,75):LINE -(200,75):LINE -(200,10):LINE -(100,10)
```

```
1190 IF I=1 THEN VV=6:HH=7
1200 IF I=2 THEN HH=28
1210 IF I=3 THEN VV=1:HH=17
1220 IF I=4 THEN VV=11
1230 LOCATE VV,HH:PRINT A$(I)
1240 LOCATE 13,1:PRINT "SPECIFY BOUNDARY CONDITION AT ";A$(I);":"
1250 LOCATE 15,1:PRINT"      1. DIRICHLET CONDITION"
1260 LOCATE 17,1:PRINT"      2. NEUMANN CONDITION"
1270 LOCATE 19,1:INPUT"ENTER YOUR CHOICE(1-2)"; SIDE(I)
1280 IF SIDE(I)<1 OR SIDE(I)>2 THEN GOTO 1270
1290 ON SIDE(I) GOTO 1300,1320
1300 LOCATE 21,1:INPUT"        BOUNDARY VALUE ="; BC(I)
1310 GOTO 1330
1320 LOCATE 21,1:INPUT"        GRADIENT="; BC(I)
1330 NEXT I
1340 'Calculate average value as starting guess
1350 SUM=0:COUNT=0
1360 FOR I=1 TO 4
1370 IF SIDE(I)=2 GOTO 1410
1380 IF I<=2 THEN WW=1/(WDTH/2)^2  ELSE  WW=1/(HEIGHT/2)^2
1390 SUM=SUM+WW*BC(I)
1400 COUNT=COUNT+WW
1410 NEXT I
1420 IF COUNT=0 THEN GOTO 1440
1430 AVER=SUM/COUNT
1440 FOR J=0 TO NY
1450 FOR I=0 TO NX
1460 U(I,J)=AVER
1470 NEXT I:NEXT J
1480 'Set conditions along boundaries
1490 FOR J=1 TO NY-1
1500 IF SIDE(1)=1 THEN U(0,J)=BC(1)
1510 IF SIDE(1)=2 AND SIDE(2)=1 THEN U(0,J)=BC(2)-WDTH*BC(1)
1520 IF SIDE(2)=1 THEN U(NX,J)=BC(2)
1530 IF SIDE(2)=2 AND SIDE(1)=1 THEN U(NX,J)=BC(1)+WDTH*BC(2)
1540 NEXT J
1550 FOR I=1 TO NX-1
1560 IF SIDE(3)=1 THEN U(I,0)=BC(3)
1570 IF SIDE(3)=2 AND SIDE(4)=1 THEN U(I,0)=BC(4)-HEIGHT*BC(3)
1580 IF SIDE(4)=1 THEN U(I,NY)=BC(4)
1590 IF SIDE(4)=2 AND SIDE(3)=1 THEN U(I,NY)=BC(3)+HEIGHT*BC(4)
1600 NEXT I
1610 'Set conditions at corners
1620 U(0,0)=(BC(1)+BC(3))/2
1630 IF SIDE(1)=1 AND SIDE(3)=2 THEN U(0,0)=BC(1)
1640 IF SIDE(1)=2 AND SIDE(3)=1 THEN U(0,0)=BC(3)
1650 IF SIDE(1)=2 AND SIDE(3)=2 THEN U(0,0)=AVER
1660 U(NX,0)=(BC(2)+BC(3))/2
1670 IF SIDE(2)=1 AND SIDE(3)=2 THEN U(NX,0)=BC(2)
1680 IF SIDE(2)=2 AND SIDE(3)=1 THEN U(NX,0)=BC(3)
1690 IF SIDE(2)=2 AND SIDE(3)=2 THEN U(NX,0)=AVER
1700 U(0,NY)=(BC(1)+BC(4))/2
1710 IF SIDE(1)=1 AND SIDE(4)=2 THEN U(0,NY)=BC(1)
1720 IF SIDE(1)=2 AND SIDE(4)=1 THEN U(0,NY)=BC(4)
1730 IF SIDE(1)=2 AND SIDE(4)=2 THEN U(0,NY)=AVER
1740 U(NX,NY)=(BC(2)+BC(4))/2
1750 IF SIDE(2)=1 AND SIDE(4)=2 THEN U(NX,NY)=BC(2)
1760 IF SIDE(2)=2 AND SIDE(4)=1 THEN U(NX,NY)=BC(4)
1770 IF SIDE(2)=2 AND SIDE(4)=2 THEN U(NX,NY)=AVER
1780 PRINT CHR$(12):TIT$="STARTING VALUES":GOSUB 3000
```

```
1790 RETURN
2000 '******** Subroutine 2: Finite difference equations ****************
2010 '
2020 'Calculate step sizes
2030 H=WDTH/NX: K=HEIGHT/NY
2040 DH=1/(H^2): DK=1/(K^2)
2050 DHDK=2*(DH+DK)
2060 ITER=0
2070 'Zero the Poisson constant when using Laplace's equation
2080 PSSN=PSSN*(RQ1-1)
2090 'Apply finite difference equations
2100 ITER=ITER+1
2110 COUNT2=0
2120 FOR J=0 TO NY
2130 FOR I=0 TO NX
2140 TEMP=U(I,J)
2150 IF I=0 AND J=0 THEN GOTO 2290
2160 IF I=NX AND J=NY THEN GOTO 2290
2170 IF I=0 AND J=NY THEN GOTO 2290
2180 IF I=NX AND J=0 THEN GOTO 2290
2190 'Sides with Neumann conditions
2200 IF I=0 AND SIDE(1)=2 THEN  U(I,J)=
     (DH*(2*U(I+1,J)-2*H*BC(1))+DK*(U(I,J+1) + U(I,J-1)))/DHDK - PSSN/DHDK
2210 IF I=NX AND SIDE(2)=2 THEN U(I,J)=
     (DH*(2*U(I-1,J)+2*H*BC(2))+DK*(U(I,J+1) + U(I,J-1)))/DHDK - PSSN/DHDK
2220 IF J=0 AND SIDE(3)=2 THEN  U(I,J)=
     (DH*(U(I+1,J)+U(I-1,J))+DK*(2*U(I,J+1) - 2*K*BC(3)))/DHDK - PSSN/DHDK
2230 IF J=NY AND SIDE(4)=2 THEN U(I,J)=
     (DH*(U(I+1,J)+U(I-1,J))+DK*(2*U(I,J-1) + 2*K*BC(4)))/DHDK - PSSN/DHDK
2240 IF I=0 OR J=0 GOTO 2330
2250 IF I=NX OR J=NY GOTO 2330
2260 'Internal points
2270 U(I,J)=(DH*(U(I+1,J)+U(I-1,J))+DK*(U(I,J+1)+U(I,J-1)))/DHDK-PSSN/DHDK
2280 'Corner points with Neumann conditions
2290 IF I=0 AND J=0 AND SIDE(1)=2 AND SIDE(3)=2 THEN   U(0,0)=
     (DH*(2*U(1,0)-2*H*BC(1))+DK*(2*U(0,1)-2*K*BC(3)))/DHDK - PSSN/DHDK
2300 IF I=NX AND J=NY AND SIDE(2)=2 AND SIDE(4)=2 THEN U(NX,NY)=
     (DH*(2*U(NX-1,NY)+2*H*BC(2))+DK*(2*U(NX,NY-1)+2*K*BC(4)))/DHDK-PSSN/DHDK
2310 IF I=0 AND J=NY AND SIDE(1)=2 AND SIDE(4)=2 THEN  U(0,NY)=
     (DH*(2*U(1,NY)-2*H*BC(1))+DK*(2*U(0,NY-1)+2*K*BC(4)))/DHDK - PSSN/DHDK
2320 IF I=NX AND J=0 AND SIDE(2)=2 AND SIDE(3)=2 THEN  U(NX,0)=
     (DH*(2*U(NX-1,0)+2*H*BC(2))+DK*(2*U(NX,1)-2*K*BC(3)))/DHDK - PSSN/DHDK
2330 'Overrelaxation
2340 U(I,J)=OMEGA*U(I,J)+(1-OMEGA)*TEMP
2350 IF ABS((U(I,J)-TEMP))<EPS THEN COUNT2=COUNT2+1
2360 NEXT I:NEXT J
2370 IF RQ2=2 THEN TIT$="INTERMEDIATE RESULTS":GOSUB 3000
2380 NTOTAL=(NX+1)*(NY+1)
2390 PRINT"ITERATION =";ITER;",  CONVERGED POINTS =";COUNT2 ;"/";NTOTAL:PRINT
2400 IF COUNT2=NTOTAL GOTO 2420
2410 GOTO 2100
2420 RETURN
3000 '**************** Subroutine 3: Print results ********************
3010 SCREEN 0:WIDTH 80
3020 PRINT TIT$:PRINT
3030 FOR J=0 TO NY
3040 FOR I=0 TO NX
3050 PRINT USING " ####.#";U(I,J);
3060 NEXT I:PRINT:NEXT J:PRINT
3070 RETURN
```

RESULTS

```
Part (a):
******************************************************************
*                                                                *
*                         EXAMPLE 6-1                            *
*                                                                *
*             ELLIPTIC PARTIAL DIFFERENTIAL EQUATIONS            *
*                                                                *
*                        (ELLIPTIC.BAS)                          *
*                                                                *
******************************************************************

                          TYPE OF EQUATION

            1. LAPLACE'S EQUATION

            2. POISSON'S EQUATION

ENTER YOUR CHOICE(1-2)? 1

                          OUTPUT OPTIONS

            1. PRINT FINAL RESULTS ONLY

            2. PRINT STEP-BY-STEP RESULTS

ENTER YOUR CHOICE(1-2)? 1

                          PARAMETERS

CONVERGENCE CRITERION (NEW VALUE - PREVIOUS VALUE)=? 0.1

OVERRELAXATION PARAMETER=? 1.5
```

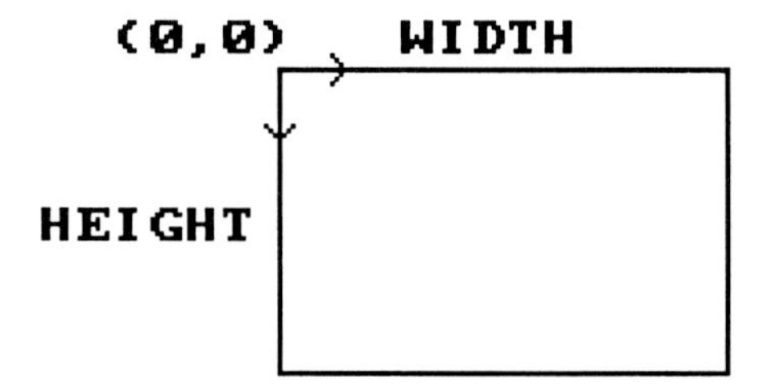


```
SPECIFY DIMENSIONS:
      WIDTH  (X-DIRECTION) =? 1
      HEIGHT (Y-DIRECTION) =? 1
NUMBER OF DIVISIONS:
      IN X-DIRECTION =? 10
      IN Y-DIRECTION =? 10
```

LEFT

```
SPECIFY BOUNDARY CONDITION AT  LEFT :

      1. DIRICHLET CONDITION

      2. NEUMANN CONDITION

ENTER YOUR CHOICE(1-2)? 1

         BOUNDARY VALUE =? 500
```

```
SPECIFY BOUNDARY CONDITION AT RIGHT :

      1. DIRICHLET CONDITION

      2. NEUMANN CONDITION

ENTER YOUR CHOICE(1-2)? 1

         BOUNDARY VALUE =? 200
```

TOP

```
SPECIFY BOUNDARY CONDITION AT  TOP  :

       1. DIRICHLET CONDITION

       2. NEUMANN CONDITION

ENTER YOUR CHOICE(1-2)? 1

          BOUNDARY VALUE =? 1000
```

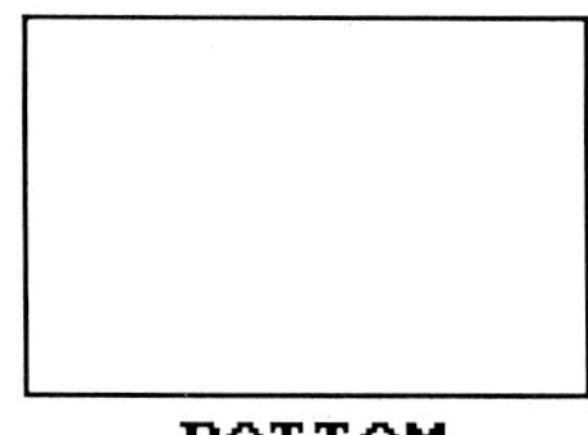

BOTTOM

```
SPECIFY BOUNDARY CONDITION AT BOTTOM:

       1. DIRICHLET CONDITION

       2. NEUMANN CONDITION

ENTER YOUR CHOICE(1-2)? 1

          BOUNDARY VALUE =? 100
```

```
STARTING VALUES

 750.0 1000.0 1000.0 1000.0 1000.0 1000.0 1000.0 1000.0 1000.0 1000.0  600.0
 500.0  450.0  450.0  450.0  450.0  450.0  450.0  450.0  450.0  450.0  200.0
 500.0  450.0  450.0  450.0  450.0  450.0  450.0  450.0  450.0  450.0  200.0
 500.0  450.0  450.0  450.0  450.0  450.0  450.0  450.0  450.0  450.0  200.0
 500.0  450.0  450.0  450.0  450.0  450.0  450.0  450.0  450.0  450.0  200.0
 500.0  450.0  450.0  450.0  450.0  450.0  450.0  450.0  450.0  450.0  200.0
 500.0  450.0  450.0  450.0  450.0  450.0  450.0  450.0  450.0  450.0  200.0
 500.0  450.0  450.0  450.0  450.0  450.0  450.0  450.0  450.0  450.0  200.0
 500.0  450.0  450.0  450.0  450.0  450.0  450.0  450.0  450.0  450.0  200.0
 500.0  450.0  450.0  450.0  450.0  450.0  450.0  450.0  450.0  450.0  200.0
 300.0  100.0  100.0  100.0  100.0  100.0  100.0  100.0  100.0  100.0  150.0

ITERATION = 1 ,  CONVERGED POINTS = 40 / 121
ITERATION = 2 ,  CONVERGED POINTS = 40 / 121
ITERATION = 3 ,  CONVERGED POINTS = 40 / 121
ITERATION = 4 ,  CONVERGED POINTS = 40 / 121
ITERATION = 5 ,  CONVERGED POINTS = 41 / 121
ITERATION = 6 ,  CONVERGED POINTS = 40 / 121
ITERATION = 7 ,  CONVERGED POINTS = 40 / 121
ITERATION = 8 ,  CONVERGED POINTS = 40 / 121
ITERATION = 9 ,  CONVERGED POINTS = 40 / 121
ITERATION = 10 ,  CONVERGED POINTS = 40 / 121
ITERATION = 11 ,  CONVERGED POINTS = 43 / 121
ITERATION = 12 ,  CONVERGED POINTS = 41 / 121
ITERATION = 13 ,  CONVERGED POINTS = 43 / 121
ITERATION = 14 ,  CONVERGED POINTS = 46 / 121
ITERATION = 15 ,  CONVERGED POINTS = 51 / 121
ITERATION = 16 ,  CONVERGED POINTS = 61 / 121
ITERATION = 17 ,  CONVERGED POINTS = 66 / 121
ITERATION = 18 ,  CONVERGED POINTS = 78 / 121
ITERATION = 19 ,  CONVERGED POINTS = 88 / 121
ITERATION = 20 ,  CONVERGED POINTS = 107 / 121
ITERATION = 21 ,  CONVERGED POINTS = 121 / 121


CONVERGED RESULTS

 750.0 1000.0 1000.0 1000.0 1000.0 1000.0 1000.0 1000.0 1000.0 1000.0  600.0
 500.0  736.8  822.1  854.4  864.0  860.5  844.9  812.1  744.8  593.4  200.0
 500.0  624.9  697.2  731.3  741.0  732.9  707.2  658.5  573.6  428.7  200.0
 500.0  565.7  610.4  632.6  635.9  622.8  592.5  541.3  462.3  347.9  200.0
 500.0  527.4  545.9  552.7  547.3  529.7  498.7  451.8  386.3  300.7  200.0
 500.0  497.8  493.3  484.9  471.0  450.1  420.7  381.1  330.2  268.7  200.0
 500.0  470.3  444.6  422.5  401.5  379.0  352.9  321.6  284.9  243.7  200.0
 500.0  438.9  392.2  358.9  333.6  311.7  290.2  267.6  244.2  221.2  200.0
 500.0  393.0  326.5  287.2  262.2  243.9  228.4  214.5  203.0  196.9  200.0
 500.0  306.7  233.6  201.3  184.2  173.3  165.2  159.0  156.3  163.3  200.0
 300.0  100.0  100.0  100.0  100.0  100.0  100.0  100.0  100.0  100.0  150.0
```

```
                    RERUN OPTIONS
        1. SHOW SAME RESULTS AGAIN
        2. RESET INTEGRATION PARAMETERS AND RECALCULATE
        3. END PROGRAM

ENTER YOUR CHOICE(1-3)? 2


Part (b):

                    TYPE OF EQUATION

        1. LAPLACE'S EQUATION

        2. POISSON'S EQUATION

ENTER YOUR CHOICE(1-2)? 1

                    OUTPUT OPTIONS

        1. PRINT FINAL RESULTS ONLY

        2. PRINT STEP-BY-STEP RESULTS

ENTER YOUR CHOICE(1-2)? 1

                    PARAMETERS

CONVERGENCE CRITERION (NEW VALUE - PREVIOUS VALUE)=? 0.1

OVERRELAXATION PARAMETER=? 1.5
```

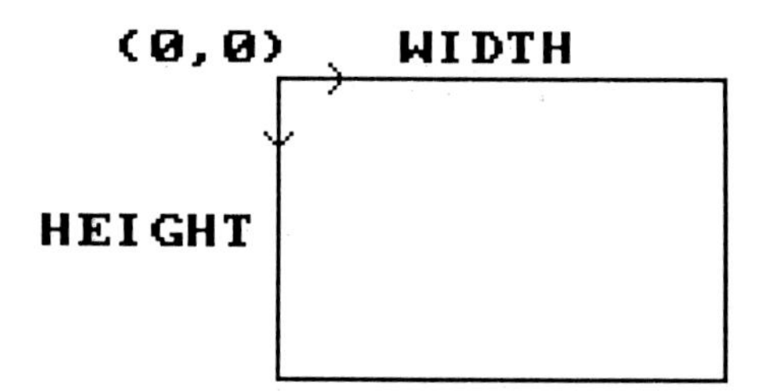


```
SPECIFY DIMENSIONS:
     WIDTH  (X-DIRECTION) =? 1
     HEIGHT (Y-DIRECTION) =? 1
NUMBER OF DIVISIONS:
     IN X-DIRECTION =? 10
     IN Y-DIRECTION =? 10
```

LEFT

```
SPECIFY BOUNDARY CONDITION AT  LEFT :

      1. DIRICHLET CONDITION

      2. NEUMANN CONDITION

ENTER YOUR CHOICE(1-2)? 1

         BOUNDARY VALUE =? 500
```

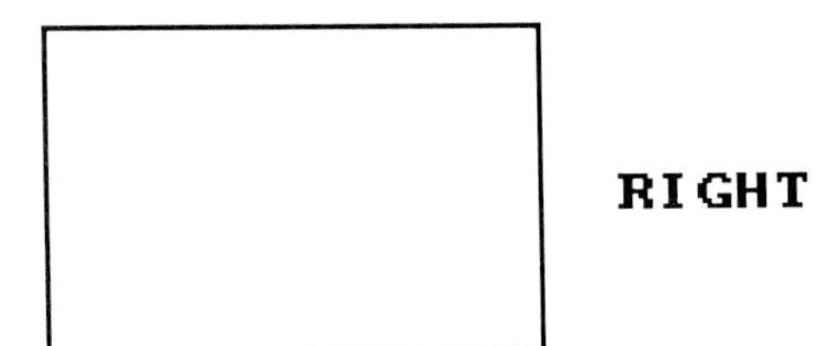


```
SPECIFY BOUNDARY CONDITION AT RIGHT :

      1. DIRICHLET CONDITION

      2. NEUMANN CONDITION

ENTER YOUR CHOICE(1-2)? 2

         GRADIENT=? 0
```

TOP

SPECIFY BOUNDARY CONDITION AT TOP :

1. DIRICHLET CONDITION

2. NEUMANN CONDITION

ENTER YOUR CHOICE(1-2)? 1

BOUNDARY VALUE =? 1000

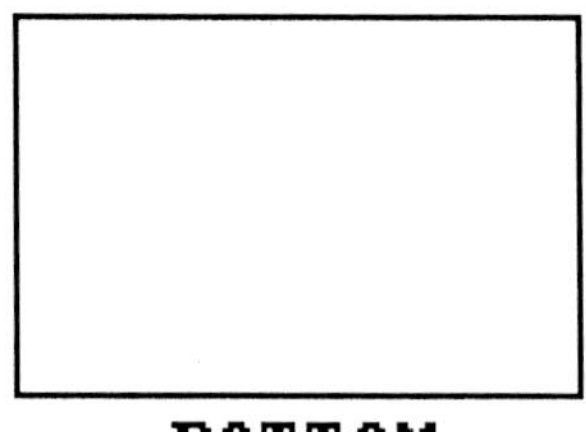


SPECIFY BOUNDARY CONDITION AT BOTTOM:

1. DIRICHLET CONDITION

2. NEUMANN CONDITION

ENTER YOUR CHOICE(1-2)? 2

GRADIENT=? 0

```
STARTING VALUES

  750.0 1000.0 1000.0 1000.0 1000.0 1000.0 1000.0 1000.0 1000.0 1000.0 1000.0
  500.0  750.0  750.0  750.0  750.0  750.0  750.0  750.0  750.0  750.0  500.0
  500.0  750.0  750.0  750.0  750.0  750.0  750.0  750.0  750.0  750.0  500.0
  500.0  750.0  750.0  750.0  750.0  750.0  750.0  750.0  750.0  750.0  500.0
  500.0  750.0  750.0  750.0  750.0  750.0  750.0  750.0  750.0  750.0  500.0
  500.0  750.0  750.0  750.0  750.0  750.0  750.0  750.0  750.0  750.0  500.0
  500.0  750.0  750.0  750.0  750.0  750.0  750.0  750.0  750.0  750.0  500.0
  500.0  750.0  750.0  750.0  750.0  750.0  750.0  750.0  750.0  750.0  500.0
  500.0  750.0  750.0  750.0  750.0  750.0  750.0  750.0  750.0  750.0  500.0
  500.0  750.0  750.0  750.0  750.0  750.0  750.0  750.0  750.0  750.0  500.0
  500.0 1000.0 1000.0 1000.0 1000.0 1000.0 1000.0 1000.0 1000.0 1000.0  750.0


ITERATION = 1 ,  CONVERGED POINTS = 31 / 121
ITERATION = 2 ,  CONVERGED POINTS = 31 / 121
ITERATION = 3 ,  CONVERGED POINTS = 31 / 121
ITERATION = 4 ,  CONVERGED POINTS = 31 / 121
ITERATION = 5 ,  CONVERGED POINTS = 31 / 121
ITERATION = 6 ,  CONVERGED POINTS = 31 / 121
ITERATION = 7 ,  CONVERGED POINTS = 31 / 121
ITERATION = 8 ,  CONVERGED POINTS = 35 / 121
ITERATION = 9 ,  CONVERGED POINTS = 33 / 121
ITERATION = 10 ,  CONVERGED POINTS = 39 / 121
ITERATION = 11 ,  CONVERGED POINTS = 35 / 121
ITERATION = 12 ,  CONVERGED POINTS = 49 / 121
ITERATION = 13 ,  CONVERGED POINTS = 59 / 121
ITERATION = 14 ,  CONVERGED POINTS = 77 / 121
ITERATION = 15 ,  CONVERGED POINTS = 95 / 121
ITERATION = 16 ,  CONVERGED POINTS = 109 / 121
ITERATION = 17 ,  CONVERGED POINTS = 113 / 121
ITERATION = 18 ,  CONVERGED POINTS = 117 / 121
ITERATION = 19 ,  CONVERGED POINTS = 119 / 121
ITERATION = 20 ,  CONVERGED POINTS = 121 / 121


CONVERGED RESULTS

  750.0 1000.0 1000.0 1000.0 1000.0 1000.0 1000.0 1000.0 1000.0 1000.0 1000.0
  500.0  750.0  848.8  895.1  920.2  935.3  945.0  951.3  955.4  957.6  958.3
  500.0  651.2  750.0  811.4  850.4  876.0  893.3  905.0  912.5  916.7  918.1
  500.0  604.9  688.6  750.0  793.9  825.1  847.3  862.7  872.9  878.7  880.6
  500.0  579.8  649.6  706.1  750.0  783.3  808.0  825.8  837.7  844.6  846.9
  500.0  564.7  624.0  674.9  716.7  750.0  775.6  794.6  807.6  815.2  817.7
  500.0  555.0  606.7  652.7  692.0  724.4  750.0  769.4  782.8  790.8  793.4
  500.0  548.7  595.0  637.3  674.2  705.4  730.6  750.0  763.7  771.8  774.5
  500.0  544.6  587.5  627.1  662.3  692.4  717.2  736.3  750.0  758.2  760.9
  500.0  542.4  583.3  621.3  655.4  684.8  709.2  728.2  741.8  750.0  752.7
  500.0  541.7  581.9  619.4  653.1  682.3  706.6  725.5  739.1  747.3  750.0
```

```
                          RERUN OPTIONS

            1. SHOW SAME RESULTS AGAIN

            2. RESET INTEGRATION PARAMETERS AND RECALCULATE

            3. END PROGRAM
ENTER YOUR CHOICE(1-3)? 2
```

Part (c):

```
                          TYPE OF EQUATION

            1. LAPLACE'S EQUATION

            2. POISSON'S EQUATION

ENTER YOUR CHOICE(1-2)? 2
                          OUTPUT OPTIONS

            1. PRINT FINAL RESULTS ONLY

            2. PRINT STEP-BY-STEP RESULTS

ENTER YOUR CHOICE(1-2)? 1
                          PARAMETERS

CONVERGENCE CRITERION (NEW VALUE - PREVIOUS VALUE)=? 0.1

OVERRELAXATION PARAMETER=? 1.5

POISSON CONSTANT(UNITS/AREA)=? -20000
```

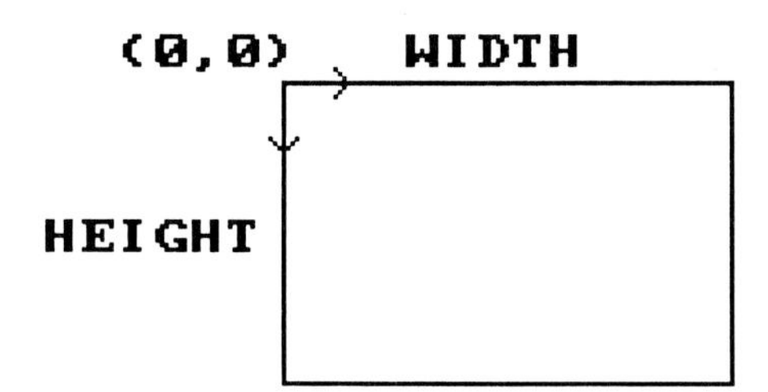


```
SPECIFY DIMENSIONS:
      WIDTH  (X-DIRECTION) =? 1
      HEIGHT (Y-DIRECTION) =? 1
NUMBER OF DIVISIONS:
      IN X-DIRECTION =? 10
      IN Y-DIRECTION =? 10
```

```
LEFT

SPECIFY BOUNDARY CONDITION AT  LEFT :

      1. DIRICHLET CONDITION

      2. NEUMANN CONDITION

ENTER YOUR CHOICE(1-2)? 1

         BOUNDARY VALUE =? 100
```

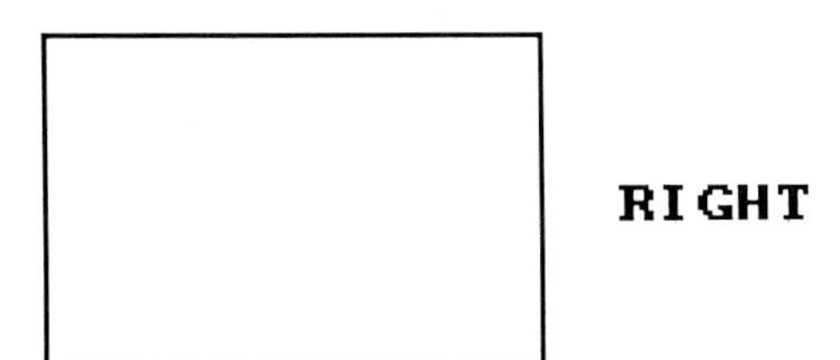


```
SPECIFY BOUNDARY CONDITION AT RIGHT :

      1. DIRICHLET CONDITION

      2. NEUMANN CONDITION

ENTER YOUR CHOICE(1-2)? 1

         BOUNDARY VALUE =? 100
```

```
                TOP

SPECIFY BOUNDARY CONDITION AT  TOP  :

      1. DIRICHLET CONDITION

      2. NEUMANN CONDITION

ENTER YOUR CHOICE(1-2)? 1

         BOUNDARY VALUE =? 100
```

```
               BOTTOM

SPECIFY BOUNDARY CONDITION AT BOTTOM:

      1. DIRICHLET CONDITION

      2. NEUMANN CONDITION

ENTER YOUR CHOICE(1-2)? 1

         BOUNDARY VALUE =? 100
```

```
STARTING VALUES

  100.0  100.0  100.0  100.0  100.0  100.0  100.0  100.0  100.0  100.0  100.0
  100.0  100.0  100.0  100.0  100.0  100.0  100.0  100.0  100.0  100.0  100.0
  100.0  100.0  100.0  100.0  100.0  100.0  100.0  100.0  100.0  100.0  100.0
  100.0  100.0  100.0  100.0  100.0  100.0  100.0  100.0  100.0  100.0  100.0
  100.0  100.0  100.0  100.0  100.0  100.0  100.0  100.0  100.0  100.0  100.0
  100.0  100.0  100.0  100.0  100.0  100.0  100.0  100.0  100.0  100.0  100.0
  100.0  100.0  100.0  100.0  100.0  100.0  100.0  100.0  100.0  100.0  100.0
  100.0  100.0  100.0  100.0  100.0  100.0  100.0  100.0  100.0  100.0  100.0
  100.0  100.0  100.0  100.0  100.0  100.0  100.0  100.0  100.0  100.0  100.0
  100.0  100.0  100.0  100.0  100.0  100.0  100.0  100.0  100.0  100.0  100.0
  100.0  100.0  100.0  100.0  100.0  100.0  100.0  100.0  100.0  100.0  100.0
ITERATION = 1 ,  CONVERGED POINTS = 40 / 121

ITERATION = 2 ,  CONVERGED POINTS = 40 / 121

ITERATION = 3 ,  CONVERGED POINTS = 40 / 121

ITERATION = 4 ,  CONVERGED POINTS = 40 / 121

ITERATION = 5 ,  CONVERGED POINTS = 40 / 121

ITERATION = 6 ,  CONVERGED POINTS = 40 / 121

ITERATION = 7 ,  CONVERGED POINTS = 40 / 121

ITERATION = 8 ,  CONVERGED POINTS = 40 / 121

ITERATION = 9 ,  CONVERGED POINTS = 40 / 121

ITERATION = 10 ,  CONVERGED POINTS = 40 / 121

ITERATION = 11 ,  CONVERGED POINTS = 40 / 121

ITERATION = 12 ,  CONVERGED POINTS = 40 / 121

ITERATION = 13 ,  CONVERGED POINTS = 40 / 121

ITERATION = 14 ,  CONVERGED POINTS = 41 / 121

ITERATION = 15 ,  CONVERGED POINTS = 41 / 121

ITERATION = 16 ,  CONVERGED POINTS = 43 / 121

ITERATION = 17 ,  CONVERGED POINTS = 45 / 121

ITERATION = 18 ,  CONVERGED POINTS = 51 / 121

ITERATION = 19 ,  CONVERGED POINTS = 60 / 121

ITERATION = 20 ,  CONVERGED POINTS = 64 / 121

ITERATION = 21 ,  CONVERGED POINTS = 77 / 121

ITERATION = 22 ,  CONVERGED POINTS = 87 / 121

ITERATION = 23 ,  CONVERGED POINTS = 105 / 121
```

```
ITERATION = 24 ,   CONVERGED POINTS = 120 / 121

ITERATION = 25 ,   CONVERGED POINTS = 121 / 121


CONVERGED RESULTS

  100.0  100.0  100.0  100.0  100.0  100.0  100.0  100.0  100.0  100.0  100.0
  100.0  356.2  512.5  607.9  659.9  676.5  659.9  607.9  512.5  356.3  100.0
  100.0  512.5  785.8  959.1 1055.3 1086.2 1055.4  959.2  785.9  512.5  100.0
  100.0  607.9  959.1 1187.6 1316.1 1357.6 1316.2 1187.7  959.2  607.9  100.0
  100.0  659.9 1055.3 1316.1 1464.0 1511.9 1464.0 1316.2 1055.4  660.0  100.0
  100.0  676.5 1086.2 1357.6 1511.9 1561.9 1511.9 1357.6 1086.3  676.6  100.0
  100.0  659.9 1055.4 1316.2 1464.0 1511.9 1464.1 1316.2 1055.4  660.0  100.0
  100.0  607.9  959.2 1187.7 1316.2 1357.6 1316.2 1187.7  959.2  607.9  100.0
  100.0  512.5  785.9  959.2 1055.4 1086.3 1055.4  959.2  785.9  512.5  100.0
  100.0  356.3  512.5  607.9  660.0  676.6  660.0  607.9  512.5  356.3  100.0
  100.0  100.0  100.0  100.0  100.0  100.0  100.0  100.0  100.0  100.0  100.0


                        RERUN OPTIONS

            1. SHOW SAME RESULTS AGAIN

            2. RESET INTEGRATION PARAMETERS AND RECALCULATE

            3. END PROGRAM

ENTER YOUR CHOICE(1-3)? 3
Ok
```

### Discussion of results

*Part a* The Laplace equation is solved with a convergence criterion of 0.1°F and an overrelaxation parameter of 1.5 (see part *d* for determination of the optimum relaxation parameter). The grid was chosen with ten divisions in each direction. A finer grid would have resulted in more accurate profiles, but this was deemed unnecessary for this example.

The results show that starting with a weighted average of 450°F for all internal grid points, the program converges to the final temperature profiles within 21 iterations.

*Part b* The effect of insulation on the right and bottom edges of the plate is evident in these results. The gradient of the temperature near these boundaries approaches zero to satisfy the imposed boundary conditions. Since the insulation stops the flow of heat through these boundaries, the temperature along the insulated edges is higher than that of part *a*.

*Part c* The Poisson equation is solved with a Poisson constant determined from Eq. (6-59):

$$f = -\frac{Q'}{k} = -\frac{200{,}000}{10} = -20{,}000\text{°F/ft}^2$$

The temperature within the plate rises sharply to its highest value of 1561.9°F at the center point. Under these circumstances the metal will begin to melt at the center core.

*Part d* The number of iterations needed for convergence for different values of the overrelaxation parameter are tabulated in Table E6-1*b* and plotted in Fig. E6-1. The value of 1.5 appears to be the optimum level of the overrelaxation parameter for this case.

**Table E6-1*b***

| Parameter | Iterations |
|---|---|
| 1.0 | 47 |
| 1.1 | 42 |
| 1.2 | 37 |
| 1.3 | 32 |
| 1.4 | 27 |
| 1.5 | 21 |
| 1.6 | 22 |
| 1.7 | 28 |
| 1.8 | 41 |
| 1.9 | 81 |

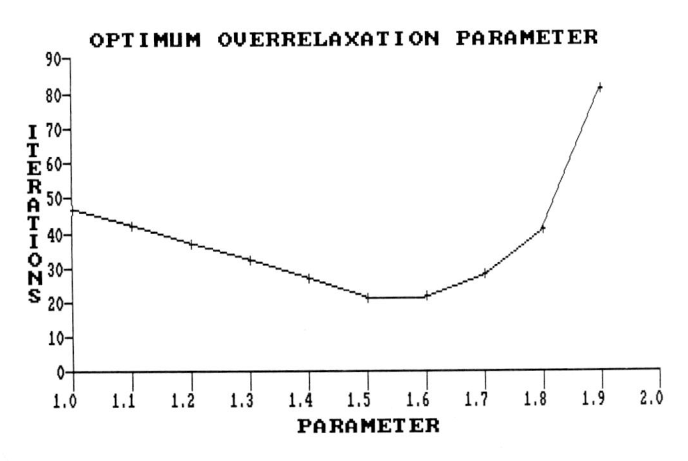


**Figure E6-1.**

### 6-4.3 Solution of Parabolic Partial Differential Equations

Classic examples of parabolic differential equations are the one-dimensional unsteady-state heat conduction equation

$$\alpha \frac{\partial^2 T}{\partial x^2} = \frac{\partial T}{\partial t} \tag{6-61}$$

and Fick's second law of diffusion

$$D_{AB} \frac{\partial^2 C_A}{\partial x^2} = \frac{\partial C_A}{\partial t} \tag{6-62}$$

with Dirichlet, Neumann, or Cauchy boundary conditions.

Let us consider this class of equations in the general form

$$\alpha \frac{\partial^2 u}{\partial x^2} = \frac{\partial u}{\partial t} \tag{6-63}$$

and express the derivatives in terms of central differences around the point $(i, j)$, using the counter $i$ for the $x$-direction and $j$ for the $t$-direction:

$$\left.\frac{\partial^2 u}{\partial x^2}\right|_{i,j} = \frac{1}{h^2}(u_{i+1,j} - 2u_{i,j} + u_{i-1,j}) + O(h^2) \tag{6-64}$$

$$\left.\frac{\partial u}{\partial t}\right|_{i,j} = \frac{1}{2k}(u_{i,j+1} - u_{i,j-1}) + O(k^2) \tag{6-65}$$

Combining Eqs. (6-63) to (6-65) and rearranging,

$$u_{i,j+1} = u_{i,j-1} + \frac{2\alpha k}{h^2}(u_{i+1,j} - 2u_{i,j} + u_{i-1,j}) + O(h^2 + k^2) \tag{6-66}$$

This is an *explicit* algebraic formula which calculates the value of the dependent variable at the next time step ($u_{i,j+1}$) from values at the current and earlier time steps. Once the initial and boundary conditions of the problem are specified, solution of an explicit formula is usually straightforward. However, this particular explicit formula is *unstable*, because it contains negative terms on the right side.† As a rule of thumb, when all the known values are arranged on the right side of the finite difference formulation, if there are any negative coefficients, the solution is unstable. This is stated more precisely by the positivity rule:‡ "For

$$u_{i,j+1} = Au_{i+1,j} + Bu_{i,j} + Cu_{i-1,j} \tag{6-67}$$

† A rigorous discussion of stability analysis is given in Sec. 6-5.

‡ For proof see Finlayson [6, p. 216].

if $A$, $B$, $C$ are positive and $A + B + C \leq 1$, then the numerical scheme is stable."

In order to eliminate the instability problem, we replace the first-order derivative in Eq. (6-63) with the forward difference:

$$\left.\frac{\partial u}{\partial t}\right|_{i,j} = \frac{1}{k}(u_{i,j+1} - u_{i,j}) + O(k) \tag{6-68}$$

Combining (6-63), (6-64), and (6-68) we obtain the explicit formula

$$u_{i,j+1} = \left(\frac{\alpha k}{h^2}\right)u_{i+1,j} + \left(1 - 2\frac{\alpha k}{h^2}\right)u_{i,j} + \left(\frac{\alpha k}{h^2}\right)u_{i-1,j} + O(h^2 + k) \tag{6-69}$$

For a stable solution, the positivity rule requires that

$$1 - 2\frac{\alpha k}{h^2} \geq 0 \tag{6-70}$$

Rearranging

$$\frac{\alpha k}{h^2} \leq \frac{1}{2} \tag{6-71}$$

This inequality determines the relationship between the two integration steps, $h$ in the $x$-direction and $k$ in the $t$-direction. As $h$ gets smaller, $k$ becomes much smaller, thus requiring longer computation times.

If we choose to work with the equality part of (6-70) or (6-71), i.e.,

$$1 - 2\frac{\alpha k}{h^2} = 0 \qquad \text{or} \qquad \frac{\alpha k}{h^2} = \frac{1}{2} \tag{6-72}$$

then Eq. (6-69) simplifies to

$$u_{i,j+1} = \tfrac{1}{2}(u_{i+1,j} + u_{i-1,j}) + O(h^2 + k) \tag{6-73}$$

This explicit formula calculates the value of the dependent variable at position $i$ of the next time step $j + 1$ from values to the right and left of $i$ at the present time step $j$. The computational molecule for this equation is shown in Fig. 6-6.

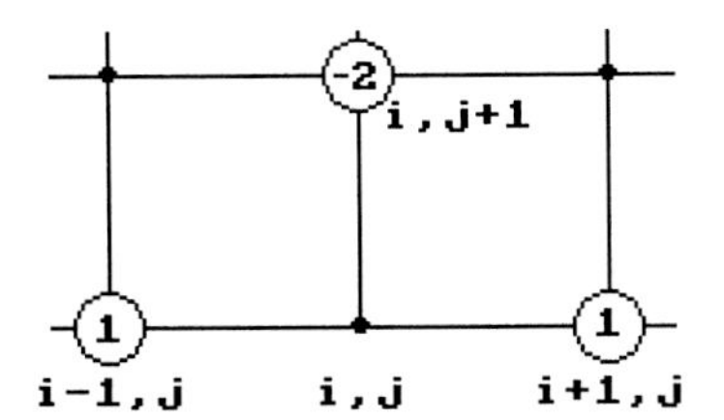


**Figure 6-6** Computational molecule for Eq. (6-73).

It should be emphasized that using the forward difference for the first-order derivatives introduces the error of order $(k)$; therefore, Eq. (6-69) is of order $O(k)$ in the time direction and $O(h^2)$ in the $x$-direction. However, the advantage of gaining stability outweighs the loss of accuracy in this case.

Parabolic partial differential equations can have initial and boundary conditions of the Dirichlet, Neumann, Cauchy, or Robbins type. These were discussed in Sec. 6-3. Examples of these conditions for the heat conduction problem are demonstrated in Fig. 6-2. The boundary conditions must be discretized using the same finite difference grid as used for the differential equation. For Dirichlet conditions this simply involves setting the values of the dependent variable along the appropriate boundary equal to the given boundary condition. For Neumann and Robbins conditions, the gradient at the boundaries must be replaced by finite difference approximations, resulting in additional algebraic equations which must be incorporated into the overall scheme of solution of the resulting set of algebraic equations.

The *explicit* method of solution of a parabolic differential equation (6-63), and the incorporation of boundary conditions, is demonstrated in Example 6-2.

**Implicit methods** Let us now consider some implicit methods for the solution of parabolic equations. We utilize the grid of Fig. 6-7, in which the half point in the $j$-direction $(i, j + \frac{1}{2})$ is shown [7]. Instead of expressing $\partial u/\partial t$ in terms of forward difference around $(i, j)$ as it was done in the explicit form, we express this partial derivative in terms of *central difference* around the half point:

$$\left.\frac{\partial u}{\partial t}\right|_{i,\, j+1/2} = \frac{1}{k}\,(u_{i,\, j+1} - u_{i,\, j}) \tag{6-74}$$

In addition, the second-order partial derivative is expressed at the half point as a weighted average of the central differences at points $(i, j + 1)$ and $(i, j)$:

$$\begin{aligned} \left.\frac{\partial^2 u}{\partial x^2}\right|_{i,\, j+1/2} &= \theta\delta^2 u_{i,\, j+1} + (1-\theta)\delta^2 u_{i,\, j} \\ &= \theta\left[\frac{1}{h^2}\,(u_{i+1,\, j+1} - 2u_{i,\, j+1} + u_{i-1,\, j+1})\right] \\ &\qquad + (1-\theta)\left[\frac{1}{h^2}\,(u_{i+1,\, j} - 2u_{i,\, j} + u_{i-1,\, j})\right] \end{aligned} \tag{6-75}$$

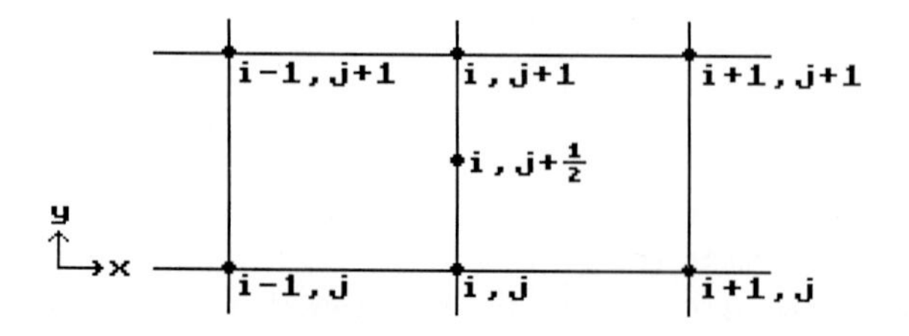


**Figure 6-7** Finite difference grid for derivation of implicit formulas.

where $\theta$ is in the range $0 \le \theta \le 1$. A combination of (6-63), (6-74), and (6-75) results in the *variable-weighted implicit* approximation of the parabolic partial differential equation:

$$\alpha\theta\left[\frac{1}{h^2}(u_{i+1,\,j+1} - 2u_{i,\,j+1} + u_{i-1,\,j+1})\right] - \frac{1}{k}u_{i,\,j+1}$$

$$= -\alpha(1-\theta)\left[\frac{1}{h^2}(u_{i+1,\,j} - 2u_{i,\,j} + u_{i-1,\,j})\right] - \frac{1}{k}u_{i,\,j} \qquad (6\text{-}76)$$

This formula is implicit because the left-hand side involves more than one value at the $j+1$ position of the difference grid (i.e., more than one unknown at any step in the time domain).

When $\theta = 0$, Eq. (6-76) becomes identical to the classic explicit formula (6-69). When $\theta = 1$, Eq. (6-76) becomes

$$-\left(\frac{\alpha k}{h^2}\right)u_{i-1,\,j+1} + \left(1 + \frac{2\alpha k}{h^2}\right)u_{i,\,j+1} - \left(\frac{\alpha k}{h^2}\right)u_{i+1,\,j+1} = u_{i,\,j} \qquad (6\text{-}77)$$

This is called the *backward implicit* approximation, which can also be obtained by approximating the first-order partial derivative using the backward difference at $(i, j+1)$ and the second-order partial derivative by the central difference at $(i, j+1)$.

Finally, when $\theta = \frac{1}{2}$, Eq. (6-76) yields the well-known *Crank-Nicolson implicit formula*:

$$-\left(\frac{\alpha k}{h^2}\right)u_{i-1,\,j+1} + 2\left(1 + \frac{\alpha k}{h^2}\right)u_{i,\,j+1} - \left(\frac{\alpha k}{h^2}\right)u_{i+1,\,j+1}$$

$$= \left(\frac{\alpha k}{h^2}\right)u_{i-1,\,j} + 2\left(1 - \frac{\alpha k}{h^2}\right)u_{i,\,j} + \left(\frac{\alpha k}{h^2}\right)u_{i+1,\,j} \qquad (6\text{-}78)$$

When written for the entire difference grid, implicit formulas generate sets of simultaneous linear algebraic equations whose matrix of coefficients is usually a tridiagonal matrix. This type of problem may be solved using a Gauss elimination procedure, or more efficiently using the Thomas algorithm [2, p. 216], which is a variation of Gauss elimination.

Implicit formulas of the type described above have been found to be unconditionally stable. It can be generalized that most explicit finite difference approximations are conditionally stable whereas most implicit approximations are unconditionally stable [2]. The explicit methods, however, are computationally easier to solve than the implicit techniques.

**Method of lines** Another technique for the solution of parabolic partial differential equations is the *method of lines*. This is based on the concept of converting the partial differential equation into a set of ordinary differential equations by discretizing only the spatial derivatives using finite differences, and leaving the time derivatives unchanged [8]:

$$\frac{du_i}{dt} = \frac{\alpha}{h^2}(u_{i+1} - 2u_i + u_{i-1})$$

There will be as many of these ordinary differential equations as there are grid points in the $x$-direction (Fig. 6-8). The complete set of differential equations for $0 \le i \le N$ would be

$$\frac{du_0}{dt} = \frac{\alpha}{h^2}(u_1 - 2u_0 + u_{-1}) \tag{6-79a}$$

$$\vdots$$

$$\frac{du_i}{dt} = \frac{\alpha}{h^2}(u_{i+1} - 2u_i + u_{i-1}) \tag{6.79b}$$

$$\vdots$$

$$\frac{du_N}{dt} = \frac{\alpha}{h^2}(u_{N+1} - 2u_N + u_{N-1}) \tag{6-79c}$$

The two equations at the boundaries, (6-79$a$) and (6-79$c$), would have to be modified according to the boundary conditions that are specified in the particular problem. For example, if a Dirichlet condition is given at $x = 0$ and $t > 0$, that is,

$$u_0 = \beta(\text{constant}) \quad \text{for } t > 0$$

then Eq. (6-79$a$) is modified to

$$\frac{du_0}{dt} = 0 \qquad u_0(0) = \beta \tag{6-80}$$

On the other hand, if a Neumann condition is given at this boundary, i.e.,

$$\left.\frac{\partial u}{\partial x}\right|_{0,\,t} = 0 \qquad \text{at } x = 0 \text{ and } t > 0$$

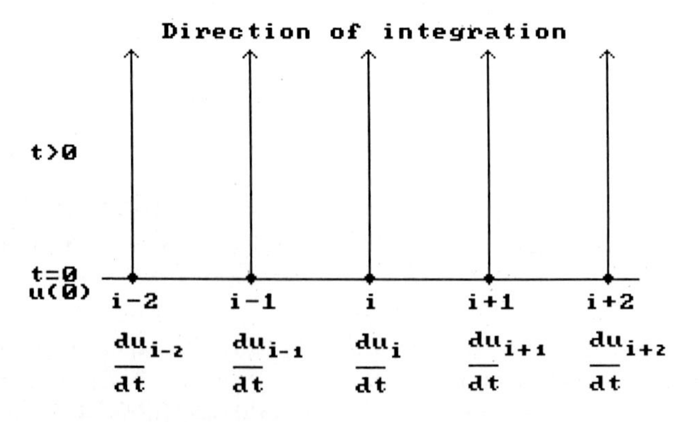


**Figure 6-8** Method of lines.

the partial is replaced by a central difference approximation

$$\left.\frac{\partial u}{\partial x}\right|_{0,\,t} = \frac{u_1 - u_{-1}}{2h} = 0$$

Then Eq. (6-79*a*) becomes

$$\frac{du_0}{dt} = \frac{\alpha}{h^2}(2u_1 - 2u_0) \tag{6-81}$$

The complete set of simultaneous differential equations must be integrated forward in time (the *j*-direction) starting with the initial conditions of the problem. This method gives stable solutions for parabolic partial differential equations.

**Example 6-2 Parabolic partial differential equations** Write a computer program to determine the numerical solution of the parabolic partial differential equation

$$\alpha \frac{\partial^2 u}{\partial x^2} = \frac{\partial u}{\partial t} \tag{6-63}$$

for a semi-infinite solid slab of variable width. The solid slab could have Dirichlet, Neumann, or Cauchy boundary conditions. Use this program to develop the solution to the following problems. Show the integrated results in tabular and graphical form.

(*a*) The wall of a furnace is 1 ft thick and is made of brick, which has a thermal diffusivity of 0.01 $ft^2/h$. The temperature of the wall is 100°F when the furnace is off. When the furnace is fired, the temperature on the inside face of the wall reaches 1000°F quite rapidly. The temperature of the outside face of the wall is maintained at 100°F by natural convection. Determine the evolution of temperature profiles within the brick wall.

(*b*) Insulation is placed on the outside surface of the wall. Assume this is perfect insulation and show the evolution of the temperature profiles within the wall when the furnace is fired to 1000°F.

(*c*) The furnace wall of part *a* is initially at a uniform temperature of 1000°F. Both sides of the wall are exposed to forced air circulation which maintains the temperature of both surfaces at 70°F. Show the temperature profiles within the wall and estimate how long it takes to cool the wall to temperatures below 200°F.

METHOD OF SOLUTION The explicit formula (6-73) is used for the solution of this problem:

$$u_{i,\,j+1} = \tfrac{1}{2}(u_{i+1,\,j} + u_{i-1,\,j}) + O(h^2 + k) \tag{6-73}$$

The value of the time increment $k$ for a stable solution is calculated from Eq. (6-72):

$$\frac{\alpha k}{h^2} = \frac{1}{2} \tag{6-72}$$

When Neumann conditions are specified, e.g.,

$$\left.\frac{\partial u}{\partial x}\right|_{0,\,t} = \beta$$

a central difference approximation of the condition is used at this boundary

$$\frac{u_{1,\,j} - u_{-1,\,j}}{2h} = \beta$$

to obtain the fictitious value of $u$ as

$$u_{-1,\,j} = u_{1,\,j} - 2h\beta$$

This modifies Eq. (6-73), at this boundary, to

$$u_{0,\,j+1} = \tfrac{1}{2}(u_{1,\,j} + u_{1,\,j} - 2h\beta)$$

Similarly, if the Neumann condition is at the right-hand boundary, Eq. (6-73) becomes

$$u_{N,\,j+1} = \tfrac{1}{2}(u_{N-1,\,j} + u_{N-1,\,j} + 2h\beta)$$

where $N$ = number of divisions in the $x$-direction.

PROGRAM DESCRIPTION This program, which is called PARABOL.BAS, consists of the sections shown in Table E6-2.

The program follows closely the method of solution described above; therefore, no further discussion is necessary.

**Table E6-2**

| Program section | Line numbers |
|---|---|
| Title | 1–90 |
| Main program | 100–270 |
| Parameter input | 110–130 |
| Subroutine calls | 140–170 |
| Rerun options | 180–260 |
| Subroutines | |
| 1. Boundary conditions | 1000–1310 |
| 2. Finite difference equations | 2000–2110 |
| 3. Print results | 3000–3080 |
| 4. Plotting | 4000–4370 |

PROGRAM

```
1 SCREEN 0:WIDTH 80:CLS:KEY OFF
5 PRINT "*********************************************************************"
10 PRINT"*                                                                   *"
15 PRINT"*                          EXAMPLE 6-2                              *"
20 PRINT"*                                                                   *"
25 PRINT"*             PARABOLIC PARTIAL DIFFERENTIAL EQUATIONS              *"
30 PRINT"*                                                                   *"
35 PRINT"*                          (PARABOL.BAS)                            *"
40 PRINT"*                                                                   *"
90 PRINT"*********************************************************************"
100 '***************************** Main Program ***************************
110 PRINT:PRINT"      ONE-DIMENSIONAL, UNSTEADY-STATE TRANSFER EQUATION"
120 PRINT:PRINT"                              PARAMETERS"
130 PRINT:INPUT"DIFFUSIVITY (AREA/TIME)=";ALPHA
140 GOSUB 1000
150 GOSUB 2000
160 TIT$="RESULTS":PRINT CHR$(12):GOSUB 3000
170 PRINT CHR$(12):GOSUB 4000
180 SCREEN 0:WIDTH 80
190 PRINT:PRINT"                            RERUN OPTIONS"
200 PRINT:PRINT"              1. SHOW SAME RESULTS AGAIN"
210 PRINT:PRINT"              2. RESET INTEGRATION PARAMETERS AND RECALCULATE"
220 PRINT:PRINT"              3. END PROGRAM"
230 PRINT:PRINT:INPUT"ENTER YOUR CHOICE(1-3)"; RQ3
240 IF RQ3<1 OR RQ3>3 THEN GOTO 230
250 ON RQ3 GOTO 160,260,270
260 PRINT:PRINT:RUN 100
270  END
1000 '************* Subroutine 1: Boundary conditions *******************
1010 '
1020 A$(1)="LEFT ":A$(2)="RIGHT"
1030 PRINT:INPUT"INITIAL CONDITION =";IC
1040 PRINT:INPUT"WIDTH (X-DIRECTION) =";WDTH
1050 PRINT:INPUT"NUMBER OF DIVISIONS IN X-DIRECTION =";NX
1060 'Calculate step sizes
1070 H=WDTH/NX:KK=H^2/(2*ALPHA)
1080 PRINT:PRINT"INCREMENT IN TIME FOR STABLE SOLUTION IS";KK;"TIME UNITS"
1090 PRINT:INPUT"HOW MANY INCREMENTS IN TIME DO YOU WANT TO CALCULATE";NY
1100 DIM U(NX,NY),XX(NX)
1110 FOR I=1 TO 2
1120 PRINT:PRINT "SPECIFY BOUNDARY CONDITION AT ";A$(I);":"
1130 PRINT:PRINT"      1. DIRICHLET CONDITION"
1140 PRINT:PRINT"      2. NEUMANN CONDITION"
1150 PRINT:INPUT"ENTER YOUR CHOICE(1-2)"; SIDE(I)
1160 IF SIDE(I)<1 OR SIDE(I)>2 THEN GOTO 1150
1170 ON SIDE(I) GOTO 1180,1200
1180 PRINT:INPUT"         BOUNDARY VALUE ="; BC(I)
1190 GOTO 1210
1200 PRINT:INPUT"         GRADIENT="; BC(I)
1210 NEXT I
1220 'Set initial conditions
1230 FOR I=0 TO NX
1240 XX(I)=I*H
1250 U(I,0)=IC
1260 NEXT I
1270 'Set conditions along boundaries
1280 FOR J=1 TO NY
1290 U(0,J)=BC(1):U(NX,J)=BC(2)
```

```
1300 NEXT J
1310 RETURN
2000 '******** Subroutine 2: Finite difference equations *****************
2010 'Apply finite difference equations
2020 FOR J=0 TO NY-1
2030 FOR I=0 TO NX
2040 'Sides with Neumann conditions
2050 IF I=0  AND SIDE(1)=2 THEN  U(I,J+1)=(U(I+1,J)+U(I+1,J)-BC(1)*2*H)/2
2060 IF I=NX AND SIDE(2)=2 THEN  U(I,J+1)=(U(I-1,J)+U(I-1,J)+BC(2)*2*H)/2
2070 IF I=0 OR I=NX GOTO 2100
2080 'Internal points
2090 U(I,J+1)=(U(I+1,J)+U(I-1,J))/2
2100 NEXT I:NEXT J
2110 RETURN
3000 '***************** Subroutine 3: Print results **********************
3010 SCREEN 0:WIDTH 80
3020 PRINT TIT$:PRINT
3030 FOR J=0 TO NY
3040 FOR I=0 TO NX
3050 PRINT USING " ####.#";U(I,J);
3060 NEXT I:PRINT:NEXT J:PRINT
3070 LOCATE 25,1:INPUT"PRESS THE ENTER KEY TO CONTINUE",KK$
3080 RETURN
4000 '************* Subroutine 4: Plotting ******************************
4010 'Locate maxima and minima for scaling axes
4020 MINX=XX(0):MAXX=XX(NX):MINY =U(0,0):MAXY =U(0,0)
4030 FOR K = 0 TO NX
4040 FOR J = 0 TO NY
4050 IF U(K ,J) < MINY THEN MINY = U(K ,J)
4060 IF U(K ,J) > MAXY THEN MAXY = U(K ,J)
4070 NEXT J:NEXT K
4080 CLS:SCREEN 1
4090 'Draw the Newmann boundaries as colored boxes
4100 IF SIDE(1)=1 AND SIDE(2)=1 GOTO 4160
4110 VIEW(0,176)-(9,182),1,1:LOCATE 23,3:PRINT"= NEUMANN BOUNDARY CONDITION"
4120 IF SIDE(1)=1 GOTO 4140
4130 VIEW(40,11)-(49,154),1,1
4140 IF SIDE(2)=1 GOTO 4160
4150 VIEW(271,11)-(280,154),1,1
4160 'Draw the axes
4170 VIEW (50,10)-(270,155)
4180 IF MINY=MAXY THEN MINY=MINY/2:MAXY=2*MAXY
4190 IF MINY=MAXY AND MINY=0 THEN MINY=-1:MAXY=1
4200 WINDOW (MINX,MINY)-(MAXX,MAXY)
4210 LINE(MINX,MINY)-(MINX,MAXY) : LINE(MAXX,MINY)-(MAXX,MAXY)
4220 'Center horizontal title
4230 LOCATE 21,17 :PRINT "X-DIRECTION"
4240 'Show maximum and minimum values
4250 LOCATE 1,2:PRINT MAXY
4260 LOCATE 21,2:PRINT MINY
4270 'Draw the profiles
4280 FOR J=0 TO NY
4290 X=XX(0):Y=U(0,J)
4300 PSET(X,Y)
4310 FOR I=1 TO NX
4320 X=XX(I) :Y=U(I,J)
4330 LINE -(X,Y)
4340 NEXT I
4350 NEXT J
4360 LOCATE 25,1:INPUT"PRESS THE ENTER KEY TO CONTINUE",KK$
4370 RETURN
```

## RESULTS

```
*******************************************************************
*                                                                 *
*                         EXAMPLE 6-2                             *
*                                                                 *
*            PARABOLIC PARTIAL DIFFERENTIAL EQUATIONS             *
*                                                                 *
*                        (PARABOL.BAS)                            *
*                                                                 *
*******************************************************************


       ONE-DIMENSIONAL, UNSTEADY-STATE TRANSFER EQUATION


                          PARAMETERS


DIFFUSIVITY (AREA/TIME)=? 0.01

INITIAL CONDITION =? 100

WIDTH (X-DIRECTION) =? 1

NUMBER OF DIVISIONS IN X-DIRECTION =? 7

INCREMENT IN TIME FOR STABLE SOLUTION IS 1.020408 TIME UNITS

HOW MANY INCREMENTS IN TIME DO YOU WANT TO CALCULATE? 25


SPECIFY BOUNDARY CONDITION AT LEFT :

      1. DIRICHLET CONDITION

      2. NEUMANN CONDITION


ENTER YOUR CHOICE(1-2)? 1

         BOUNDARY VALUE =? 1000


SPECIFY BOUNDARY CONDITION AT RIGHT:

      1. DIRICHLET CONDITION

      2. NEUMANN CONDITION


ENTER YOUR CHOICE(1-2)? 1

         BOUNDARY VALUE =? 100
```

```
RESULTS

  100.0  100.0  100.0  100.0  100.0  100.0  100.0  100.0
 1000.0  100.0  100.0  100.0  100.0  100.0  100.0  100.0
 1000.0  550.0  100.0  100.0  100.0  100.0  100.0  100.0
 1000.0  550.0  325.0  100.0  100.0  100.0  100.0  100.0
 1000.0  662.5  325.0  212.5  100.0  100.0  100.0  100.0
 1000.0  662.5  437.5  212.5  156.3  100.0  100.0  100.0
 1000.0  718.8  437.5  296.9  156.3  128.1  100.0  100.0
 1000.0  718.8  507.8  296.9  212.5  128.1  114.1  100.0
 1000.0  753.9  507.8  360.2  212.5  163.3  114.1  100.0
 1000.0  753.9  557.0  360.2  261.7  163.3  131.6  100.0
 1000.0  778.5  557.0  409.4  261.7  196.7  131.6  100.0
 1000.0  778.5  593.9  409.4  303.0  196.7  148.3  100.0
 1000.0  797.0  593.9  448.5  303.0  225.7  148.3  100.0
 1000.0  797.0  622.7  448.5  337.1  225.7  162.8  100.0
 1000.0  811.4  622.7  479.9  337.1  250.0  162.8  100.0
 1000.0  811.4  645.6  479.9  364.9  250.0  175.0  100.0
 1000.0  822.8  645.6  505.3  364.9  270.0  175.0  100.0
 1000.0  822.8  664.1  505.3  387.6  270.0  185.0  100.0
 1000.0  832.0  664.1  525.8  387.6  286.3  185.0  100.0
 1000.0  832.0  678.9  525.8  406.1  286.3  193.2  100.0
 1000.0  839.5  678.9  542.5  406.1  299.6  193.2  100.0
 1000.0  839.5  691.0  542.5  421.1  299.6  199.8  100.0
 1000.0  845.5  691.0  556.0  421.1  310.4  199.8  100.0
 1000.0  845.5  700.8  556.0  433.2  310.4  205.2  100.0
 1000.0  850.4  700.8  567.0  433.2  319.2  205.2  100.0
 1000.0  850.4  708.7  567.0  443.1  319.2  209.6  100.0

PRESS THE ENTER KEY TO CONTINUE
```

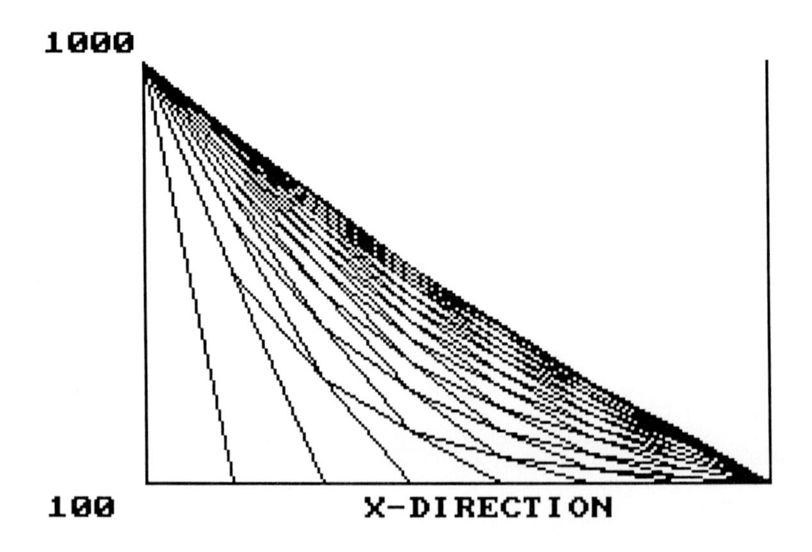


PRESS THE ENTER KEY TO CONTINUE

```
                         RERUN OPTIONS

            1. SHOW SAME RESULTS AGAIN

            2. RESET INTEGRATION PARAMETERS AND RECALCULATE

            3. END PROGRAM

ENTER YOUR CHOICE(1-3)? 2



       ONE-DIMENSIONAL, UNSTEADY-STATE TRANSFER EQUATION

                           PARAMETERS

DIFFUSIVITY (AREA/TIME)=? 0.01

INITIAL CONDITION =? 100

WIDTH (X-DIRECTION) =? 1

NUMBER OF DIVISIONS IN X-DIRECTION =? 7

INCREMENT IN TIME FOR STABLE SOLUTION IS 1.020408 TIME UNITS

HOW MANY INCREMENTS IN TIME DO YOU WANT TO CALCULATE? 25


SPECIFY BOUNDARY CONDITION AT LEFT :

      1. DIRICHLET CONDITION

      2. NEUMANN CONDITION


ENTER YOUR CHOICE(1-2)? 1

         BOUNDARY VALUE =? 1000


SPECIFY BOUNDARY CONDITION AT RIGHT:

      1. DIRICHLET CONDITION

      2. NEUMANN CONDITION


ENTER YOUR CHOICE(1-2)? 2

         GRADIENT=? 0
```

RESULTS

| | | | | | | | |
|---|---|---|---|---|---|---|---|
| 100.0 | 100.0 | 100.0 | 100.0 | 100.0 | 100.0 | 100.0 | 100.0 |
| 1000.0 | 100.0 | 100.0 | 100.0 | 100.0 | 100.0 | 100.0 | 100.0 |
| 1000.0 | 550.0 | 100.0 | 100.0 | 100.0 | 100.0 | 100.0 | 100.0 |
| 1000.0 | 550.0 | 325.0 | 100.0 | 100.0 | 100.0 | 100.0 | 100.0 |
| 1000.0 | 662.5 | 325.0 | 212.5 | 100.0 | 100.0 | 100.0 | 100.0 |
| 1000.0 | 662.5 | 437.5 | 212.5 | 156.3 | 100.0 | 100.0 | 100.0 |
| 1000.0 | 718.8 | 437.5 | 296.9 | 156.3 | 128.1 | 100.0 | 100.0 |
| 1000.0 | 718.8 | 507.8 | 296.9 | 212.5 | 128.1 | 114.1 | 100.0 |
| 1000.0 | 753.9 | 507.8 | 360.2 | 212.5 | 163.3 | 114.1 | 114.1 |
| 1000.0 | 753.9 | 557.0 | 360.2 | 261.7 | 163.3 | 138.7 | 114.1 |
| 1000.0 | 778.5 | 557.0 | 409.4 | 261.7 | 200.2 | 138.7 | 138.7 |
| 1000.0 | 778.5 | 593.9 | 409.4 | 304.8 | 200.2 | 169.4 | 138.7 |
| 1000.0 | 797.0 | 593.9 | 449.4 | 304.8 | 237.1 | 169.4 | 169.4 |
| 1000.0 | 797.0 | 623.2 | 449.4 | 343.2 | 237.1 | 203.3 | 169.4 |
| 1000.0 | 811.6 | 623.2 | 483.2 | 343.2 | 273.3 | 203.3 | 203.3 |
| 1000.0 | 811.6 | 647.4 | 483.2 | 378.2 | 273.3 | 238.3 | 203.3 |
| 1000.0 | 823.7 | 647.4 | 512.8 | 378.2 | 308.2 | 238.3 | 238.3 |
| 1000.0 | 823.7 | 668.3 | 512.8 | 410.5 | 308.2 | 273.3 | 238.3 |
| 1000.0 | 834.1 | 668.3 | 539.4 | 410.5 | 341.9 | 273.3 | 273.3 |
| 1000.0 | 834.1 | 686.8 | 539.4 | 440.6 | 341.9 | 307.6 | 273.3 |
| 1000.0 | 843.4 | 686.8 | 563.7 | 440.6 | 374.1 | 307.6 | 307.6 |
| 1000.0 | 843.4 | 703.5 | 563.7 | 468.9 | 374.1 | 340.8 | 307.6 |
| 1000.0 | 851.8 | 703.5 | 586.2 | 468.9 | 404.9 | 340.8 | 340.8 |
| 1000.0 | 851.8 | 719.0 | 586.2 | 495.5 | 404.9 | 372.9 | 340.8 |
| 1000.0 | 859.5 | 719.0 | 607.3 | 495.5 | 434.2 | 372.9 | 372.9 |
| 1000.0 | 859.5 | 733.4 | 607.3 | 520.7 | 434.2 | 403.5 | 372.9 |

PRESS THE ENTER KEY TO CONTINUE

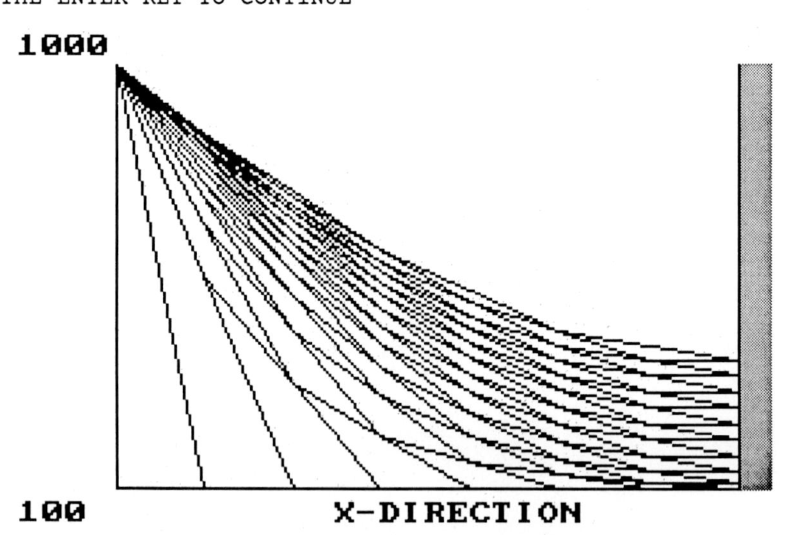

```
                          RERUN OPTIONS

            1. SHOW SAME RESULTS AGAIN

            2. RESET INTEGRATION PARAMETERS AND RECALCULATE

            3. END PROGRAM

ENTER YOUR CHOICE(1-3)? 2



       ONE-DIMENSIONAL, UNSTEADY-STATE TRANSFER EQUATION

                           PARAMETERS

DIFFUSIVITY (AREA/TIME)=? 0.01

INITIAL CONDITION =? 1000

WIDTH (X-DIRECTION) =? 1

NUMBER OF DIVISIONS IN X-DIRECTION =? 7

INCREMENT IN TIME FOR STABLE SOLUTION IS 1.020408 TIME UNITS

HOW MANY INCREMENTS IN TIME DO YOU WANT TO CALCULATE? 25


SPECIFY BOUNDARY CONDITION AT LEFT :

      1. DIRICHLET CONDITION

      2. NEUMANN CONDITION


ENTER YOUR CHOICE(1-2)? 1

         BOUNDARY VALUE =? 70


SPECIFY BOUNDARY CONDITION AT RIGHT:

      1. DIRICHLET CONDITION

      2. NEUMANN CONDITION


ENTER YOUR CHOICE(1-2)? 1

         BOUNDARY VALUE =? 70
```

RESULTS

```
 1000.0 1000.0 1000.0 1000.0 1000.0 1000.0 1000.0 1000.0
   70.0 1000.0 1000.0 1000.0 1000.0 1000.0 1000.0   70.0
   70.0  535.0 1000.0 1000.0 1000.0 1000.0  535.0   70.0
   70.0  535.0  767.5 1000.0 1000.0  767.5  535.0   70.0
   70.0  418.8  767.5  883.8  883.8  767.5  418.8   70.0
   70.0  418.8  651.3  825.6  825.6  651.3  418.8   70.0
   70.0  360.6  622.2  738.4  738.4  622.2  360.6   70.0
   70.0  346.1  549.5  680.3  680.3  549.5  346.1   70.0
   70.0  309.8  513.2  614.9  614.9  513.2  309.8   70.0
   70.0  291.6  462.3  564.1  564.1  462.3  291.6   70.0
   70.0  266.2  427.8  513.2  513.2  427.8  266.2   70.0
   70.0  248.9  389.7  470.5  470.5  389.7  248.9   70.0
   70.0  229.8  359.7  430.1  430.1  359.7  229.8   70.0
   70.0  214.9  330.0  394.9  394.9  330.0  214.9   70.0
   70.0  200.0  304.9  362.4  362.4  304.9  200.0   70.0
   70.0  187.4  281.2  333.7  333.7  281.2  187.4   70.0
   70.0  175.6  260.6  307.4  307.4  260.6  175.6   70.0
   70.0  165.3  241.5  284.0  284.0  241.5  165.3   70.0
   70.0  155.8  224.6  262.8  262.8  224.6  155.8   70.0
   70.0  147.3  209.3  243.7  243.7  209.3  147.3   70.0
   70.0  139.6  195.5  226.5  226.5  195.5  139.6   70.0
   70.0  132.8  183.1  211.0  211.0  183.1  132.8   70.0
   70.0  126.5  171.9  197.0  197.0  171.9  126.5   70.0
   70.0  120.9  161.8  184.4  184.4  161.8  120.9   70.0
   70.0  115.9  152.7  173.1  173.1  152.7  115.9   70.0
   70.0  111.3  144.5  162.9  162.9  144.5  111.3   70.0
```

PRESS THE ENTER KEY TO CONTINUE

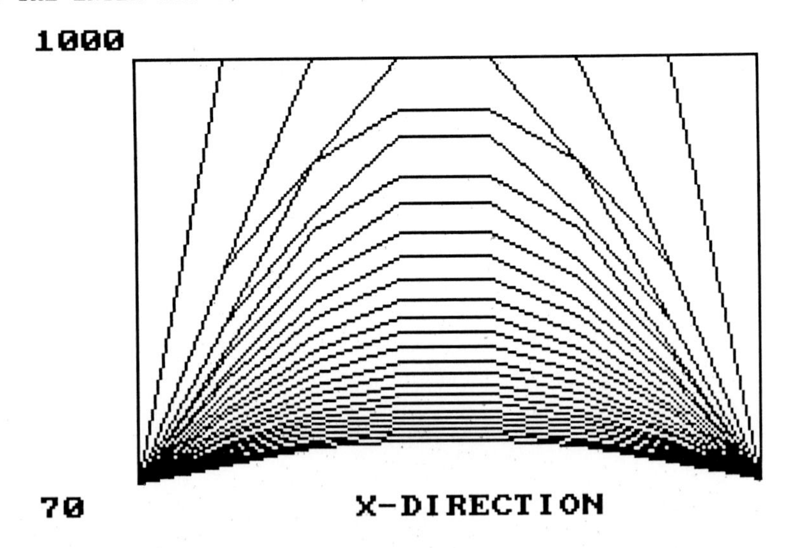


PRESS THE ENTER KEY TO CONTINUE

```
                    RERUN OPTIONS

          1. SHOW SAME RESULTS AGAIN

          2. RESET INTEGRATION PARAMETERS AND RECALCULATE

          3. END PROGRAM

ENTER YOUR CHOICE(1-3)? 3
Ok
```

### Discussion of results

*Part a* Given a thermal diffusivity of 0.01 ft$^2$/h, a width of 1 ft, and the number of divisions in the $x$-direction = 7, the program calculates the time increment for stable solution as follows:

$$k = \frac{h^2}{2\alpha} = \frac{(\frac{1}{7})^2}{(2)(0.01)} = 1.02 \text{ h}$$

Taking 25 time increments, the temperature profile progresses from the lower left of the figure toward the center. If the integration is continued for a sufficiently long time, the profile will reach the steady state, which for this case is a straight line connecting the two Dirichlet conditions. This is easily verified from the analytical solution of the steady-state problem,

$$\frac{\partial^2 u}{\partial x^2} = 0$$

which is the equation of the straight line,

$$u = c_1 x + c_2$$

where the constants, $c_1 = -900$ and $c_2 = 1000$, are calculated from the Dirichlet conditions.

*Part b* In this case, the insulation installed on the outside surface of the furnace wall causes the temperature within the wall to continue rising. The steady-state temperature profile would be $u = 1000$ throughout the solid wall. This is also verifiable from the analytical solution of the steady-state problem in conjunction with the imposed boundary conditions.

*Part c* The cooling of the wall occurs from both sides, and the temperature profile moves, symmetrically, from the initial temperature of 1000°F to the limit of 70°F. It takes 22 time increments (approximately 22.5 h) for the temperature to drop below 200°F at the center point of the wall.

### 6-4.4 Solution of Hyperbolic Partial Differential Equations

Second-order partial differential equations of the hyperbolic type occur principally in physical problems connected with vibration processes. For example, the *one-dimensional* wave equation†

$$\rho \frac{\partial^2 u}{\partial t^2} = T_0 \frac{\partial^2 u}{\partial x^2} + f(x, t) \tag{6-82}$$

describes the transverse motion of a vibrating string which is subjected to tension $T_0$ and external force $f(x, t)$. In the case of constant density $\rho$, the equation is written in the form

$$\frac{\partial^2 u}{\partial t^2} = a^2 \frac{\partial^2 u}{\partial x^2} + F(x, t) \tag{6-83}$$

where

$$a^2 = \frac{T_0}{\rho}$$

$$F(x, t) = \frac{1}{\rho} f(x, t)$$

If no external force acts on the string, (6-83) becomes a homogeneous equation:

$$\frac{\partial^2 u}{\partial t^2} = a^2 \frac{\partial^2 u}{\partial x^2} \tag{6-84}$$

The two-dimensional extension of equation (6-83) is†

$$\frac{\partial^2 u}{\partial t^2} = a^2\left(\frac{\partial^2 u}{\partial x^2} + \frac{\partial^2 u}{\partial y^2}\right) + F(x, y, t) \tag{6-85}$$

which describes the vibrations of a membrane subjected to tension $T_0$ and external force $f(x, y, t)$.

To find the numerical solution of Eq. (6-84) we expand each second-order derivative in terms of central finite differences to obtain

$$\frac{u_{i,t+1} - 2u_{i,t} + u_{i,t-1}}{k^2} = a^2\left(\frac{u_{i+1,t} - 2u_{i,t} + u_{i-1,t}}{h^2}\right) + O(h^2 + k^2) \tag{6-86}$$

where the counter $i$ is used for the $x$-direction and $t$ for the $t$-direction. Rearranging to solve for $u_{i,t+1}$,

† For derivations of these equations see Tychonov and Samarski [3, p. 12].

$$u_{i,\,t+1} = 2u_{i,\,t} - u_{i,\,t-1} + \frac{a^2k^2}{h^2}(u_{i+1,\,t} - 2u_{i,\,t} + u_{i-1,\,t}) + O(h^2 + k^2) \tag{6-87}$$

This is an *explicit* numerical solution of the hyperbolic equation (6-84).

Its stability depends on the value of $a^2k^2/h^2$. It is shown in Sec. 6-5 that a stable solution can be obtained for

$$\frac{a^2k^2}{h^2} \leq 1 \tag{6-88}$$

Similarly, the homogeneous form of the two-dimensional hyperbolic equation

$$\frac{\partial^2 u}{\partial t^2} = a^2\left(\frac{\partial^2 u}{\partial x^2} + \frac{\partial^2 u}{\partial y^2}\right) \tag{6-89}$$

is expanded using central finite difference approximations to yield

$$\begin{aligned} \frac{u_{i,\,j,\,t+1} - 2u_{i,\,j,\,t} + u_{i,\,j,\,t-1}}{k^2} &= a^2\left(\frac{u_{i+1,\,j,\,t} - 2u_{i,\,j,\,t} + u_{i-1,\,j,\,t}}{h^2}\right) + a^2\left(\frac{u_{i,\,j+1,\,t} - 2u_{i,\,j,\,t} + u_{i,\,j-1,\,t}}{h^2}\right) \\ &\quad + O(h^2 + h^2 + k^2) \end{aligned} \tag{6-90}$$

An equidistant grid in the $x$- and $y$-directions is used. This equation rearranges to the explicit form

$$\begin{aligned} u_{i,\,j,\,t+1} &= 2\left[1 - 2\left(\frac{a^2k^2}{h^2}\right)\right]u_{i,\,j,\,t} - u_{i,\,j,\,t-1} \\ &\quad + \frac{a^2k^2}{h^2}(u_{i+1,\,j,\,t} + u_{i-1,\,j,\,t} + u_{i,\,j+1,\,t} + u_{i,\,j-1,\,t}) \end{aligned} \tag{6-91}$$

This solution is stable when

$$\frac{a^2k^2}{h^2} \leq \frac{1}{2} \tag{6-92}$$

*Implicit* methods for the solution of hyperbolic partial differential equations can be developed using the *variable-weight* approach, where the space partial derivatives are weighted at $t+1$, $t$, and $t-1$. The implicit formulation of (6-84) is

$$\frac{u_{i,t+1} - 2u_{i,t} + u_{i,t-1}}{k^2} = \frac{a^2}{h^2}[\theta(u_{i+1,t+1} - 2u_{i,t+1} + u_{i-1,t+1}) + (1-2\theta)(u_{i+1,t} - 2u_{i,t} + u_{i-1,t}) + \theta(u_{i+1,t-1} - 2u_{i,t-1} + u_{i-1,t-1})] \tag{6-93}$$

where $0 \leq \theta \leq 1$. When $\theta = 0$, Eq. (6-93) reverts back to the explicit method (6-87). When $\theta = \frac{1}{2}$, (6-93) is a Crank-Nicolson-type approximation. Implicit methods yield tridiagonal sets of linear algebraic equations whose solutions can be obtained using Gauss elimination methods.

### 6-4.5 Irregular Boundaries and Polar Coordinate Systems

The finite difference approximations of partial differential equations developed in this chapter so far were based on regular cartesian coordinate systems. Quite often, however, the objects, whose properties are being modeled by the partial differential equations, may have circular, cylindrical, or spherical shapes, or may have altogether irregular boundaries. The finite difference approximations may be modified to handle such cases.

Let us first consider an object which is well described by cartesian coordinates everywhere except near the boundary which is of irregular shape, as shown in Fig. 6-9. The second partial derivative in the $x$-direction evaluated at point $(i, j)$ is given by†

$$\left.\frac{\partial^2 u}{\partial x^2}\right|_{i,j} = \left(\frac{1}{h^2}\right)\left[\frac{2}{\alpha(\alpha+1)}\right][u_{i+1,j} - (1+\alpha)u_{i,j} + \alpha u_{i-1,j}] \tag{6-94}$$

Similarly, in the $y$-direction,

$$\left.\frac{\partial^2 u}{\partial y^2}\right|_{i,j} = \left(\frac{1}{k^2}\right)\left[\frac{2}{\beta(\beta+1)}\right][u_{i,j+1} - (1+\beta)u_{i,j} + \beta u_{i,j-1}] \tag{6-95}$$

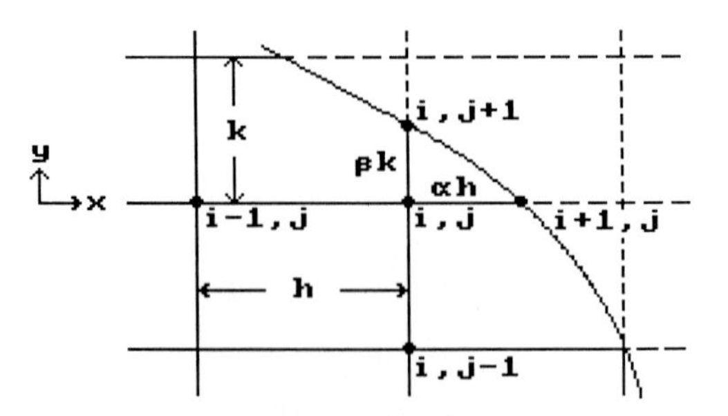


**Figure 6-9** Finite difference grid for irregular boundaries.

† For derivation see Salvadori and Baron [9, p. 67].

When $\alpha = \beta = 1$, Eqs. (6-94) and (6-95) become identical to those developed earlier in this chapter for regular cartesian coordinate systems. Therefore, for objects with irregular boundaries, the partial differential equations would be converted to algebraic equations using (6-94) and (6-95). For points adjacent to the boundary the parameters $\alpha$ and $\beta$ would assume values which reflect the irregular shape of the boundary and for internal points away from the boundary the value of $\alpha$ and $\beta$ would be unity.

Cylindrical shaped objects are more conveniently expressed in polar coordinates. The transformation from cartesian coordinate systems to polar coordinate systems is performed using the following relationships, which are based on Fig. 6-10:

$$x = r\cos\theta \qquad y = r\sin\theta$$
$$r = \sqrt{x^2 + y^2} \qquad \theta = \tan^{-1}\frac{y}{x} \tag{6-96}$$

The Laplacian operator in polar coordinates becomes

$$\frac{\partial^2 u}{\partial x^2} + \frac{\partial^2 u}{\partial y^2} = \frac{\partial^2 u}{\partial r^2} + \frac{1}{r}\frac{\partial u}{\partial r} + \frac{1}{r^2}\frac{\partial^2 u}{\partial \theta^2} \tag{6-97}$$

Fick's second law of diffusion [Eq. (6-10)] in polar coordinates is

$$\frac{\partial c_A}{\partial t} = D_{AB}\left(\frac{\partial^2 c_A}{\partial r^2} + \frac{1}{r}\frac{\partial c_A}{\partial r} + \frac{1}{r^2}\frac{\partial^2 c_A}{\partial \theta^2} + \frac{\partial^2 c_A}{\partial z^2}\right) \tag{6-98}$$

Using the finite difference grid for polar coordinates shown in Fig. 6-11, the partial derivatives are approximated by

$$\left.\frac{\partial^2 u}{\partial r^2}\right|_{i,j} = \frac{1}{h^2}\left(u_{i,j+1} - 2u_{i,j} + u_{i,j-1}\right) \tag{6-99}$$

$$\left.\frac{\partial^2 u}{\partial \theta^2}\right|_{i,j} = \frac{1}{\gamma^2}\left(u_{i+1,j} - 2u_{i,j} + u_{i-1,j}\right) \tag{6-100}$$

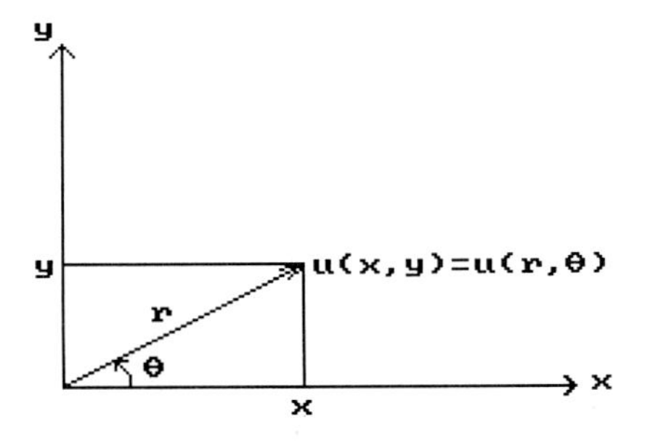


**Figure 6-10** Transformation to polar coordinates.

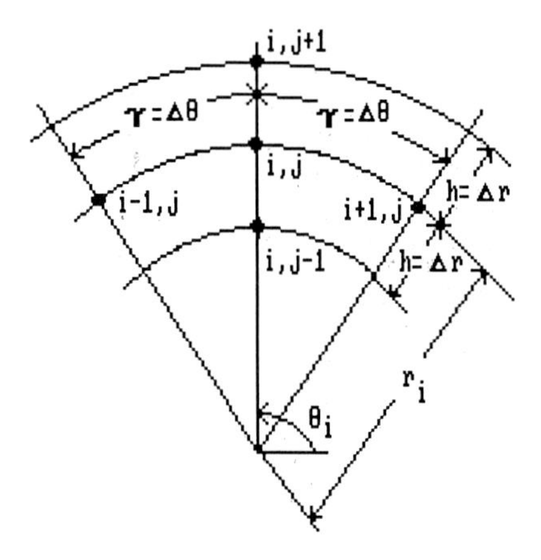


**Figure 6-11** Finite difference grid for polar coordinates.

$$\left.\frac{\partial u}{\partial r}\right|_{i,j} = \frac{1}{2h}(u_{i,j+1} - u_{i,j-1}) \tag{6-101}$$

Partial derivatives in the $z$- and $t$-directions (not shown in Fig. 6-11) would be similarly expressed, with the use of additional subscripts.

### 6-4.6 Nonlinear Partial Differential Equations

The discussion in this chapter has focused on linear partial differential equations which yield sets of linear algebraic equations when expressed in finite difference approximations. On the other hand, if the partial differential equation is nonlinear, e.g.,

$$u\frac{\partial^2 u}{\partial x^2} + u\frac{\partial^2 u}{\partial y^2} = f(u) \tag{6-102}$$

the resulting finite difference discretization would generate sets of nonlinear algebraic equations. The solution of this problem would require the application of Newton's method for simultaneous nonlinear algebraic equations (see Chap. 2).

## 6-5 STABILITY ANALYSIS

In this section we discuss the stability of finite difference approximations using the well-known von Neumann procedure. This method introduces an initial error represented by a finite Fourier series and examines how this error propagates during the solution. The von Neumann method applies to initial-value problems; for this reason it is used to analyze the stability of the explicit method for parabolic equations developed in Sec. 6-4.3 and the explicit method for hyperbolic equations developed in Sec. 6-4.4.

Define the error $\epsilon_{n,m}$ as the difference between the solution $u_{n,m}$ of the finite difference approximation and the exact solution $\bar{u}_{n,m}$ of the differential equation at step $(n, m)$:

$$\epsilon_{n,m} \equiv u_{n,m} - \bar{u}_{n,m} \tag{6-103}$$

The explicit finite difference solution (6-69) of the parabolic partial differential equation (6-63) can be written for $u_{n,m+1}$ and $\bar{u}_{n,m+1}$ as follows:

$$u_{n,m+1} = \left(\frac{\alpha k}{h^2}\right)u_{n+1,m} + \left(1 - \frac{2\alpha k}{h^2}\right)u_{n,m} + \left(\frac{\alpha k}{h^2}\right)u_{n-1,m} + R_{E_{n,m+1}} \tag{6-104}$$

$$\bar{u}_{n,m+1} = \left(\frac{\alpha k}{h^2}\right)\bar{u}_{n+1,m} + \left(1 - \frac{2\alpha k}{h^2}\right)\bar{u}_{n,m} + \left(\frac{\alpha k}{h^2}\right)\bar{u}_{n-1,m} + T_{E_{n,m+1}} \tag{6-105}$$

where $R_{E_{n,m+1}}$ and $T_{E_{n,m+1}}$ are the roundoff and truncation errors, respectively, at step $(n, m+1)$.

Combining Eqs. (6-103) to (6-105) we obtain

$$\epsilon_{n,m+1} - \left(\frac{\alpha k}{h^2}\right)\epsilon_{n+1,m} - \left(1 - 2\frac{\alpha k}{h^2}\right)\epsilon_{n,m} - \left(\frac{\alpha k}{h^2}\right)\epsilon_{n-1,m} = R_{E_{n,m+1}} + T_{E_{n,m+1}} \tag{6-106}$$

This is a *nonhomogeneous finite difference equation in two dimensions*, representing the propagation of error during the numerical solution of the parabolic partial differential equation (6-63). The solution of this finite difference equation is rather difficult to obtain. For this reason the von Neumann analysis considers the *homogeneous* part of Eq. (6-106):

$$\epsilon_{n,m+1} - \left(\frac{\alpha k}{h^2}\right)\epsilon_{n+1,m} - \left(1 - 2\frac{\alpha k}{h^2}\right)\epsilon_{n,m} - \left(\frac{\alpha k}{h^2}\right)\epsilon_{n-1,m} = 0 \tag{6-107}$$

which represents the propagation of the error introduced at the initial point $(m = 0)$ only and ignores truncation and roundoff errors that enter the solution at $m > 0$.

The solution of the homogeneous finite difference equation may be written in the separable form

$$\epsilon_{n,m} = ce^{\gamma mk}e^{i\beta nh} \tag{6-108}$$

where $i = \sqrt{-1}$ and $c$, $\gamma$, $\beta$ are constants. At $m = 0$

$$\epsilon_{n,0} = ce^{i\beta nh} \tag{6-109}$$

which is the error at the initial point. Therefore, the term $\epsilon^{\gamma k}$ is the *amplifica-*

*tion factor* of the initial error. In order for the original error not to grow as $m$ increases, the amplification factor must satisfy the *von Neumann condition for stability*:

$$|e^{\gamma k}| \le 1 \tag{6-110}$$

The amplification factor can have complex values. In that case, the modulus of the complex numbers must satisfy the above inequality, i.e.,

$$|r| \le 1 \tag{6-111}$$

Therefore the stability region in the complex plane is a circle of radius = 1, as shown in Fig. 6-12.†

The amplification factor is determined by substituting (6-108) into (6-107) and rearranging to obtain

$$e^{\gamma k} = \left(1 - 2\frac{\alpha k}{h^2}\right) + \frac{\alpha k}{h^2}\,(e^{i\beta h} + e^{-i\beta h}) \tag{6-112}$$

Using the trigonometric identities

$$\frac{e^{i\beta h} + e^{-i\beta h}}{2} = \cos \beta h \tag{6-113}$$

and

$$1 - \cos \beta h = 2 \sin^2 \frac{\beta h}{2} \tag{6-114}$$

Eq. (6-112) becomes

$$e^{\gamma k} = 1 - \left(4\frac{\alpha k}{h^2}\right)\left(\sin^2 \frac{\beta h}{2}\right) \tag{6-115}$$

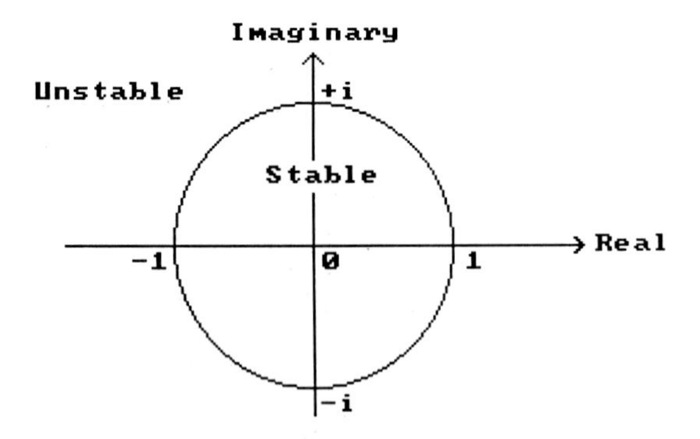


**Figure 6-12** Stability region in the complex plane.

† For a complete discussion of stability analysis of hyperbolic equations see Vichnevetsky and Bowles [10].

Combining this with the von Neumann condition for stability, we obtain the stability bound

$$0 < \frac{\alpha k}{h^2}\left(\sin^2 \frac{\beta h}{2}\right) \leq \frac{1}{2} \tag{6-116}$$

The $\sin^2(\beta h/2)$ term has its highest value equal to unity; therefore,

$$0 < \frac{\alpha k}{h^2} \leq \frac{1}{2} \tag{6-117}$$

is the limit for conditional stability for this method. It should be noted that this limit is identical to that obtained by using the positivity rule (Sec. 6-4.3).

The stability of the explicit solution (6-87) of the hyperbolic equation (6-84) can be similarly analyzed using the von Neumann method. The homogeneous equation for the error propagation of that solution is

$$\epsilon_{n,\,m+1} - 2\left(1 - \frac{a^2k^2}{h^2}\right)\epsilon_{n,\,m} - \frac{a^2k^2}{h^2}(\epsilon_{n+1,\,m} + \epsilon_{n-1,\,m}) + \epsilon_{n,\,m-1} = 0 \tag{6-118}$$

Substitution of the solution (6-108) into (6-118) and use of the trigonometric identities (6-113) and (6-114) give the amplification factor as

$$\epsilon^{\gamma\kappa} = \left(1 - 2\frac{a^2k^2}{h^2}\sin^2\frac{\beta h}{2}\right) \pm \sqrt{\left(1 - 2\frac{a^2k^2}{h^2}\sin^2\frac{\beta h}{2}\right)^2 - 1} \tag{6-119}$$

The above amplification factor satisfies inequality (6-111) in the complex plane, i.e., when

$$\left(1 - 2\frac{a^2k^2}{h^2}\sin^2\frac{\beta h}{2}\right)^2 - 1 \leq 0 \tag{6-120}$$

which converts to the following inequality,

$$\frac{a^2k^2}{h^2} \leq \frac{1}{\sin^2(\beta h/2)} \tag{6-121}$$

The $\sin^2(\beta h/2)$ term has its highest value equal to unity; therefore,

$$\frac{a^2k^2}{h^2} \leq 1 \tag{6-122}$$

is the conditional stability limit for this method.

In a similar manner the stability of other explicit and implicit finite difference methods may be examined. This has been done by Lapidus and Pinder [2], who conclude that "most explicit finite difference approximations are conditionally stable, whereas most implicit approximations are unconditionally stable."

## 6-6 INTRODUCTION TO FINITE ELEMENT METHODS

The finite element methods are powerful techniques for the numerical solution of differential equations. Their theoretical formulation is based on the variational principle. The minimization of the functional of the form

$$J(U) = \int_D \phi\left(U, \frac{\partial U}{\partial x}, \frac{\partial U}{\partial y}, x, y\right) dD \tag{6-123}$$

must satisfy the Euler-Lagrange equation

$$\frac{\partial}{\partial x}\left[\frac{\partial \phi}{\partial(\partial U/\partial x)}\right] + \frac{\partial}{\partial y}\left[\frac{\partial \phi}{\partial(\partial U/\partial y)}\right] - \frac{\partial \phi}{\partial U} = 0 \tag{6-124}$$

which is a partial differential equation with certain natural boundary conditions.

It has been shown that many differential equations which originate from the physical sciences have equivalent variational formulations.† This is the basis for the well-known Rayleigh-Ritz procedure which in turn forms the basis for the finite element methods.

An equivalent formulation of finite element methods can be developed using the concept of weighted residuals. In Sec. 5-7.3 we discussed the method of weighted residuals in connection with the solution of the two-point boundary-value problem. In that case we chose the solution of the ordinary differential equation as a polynomial *basis function* and caused the integral of weighted residuals to vanish:

$$\int_0^1 W_k R(x)\, dx = 0 \tag{5-207}$$

We now extend this method to the solution of partial differential equations where the desired solution $u(\cdot)$ is replaced by a piecewise polynomial approximation of the form

$$u(\cdot) = \sum_{j=1}^{N} a_j \phi_j(\cdot) \tag{6-125}$$

The set of functions $\{\phi_j(\cdot) | j = 1, 2, \ldots, N\}$ are the basis functions and the $\{a_j | j = 1, 2, \ldots, N\}$ are undetermined coefficients. The integral of weighted residuals is made to vanish.

$$\int_t \int_V W_j(\cdot) R(\cdot)\, dV\, dt = 0 \tag{6-126}$$

† For a complete discussion of the variational formulation of the finite element method, see Vichnevetsky [4, p. 179], and Vemuri and Karplus [11, p. 230].

The choice of basis functions $\phi_j(\cdot)$ and weighting functions $W_j(\cdot)$ determines the particular finite element method. The *Galerkin* method chooses the basis and weighting functions to be identical to each other. The *orthogonal collocation* method uses the Dirac delta functions for weights and orthogonal polynomials for basis functions. The *subdomain* method chooses the weighting function to be unity in the subregion $V_i$, for which it is defined, and zero elsewhere. A complete discussion of the finite element methods is outside the scope of this book. The interested reader is referred to Lapidus and Pinder [2], Finlayson [6], Fairweather [12], Vichnevetsky [4], Norrie and de Vries [13], and Vemuri and Karplus [11] for detailed development of these methods.

# PROBLEMS

**6-1** Modify the program in Example 6-1 to solve for the three-dimensional problem

$$\frac{\partial^2 u}{\partial x^2} + \frac{\partial^2 u}{\partial y^2} + \frac{\partial^2 u}{\partial z^2} = 0$$

Apply this program to calculate the distribution of the dependent variable within a solid body which is subjected to the following boundary conditions:

$$u(0, y, z) = 100 \qquad u(1, y, z) = 100$$

$$u(x, 0, z) = 0 \qquad u(x, 1, z) = 0$$

$$u(x, y, 0) = 50 \qquad u(x, y, 1) = 50$$

**6-2** Solve Laplace's equation with the following boundary conditions:

$$u(0, y) = 100$$

$$\left.\frac{\partial u}{\partial x}\right|_{10,\, y} = 10$$

$$\left.\frac{\partial u}{\partial y}\right|_{x,\, 0} = 0$$

$$\left.\frac{\partial u}{\partial y}\right|_{x,\, 1} = 0$$

Discuss the results. Determine the optimum value of the overrelaxation parameter for this problem.

**6-3** Extend the program in Example 6-2 to include Robbins boundary conditions of the type shown below:

$$k\,\frac{\partial u}{\partial x} = h(u - u_f) \qquad \text{at } x = 0 \text{ and } t \geq 0$$

and

$$-k\,\frac{\partial u}{\partial x} = h(u - u_f) \qquad \text{at } x = X \text{ and } t \geq 0$$

where $u$ is the value of the dependent variable at the boundary and $u_f$ is a known value of the dependent variable in the fluid next to the boundary; $k$ and $h$ are known constants.

Apply this program to solve the following problem: The ambient temperature surrounding a house is 50°F. The heat in the house had been turned off; therefore, the internal temperature is also at 50°F at $t = 0$. The heating system is turned on and raises the internal temperature to 70°F at the rate of 4°F/h. The ambient temperature remains at 50°F. The wall of the house is 0.5 ft thick and is made of material which has an average thermal diffusivity $\alpha = 0.01\ \text{ft}^2/\text{h}$ and a thermal conductivity $k = 0.2\ \text{Btu}/(\text{h} \cdot \text{ft} \cdot {}^\circ\text{F})$. The heat transfer coefficient on the inside of the wall is $h_{\text{in}} = 1.0\ \text{Btu}/(\text{h} \cdot \text{ft}^2 \cdot {}^\circ\text{F})$, and the heat transfer coefficient on the outside is $h_{\text{out}} = 2.0\ \text{Btu}/(\text{h} \cdot \text{ft}^2 \cdot {}^\circ\text{F})$. Estimate how long it will take to reach a steady-state temperature distribution across the wall.

**6-4** Develop the finite difference approximation of Fick's second law of diffusion in polar coordinates. Write a computer program which can be used to solve the following problem [14, p. 330]:

> A wet cylinder of agar gel at 278 K with a uniform concentration of urea of $0.1\ \text{kg} \cdot \text{mol}/\text{m}^3$ has a diameter of 30.48 mm and is 38.1 mm long with flat parallel ends. The diffusivity is $4.72 \times 10^{-10}\ \text{m}^2/\text{s}$. Calculate the concentration at the midpoint of the cylinder after 100 h for the following cases if the cylinder is suddenly immersed in turbulent pure water:
> (*a*) For radial diffusion only
> (*b*) Diffusion that occurs radially and axially

**6-5** Express the two-dimensional parabolic partial differential equation

$$\frac{\partial u}{\partial t} = \alpha\left(\frac{\partial^2 u}{\partial x^2} + \frac{\partial^2 u}{\partial y^2}\right)$$

in an explicit finite difference formulation. Determine the limits of conditional stability for this method using
(*a*) The von Neumann stability analysis
(*b*) The positivity rule

**6-6** Consider a first-order chemical reaction being carried out under isothermal steady-state conditions in a tubular-flow reactor. On the assumptions of laminar flow and negligible axial diffusion, the material balance equation is

$$-v_0\left[1 - \left(\frac{r}{R}\right)^2\right]\frac{\partial c}{\partial z} + D\left(\frac{\partial^2 c}{\partial r^2} + \frac{1}{r}\frac{\partial c}{\partial r}\right) - kc = 0$$

where $v_0$ = velocity of central stream line
$R$ = tube radius
$k$ = reaction-velocity constant
$c$ = concentration of reactant
$D$ = radial diffusion constant
$z$ = axial distance down tube
$r$ = radial distance from center

Upon defining the following dimensionless variables,

$$\lambda = \frac{kz}{v_0} \qquad C = \frac{c}{c_0} \qquad \alpha = \frac{D}{kR^2} \qquad U = \frac{r}{R}$$

the equation becomes

$$(1 - U^2)\frac{\partial C}{\partial \lambda} = \alpha\left(\frac{\partial^2 C}{\partial U^2} + \frac{1}{U}\frac{\partial C}{\partial U}\right) - C$$

where $c_0$ is the entering concentration of the reactant to the reactor.

(*a*) Choose a set of appropriate boundary conditions for this problem. Explain your choice.

(*b*) What class of PDE is the above equation (hyperbolic, parabolic, or elliptic)?

(*c*) Set up the equation for numerical solution using finite difference approximations.

(*d*) Does your choice of finite differences result in an explicit or implicit set of equations? Give details of the procedure for the solution of this set of equations.

(*e*) Discuss stability considerations with respect to the method you have chosen.

**6-7** A 12-in-square membrane (no bending or shear stresses), with a 4-in-square hole in the middle, is fastened at the outside and inside boundaries as shown in Fig. P6-7. If a highly stretched membrane is subjected to a pressure $p$, the partial differential equation for the deflection $w$ in the $z$-direction is

$$\frac{\partial^2 w}{\partial x^2} + \frac{\partial^2 w}{\partial y^2} = -\frac{p}{T}$$

where $T$ is the tension (pounds per linear inch). For a highly stretched membrane, the tension $T$ may be assumed constant for small deflections [5, p. 562]. Utilizing the following values of pressure and tension,

$$p = 5 \text{ psi} \qquad \text{(uniformly distributed)}$$

$$T = 100 \text{ lb/in}$$

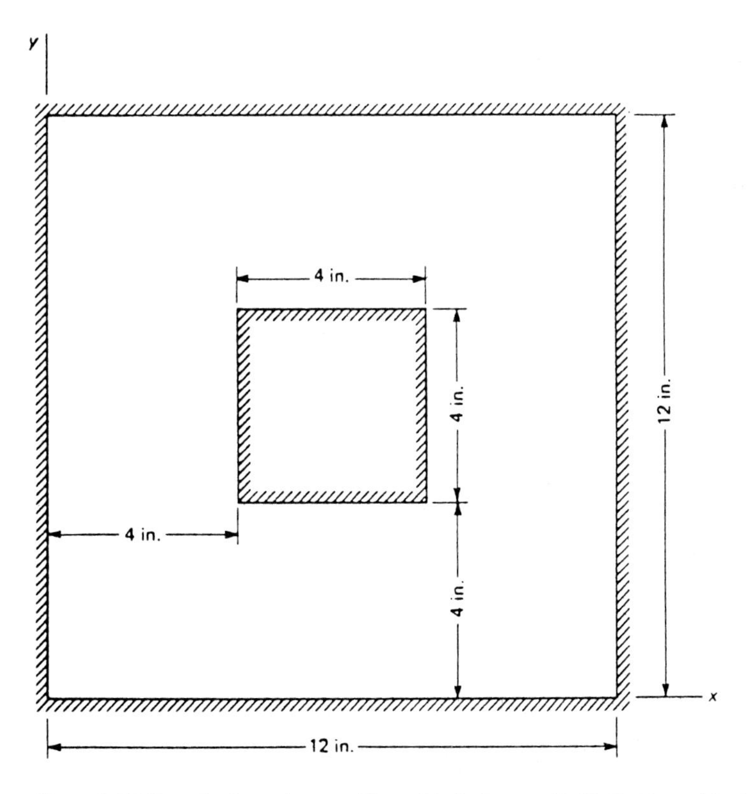


**Figure P6-7** Stretched membrane. *(From M. L. James, G. M. Smith, and J. C. Wolford, Ref. 5, by permission of the publisher.)*

(*a*) Express the differential equation in finite difference form to obtain the deflections $w$ of the membrane.

(*b*) List all the boundary conditions needed for the numerical solution of the problem. Utilize some or all of these boundary conditions to simplify the finite difference equations of part *a*.

(*c*) Write a computer program for the numerical solution of this problem.

**6-8** Figure P6.8 shows a cross section of a long cooling fin of width $W$, thickness $t$, and thermal conductivity $k$ that is bonded to a hot wall, maintaining its base (at $x = 0$) at a temperature $T_w$ [7, p. 528]. Heat is conducted steadily through the fin in the plane of Fig. P6-8 so that the fin temperature $T$ obeys Laplace's equation, $\partial^2 T/\partial x^2 + \partial^2 T/\partial y^2 = 0$. (Temperature variations along the length of the fin in the $z$-direction are ignored.)

Heat is lost from the sides and tip of the fin by convection to the surrounding air (radiation is neglected at sufficiently low temperatures) at a local rate $q = h(T_s - T_a)$ Btu/(h · ft$^2$). Here, $T_s$ and $T_a$, in degrees Fahrenheit, are the temperatures at a point on the fin surface and of the air, respectively. If the surface of the fin is vertical, the heat transfer coefficient $h$ obeys the dimensional correlation $h = 0.21(T_s - T_a)^{1/3}$.

(*a*) Set up the equations for a numerical solution of this problem to determine the temperatures at a finite number of points within the fin and at the surface of the fin.

(*b*) Describe in detail the step-by-step procedure for solving the equations of part *a* and evaluating the temperatures within the fin and at the surface.

(*c*) Write a computer program to find the numerical solution of this problem using the following quantities:

$$T_w = 200°\text{F} \qquad T_a = 70°\text{F}$$

$$t = 0.25 \text{ in} \qquad k = 25.9 \text{ Btu/(h} \cdot \text{ft} \cdot °\text{F)}$$

$$W = 0.5 \text{ in}$$

**6-9** Consider a steady-state plug flow reactor of length $z$ through which a substrate is flowing with a constant velocity $v$ with no dispersion effects. The reactor is made up of a series of collagen membranes, each impregnated with two enzymes catalyzing the sequential reaction:

$$\text{A} \xrightarrow{\text{Enzyme 1}} \text{B} \xrightarrow{\text{Enzyme 2}} \text{C}$$

The membranes in the reactor are arranged in parallel, as shown in Fig. P6-9 [15]. The nomenclature for this problem is shown in Table P6-9.

For a substrate molecule to encounter the immobilized enzymes, it must diffuse across a Nernst diffusion layer on the surface of the support and then some distance into the membrane.

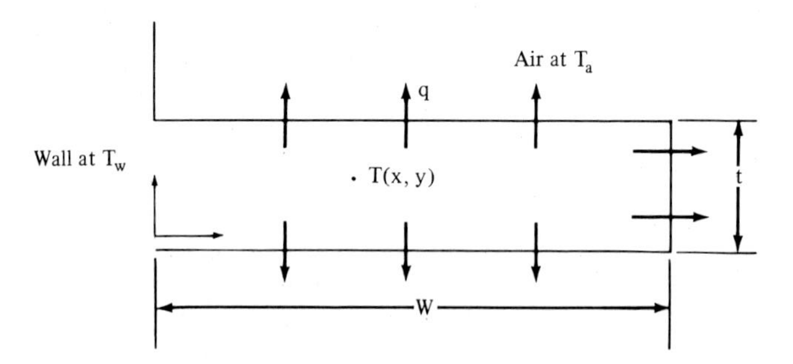


**Figure P6-8** Cooling fin. (*Reproduced from B. Carnahan, H. A. Luther, and J. O. Wilkes, Ref. 7, by permission of the publisher.*)

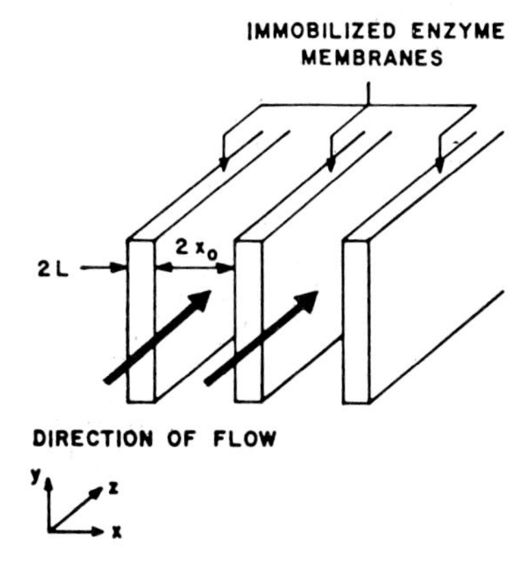


**Figure P6-9** Biocatalytic reactor.

The coupled reaction takes place in the membrane and the product, the unreacted intermediate, and substrate diffuse back into the bulk fluid phase. No inactivation of the enzymes occurs and it is assumed that the enzymes behave independently of each other.

Since the membrane can accommodate only a finite number of enzyme molecules per unit weight, it becomes necessary to introduce a control parameter $\epsilon$ which measures the ratio of molar concentration of enzyme 1 to molar concentration of enzymes 1 and 2. It is implicitly assumed that the binding sites on collagen do not discriminate between the enzymes. Thus, when both enzymes are present, the maximum reaction velocities reduce to $\epsilon V_1$ and $(1-\epsilon)V_2$. The control $\epsilon$ is constrained between the bounds of 0 (only enzyme 2 present) and 1 (only enzyme 1 present).

The reaction rates for the two sequential reactions are given by the Michaelis-Menten relationship:

## Table P6-9 Nomenclature

| | |
|---|---|
| $C_{\mathrm{Af}}$ = | concentation of A in feed, mol/L |
| $D$ = | molecular diffusivity of reactants or products in membrane, $\mathrm{cm^2/s}$ |
| $k_L$ = | overall mass transfer coefficient in the fluid phase, cm/s |
| $K_{\mathrm{M1}}, K_{\mathrm{M2}}$ = | Michaelis-Menten constant for enzymes 1 and 2, mol/L |
| $L$ = | half thickness of membrane, mils |
| $\nu$ = | superficial fluid velocity in reactor, cm/s |
| $V_1, V_2$ = | maximum reaction velocity for enzymes 1 and 2, $\mathrm{mol/(L \cdot s)}$ |
| $x$ = | variable axial distance from center of membrane to surface, cm |
| $x_0$ = | half distance between two consecutive membranes, cm |
| $X$ = | $x/L$, dimensionless |
| $Y_{\mathrm{Ab}}, Y_{\mathrm{Bb}}, Y_{\mathrm{Cb}}$ = | bulk concentration of species A, B, or C divided by the feed concentration of A ($C_{\mathrm{Af}}$), dimensionless |
| $Y_{\mathrm{Am}}, Y_{\mathrm{Bm}}, Y_{\mathrm{Cm}}$ = | membrane concentration of species A, B, or C divided by the feed concentration of A ($C_{\mathrm{Af}}$), dimensionless |
| $Y_{\mathrm{As}}, Y_{\mathrm{Bs}}, Y_{\mathrm{Cs}}$ = | surface concentration of species A, B, or C divided by the feed concentration of A ($C_{\mathrm{Af}}$), dimensionless |
| $z$ = | variable longitudinal distance from entrance of reactor, cm |
| $\epsilon$ = | control, ratio of molar concentration of enzyme 1 to total molar concentration of enzymes 1 and 2, dimensionless |
| $\theta$ = | space time, $z/\nu$, s |

$$R_1 = \frac{\epsilon V_1 Y_{\text{Am}}}{K_{\text{M1}} + C_{\text{Af}} Y_{\text{Am}}}$$

$$R_2 = \frac{(1-\epsilon) V_2 Y_{\text{Bm}}}{K_{\text{M2}} + C_{\text{Af}} Y_{\text{Bm}}}$$

Material balances for the species A, B, and C in the membrane yield the following differential equations:

$$\frac{D}{L^2}\frac{\partial^2 Y_{\text{Am}}}{\partial X^2} - R_1 = 0$$

$$\frac{D}{L^2}\frac{\partial^2 Y_{\text{Bm}}}{\partial X^2} + R_2 - R_1 = 0$$

$$\frac{D}{L^2}\frac{\partial^2 Y_{\text{Cm}}}{\partial X^2} + R_2 = 0$$

In the bulk fluid phase, the material balances for species A, B, and C can be defined as follows:

$$\frac{dY_{\text{Ab}}}{d\theta} + \frac{k_L}{x_0}(Y_{\text{Ab}} - Y_{\text{As}}) = 0$$

$$\frac{dY_{\text{Bb}}}{d\theta} + \frac{k_L}{x_0}(Y_{\text{Bb}} - Y_{\text{Bs}}) = 0$$

$$\frac{dY_{\text{Cb}}}{d\theta} + \frac{k_L}{x_0}(Y_{\text{Cb}} - Y_{\text{Cs}}) = 0$$

Since each membrane is symmetric about $X = 0$, the boundary conditions at $X = 0$ and $X = 1$ become

$$\frac{\partial Y_{\text{Am}}}{\partial X} = \frac{\partial Y_{\text{Bm}}}{\partial X} = \frac{\partial Y_{\text{Cm}}}{\partial X} = 0 \qquad \text{at } X = 0$$

$$\left.\begin{aligned} Y_{\text{Am}} &= Y_{\text{As}} \\ Y_{\text{Bm}} &= Y_{\text{Bs}} \\ Y_{\text{Cm}} &= Y_{\text{Cs}} \end{aligned}\right\} \qquad \text{at } X = 1$$

The surface concentrations are determined by equating the surface flux to the bulk transport flux, i.e.,

$$\frac{D}{L}\left(\frac{\partial Y_{\text{Am}}}{\partial X}\right)_{X=1} = k_L(Y_{\text{Ab}} - Y_{\text{As}})$$

$$\frac{D}{L}\left(\frac{\partial Y_{\text{Bm}}}{\partial X}\right)_{X=1} = k_L(Y_{\text{Bb}} - Y_{\text{Bs}})$$

$$\frac{D}{L}\left(\frac{\partial Y_{\text{Cm}}}{\partial X}\right)_{X=1} = k_L(Y_{\text{Cb}} - Y_{\text{Cs}})$$

Finally at the entrance of the reactor, i.e., at $\theta = 0$,

$$Y_{\text{Ab}} = 1 \qquad Y_{\text{Bb}} = 0 \qquad Y_{\text{C}} = 0$$

Develop numerical procedures for solving the above set of equations and write a computer program to calculate the concentration profiles in the membranes and in the bulk fluid for the following set of kinetic and transport parameters:

$$V_1 = 4.4 \times 10^{-3}\ \text{mol/L}\cdot\text{s} \qquad V_2 = 12.0 \times 10^{-3}\ \text{mol/L}\cdot\text{s}$$

$$K_{M1} = 0.022\ (\text{mol/L}) \qquad K_{M2} = 0.010\ (\text{mol/L})$$

$$D = 5.7 \times 10^{-8}\ \text{cm}^2/\text{s} \qquad C_{Af} = 1.0\ \text{mol/L}$$

$$k_L = 1.2 \times 10^{-4}\ \text{cm/s} \qquad x_0 = 23\ \text{mils}$$

$$\epsilon = 0.75$$

$$2L = 3\ \text{mils}$$

**6-10** Coulet et al. [16] have developed a glucose sensor that has glucose oxidase enzyme immobilized as a surface layer on a highly polymerized collagen membrane. In this system, glucose (analyte) is converted to hydrogen peroxide which is subsequently detected on the membrane face (that is not exposed to the analyte solution) by an amperometric electrode. The hydrogen peroxide flux is a direct measure of the sensor response [17].

The physical model and coordinate system are shown in Fig. P6-10. The local analyte concentration at the enzyme surface is low so that the reaction kinetics are adequately described by a first-order law. This latter assumption ensures that the electrode response is proportional to the analyte concentration.

The governing dimensionless equation describing analyte transport within the membrane is

$$\frac{\partial C}{\partial \xi} = \frac{\partial^2 C}{\partial \zeta^2}$$

where the dimensionless time $\xi$ and penetration $\zeta$ variables are defined as

$$\xi = \frac{Dt}{\delta^2}$$

$$\zeta = \frac{x}{\delta}$$

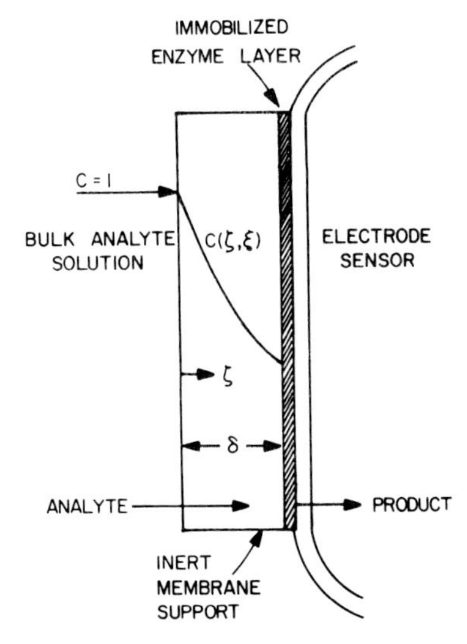


**Figure P6-10** Schematic description of an anisotropic enzyme electrode. The membrane (exaggerated) has active enzyme deposited as a surface layer at the electrode sensor interface. The product flux is the result of the reaction involving analyte diffusing through the membrane.

$\delta$ is the membrane thickness and $D$ the diffusion coefficient. The initial and boundary conditions are

$$C = 0 \qquad \xi = 0 \qquad 0 \le \zeta \le 1$$

$$C = 1 \qquad \xi > 0 \qquad \zeta = 0$$

$$\frac{\partial C}{\partial \zeta} = -\mu C \qquad \xi > 0 \qquad \zeta = 1$$

where $\mu$ is the Damkoehler number, defined as

$$\mu = \frac{k''\delta}{D}$$

The surface rate constant $k''$ is related to the surface concentration of the enzyme $[E'']$, the turnover number $k_{\text{cat}}$, and the intrinsic Michaelis-Menten constant $K_m$ by

$$k'' = \frac{k_{\text{cat}}[E'']}{K_m}$$

(*a*) Predict the electrode response as a function of the dimensionless time $\xi$ for a 0.3-mm-thick membrane with the analyte diffusion coefficient $D = 2 \times 10^{-6}\ \text{cm}^2/\text{s}$ and immobilized enzyme with the surface rate constant $k'' = 0.24\ \text{cm/h}$.

(*b*) Repeat part *a* for the reaction kinetics defined by the Michaelis-Menten law.

# REFERENCES


1. Bird, R. B., Stewart, W. E., and Lightfoot, E. N.: *Transport Phenomena*, John Wiley & Sons, Inc., New York, 1960.
2. Lapidus, L., and Pinder, G. F.: *Numerical Solution of Partial Differential Equations in Science and Engineering*, John Wiley & Sons, Inc., New York, 1982.
3. Tychonov, A. N., and Samarski, A. A.: *Partial Differential Equations of Mathematical Physics*, Holden-Day, Inc., Publisher, San Francisco, 1964.
4. Vichnevetsky, R.: *Computer Methods for Partial Differential Equations*, vol. I, Prentice-Hall, Inc., Englewood Cliffs, N.J. 1981.
5. James, M. L., Smith, G. M., and Wolford, J. C.: *Applied Numerical Methods for Digital Computation with FORTRAN and CSMP*, 2d ed., Harper & Row, Publishers, Incorporated, New York, 1977.
6. Finlayson, B. A.: *Nonlinear Analysis in Chemical Engineering*, McGraw-Hill Book Company, New York, p. 216.
7. Carnahan, B., Luther, H. A. and Wilkes, J. O.: *Applied Numerical Methods*, John Wiley & Sons, Inc., New York, 1969, p. 451.
8. Davis, M. E.: *Numerical Methods and Modeling for Chemical Engineers*, John Wiley & Sons, Inc., New York, 1984, p. 128.
9. Salvadori, M. G., and Baron, M. L.: *Numerical Methods in Engineering*, Prentice-Hall, Inc., Englewood Cliffs, N.J., 1961.
10. Vichnevetsky, R., and Bowles, J. B.: *Fourier Analysis of Numerical Approximations of Hyperbolic Equations*, SIAM, Philadelphia, 1982.
11. Vemuri, V., and Karplus, W. J.: *Digital Computer Treatment of Partial Differential Equations*, Prentice-Hall, Inc., Englewood Cliffs, N.J., 1981.
12. Fairweather, G.: *Finite Element Galerkin Methods for Differential Equations*, Marcel Dekker, Inc., New York, 1978.

13. Norrie, D. H., and de Vries, G.: *An Introduction to Finite Element Analysis*, Academic Press, Inc., New York, 1978.
14. Geankoplis, C. J.: *Transport Processes and Unit Operations*, Allyn and Bacon, Inc., Boston, 1978.
15. Fernandes, P. M., Constantinides, A., Vieth, W. R., and Venkatasubramanian, K.: "Enzyme Engineering: Part V. Modeling and Optimizing Multi-Enzyme Reactor Systems," *Chemtech*, July 1975, p. 438.
16. Coulet, P. R., Sternberg, R., and Thevenot, D. R.: "Electrochemical Study of Reactions at Interfaces of Glucose Oxidase Collagen Membranes," *Biochim. Biophys. Acta*, vol. 612, 1980, p. 317.
17. Pedersen, H., and Chotani, G. K.,: "Analysis of a Theoretical Model for Anisotropic Enzyme Membranes: Application to Enzyme Electrodes," *Appl. Biochem. Biotech.*, vol. 6, 1981, p. 309.

CHAPTER

# SEVEN

# LINEAR AND NONLINEAR REGRESSION ANALYSIS

## 7-1 PROCESS ANALYSIS, MATHEMATICAL MODELING, AND REGRESSION ANALYSIS

Engineers and scientists are often required to analyze complex physical or chemical systems and to develop mathematical models which simulate the behavior of such systems. *Process analysis* is a term commonly used by chemical engineers to describe the study of complex chemical, biochemical, or petrochemical processes. More recently coined phrases such as *systems engineering* and *systems analysis* are used by electrical engineers and computer scientists to refer to the analysis of electric networks and computer systems. No matter what the phraseology is, the principles applied are the same. According to Himmelblau and Bischoff [1]: "Process analysis is the application of scientific methods to the recognition and definition of problems and the development of procedures for their solution. In more detail, this means (1) mathematical specification of the problem for the given physical situation, (2) detailed analysis to obtain mathematical models, and (3) synthesis and presentation of results to ensure full comprehension."

In the heart of successful process analysis is the step of *mathematical modeling*. The objective of modeling is to construct, from theoretical and empirical knowledge of a process, a mathematical formulation which can be used to predict the behavior of this process. Complete understanding of the mechanism of the chemical, physical, or biological aspects of the process under investigation is not usually possible. However, some information on the mechanism of the system may be available; therefore, a combination of empirical and theoretical methods can be used. According to Box and Hunter [2]: "No model can give a precise description of what happens. A working theoretical model, however, supplies information on the system under study over important ranges of the variables by means of equations which reflect at least the major features of the mechanism."

The engineer in the process industries is usually concerned with the operation of existing plants and the development of new processes. In the first case, the control, improvement, and optimization of the operation are the engineer's main objectives. In order to achieve this, a quantitative representation of the process, a model, is needed which would give the relationship between the various parts of the system. In the design of new processes, the engineer draws information from theory and the literature to construct mathematical models which may be used to simulate the process (see Fig. 7-1). The development of mathematical models often requires the implementation of an experimental program in order to obtain the necessary information for the verification of the models. The experimental program is originally designed based on the theoretical considerations coupled with a priori knowledge of the process and is subsequently modified based on the results of regression analysis.

*Regression analysis* is the application of mathematical and statistical methods for the analysis of the experimental data, and the fitting of the mathematical models to these data by the estimation of the unknown parameters of the models. The series of statistical tests, which normally accompany regression analysis, serve in model identification, model verification, and efficient design of the experimental program.

Strictly speaking, a mathematical model of a dynamic system is a set of equations that can be used to calculate how the state of the system evolves

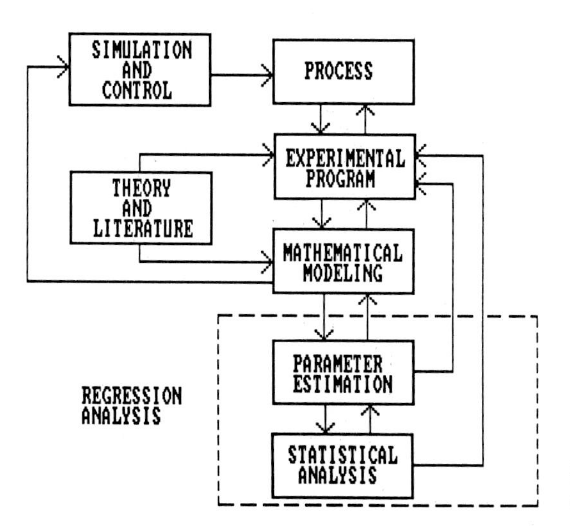


**Figure 7-1** Mathematical modeling and regression analysis.

through time under the action of the control variables, given the state of the system at some initial time. The state of the system is described by a set of variables known as state variables. The first stage in the development of a mathematical model is to identify the state and control variables.

The control variables are those which can be directly controlled by the experimenter and which influence the way the system changes from its initial state to that of any later time. Examples of control variables in a chemical reaction system may be the temperature, pressure, and/or concentration of some of the components. The state variables are those which describe the state of the system and which are not under direct control. The concentrations of reactants and products are state variables in chemical systems. The distinction between state and control variables is not always fixed but can change when the method of operating the system changes. For example, if temperature is not directly controlled, it becomes a state variable.

The equations comprising the mathematical model of the process are called the *performance equations*. These equations should show the effect of the control variables on the evolution of the state variables. The performance equations may be a set of differential equations and/or a set of algebraic equations. For example, a set of ordinary differential equations describing the dynamics of a process may have the general form

$$\mathbf{y}' = \mathbf{g}(x, \mathbf{y}, \boldsymbol{\theta}, \mathbf{b}) \tag{7-1}$$

where $x$ = independent variable
$\mathbf{y}$ = vector of state (dependent) variables
$\boldsymbol{\theta}$ = vector of control variables
$\mathbf{b}$ = vector of parameters whose values must be determined.

In this chapter, we concern ourselves with the methods of estimating the parameter vector **b** using regression analysis. For this purpose, we assume that the vector of control variables $\boldsymbol{\theta}$ is fixed; therefore, the mathematical model simplifies to

$$\mathbf{y}' = \mathbf{g}(x, \mathbf{y}, \mathbf{b}) \tag{7-2}$$

In their integrated form, the above set of performance equations convert to

$$\mathbf{y} = \mathbf{f}(x, \mathbf{b}) \tag{7-3}$$

For regression analysis, mathematical models are classified as *linear* or *nonlinear* with respect to the *unknown parameters*. For example, the following differential equation

$$\frac{dy}{dt} = ky \tag{7-4}$$

which we classified earlier as linear with respect to the dependent variable (see Chap. 5), is *nonlinear* with respect to the parameter $k$. This is clearly shown by the integrated form of Eq. (7-4),

$$y = y_0 e^{kt} \tag{7-5}$$

where $y$ is highly nonlinear with respect to $k$.

Most mathematical models encountered in engineering and the sciences are nonlinear in the parameters. Attempts at linearizing these models, by rearranging the equations and regrouping the variables, were common practice in the precomputer era when graph paper and the straightedge were the tools for fitting models to experimental data. Such primitive techniques have been replaced by the implementation of *linear* and *nonlinear regression methods* on the computer.

The theory of linear regression has been expounded by statisticians and econometricians, and a rigorous statistical analysis of the regression results has been developed. Nonlinear regression is an extension of the linear regression methods used iteratively to arrive at the values of the parameters of the nonlinear models. The statistical analysis of the nonlinear regression results is also an extension of that applied in linear analysis but does not possess the rigorous theoretical basis of the latter.

In this chapter, after giving a brief review of statistical terminology, we develop the basic algorithm of linear regression and then show how this is extended to nonlinear regression. We develop the methods in matrix notation so that the algorithms are equally applicable to fitting single or multiple variables, and to using single or multiple sets of experimental data.

## 7-2 REVIEW OF STATISTICAL TERMINOLOGY USED IN REGRESSION ANALYSIS

It is assumed that the reader has a rudimentary knowledge of statistics. This section serves as a review of the statistical definitions and terminology needed for understanding the application of linear and nonlinear regression analysis, and the statistical treatment of the results of this analysis. For a more complete discussion of statistics the reader should consult a standard text on statistics, such as Brownlee [3] and Lipson and Sheth [4].

### 7-2.1 Population and Sample Statistics

A *population* is defined as a group of similar items, or events, from which a sample is drawn for test purposes; the population is usually assumed to be very large, sometimes infinite. A *sample* is a random selection of items from a population, usually made for evaluating a variable of that population. The *variable* under investigation is a characteristic property of the population.

A *random variable* is defined as a variable which can assume any value from a set of possible values. A *statistic* or *statistical parameter* is any quantity computed from a sample; it is characteristic of the sample and it is used to estimate the characteristics of the population variable.

*Degrees of freedom* can be defined as the number of observations made in excess of the minimum theoretically necessary to estimate a statistical parameter or any unknown quantity [5].

Let us use the notation $N$ to designate the total number of items in the population under study, where $0 \leq N \leq \infty$, and $n$ to specify the number of items contained in the sample taken from that population, where $0 \leq n \leq N$. The variable being investigated will be designated as $X$; it may have discrete values, or it may be a continuous function, in the range $-\infty < x < \infty$. For specific populations these ranges may be more limited, as will be mentioned below.

For the sake of example and in order to clarify these terms, let us consider for study the entire human population and examine the age of this population. The value of $N$, in this case, would be approximately 5 billion. The age variable would range from 0 to possibly 150 years. Age can be considered either as a continuous variable, since all ages are possible, or more commonly as a discrete variable, since ages are usually grouped by year. In the continuous case, the age variable takes an infinite number of values in the range $0 < x \leq 150$, and in the discrete case it takes a finite number of values $x_j$, where $j = 1, 2, 3, \ldots, M$ and $M \leq 150$. Assume that a random sample of $n$ persons is chosen from the total population (say $n = 1$ million) and the age of each person in the sample is recorded.

The frequency at which each value of the variable (age, in the above example) may occur in the population is not the same; some values (ages) will occur more frequently than others. Designating $m_j$ as the number of times the value of $x_j$ occurs, we can define the concept of *probability of occurrence* as

$$\Pr\{X = x_j\} = \left\{\begin{matrix}\text{Probability of} \\ \text{occurrence of} \\ x_j\end{matrix}\right\} = \frac{\text{number of occurrences of } x_j}{\text{total number of observations}}$$

$$= \lim_{n \to N} \frac{m_j}{n}$$

$$= p(x_j) \tag{7-6}$$

For a discrete random variable, $p(x_j)$ is called the *probability function* and it has the following properties:

$$0 \leq p(x_j) \leq 1$$

$$\sum_{j=1}^{M} p(x_j) = 1 \tag{7-7}$$

The shape of a typical probability function is shown in Fig. 7-2$a$.

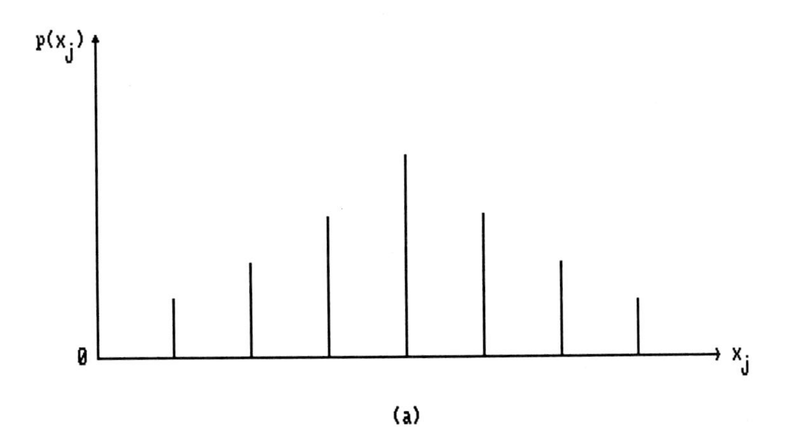


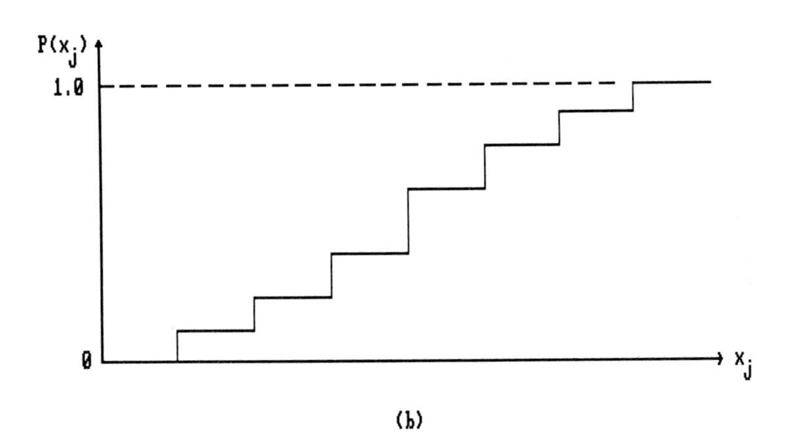


**Figure 7-2** (*a*) Probability function and (*b*) cumulative distribution function for a discrete random variable.

For a continuous random variable, the probability of occurrence is measured by the continuous function $p(x)$, which is called the *probability density function*, so that

$$\Pr\{x < X \leq x + dx\} = p(x)\, dx \tag{7-8}$$

The probability density function has the following properties:

$$\begin{gathered} 0 \leq p(x) \leq 1 \\ \int_{-\infty}^{\infty} p(x)\, dx = 1 \end{gathered} \tag{7-9}$$

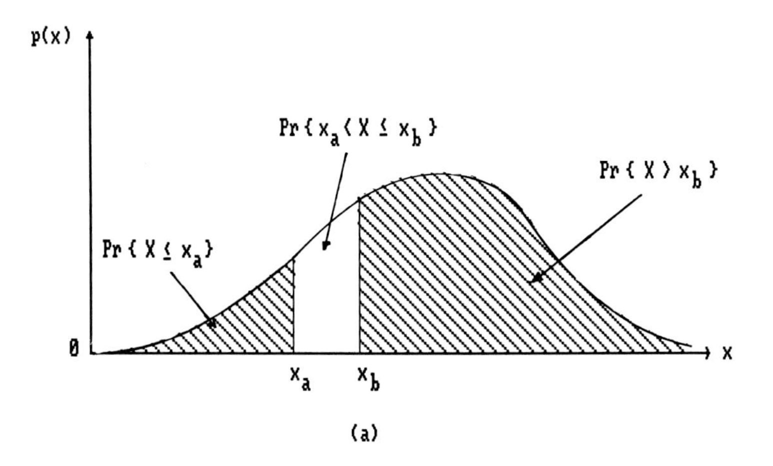


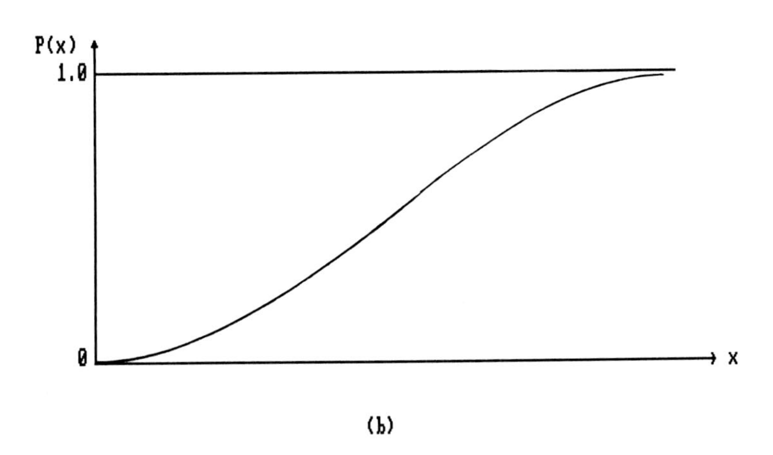


**Figure 7-3** (*a*) Probability density function and (*b*) cumulative distribution function for a continuous random variable.

The smooth curve obtained from plotting $p(x)$ versus $x$ (Fig. 7-3*a*) is called the *continuous probability density distribution*.

The *cumulative distribution function* is defined as the probability that a random variable $X$ will not exceed a given value $x$, that is,

$$\Pr\{X \leq x\} = P(x) = \int_{-\infty}^{x} p(x)\, dx \tag{7-10}$$

The equivalent of (7-10) for a discrete random variable is

$$\Pr\{X \leq x_i\} = P(x_i) = \sum_{j=1}^{i} p(x_j) \tag{7-11}$$

The cumulative distribution functions for discrete and continuous random variables are illustrated in Figs. 7-2*b* and 7-3*b*, respectively.

It is obvious from the integral of Eq. (7-10) that the cumulative distribution function is obtained from calculating the area under the density distribution function. The three area segments shown in Fig. 7-3*a* correspond to the following three probabilities:

$$\Pr\{X \le x_a\} = \int_{-\infty}^{x_a} p(x)\, dx \tag{7-12}$$

$$\Pr\{x_a < X \le x_b\} = \int_{x_a}^{x_b} p(x)\, dx \tag{7-13}$$

$$\Pr\{X > x_b\} = \int_{x_b}^{\infty} p(x)\, dx \tag{7-14}$$

The *population mean*, or *expected value*, of a discrete random variable is defined as

$$\mu = E[X] = \sum_{j=1}^{M} x_j p(x_j) \tag{7-15}$$

and that of a continuous random variable as

$$\mu = E[X] = \int_{-\infty}^{\infty} x p(x)\, dx \tag{7-16}$$

The usefulness of the concept of *expectation*, as defined above, is that it corresponds to our intuitive idea of *average*, or equivalently to the center of gravity of the probability density distribution along the $x$-axis. It is easy to show that combining (7-15) and (7-6) yields the arithmetic average of the random variable for the entire population:

$$\mu = E[X] = \frac{\sum_{i=1}^{N} x_i}{N} \tag{7-17}$$

In addition, the integral of Eq. (7-16) can be recognized from the field of mechanics as the *first noncentral moment of X*.

The *sample mean*, or *arithmetic average*, of a sample of observations is the value obtained by dividing the sum of observations by their total number:

$$\bar{x} = \frac{\sum_{i=1}^{n} x_i}{n} \tag{7-18}$$

The expected value of the sample mean is given by

$$E[\bar{x}] = E\left[\frac{\sum_{i=1}^{n} x_i}{n}\right] = \frac{1}{n}\sum_{i=1}^{n} E[x_i] = \frac{1}{n}\sum_{i=1}^{n} \mu = \mu \tag{7-19}$$

i.e., the sample mean is an *unbiased estimate* of the population mean.

The *population variance* is defined as the expected value of the square of the deviation of the random variable $X$ from its expectation:

$$\begin{aligned} \sigma^2 &= V[X] \\ &= E[(X - E[X])^2] \\ &= E[(X - \mu)^2] \end{aligned} \tag{7-20}$$

For a discrete random variable, (7-20) is equivalent to

$$\sigma^2 = \sum_{j=1}^{M} (x_j - \mu)^2 p(x_j) \tag{7-21}$$

When combined with (7-6), Eq. (7-21) becomes

$$\sigma^2 = \frac{\sum_{i=1}^{N} (x_i - \mu)^2}{N} \tag{7-22}$$

which is the arithmetic average of the square of the deviations of the random variable from its mean. For a continuous random variable, (7-20) is equivalent to

$$\sigma^2 = \int_{-\infty}^{\infty} (x - \mu)^2 p(x)\, dx \tag{7-23}$$

which is the *second central moment of* $X$ about its mean.

It is interesting and useful to note that (7-20) expands as follows:†

$$\begin{aligned} V[X] &= E[(X - E[X])^2] = E[X^2 + (E[X])^2 - 2XE[X]] \\ &= E[X^2] + E[(E[X])^2] - 2E[XE[X]] \\ &= E[X^2] + (E[X])^2 - 2(E[X])^2 \\ &= E[X^2] - (E[X])^2 \\ &= E[X^2] - \mu^2 \end{aligned} \tag{7-24}$$

The positive square root of the population variance is called the *population standard deviation*:

† The expected value of a constant is that constant. The expected value of $X$ is a constant; therefore, $E[E[X]] = E[X]$.

$$\sigma = +\sqrt{\sigma^2} \tag{7-25}$$

The *sample variance* is defined as the arithmetic average of the square of the deviations of $x_i$ from the population mean $\mu$:

$$s^2 = \frac{\sum_{i=1}^{n} (x_i - \mu)^2}{n} \tag{7-26}$$

However, since $\mu$ is not usually known, $\bar{x}$ is used as an estimate of $\mu$, and the sample variance is calculated from

$$s^2 = \frac{\sum_{i=1}^{n} (x_i - \bar{x})^2}{n - 1} \tag{7-27}$$

where the degrees of freedom have been reduced to $n - 1$, since the calculation of the sample mean consumes one degree of freedom. The sample variance obtained from Eq. (7-27) is an unbiased estimate of population variance,† i.e.,

$$E[s^2] = \sigma^2 \tag{7-28}$$

The positive square root of the sample variance is called the *sample standard deviation*:

$$s = +\sqrt{s^2} \tag{7-29}$$

The *covariance* of two random variables $X$ and $Y$ is defined as the expected value of the product of the deviations of $X$ and $Y$ from their expected values:

$$\text{Cov}[X, Y] = E[(X - E[X])(Y - E[Y])] \tag{7-30}$$

Equation (7-30) expands to

$$\begin{aligned} \text{Cov}[X, Y] &= E[XY - YE[X] - XE[Y] + E[X]E[Y]] \\ &= E[XY] - E[X]E[Y] \end{aligned} \tag{7-31}$$

The covariance is a measurement of the association between the two variables. If large positive deviations of $X$ are associated with large positive deviations of $Y$, and likewise large negative deviations of the two variables occur together, then the covariance will be positive. Furthermore, if positive deviations of $X$ are associated with negative deviations of $Y$, and vice versa, then the covariance will be negative. On the other hand, if positive and negative deviations of $X$ occur equally frequently with positive and negative deviations of $Y$, then the covariance will tend to zero [3].

† See Carnahan et al. [6, p. 542].

The variance of $X$, defined earlier in Eq. (7-20), is a special case of the covariance of the random variable with itself:

$$\begin{aligned} \text{Cov}\,[X, X] &= E[(X - E[X])(X - E[X])] \\ &= E[(X - E[X])^2] = V[X] \end{aligned} \tag{7-32}$$

The magnitude of the covariance depends upon the magnitude and units of $X$ and $Y$, and could conceivably range from $-\infty$ to $\infty$. To make the measurement of covariance more manageable, the two dimensionless standardized variables are formed:

$$\frac{X - E[X]}{\sqrt{V[X]}} \qquad \text{and} \qquad \frac{Y - E[Y]}{\sqrt{V[Y]}}$$

The covariance of the standardized variables is known as the *correlation coefficient*:

$$\rho_{XY} = \text{Cov}\left[\frac{X - E[X]}{\sqrt{V[X]}}, \frac{Y - E[Y]}{\sqrt{V[Y]}}\right] \tag{7-33}$$

Using the definition of covariance reduces the correlation coefficient to

$$\rho_{XY} = \frac{\text{Cov}\,[X, Y]}{\sqrt{V[X]V[Y]}} \tag{7-34}$$

If $\rho_{XY} = 0$, we say that $X$ and $Y$ are *uncorrelated*, and this implies

$$\text{Cov}\,[X, Y] = 0$$

We know from probability theory (see Brownlee [3], p. 75) that if $X$ and $Y$ are *independent variables*, then

$$p\{x, y\} = p_x(x)p_y(y)$$

from which it follows that

$$E[XY] = E[X]E[Y] \tag{7-35}$$

Combining (7-35) and (7-31) shows that

$$\text{Cov}\,[X, Y] = 0$$

and from (7-34)

$$\rho_{xy} = 0$$

Thus independent variables are uncorrelated.

The population and sample statistics discussed above are summarized in Table 7-1.

**Table 7-1 Summary of population and sample statistics**

| Statistic | Population: Continuous variable | Population: Discrete variable | Sample |
|---|---|---|---|
| Mean | $\mu = E[X] = \int_{-\infty}^{\infty} xp(x)\,dx$ | $\mu = E[X] = \sum_{j=1}^{M} x_j p(x_j)$ | $\bar{x} = \frac{\sum_{i=1}^{n} x_i}{n}$ |
| Variance | $\sigma^2 = V[X] = E[(X - E[X])^2]$ | $\sigma^2 = V[X] = E[(X - E[X])^2]$ | $s^2 = \frac{\sum_{i=1}^{n} (x_i - \mu)^2}{n}$ |
| | | | or |
| | $= \int_{-\infty}^{\infty} (x - \mu)p(x)\,dx$ | $= \sum_{j=1}^{M} (x_j - \mu)^2 p(x_j)$ | $s^2 = \frac{\sum_{i=1}^{n} (x_i - \bar{x})^2}{n - 1}$ |
| Standard deviation | $\sigma = +\sqrt{\sigma^2}$ | $\sigma = +\sqrt{\sigma^2}$ | $s = +\sqrt{s^2}$ |
| Covariance | $\mathrm{Cov}\,[X, Y] = E[(X - E[X])(Y - E[Y])]$ | | |
| Correlation coefficient | $\rho_{XY} = \frac{\mathrm{Cov}\,[X, Y]}{\sqrt{V[X]V[Y]}}$ | | |

### 7-2.2 Probability Density Functions and Probability Distributions

There are many different probability density functions encountered in statistical analysis. Of particular interest to regression analysis are the *normal*, $\chi^2$, $t$ and $F$ *distributions*, which will be discussed in this section. The *normal* or *gaussian* density function has the form

$$p(x) = \frac{1}{\sqrt{2\pi}\sigma} \exp\left[-\frac{1}{2}\left(\frac{x-\mu}{\sigma}\right)^2\right] \tag{7-36}$$

where $-\infty < x < \infty$. The cumulative distribution function of the normal density function is

$$P(x) = \frac{1}{\sqrt{2\pi}\sigma} \int_{-\infty}^{x} \exp\left[-\frac{1}{2}\left(\frac{x-\mu}{\sigma}\right)^2\right] dx \tag{7-37}$$

involving an integral which does not have an explicit form and must be integrated numerically. The normal probability distributions are illustrated in Fig. 7-4.

The expected value of a variable which has a normal distribution is†

$$E[X] = \mu \tag{7-38}$$

and the variance is

$$V[X] = \sigma^2 \tag{7-39}$$

For this reason the normal density function is usually abbreviated as $N(\mu, \sigma^2)$, and the notation

$$X \sim N(\mu, \sigma^2) \tag{7-40}$$

means that the variable $X$ has a normal distribution with expected value $\mu$ and variance $\sigma^2$.

A normal density function can be transformed to the *standard normal density function* by the substitution

$$u = \frac{x-\mu}{\sigma} \tag{7-41}$$

which transforms Eq. (7-36) to‡

† See Brownlee [3, pp. 54 and 60].
‡ See Brownlee [3, p. 47].

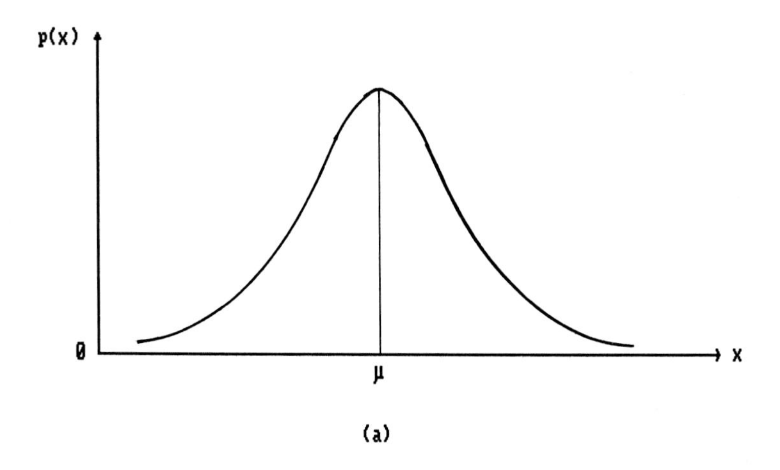


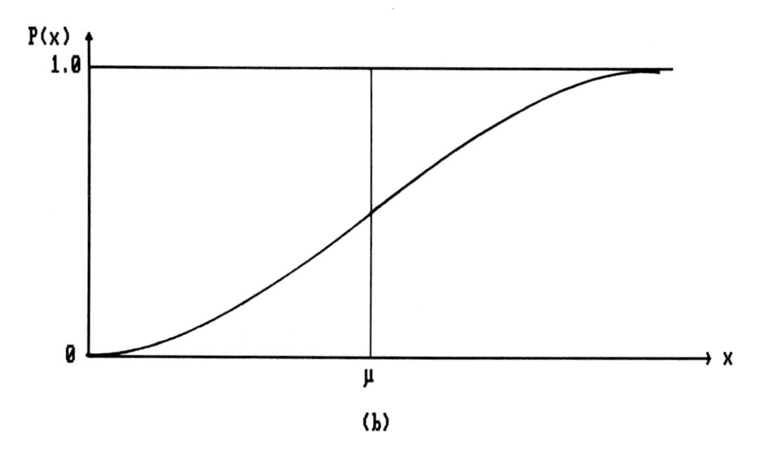


**Figure 7-4** (*a*) Normal probability density function and (*b*) normal cumulative distribution function for a continuous random variable.

$$\phi(u) = \frac{1}{\sqrt{2\pi}} \exp\left[-\frac{u^2}{2}\right] \tag{7-42}$$

The expected value of the standardized variable $u$ is

$$E[u] = 0 \tag{7-43}$$

and the variance is

$$V[u] = 1 \tag{7-44}$$

therefore

$$u \sim N(0, 1) \tag{7-45}$$

The standard normal density function $\phi(u)$ and its cumulative distribution function

$$\Phi(u) = \frac{1}{\sqrt{2\pi}} \int_{-\infty}^{u} \exp\left[-\frac{u^2}{2}\right] du \tag{7-46}$$

are shown in Fig. 7-5. Tables of the statistical distribution functions may be found in Refs. 3, 7, and 8.

The function $\phi(u)$ is symmetrical about zero; therefore, the area in the left tail, below $-u$, is equal to the area in the right tail, above $+u$ (shaded areas in Fig. 7-5). The unshaded area between $-1.960$ and $1.960$ is equivalent to 95 percent of the total area under the density function. This area is designated as $(1-\alpha)$ and the area under each tail as $\alpha/2$. Application of Eqs. (7-12) to (7-14) shows that

$$\Pr\{u \leq -1.960\} = \frac{\alpha}{2} = 0.025$$

$$\Pr\{-1.960 < u \leq 1.960\} = 1 - \alpha = 0.95$$

and

$$\Pr\{u > 1.960\} = \frac{\alpha}{2} = 0.025$$

If a set of normally distributed variables $X_k$, where

$$X_k \sim N(\mu_k, \sigma_k^2) \tag{7-47}$$

is linearly combined to form another variable $Y$, where

$$Y = \sum_k a_k X_k \tag{7-48}$$

then $Y$ is also normally distributed, i.e.,

$$Y \sim N\left(\sum_k a_k \mu_k, \sum_k a_k^2 \sigma_k^2\right) \tag{7-49}$$

The sample mean [Eq. (7-18)] of a normally distributed population is a linear combination of normally distributed variables; therefore, the sample mean itself is normally distributed:

$$\bar{x} \sim N\left(\mu, \frac{\sigma^2}{n}\right) \tag{7-50}$$

It follows then, from Eqs. (7-41) and (7-45), that

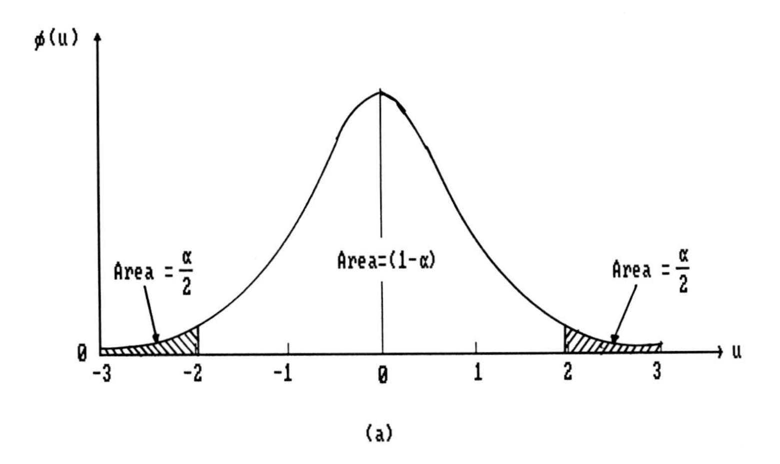


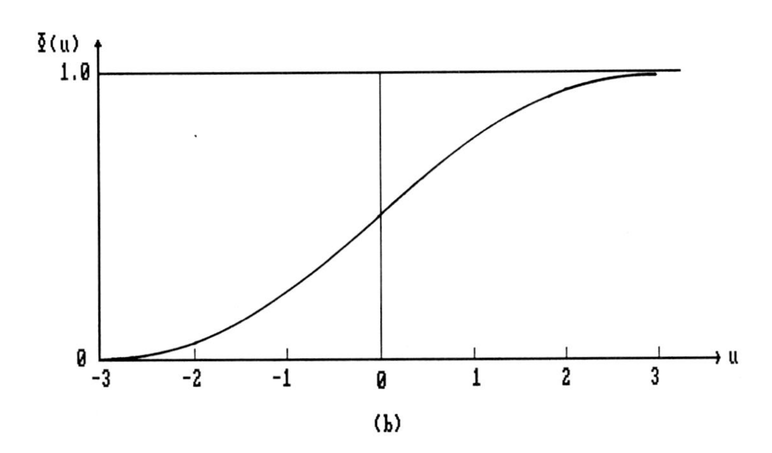


**Figure 7-5** (*a*) Standardized normal probability density function and (*b*) standardized normal cumulative distribution function.

$$\frac{\bar{x} - \mu}{\sqrt{\sigma^2/n}} \sim N(0, 1) \tag{7-51}$$

If we wish to test the hypothesis that a sample, whose mean is $\bar{x}$, could come from a normal distribution of mean $\mu$ and known variance $\sigma^2$, the procedure is easy, for the variable $(\bar{x} - \mu)/\sqrt{\sigma^2/n}$ is $N(0, 1)$ and can readily be compared with tabulated values. However, if $\sigma^2$ is unknown and must be estimated from the sample variance $s^2$, then *Student's t distribution*, which is described later in this section, is needed [6].

Now consider a sequence $X_k$ of identically distributed, independent random variables (not necessarily normally distributed) whose second-order moment exists. Let $E[X_k] = \mu$ and $E[(X_k - \mu)^2] = V[X_k] = \sigma^2$, for every $k$. Consider the random variable $Z_n$ defined by

$$Z_n = X_1 + X_2 + \cdots + X_n \tag{7-52}$$

where $E[Z_n] = n\mu$ and, by independence of the $X_k$, $E[(Z_n - n\mu)^2] = n\sigma^2$. Let

$$\hat{Z}_n = \frac{Z_n - n\mu}{\sigma\sqrt{n}} \tag{7-53}$$

Then the distribution of $\hat{Z}_n$ approaches the standard normal distribution, i.e.,

$$\lim_{n\to\infty} P_n(z) = \frac{1}{\sqrt{2\pi}} \int_{-\infty}^{z} \exp\left[-\frac{z^2}{2}\right] dz \tag{7-54}$$

This is the *central limit theorem*, a proof of which can be found in Seinfeld and Lapidus [9, p. 188]. This is a very important theorem of statistics, particularly in regression analysis where experimental data are being analyzed. The experimental error is a composite of many separate errors whose probability distributions are not necessarily normal distributions. However, as the number of components contributing to the error increases, the central limit theorem justifies the assumption of normality of the error.

Suppose we have a set of $\nu$ independent observations, $x_1, \ldots, x_\nu$ from a normal distribution $N(\mu, \sigma^2)$. The standardized variables

$$u_i = \frac{x_i - \mu}{\sigma} \tag{7-41}$$

will also be independent and have distribution $N(0, 1)$. The variable $\chi^2(\nu)$ is defined as the sum of the squares of $u_i$

$$\chi^2(\nu) = \sum_{i=1}^{\nu} u_i^2 = \sum_{i=1}^{\nu} \frac{(x_i - \mu)^2}{\sigma^2} \tag{7-55}$$

The $\chi^2(\nu)$ variable has the so-called $\chi^2$ (*chi-square*) *distribution function* which is given by

$$p(\chi^2) = \frac{1}{2^{\nu/2}\Gamma(\nu/2)} e^{-\chi^2/2} (\chi^2)^{(\nu/2)-1} \tag{7-56}$$

where $\chi^2 \geq 0$ and

$$\Gamma\left(\frac{\nu}{2}\right) = \int_0^\infty e^{-x} x^{(\nu/2)-1}\, dx \qquad \text{for } \frac{\nu}{2} > 0 \tag{7-57}$$

The $\chi^2$ distribution is a function of the degrees of freedom $\nu$, as shown in Fig. 7-6. The distribution is confined to the positive half of the $\chi^2$-axis since the $u_i^2$ quantities are always positive.

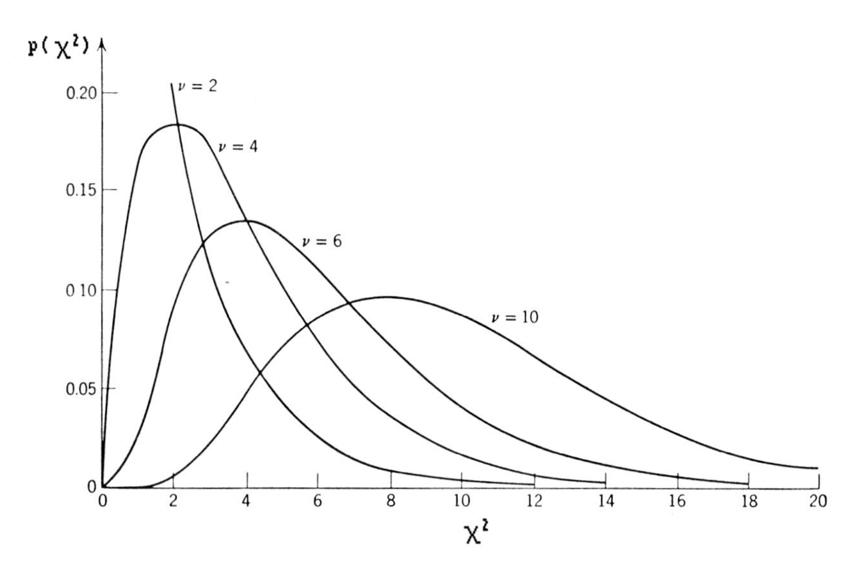


**Figure 7-6** The $\chi^2$ distribution function.

The expected value of the $\chi^2$ variable is

$$\mu = E[\chi^2] = \int_0^\infty \chi^2 p(\chi^2)\, d\chi^2 = \nu \tag{7-58}$$

and its variance is

$$\sigma^2 = V[\chi^2] = \int_0^\infty (\chi^2)^2 p(\chi^2)\, d\chi^2 = 2\nu \tag{7-59}$$

The $\chi^2$ distribution tends toward the normal distribution $N(\nu, 2\nu)$ as $\nu$ becomes large. The $\chi^2$ distribution is widely used in statistical analysis for testing independence of variables and the fit of probability distributions to experimental data.

We saw earlier that the sample variance was obtained from Eq. (7-27):

$$s^2 = \frac{\sum_{i=1}^{n} (x_i - \bar{x})^2}{n - 1} \tag{7-27}$$

with $n - 1$ degrees of freedom. When $\bar{x}$ is assumed to be equal to $\mu$ then

$$s^2 = \frac{\sum_{i=1}^{n} (x_i - \mu)^2}{n - 1} \tag{7-60}$$

Combining Eqs. (7-55) and (7-60) shows that

$$\chi^2 = (n-1)\frac{s^2}{\sigma^2} \tag{7-61}$$

with $\nu = n - 1$ degrees of freedom. This equation will be very useful in Sec. 7-2.3 in constructing confidence intervals for the population variance.

Let us define a new random variable $t$, so that

$$t = \frac{u}{\sqrt{\chi^2/\nu}} \tag{7-62}$$

where $u \sim N(0, 1)$ and $\chi^2$ is distributed as chi-square with $\nu$ degrees of freedom. It is assumed that $u$ and $\chi^2$ are independent of each other. The variable $t$ is called *Student's t* and has the probability density function

$$p(t) = \frac{1}{\sqrt{\nu\pi}}\,\frac{\Gamma[(\nu+1)/2]}{\Gamma(\nu/2)}\left(1+\frac{t^2}{\nu}\right)^{-(\nu+1)/2} \tag{7-63}$$

with $\nu$ degrees of freedom. The shape of the $t$ density function is shown in Fig. 7-7.

The expected value of the $t$ variable is

$$\mu_t = E[t] = \int_{-\infty}^{\infty} t\,p(t)\,dt = 0 \qquad \text{for } \nu > 1 \tag{7-64}$$

and the variance is

$$\sigma_t^2 = V[t] = \int_{-\infty}^{\infty} t^2 p(t)\,dt = \frac{\nu}{\nu-2} \qquad \text{for } \nu > 2 \tag{7-65}$$

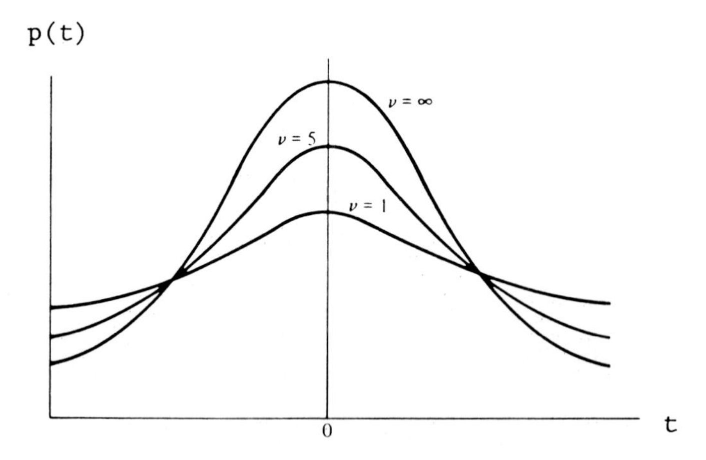


**Figure 7-7** The Student's $t$ density function.

The $t$ distribution tends toward the normal distribution as $\nu$ becomes large.

Combining Eq. (7-62) with (7-51) and (7-61) gives

$$t = \frac{(\bar{x} - \mu)/\sqrt{\sigma^2/n}}{\sqrt{s^2/\sigma^2}} = \frac{\bar{x} - \mu}{\sqrt{s^2/n}} \tag{7-66}$$

The quantity on the right-hand side of (7-66) is independent of $\sigma$ and has a $t$ distribution. Therefore, the $t$ distribution provides a test of significance for the deviation of a sample mean from its expected value when the population variance is unknown and must be estimated from the sample variance.

Finally, we define the ratio

$$F(\nu_1, \nu_2) = \frac{\chi_1^2/\nu_1}{\chi_2^2/\nu_2} \tag{7-67}$$

where $\chi_1^2$ and $\chi_2^2$ are two independent random variables of chi-square distribution with $\nu_1$ and $\nu_2$ degrees of freedom, respectively. The variable $F(\nu_1, \nu_2)$ has the $F$ distribution density function with $\nu_1$ and $\nu_2$ degrees of freedom shown below:

$$p(F) = \frac{\left(\dfrac{\nu_1}{\nu_2}\right)^{\nu_1/2} F^{(\nu_1/2)-1}\left(1 + \dfrac{\nu_1}{\nu_2} F\right)^{-(\nu_1+\nu_2)/2}}{\displaystyle\int_0^1 x^{(\nu_1/2)-1}(1 - x)^{(\nu_2/2)-1}\, dx} \tag{7-68}$$

The $F$ distribution is very useful in the analysis of variance of populations. Consider two normally distributed independent random samples

$$x_{1,1}, x_{1,2}, \ldots, x_{1,n_1}$$

and

$$x_{2,1}, x_{2,2}, \ldots, x_{2,n_2}$$

The first sample which has a sample variance $s_1^2$ is from a population with mean $\mu_1$ and variance $\sigma_1^2$. The second sample which has a sample variance of $s_2^2$ is from a population with mean $\mu_2$ and variance $\sigma_2^2$. Using Eq. (7-61), we see that

$$\chi_1^2 = (n_1 - 1)\frac{s_1^2}{\sigma_1^2} \tag{7-69}$$

and

$$\chi_2^2 = (n_2 - 1)\frac{s_2^2}{\sigma_2^2} \tag{7-70}$$

Combining (7-67) with (7-69) and (7-70) shows that

$$F(n_1-1, n_2-1) = \frac{\chi_1^2/(n_1-1)}{\chi_2^2/(n_2-1)} = \frac{s_1^2/\sigma_1^2}{s_2^2/\sigma_2^2} \tag{7-71}$$

with $n_1 - 1$ and $n_2 - 1$ degrees of freedom. Furthermore, if the two populations have the same variance, i.e., if $\sigma_1^2 = \sigma_2^2$, then

$$F(n_1-1, n_2-1) = \frac{s_1^2}{s_2^2} \tag{7-72}$$

Therefore, the $F$ distribution provides a means of comparing variances, as will be seen in Sec. 7-2.3.

### 7-2.3 Confidence Intervals and Hypothesis Testing

The concept of *confidence intervals* is of considerable importance in regression analysis. A confidence interval is a range of values defined by an upper and a lower limit, the *confidence limits*. This range is constructed in such a way that we can say with certain confidence that the true value of the statistic being examined lies within this range. The level of confidence is chosen at $100(1-\alpha)$ percent, where $\alpha$ is usually small, say, 0.05 or 0.01. For example, when $\alpha = 0.05$, the confidence level is 95 percent. We demonstrate the concept of confidence intervals by first constructing such an interval for the standard normal distribution, extending the concept to other distributions, and then calculating specific confidence intervals for the mean and variance.

We saw earlier that the standard normal variable $u$ has a density function $\phi(u)$ [Eq. (7-42)] and a cumulative distribution function $\Phi(u)$ [Eq. (7-46)] and is distributed with $N(0, 1)$. Applying Eqs. (7-12) and (7-13) to the standard normal distribution,

$$\Pr\{u \le u_{\alpha/2}\} = \int_{-\infty}^{u_{\alpha/2}} \phi(u)\, du = \Phi(u_{\alpha/2}) = \frac{\alpha}{2} \tag{7-73}$$

$$\Pr\{u \le u_{1-\alpha/2}\} = \int_{-\infty}^{u_{1-\alpha/2}} \phi(u)\, du = \Phi(u_{1-\alpha/2}) = 1 - \frac{\alpha}{2} \tag{7-74}$$

and

$$\begin{aligned}\Pr\{u_{\alpha/2} < u \le u_{1-\alpha/2}\} &= \int_{u_{\alpha/2}}^{u_{1-\alpha/2}} \phi(u)\, du \\ &= \Phi(u_{1-\alpha/2}) - \Phi(u_{\alpha/2}) \\ &= 1 - \alpha \end{aligned} \tag{7-75}$$

The inequality

$$u_{\alpha/2} < u \le u_{1-\alpha/2} \tag{7-76}$$

defines the $100(1-\alpha)$ percent interval for the variable $u$. If $\alpha = 0.05$, then the 95 percent confidence interval for the standard normal variable is

$$-1.96 < u \leq 1.96 \tag{7-77}$$

where the values of the confidence limits were obtained from Table I of Ref. 3.

Let us now determine a confidence interval for the mean of a normally distributed population. We saw earlier [Eq. (7-50)] that the sample mean $\bar{x}$ of a normally distributed population is also normally distributed,

$$\bar{x} \sim N\left(\mu, \frac{\sigma^2}{n}\right) \tag{7-50}$$

and that this can be converted to the standard normal distribution so

$$\frac{\bar{x}-\mu}{\sqrt{\sigma^2/n}} \sim N(0, 1) \tag{7-51}$$

Since the quantity $[(\bar{x}-\mu)/\sqrt{\sigma^2/n}]$ is equivalent to $u$, Eq. (7-75) can be written as

$$\Pr\left\{u_{\alpha/2} < \frac{\bar{x}-\mu}{\sqrt{\sigma^2/n}} \leq u_{1-\alpha/2}\right\} = 1-\alpha \tag{7-78}$$

or rearranged to

$$\Pr\left\{\bar{x} - u_{1-\alpha/2}\sqrt{\frac{\sigma^2}{n}} \leq \mu < \bar{x} - u_{\alpha/2}\sqrt{\frac{\sigma^2}{n}}\right\} = 1-\alpha \tag{7-79}$$

The inequality†

$$\bar{x} - u_{1-\alpha/2}\sqrt{\frac{\sigma^2}{n}} \leq \mu < \bar{x} + u_{1-\alpha/2}\sqrt{\frac{\sigma^2}{n}} \tag{7-80}$$

is the $100(1-\alpha)$ percent confidence interval for the population mean. For $\alpha = 0.05$, the 95 percent confidence interval of the mean of a normally distributed population is

$$\bar{x} - 1.96\sqrt{\frac{\sigma^2}{n}} \leq \mu < \bar{x} + 1.96\sqrt{\frac{\sigma^2}{n}} \tag{7-81}$$

where $\bar{x}$ is the sample mean and $\sigma^2$ is the population variance. This simply says that we can state with 95 percent confidence that the true value of the population mean is in the range defined by the inequality (7-81).

† Note that the density distribution of $u$ is symmetrical around $u = 0$, so that $u_{\alpha/2} = -u_{1-\alpha/2}$. This substitution has been made in obtaining (7-80).

If the population variance $\sigma^2$ is not known, it will be estimated from the sample variance $s^2$. Replacing $\sigma^2$ with $s^2$ in the quantity $[(\bar{x} - \mu)/\sqrt{\sigma^2/n}]$, we obtain the variable

$$t = \frac{\bar{x} - \mu}{\sqrt{s^2/n}} \tag{7-66}$$

which has a Student's $t$ distribution with $\nu = n - 1$ degrees of freedom, as shown in Sec. 7-2.2. The confidence interval in this case is obtained from

$$\Pr\left\{t_{\alpha/2} < \frac{\bar{x} - \mu}{\sqrt{s^2/n}} \leq t_{1-\alpha/2}\right\} = 1 - \alpha \tag{7-82}$$

which rearranges to

$$\Pr\left\{\bar{x} - t_{1-\alpha/2}\sqrt{\frac{s^2}{n}} \leq \mu < \bar{x} - t_{\alpha/2}\sqrt{\frac{s^2}{n}}\right\} = 1 - \alpha \tag{7-83}$$

to yield the $100(1 - \alpha)$ percent confidence interval†

$$\bar{x} - t_{1-\alpha/2}\sqrt{\frac{s^2}{n}} \leq \mu < \bar{x} + t_{1-\alpha/2}\sqrt{\frac{s^2}{n}} \tag{7-84}$$

In Sec. 7-2.2 we showed that the sample variance $s^2$ and the population variance $\sigma^2$ were related through the $\chi^2$ distribution:

$$\chi^2 = (n - 1)\frac{s^2}{\sigma^2} \tag{7-61}$$

with $\nu = n - 1$ degrees of freedom. This relation can now be used to construct the confidence interval for the variance from

$$\Pr\left\{\chi^2_{\alpha/2} < (n - 1)\frac{s^2}{\sigma^2} \leq \chi^2_{1-\alpha/2}\right\} = 1 - \alpha \tag{7-85}$$

which gives the $100(1 - \alpha)$ percent confidence interval for $\sigma^2$ as

$$\frac{(n - 1)s^2}{\chi^2_{1-\alpha/2}} \leq \sigma^2 < \frac{(n - 1)s^2}{\chi^2_{\alpha/2}} \tag{7-86}$$

This discussion leads us to the concept of *hypothesis testing*. This consists of making an assumption about the distribution function of a random variable, very often about the numerical values of the statistical parameters of the

† The density distribution of the $t$ variable is symmetrical around $t = 0$, so that $t_{\alpha/2} = -t_{1-\alpha/2}$. This substitution has been made in obtaining (7-84).

distribution function (mean and variance), and deciding whether those values of the parameters are consistent with our sample of observations on that random variable.

For example, suppose that a sample of $n_1 = 10$ observations has a sample mean $\bar{x}_1 = 2.0$ and a sample variance $s_1^2 = 4$. Let us make the assumption that this sample came from a population which has a normal distribution with $\mu = 0$ and $\sigma_1^2$ unknown, that is, $X \sim N(0, \sigma_1^2)$. In order to test this assumption, we formalize it by stating the *null hypothesis*,

$$H_0: \quad \mu = \mu_0 = 0$$

and the *alternative hypothesis*,

$$H_A: \quad \mu = \mu_A \neq 0$$

We recall from (7-66) that the quantity $[(\bar{x}_1 - \mu)/\sqrt{s_1^2/n_1}]$ has a $t$ distribution, and from (7-82) that

$$\Pr\{t_{\alpha/2} < t \leq t_{1-\alpha/2}\} = 1 - \alpha \tag{7-87}$$

For 95 percent probability and $\nu = n_1 - 1 = 9$ degrees of freedom, the above equation is

$$\Pr\{-2.262 < t \leq 2.262\} = 0.95 \tag{7-88}$$

The region defined by Eq. (7-88) is shown in Fig. 7-8 as the *region of acceptance*, while the regions outside this range are labeled *regions of rejection*. Based on the assumption that the null hypothesis is true, if the statistic calculated from the experimental sample falls outside the region of acceptance,

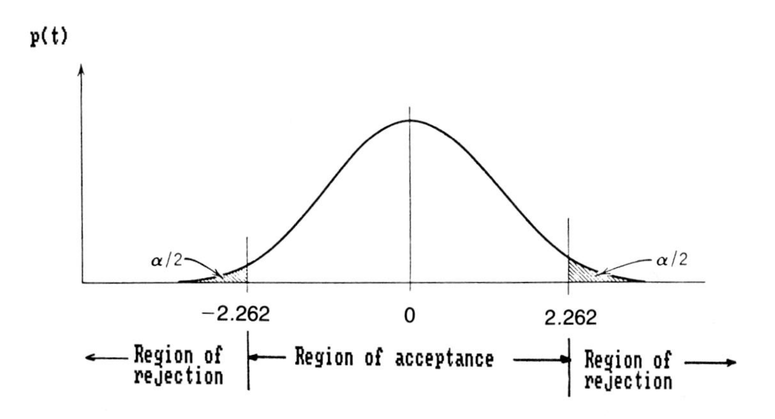


**Figure 7-8** Hypothesis test for the mean.

the null hypothesis is rejected and $H_A$ is accepted. Otherwise, $H_0$ is accepted and $H_A$ is rejected. In this example we calculate $t_0$:

$$t_0 = \frac{\bar{x}_1 - \mu_0}{\sqrt{s_1^2/n_1}} = \frac{2.0 - 0}{\sqrt{\frac{4}{10}}} = 3.16$$

We see that $t_0$ is outside the region of acceptance defined by (7-88); therefore we reject the null hypothesis.

We can generalize this test by saying that if

$$\left| \frac{\bar{x} - \mu_0}{\sqrt{s^2/n}} \right| > t_{1-\alpha/2} \tag{7-89}$$

then the null hypothesis that $\mu = \mu_0$ must be rejected. This is the well-known *two-sided t test*, which is used extensively in regression analysis to test the values of regression parameters.

Let us now examine the variance of the sample. We draw a second sample of $n_2 = 21$ observations and find that $\bar{x}_2 = 2.0$ and $s_2^2 = 3$. We ask the question: "Is the second sample taken from the same population as the first sample, or from one which has a different variance than the first?" We state the null hypothesis,

$$H_0: \quad \frac{\sigma_1^2}{\sigma_2^2} = 1$$

and the alternative hypothesis,

$$H_A: \quad \frac{\sigma_1^2}{\sigma^2} \neq 1$$

We recall from Eq. (7-71) that the ratio $(s_1^2/\sigma_1^2)/(s_2^2/\sigma_2^2)$ has an $F$ distribution with $(\nu_1, \nu_2)$ degrees of freedom. From the probability distribution function

$$\Pr\{F_{\alpha/2}(\nu_1, \nu_2) < F \leq F_{1-\alpha/2}(\nu_1, \nu_2)\} = 1 - \alpha \tag{7-90}$$

To test the null hypothesis at the 95 percent confidence level, obtain the values of $F_{0.025}(9, 20)$ and $F_{0.975}(9, 20)$ for this example from the $F$ distribution tables.† Therefore the interval of acceptance is given by

$$\Pr\left\{0.272 < \frac{s_1^2/\sigma_1^2}{s_2^2/\sigma_2^2} \leq 2.84\right\} = 0.95 \tag{7-91}$$

The null hypothesis assumes that $\sigma_1^2 = \sigma_2^2$; therefore the inequality becomes

†The value of $F_{\alpha/2}(\nu_1, \nu_2)$ is obtained from the relationship $F_{\alpha/2}(\nu_1, \nu_2) = 1/F_{1-\alpha/2}(\nu_2, \nu_1)$.

$$0.272 < \frac{s_1^2}{s_2^2} \leq 2.84 \tag{7-92}$$

For this example

$$\frac{s_1^2}{s_2^2} = \tfrac{4}{3} = 1.33$$

therefore, the null hypothesis can be accepted.

This is the *two-sided F test* which is used in the analysis of variance of regression results to test the adequacy of a model in fitting the experimental data (see Sec. 7-5).

Hypothesis testing is an involved procedure, which we have briefly introduced here. It is outside the scope of this chapter to discuss hypothesis testing in more depth. The interested reader to referred to Brownlee [3] and Lipson and Sheth [4] for further discussion.

## 7-3 LINEAR REGRESSION ANALYSIS

### 7-3.1 Derivation of the Normal Equations

Most mathematical models in engineering and science are nonlinear in the parameters. However, for a complete understanding of nonlinear regression methods it is necessary to develop first the linear regression case and show how this extends to nonlinear models.

The exact representation of a linear relationship may be shown as

$$y = \alpha + \beta x \tag{7-93}$$

where $y$ represents the true value of the dependent variable, $x$ is the true value of the independent variable, $\beta$ is the slope of the line, and $\alpha$ is the $y$-intercept of the line. This relationship is not useful in this form because it requires knowledge of the true values of $y$ and $x$. Instead, the linear model is rewritten in terms of the observations of the values of the variables

$$\mathbf{Y}^* = \alpha + \beta \mathbf{X} + \mathbf{u} \tag{7-94}$$

where $\mathbf{Y}^*$ is the vector of observations of the dependent variable, $\mathbf{X}$ is the vector of observations of the independent variable, and $\mathbf{u}$ is the vector of *disturbance terms*. The purpose of the $\mathbf{u}$ term is to characterize the discrepancies that emerge between the true values and the observed values of the variables. These discrepancies can be attributed mainly to experimental error. Later in this section, $\mathbf{u}$ will be assumed to be a stochastic variable with some specified probability distribution.

Equation (7-94) can be extended to include more than one independent variable:

$$\mathbf{Y}^* = \beta_1 \mathbf{X}_1 + \beta_2 \mathbf{X}_2 + \cdots + \beta_k \mathbf{X}_k + \mathbf{u} \tag{7-95}$$

where $\mathbf{X}_1, \mathbf{X}_2, \ldots, \mathbf{X}_k$ are the vectors of observations of $k$ independent variables. (To allow for a $y$-intercept, the vector $\mathbf{X}_1$ can be taken as a vector whose components are all unity; thus $\beta_1$ becomes the parameter specifying the value of the $y$-intercept.)

Equation (7-95) can be condensed to matrix form

$$\mathbf{Y}^* = \mathbf{X}\boldsymbol{\beta} + \mathbf{u} \tag{7-96}$$

where $\mathbf{Y}^* = (n \times 1)$ vector of observations of the dependent variable
$\mathbf{X} = (n \times k)$ matrix of observations of the independent variables
$\boldsymbol{\beta} = (k \times 1)$ vector of parameters
$\mathbf{u} = (n \times 1)$ vector of disturbance terms
$n$ = number of observations

Given a set of $n$ observations in the $Y$ variable, and in each of the $k$ independent variables, the problem now is to obtain an estimate of the $\boldsymbol{\beta}$ vector.

The basic assumptions made in the derivation of the method for estimating the parameters are the following:

1. The disturbance terms, represented by the vector $\mathbf{u}$, are random variables with zero expectation, i.e.,

$$E[\mathbf{u}] = \bar{\mathbf{u}} = 0 \tag{7-97}$$

Since the value of $\mathbf{u}$ is the sum of errors from several sources, the central limit theorem implies that the distribution of $\mathbf{u}$ tends toward the normal distribution as the number of factors contributing to $\mathbf{u}$ increases.

2. The variance of the distribution of $\mathbf{u}$ is constant and independent of $\mathbf{X}$, that is,

$$\begin{aligned} V[\mathbf{u}_1] &= E[(\mathbf{u}_1 - \bar{\mathbf{u}}_1)^2] = \sigma^2 \\ V[\mathbf{u}_2] &= E[(\mathbf{u}_2 - \bar{\mathbf{u}}_2)^2] = \sigma^2 \\ &\vdots \\ V[\mathbf{u}_n] &= E[(\mathbf{u}_n - \bar{\mathbf{u}}_n)^2] = \sigma^2 \end{aligned} \tag{7-98}$$

In addition, the values of $\mathbf{u}$ for each set of observations are independent of one another, i.e.,

$$E[\mathbf{u}_i \mathbf{u}_j] = E[\mathbf{u}_i]E[\mathbf{u}_j] \qquad i \neq j \tag{7-99}$$

From assumption 1 and from Eqs. (7-31) and (7-35), we also conclude that the covariance of $\mathbf{u}$ is zero:

$$\text{Cov}\,[\mathbf{u}_i, \mathbf{u}_j] = 0 \qquad i \neq j \tag{7-100}$$

The variance-covariance matrix is defined as

$$\text{Var-Cov}\,[\mathbf{u}] = \begin{bmatrix} V[\mathbf{u}_1] & \text{Cov}\,[\mathbf{u}_1, \mathbf{u}_2] & \cdots & \text{Cov}\,[\mathbf{u}_1, \mathbf{u}_n] \\ \text{Cov}\,[\mathbf{u}_2, \mathbf{u}_1] & V[\mathbf{u}_2] & \cdots & \text{Cov}\,[\mathbf{u}_2, \mathbf{u}_n] \\ \cdots & \cdots & \cdots & \cdots \\ \text{Cov}\,[\mathbf{u}_n, \mathbf{u}_1] & \text{Cov}\,[\mathbf{u}_n, \mathbf{u}_2] & \cdots & V[\mathbf{u}_n] \end{bmatrix}$$

$$= E[(\mathbf{u} - E[\mathbf{u}])(\mathbf{u} - E[\mathbf{u}])^T] \tag{7-101}$$

Combining (7-97), (7-98), (7-100) and (7-101), we obtain

$$\text{Var-Cov}\,[\mathbf{u}] = E[\mathbf{u}\mathbf{u}^T]$$

$$= \begin{bmatrix} \sigma^2 & 0 & \cdots & 0 \\ 0 & \sigma^2 & \cdots & 0 \\ & & & \\ 0 & 0 & \cdots & \sigma^2 \end{bmatrix} = \sigma^2 \mathbf{I} \tag{7-102}$$

In summary, (7-102) says that each $\mathbf{u}$ distribution has the same variance, and that all disturbances are pairwise uncorrelated [8].

3. The matrix $\mathbf{X}$ is a set of fixed numbers, i.e., the values of $\mathbf{X}$ do not contain error.
4. The rank of the matrix $\mathbf{X}$ is equal to $k$, and $k < n$. The first part of this assumption ensures that the $k$ variables are linearly independent. The second part requires that the number of observations exceeds the number of parameters to be estimated. This is essential in order to have the necessary degrees of freedom for the parameters estimation.
5. The vector $\mathbf{u}$ has a multivariate normal distribution:

$$\mathbf{u} \sim N(0, \sigma^2 \mathbf{I}) \tag{7-103}$$

These assumptions are not overly restrictive. Since the value of $\mathbf{u}$ is due to many factors acting in opposite directions, it should be expected that small values of $\mathbf{u}$ occur more frequently than large values, and that $\mathbf{u}$ is a variable with a probability distribution centered at zero and having a finite variance $\sigma^2$. This is true when the form of Eq. (7-96) is close to the correct relationship. Because of the many factors involved, the central limit theorem would further suggest that $\mathbf{u}$ has a normal distribution, which gives the parameter estimates the desirable property of being maximum-likelihood estimates. Later on in the discussion, it will be shown that the regression method can handle cases where $\sigma^2$ is not constant, and where $\mathbf{u}$ is not independent of $\mathbf{X}$.

Let us now return to the hypothesized linear model

$$\mathbf{Y}^* = \mathbf{X}\boldsymbol{\beta} + \mathbf{u} \tag{7-96}$$

Let **b** denote a $k$-element vector which is an estimate of the parameter vector $\boldsymbol{\beta}$. We use this estimate to define a vector of residuals

$$\boldsymbol{\epsilon} = \mathbf{Y}^* - \mathbf{X}\mathbf{b} = \mathbf{Y}^* - \mathbf{Y} \tag{7-104}$$

These residuals are the differences between the experimental observations $\mathbf{Y}^*$ and the calculated values of **Y** using the estimated vector **b**. Now the *least squares method* can be used to find the vector **b** which minimizes the sum of the squared residuals $\Phi$:

$$\Phi = \boldsymbol{\epsilon}^T\boldsymbol{\epsilon} = (\mathbf{Y}^* - \mathbf{X}\mathbf{b})^T(\mathbf{Y}^* - \mathbf{X}\mathbf{b}) \tag{7-105}$$

In order to calculate the vector **b** which minimizes $\Phi$, we take the partial derivative of $\Phi$ with respect to **b** and set it equal to zero:

$$\frac{\partial \Phi}{\partial \mathbf{b}} = (-\mathbf{X})^T(\mathbf{Y}^* - \mathbf{X}\mathbf{b}) + (\mathbf{Y}^* - \mathbf{X}\mathbf{b})^T(-\mathbf{X}) = 0 \tag{7-106}$$

We simplify this utilizing the matrix-vector identity $\mathbf{A}^T\mathbf{y} = \mathbf{y}^T\mathbf{A}$:

$$-2\mathbf{X}^T(\mathbf{Y}^* - \mathbf{X}\mathbf{b}) = 0 \tag{7-107}$$

Equation (7-107) can be further rearranged to yield

$$(\mathbf{X}^T\mathbf{X})\mathbf{b} = \mathbf{X}^T\mathbf{Y}^* \tag{7-108}$$

The above constitutes a set of simultaneous linear algebraic equations, called the *normal equations*. The matrix $(\mathbf{X}^T\mathbf{X})$ is a $(k \times k)$ symmetric matrix. Assumption 4 made earlier guarantees that $(\mathbf{X}^T\mathbf{X})$ is nonsingular; therefore its inverse exists. Thus, the normal equations can be solved for the vector **b**:

$$\mathbf{b} = (\mathbf{X}^T\mathbf{X})^{-1}\mathbf{X}^T\mathbf{Y}^* \tag{7-109}$$

The values of the elements of vector **b** can be obtained readily from (7-109) since the right-hand side of this equation contains the matrix of observations of the independent variables **X** and the vector of observations of the dependent variable $\mathbf{Y}^*$, all of which are known.

### 7-3.2 Properties of the Estimated Vector of Parameters

The vector **b** is an estimate of $\boldsymbol{\beta}$ which minimizes the sum of the squared residuals, irrespective of any distribution properties of the residuals. In addition, $\mathbf{b}$ is an unbiased estimate of $\boldsymbol{\beta}$. To show this, we combine Eqs. (7-109) and (7-96),

$$\begin{aligned}\mathbf{b} &= (\mathbf{X}^T\mathbf{X})^{-1}\mathbf{X}^T(\mathbf{X}\boldsymbol{\beta} + \mathbf{u}) \\ &= (\mathbf{X}^T\mathbf{X})^{-1}(\mathbf{X}^T\mathbf{X})\boldsymbol{\beta} + (\mathbf{X}^T\mathbf{X})^{-1}\mathbf{X}^T\mathbf{u} \\ &= \boldsymbol{\beta} + (\mathbf{X}^T\mathbf{X})^{-1}\mathbf{X}^T\mathbf{u}\end{aligned} \tag{7-110}$$

and take the expected value of $\mathbf{b}$,

$$E[\mathbf{b}] = E[\boldsymbol{\beta}] + (\mathbf{X}^T\mathbf{X})^{-1}\mathbf{X}^T E[\mathbf{u}] \tag{7-111}$$

but since $E[\mathbf{u}] = 0$ (assumption 1) and $\boldsymbol{\beta}$ is constant, then

$$E[\mathbf{b}] = \boldsymbol{\beta} \tag{7-112}$$

i.e., the expected value of $\mathbf{b}$ is $\boldsymbol{\beta}$.

Furthermore, the variance of $\mathbf{b}$ can be obtained as follows. Rearranging Eq. (7-110),

$$\mathbf{b} - \boldsymbol{\beta} = (\mathbf{X}^T\mathbf{X})^{-1}\mathbf{X}^T\mathbf{u} \tag{7-113}$$

and utilizing Eq. (7-112), we obtain

$$\mathbf{b} - E[\mathbf{b}] = (\mathbf{X}^T\mathbf{X})^{-1}\mathbf{X}^T\mathbf{u} \tag{7-114}$$

From the definition of the variance-covariance matrix [Eq. (7-101)],

$$\text{Var-Cov}\,[\mathbf{b}] = E[(\mathbf{b} - E[\mathbf{b}])(\mathbf{b} - E[\mathbf{b}])^T] \tag{7-115}$$

Using (7-114) in (7-115),

$$\begin{aligned}\text{Var-Cov}\,[\mathbf{b}] &= E[(\mathbf{X}^T\mathbf{X})^{-1}\mathbf{X}^T\mathbf{u}\mathbf{u}^T\mathbf{X}(\mathbf{X}^T\mathbf{X})^{-1}]\dagger \\ &= (\mathbf{X}^T\mathbf{X})^{-1}\mathbf{X}^T E[\mathbf{u}\mathbf{u}^T]\mathbf{X}(\mathbf{X}^T\mathbf{X})^{-1}\end{aligned} \tag{7-116}$$

but from Eq. (7-102) $E[\mathbf{u}\mathbf{u}^T] = \sigma^2\mathbf{I}$; therefore the variance-covariance of $\mathbf{b}$ simplifies to

$$\text{Var-Cov}\,[\mathbf{b}] = \sigma^2(\mathbf{X}^T\mathbf{X})^{-1} \tag{7-117}$$

where $\sigma^2$ is the variance of $\mathbf{u}$, as defined by Eq. (7-98).

The elements of the matrix $(\mathbf{X}^T\mathbf{X})^{-1}$ are designated as $a_{ij}$. Therefore the

† Note that $(\mathbf{X}^T\mathbf{X})$ is a symmetric matrix; therefore its inverse $(\mathbf{X}^T\mathbf{X})^{-1}$ is also symmetric. The transpose of a symmetric matrix is the same as the original matrix.

variance of $\mathbf{b}_i$ is given by

$$V[\mathbf{b}_i] = \sigma^2 a_{ii} \tag{7-118}$$

and the covariance of $\mathbf{b}_i$ with $\mathbf{b}_j$ by

$$\text{Cov}\,[\mathbf{b}_i, \mathbf{b}_j] = \sigma^2 a_{ij} \tag{7-119}$$

Therefore, if the variance of $\mathbf{u}$ is known, or can be estimated, then the variance-covariance of the estimated parameter vector $\mathbf{b}$ can be calculated.

It can be seen from Eq. (7-116) that the variance-covariance matrix of $\mathbf{b}$ can still be calculated even if assumption 2 is not made. In that case, the matrix $E[\mathbf{uu}^T]$ would not be a diagonal matrix.

We can now draw an important conclusion regarding the distribution of $\mathbf{b}$. Equation (7-110) shows that $\mathbf{b}$ is a linear combination of $\mathbf{u}$. If $\mathbf{u}$ is a multivariate normal distribution (assumption 5, Sec. 7-3.1), then $\mathbf{b}$ is also a multivariate normal distribution, i.e.,

$$\mathbf{b} \sim N(\boldsymbol{\beta}, \sigma^2(\mathbf{X}^T\mathbf{X})^{-1}) \tag{7-120}$$

For each individual parameter

$$b_i \sim N(\beta_i, \sigma^2 a_{ii}) \tag{7-121}$$

where $a_{ii}$ is the $i$th element on the principal diagonal of $(\mathbf{X}^T\mathbf{X})^{-1}$. The normal distribution can be converted to the standard normal distribution

$$\frac{b_i - \beta_i}{\sigma\sqrt{a_{ii}}} \sim N(0, 1) \tag{7-122}$$

The variance $\sigma^2$ of the disturbance terms is not usually known unless a large number of repetitive experiments have been performed. The value of $\sigma^2$ can be estimated from

$$s^2 = \frac{\boldsymbol{\epsilon}^T\boldsymbol{\epsilon}}{n-k} \tag{7-123}$$

where $\boldsymbol{\epsilon}^T\boldsymbol{\epsilon}$ is the sum of squared residuals [see Eq. (7.105)], and $n - k$ is the number of degrees of freedom. If there is no lack of fit of the model to the data (see analysis of variances, Sec. 7-4.5), then $s^2$ is an unbiased estimate of $\sigma^2$, i.e.,

$$E[s^2] = \sigma^2 \tag{7-124}$$

If lack of fit cannot be tested, use of $s^2$ as an estimate of $\sigma^2$ implies an assumption that the model is correct [3, p. 92].

We saw earlier that the ratio of $s^2/\sigma^2$ has a chi-square distribution,

$$\chi^2 = \nu \frac{s^2}{\sigma^2} \tag{7-61}$$

and that the $t$ variable is given by

$$t = \frac{N(0, 1)}{\sqrt{\chi^2/\nu}} \tag{7-62}$$

We can, therefore, combine (7-122), (7-61), and (7-62) to form the $t$ variable:

$$t = \frac{(b_i - \beta_i)/\sigma\sqrt{a_{ii}}}{\sqrt{s^2/\sigma^2}} = \frac{b_i - \beta_i}{s\sqrt{a_{ii}}} \sim t(n-k) \tag{7-125}$$

Equation (7-125) shows that the quantity $(b_i - \beta_i)/s\sqrt{a_{ii}}$ has a $t$ distribution with $n - k$ degrees of freedom. This is a very important equation, because it enables us to construct confidence intervals of the parameters from quantities which can be calculated from the regression analysis. For example, the $100(1-\alpha)$ percent confidence interval for parameter $\beta_i$ can be obtained from

$$\Pr\left\{t_{\alpha/2} < \frac{b_i - \beta_i}{s\sqrt{a_{ii}}} \leq t_{1-\alpha/2}\right\} = 1 - \alpha \tag{7-126}$$

which yields the interval

$$b_i - t_{1-\alpha/2}s\sqrt{a_{ii}} \leq \beta_i < b_i + t_{1-\alpha/2}s\sqrt{a_{ii}} \tag{7-127}$$

The above are *individual parameter confidence intervals*. Figure 7-9*a* demonstrates these intervals for $\beta_1$ and $\beta_2$ in a two-parameter model.

Furthermore, Eq. (7-125) enables us to perform the $t$ test for hypothetical values of $\boldsymbol{\beta}$ (see Sec. 7-2.3). For example, if it is suspected that the value of $\beta_i$ is not significantly different than zero, then the null hypothesis can be stated as

$$H_0: \quad \beta_i = 0 \tag{7-128}$$

and the alternative hypothesis as

$$H_A: \quad \beta_i \neq 0 \tag{7-129}$$

When (7-128) is substituted in (7-125), the resulting expression

$$t = \frac{b_i}{s\sqrt{a_{ii}}} \tag{7-130}$$

is calculated. If this value of $t$ lies within the region of acceptance given by the $t$

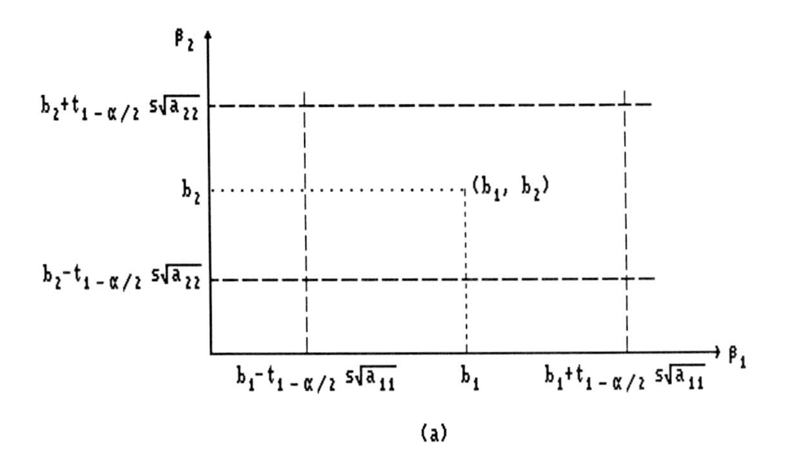


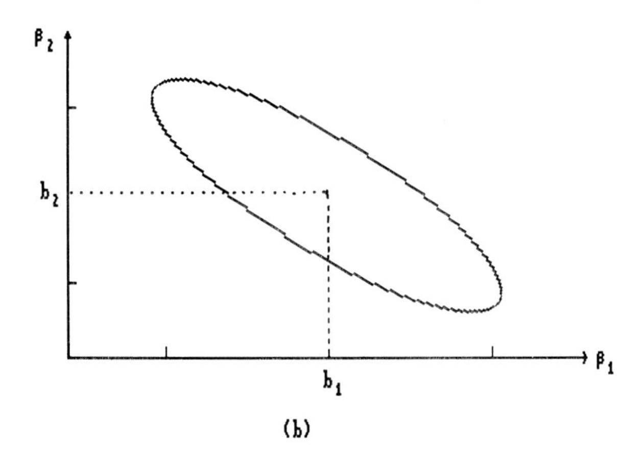


**Figure 7-9** Confidence intervals for parameters: (*a*) individual confidence intervals and (*b*) joint confidence region.

distribution for a two-sided test at the required confidence level, then the null hypothesis that $\beta_i = 0$ is accepted. This is a very useful test in deciding the significance of a parameter in a model and in helping the experimenter discriminate between competing models.

In most mathematical models, the covariance between parameters, as measured by Eq. (7-119), is nonzero, i.e., the parameters are correlated with each other. Careful experimental design may reduce, but never completely eliminate, this correlation. The individual confidence intervals calculated by (7-127) do not reflect the covariance. To do so, it is necessary to construct the *joint confidence region* of parameters. Using the multivariate normal distribution of **b** [Eq. (7-120)], we form the standardized normal variable:

$$\{\sigma^2(\mathbf{X}^T\mathbf{X})^{-1}\}^{-1/2}(\mathbf{b}-\boldsymbol{\beta}) \sim N(0, \mathbf{I}) \tag{7-131}$$

We recall that the chi-square variable is the sum of the squares of standardized normal variables [Eq. (7-55)]; therefore we can form the $\chi_1^2$ variable from (7-131):

$$\chi_1^2 = \frac{(\mathbf{b}-\boldsymbol{\beta})^T(\mathbf{X}^T\mathbf{X})(\mathbf{b}-\boldsymbol{\beta})}{\sigma^2} \tag{7-132}$$

with $k$ degrees of freedom. We can also form another chi-square variable from the ratio of $s^2$ and $\sigma^2$ [see Eq. (7-61)]:

$$\chi_2^2 = (n-k)\frac{s^2}{\sigma^2} \tag{7-133}$$

From Eq. (7-67) we recall that

$$F(\nu_1, \nu_2) = \frac{\chi_1^2/\nu_1}{\chi_2^2/\nu_2} \tag{7-67}$$

Therefore, combining (7-132), (7-133), and (7-67) we obtain

$$F(k, n-k) = \frac{[(\mathbf{b}-\boldsymbol{\beta})^T(\mathbf{X}^T\mathbf{X})(\mathbf{b}-\boldsymbol{\beta})]/k\sigma^2}{[(n-k)s^2]/[(n-k)\sigma^2]} = \frac{(\mathbf{b}-\boldsymbol{\beta})^T(\mathbf{X}^T\mathbf{X})(\mathbf{b}-\boldsymbol{\beta})}{ks^2} \tag{7-134}$$

Finally, the joint $100(1-\alpha)$ percent confidence region for all the parameters can be obtained from

$$\Pr\left\{\frac{(\mathbf{b}-\boldsymbol{\beta})^T(\mathbf{X}^T\mathbf{X})(\mathbf{b}-\boldsymbol{\beta})}{ks^2} \leq F_{1-\alpha}(k, n-k)\right\} = 1-\alpha \tag{7-135}$$

where $F_{1-\alpha}(k, n-k)$ is the $(1-\alpha)$ point of the $F$ distribution with $k$ and $n-k$ degrees of freedom. The inequality in (7-135) defines a hyperellipsoidal region in the $k$-dimensional parameter space. For a two-parameter model, this joint confidence region is shown in Fig. 7-9$b$ as an elongated tilted ellipse.

In the rare case where the parameters are uncorrelated, the matrix $(\mathbf{X}^T\mathbf{X})^{-1}$ is diagonal, the axes of the confidence ellipsoid would be parallel to the coordinates of the parameter space, and the individual parameter confidence intervals would hold for each parameter independently. However, since the parameters are usually correlated, the extent of the correlation can be measured from the *correlation coefficient matrix*, $\boldsymbol{R}$. This is obtained by applying Eq. (7-34) to the variance-covariance matrix (7-117):

$$\rho_{b_i b_j} = \frac{\text{Cov}\,[b_i, b_j]}{\sqrt{V[b_i]V[b_j]}}$$

$$= \frac{a_{ij}}{\sqrt{a_{ii}a_{jj}}} = r_{ij} \tag{7-136}$$

Matrix **R** is a $(k \times k)$ symmetric matrix and its elements $r_{ij}$ have values in the range $-1.0 \le r_{ij} \le 1.0$.

A typical correlation coefficient matrix may look like this:

$$\mathbf{R} = \begin{bmatrix} 1.0 & 0.98 & -0.56 & 0.92 \\ 0.98 & 1.0 & 0.85 & -0.97 \\ -0.56 & 0.85 & 1.0 & 0.35 \\ 0.92 & -0.97 & 0.35 & 1.0 \end{bmatrix} \tag{7-137}$$

A negative correlation between two parameters implies that the errors which cause the estimate of one to be high also cause the estimate of the other to be low. The higher the correlation between two parameters, the closer the value of $r_{ij}$ is to $|1.0|$. Consequently, the diagonal elements $r_{ii}$ are all equal to $+1.0$.

The correlation between parameters causes the axes of the confidence ellipsoids of the linear model to be at an angle to the coordinates of the parameter space. Therefore, the individual parameter confidence limits will not represent the true interval within which a parameter $b_i$ may lie and still remain within the confidence ellipsoid.

In nonlinear models, the confidence hyperspace is no longer a hyperellipsoid. The amount of distortion depends on the extent of nonlinearlity of the model. Therefore, the calculation of the confidence intervals is not as rigorous an exercise as in the linear model. Still, a lot of valuable information can be extracted from the correlation coefficient matrix which approximates the maximum-likelihood hyperspace in the vicinity of the solution where the model is nearly linear. If the values of the off-diagonal elements of **R** are close to $|1.0|$, the parameters associated with those elements are highly correlated with each other. Davies [10] tests the values of $r_{ij}$ against a normal distribution with zero mean, i.e., no correlation. He classifies the correlation as "significant" and "highly significant" if the value of $r_{ij}$ is higher than the 0.05 and 0.01 significance points of the normal distribution, respectively. High correlation between parameters implies that it is very difficult to obtain separate estimates of these parameters with the available data.

The eigenvectors $w$ of the matrix **R** give the direction of the major and minor axes of the hyperellipsoidal confidence region of the parameter space. The lengths of the axes are proportional to the square root of the eigenvalues $\lambda$

of the matrix. Box [11] calculates the values of the parameters at the ends of the axes by

$$\bar{b}_i = b_i \pm w_{ri}\{\lambda_r(s^2 a_{ii})kF_{1-\alpha}(k, n-k)\}^{1/2} \tag{7-138}$$

where
$r = 1, 2, \ldots, k$
$k$ = number of parameters
$n$ = number of points used in estimating $b_i$
$F_{1-\alpha}(k, n-k)$ = value of the $F$ distribution with $k$ and $n-k$ degrees of freedom

Subsequently, he uses these parameter values to calculate the sum of squares at each end of the axes and to compare them with the sum of squares at the center of the hyperellipsoid. This sum of squares search, which is based on a linear model, may give vital information for nonlinear models as well. In the case where the solution has only converged on a *local* minimum sum of squares, it is very likely that the search in the direction of one of the axes will produce a lower sum of squares. In such a case the regression must be repeated, starting from a different initial position, so that the local minimum may be bypassed.

## 7-4 NONLINEAR REGRESSION ANALYSIS

We have stated this earlier in the chapter, and we repeat it again: the mathematical models encountered in engineering and science are often nonlinear in their parameters. Consider, for example, the analysis of a complex chemical reaction such as

$$\mathrm{A} \xrightarrow{k_1} \mathrm{B} \xrightarrow{k_2} \mathrm{C} + \mathrm{D}$$

$$\mathrm{C} + \mathrm{A} \xrightarrow{k_3} \mathrm{E} + \mathrm{F}$$

where the rate of formation of each component may be written as

$$\begin{aligned}
\frac{dC_\mathrm{A}}{dt} &= -k_1 C_\mathrm{A} - k_3 C_\mathrm{A}^n C_\mathrm{C}^m \\
\frac{dC_\mathrm{B}}{dt} &= k_1 C_\mathrm{A} - k_2 C_\mathrm{B} \\
\frac{dC_\mathrm{C}}{dt} &= k_2 C_\mathrm{B} - k_3 C_\mathrm{A}^n C_\mathrm{C}^m \\
\frac{dC_\mathrm{E}}{dt} &= k_3 C_\mathrm{A}^n C_\mathrm{C}^m
\end{aligned} \tag{7-139}$$

This is only one possible formulation of the reaction mechanism. It contains five unknown parameters—$k_1$, $k_2$, $k_3$, $n$, and $m$—which must be calculated by fitting the model to experimental data. Suppose that experiments for this chemical system are carried out in a batch reactor and data of the form shown in Fig. 7-10 are collected. Since experimental data are available for all four dependent variables—$C_A$, $C_B$, $C_C$, and $C_E$—*multiple nonlinear regression* can be performed by simultaneously fitting all four equations of (7-139) to the data.

The methods developed in this section will enable us to fit models consisting of multiple dependent variables, such as the one described above, to

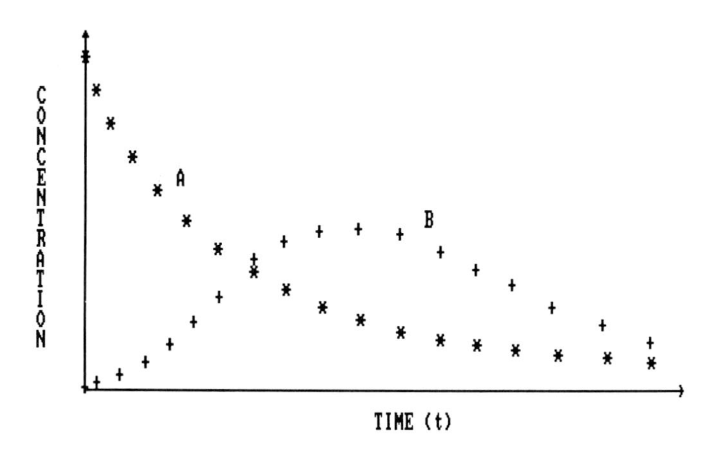


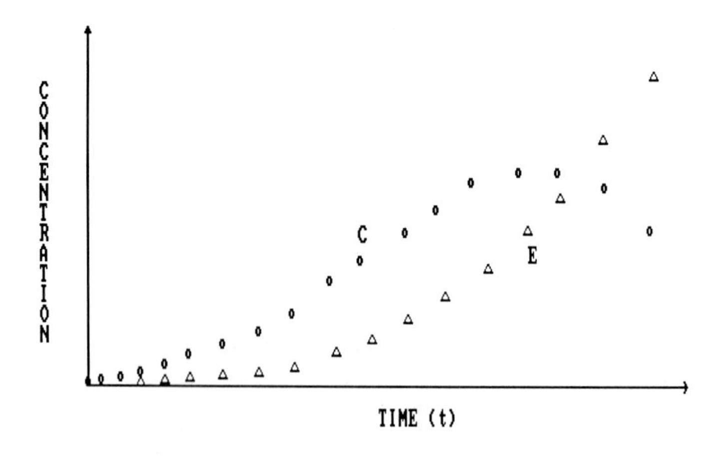


**Figure 7-10** Simulated data for batch reactor experiment.

multiresponse experimental data in order to obtain the values of the parameters of the model which minimize the *overall (weighted) sum of squared residuals*. In addition, a thorough statistical analysis of the regression results will enable us to

1. Decide whether the model gives satisfactory fit within the experimental error of the data.
2. Discriminate between competing models.
3. Measure the accuracy of the estimation of the parameters by constructing the confidence region in the parameter space.
4. Measure the correlation between parameters by examining the correlation coefficient matrix.
5. Perform tests to verify that repeated experimental data come from the same population of experiments.
6. Perform tests to verify whether the residuals between the data and the model are randomly distributed.

### 7-4.1 The Gauss-Newton Method

A model consisting of differential equations, such as (7-139), may be shown in the form:

$$\mathbf{Y}' = \mathbf{g}(x, \mathbf{Y}, \mathbf{b}) \tag{7-140}$$

where $\mathbf{Y}'$ = vector of derivatives of $Y$
$\mathbf{g}$ = vector of functions
$x$ = independent variable
$\mathbf{Y}$ = vector of dependent variables
$\mathbf{b}$ = vector of parameters

We assume that if the boundary conditions are given and if the vector $\mathbf{b}$ can be estimated, then the differential equations (7-140) can be integrated numerically or analytically to give the integrated results, which are

$$\mathbf{Y} = f(x, \mathbf{b}) \tag{7-141}$$

For the simple case where the model contains only one dependent variable, the sum of squared residuals is given by

$$\Phi = \boldsymbol{\epsilon}^T \boldsymbol{\epsilon} = (\mathbf{Y}^* - \mathbf{Y})^T (\mathbf{Y}^* - \mathbf{Y}) \tag{7-142}$$

where $\mathbf{Y}^*$ = vector of experimental observations of the dependent variable
$\mathbf{Y}$ = vector of calculated values of the dependent variables obtained from (7-141).

For multiple regression, where there are $v$ dependent variables in the model, the *weighted sum of squared residuals* is given by

$$\Phi = \sum_{j=1}^{v} w_j \boldsymbol{\epsilon}_j^T \boldsymbol{\epsilon}_j = \sum_{j=1}^{v} w_j \phi_j$$

$$= \sum_{j=1}^{v} w_j (\mathbf{Y}_j^* - \mathbf{Y}_j)^T (\mathbf{Y}_j^* - \mathbf{Y}_j) \tag{7-143}$$

where $w_j$ = weighting factor corresponding to the $j$th dependent variable
$\phi_j$ = sum of squared residuals corresponding to the $j$th dependent variable

We defer discussion of the multiple regression case to Sec. 7-4.4 and proceed here with the development of the one-variable case.

Since $\mathbf{Y}$ is nonlinear with respect to the parameters, taking the partial derivative of $\Phi$ with respect to $\mathbf{b}$ and setting it to zero will yield a nonlinear equation which would be difficult to solve for $\mathbf{b}$. This problem was alleviated by Gauss, who determined that fitting nonlinear functions by least squares can be achieved by an iterative method involving a series of linear approximations. At each stage of the iteration, linear least squares theory can be used to obtain the next approximation.

This method, known as the *Gauss-Newton method*, converts the nonlinear problem into a linear one by approximating the function $\mathbf{Y}$ by a Taylor series expansion around an estimated value of the parameter vector $\mathbf{b}$:

$$\mathbf{Y}(x, \mathbf{b} + \Delta\mathbf{b}) = \mathbf{Y}(x, \mathbf{b}) + \frac{\partial \mathbf{Y}}{\partial \mathbf{b}} \Delta\mathbf{b} \tag{7-144}$$

where the Taylor series has been truncated after the second term. Equation (7-144) is linear in $\Delta\mathbf{b}$. Therefore, the problem has been transformed from finding $\mathbf{b}$ to that of finding the correction to $\mathbf{b}$, that is, $\Delta\mathbf{b}$, which must be added to an estimate of $\mathbf{b}$ to minimize the sum of squared residuals. To do this we replace $\mathbf{Y}$ in (7-142) with the right-hand side of (7-144) to get

$$\Phi = (\mathbf{Y}^* - \mathbf{Y} - \mathbf{A}\,\Delta\mathbf{b})^T (\mathbf{Y}^* - \mathbf{Y} - \mathbf{A}\,\Delta\mathbf{b}) \tag{7-145}$$

where $\mathbf{A}$ is the Jacobian matrix of partial derivatives of $\mathbf{Y}$ with respect to $\mathbf{b}$ evaluated at all $n$ points where experimental observations are available:

$$\mathbf{A} = \begin{bmatrix} \dfrac{\partial Y_1}{\partial b_1} & \cdots & \dfrac{\partial Y_1}{\partial b_k} \\ \cdots & \cdots & \cdots \\ \dfrac{\partial Y_n}{\partial b_1} & \cdots & \dfrac{\partial Y_n}{\partial b_k} \end{bmatrix} \tag{7-146}$$

Taking the partial derivative of $\Phi$ with respect to $\Delta\mathbf{b}$, setting it equal to zero, and solving for $\Delta\mathbf{b}$, we obtain

$$\Delta\mathbf{b} = (\mathbf{A}^T\mathbf{A})^{-1}\mathbf{A}^T(\mathbf{Y}^* - \mathbf{Y}) \tag{7-147}$$

where $\Delta\mathbf{b}$ is the correction vector to be applied to the estimated value of $\mathbf{b}$ to obtain a new estimate of the parameter vector:

$$\mathbf{b}_{\text{new}} = \mathbf{b}_{\text{previous}} + \Delta\mathbf{b} \tag{7-148}$$

The Gauss-Newton method applies to both the one-variable model and the multiple regression case (see Sec. 7-4.4). The algorithm of the Gauss-Newton method involves the following steps:

1. Assume initial guesses for the parameter vector $\mathbf{b}$.
2. If the model is in the form of differential equation(s), then use the vector $\mathbf{b}$ and the boundary condition(s) to integrate the equation(s) to obtain the profile(s) of $\mathbf{Y}$. If the model is in the form of algebraic equation(s), then simply use the vector $\mathbf{b}$ to evaluate $\mathbf{Y}$ from the equation(s).
3. Evaluate the matrix $\mathbf{A}$ from the equation(s) of the model. If the model consists of algebraic equations, then $\partial\mathbf{Y}/\partial\mathbf{b}$ are easily evaluated by differentiating the model. If the model consists of differential equations, then the *variational equations* must be developed. These are obtained by taking the partial of the differential equations with respect to the parameters

$$\frac{\partial}{\partial\mathbf{b}}\left(\frac{d\mathbf{Y}}{dx}\right) = \frac{\partial\mathbf{g}}{\partial\mathbf{b}} \tag{7-149}$$

   and rearranging the order of differentiation to obtain

$$\frac{d}{dx}\left(\frac{\partial\mathbf{Y}}{\partial\mathbf{b}}\right) = \frac{\partial\mathbf{g}}{\partial\mathbf{b}} + \frac{\partial\mathbf{g}}{\partial\mathbf{Y}}\frac{\partial\mathbf{Y}}{\partial\mathbf{b}} \tag{7-150}$$

   These are a set of ordinary differential equations which can be integrated simultaneously with the model equations. Their integrated results yield the profiles of $\partial\mathbf{Y}/\partial\mathbf{b}$ which are needed to construct the matrix $\mathbf{A}$.
4. Use Eq. (7-147) to obtain the correction vector $\Delta\mathbf{b}$.
5. Evaluate the new estimate of the parameter vector from (7-148).
6. Repeat steps 2 through 5 until either
   *a*. $\Phi$ does not change anymore.
   *b*. $\Delta\mathbf{b}$ becomes very small.

The Gauss-Newton method is based on the linearization of a nonlinear model; therefore this method is expected to work well if the model is not highly nonlinear, or if the initial estimate of the parameter vector is near the minimum sum of squares. The contours of constant $\Phi$ in the parameter space of a linear model are ellipsoids (Fig. 7-11*a*). For a nonlinear model, these

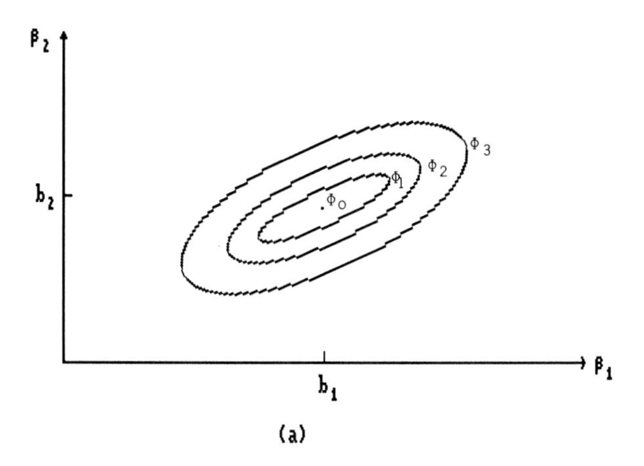


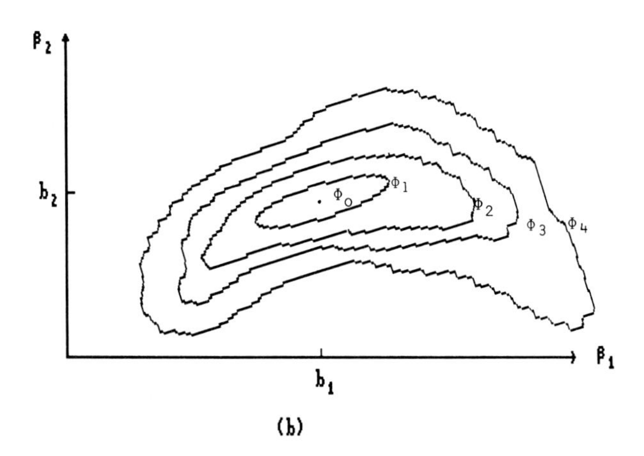


**Figure 7-11** Contours for constant sum of squares in parameter space: (*a*) linear model, (*b*) nonlinear model.

contours are distorted (Fig. 7-11*b*), but in the vicinity of the minimum $\Phi$, the contours are very nearly elliptical. Therefore the Gauss-Newton method is quite effective if the initial starting point for the search is in the nearly elliptical region. On the other hand, this method may diverge if the starting point is in the highly distorted region of the parameter hyperspace.

## 7-4.2 The Method of Steepest Descent

Another method, which has been used to arrive at the minimum sum of squares of a nonlinear model, is that of *steepest-descent*. In this method, the initial vector of parameter estimates is corrected in the direction of the negative gradient of $\Phi$:

$$\Delta\mathbf{b} = -K\left(\frac{\partial\Phi}{\partial\mathbf{b}}\right) \tag{7-151}$$

where $K$ is a suitable constant factor. It is easy to show that the negative gradient is proportional to $\mathbf{A}^T(\mathbf{Y}^* - \mathbf{Y})$ by differentiating Eq. (7-142) with respect to $\mathbf{b}$:

$$\frac{\partial\Phi}{\partial\mathbf{b}} = -2\mathbf{A}^T(\mathbf{Y}^* - \mathbf{Y}) \tag{7-152}$$

and combining with (7-151) to obtain

$$\Delta\mathbf{b} = 2K\mathbf{A}^T(\mathbf{Y}^* - \mathbf{Y}) \tag{7-153}$$

Comparison of Eqs. (7-147) and (7-153) reveals that the only difference between the correction vectors of the Gauss-Newton and steepest-descent methods is the matrix $(\mathbf{A}^T\mathbf{A})^{-1}$ and the constant factor $2K$. The steepest-descent method has an advantage over the Gauss-Newton method because it does not diverge, provided that the value of $K$, which determines the step size, is small enough. However, the rate of convergence to the minimum sum of squares decreases as the search approaches this minimum and the method loses it attractiveness because of this shortcoming.

### 7-4.3 The Marquardt Method

Marquardt [12] has developed an interpolation technique between the Gauss-Newton and the steepest-descent methods. This interpolation is achieved by adding the diagonal matrix $(\lambda\mathbf{I})$ to the matrix $(\mathbf{A}^T\mathbf{A})$ in Eq. (7-147):

$$\Delta\mathbf{b} = (\mathbf{A}^T\mathbf{A} + \lambda\mathbf{I})^{-1}\mathbf{A}^T(\mathbf{Y}^* - \mathbf{Y}) \tag{7-154}$$

The value of $\lambda$ is chosen, at each iteration, so that the corrected parameter vector will result in a lower sum of squares in the following iteration. It can be easily seen that when the value of $\lambda$ is small in comparison with the elements of matrix $(\mathbf{A}^T\mathbf{A})$, the Marquardt method approaches the Gauss-Newton method; when $\lambda$ is very large, this method is identical to the steepest-descent, with the exception of a scale factor which does not affect the direction of the parameter correction vector but which gives a small step size.

According to Marquardt, it is desired to minimize $\Phi$ in the maximum neighborhood over which the linearized function will give adequate representation of the nonlinear function. Therefore, the method for choosing $\lambda$ must give small values of $\lambda$ when the Gauss-Newton method would converge efficiently and large values of $\lambda$ when the steepest-descent method is necessary. The Marquardt algorithm is incorporated into the nonlinear regression program (NLR) which is described in Sec. 7-5.

### 7-4.4 Multiple Nonlinear Regression

In the previous three sections, the sum of squared residuals which was minimized was that given by Eq. (7-142). This was the sum of squared residuals determined from fitting one equation to measurements of one variable. However, most mathematical models may involve simultaneous equations in multiple dependent variables. For such a case, when more than one equation is fitted to multiresponse data, the weighted sum of squared residuals is given by Eq. (7-143):

$$\Phi = \sum_{j=1}^{v} w_j(\mathbf{Y}_j^* - \mathbf{Y}_j)^T(\mathbf{Y}_j^* - \mathbf{Y}_j) \tag{7-143}$$

To minimize $\Phi$, we first linearize the models using Eq. (7-144) and combine with (7-143) to obtain

$$\Phi = \sum_{j=1}^{v} w_j(\mathbf{Y}_j^* - \mathbf{Y}_j - \mathbf{A}_j\,\Delta \boldsymbol{b})^T(\mathbf{Y}_j^* - \mathbf{Y}_j - \mathbf{A}_j\,\Delta\mathbf{b}) \tag{7-155}$$

Taking the partial derivative of $\Phi$ with respect to $\Delta\mathbf{b}$, setting it equal to zero, and solving for $\Delta\mathbf{b}$ we obtain

$$\Delta\mathbf{b} = \left[\sum_{j=1}^{v} w_j \mathbf{A}_j^T \mathbf{A}_j\right]^{-1}\left[\sum_{j=1}^{v} w_j \mathbf{A}_j^T(\mathbf{Y}_j^* - \mathbf{Y}_j)\right] \tag{7-156}$$

Equation (7-156) gives the correction of the parameter vector when fitting multiple dependent variables simultaneously. Equation (7-156) becomes identical to (7-147) when $v = 1$, that is, when only one dependent variable is fitted.

The weighting factors $w_j$ are determined as follows: The basic assumption in the derivation of the regression algorithm was that the variance $\sigma^2$ of the distribution of the error in the measurements was constant throughout the profile of a single dependent variable. However, in the case of multiple regression, it is very unlikely that the variances $\sigma_j^2$ of all the curves will be the same. Therefore, in order to form an unbiased weighted sum of squared residuals, the individual sum of squares must be multiplied by a weighting factor which is proportional to $1/\sigma_j^2$. The equation for evaluating the weighting factors is given by

$$w_j = \frac{1/\sigma_j^2}{\dfrac{1}{\sum_{i=1}^{v} n_i}\left[\sum_{i=1}^{v}\sum_{l=1}^{n_i}\dfrac{1}{\sigma_i^2}\right]} \tag{7-157}$$

where $\sigma_j^2$ or $\sigma_i^2$ = variance for each curve
$n_i$ = number of experimental points available for each curve
$v$ = number of variables being fitted

The denominator of (7-157) accounts for the possibility that each curve may have a different number of experimental points $n_i$ and weighs that accordingly. If the assumption that $\sigma_j^2$ is constant within one curve does not hold, then Eq. (7-157) can be extended so that the weighting factor can be calculated at each point with the appropriate value of $\sigma_j^2$.

In most cases, the values of $\sigma_j^2$ would not be known; however, the estimates of these variances $s_j^2$ can be obtained from repeated experiments, and the values of $s_j^2$ are then used in Eq. (7-157) to calculate the weighting factors. In the worst case, where no repeated experiments are made and no a priori knowledge of $\sigma_j^2$ is available, then the values of $w_j$ must be guessed. Otherwise the nonlinear regression algorithm would introduce a bias toward fitting more satisfactorily the curve with the highest $\phi_j$ and partially ignoring the curves with low $\phi_j$.

The nonlinear regression can also be extended to fit multiple experimental values of the dependent variable at each value of the independent variable. This can be done by changing Eq. (7-143) so that the squared residuals are also summed within each group of points. Finally, if the value of the variance of the error is proportional to the value of the dependent variable, the residual in the sum of squares calculation must be divided by the theoretical (calculated) value of the dependent variable at each point in the calculation.

### 7-4.5 Analysis of Variance and Other Statistical Tests of the Regression Results

The $t$ test on parameters, described in Sec. 7-3.2, is useful in establishing whether a model contains an insignificant parameter. This information can be used to make small adjustments to models and thus discriminate between models which vary from each other by one or two parameters. This test, however, does not give a criterion for testing the adequacy of this model. The residual sum of squares, calculated by Eq. (7-105), contains two components. One is due to the scatter in the experimental data and the other is due to the lack of fit of the model. In order to test the adequacy of the fit of a model, the sum of squares must be partitioned into its components. This procedure is called *analysis of variance*, which is summarized on Table 7-2. To maintain generality, we examine a set of nonlinear data and assume the availability of multiple values of the dependent variable $y_{ij}$ at each point of the independent variable $x_i$ (see Fig. 7-12).

In Table 7-2 $p$ is the number of points of the independent variable at which there are experimental (observed) values of the dependent variable; $n_i$ are the numbers of repeated experiments available at each point of the independent variable; $\bar{y}_i$ is the mean value of each group of repeated experiments; $y_i$ are the

**Table 7-2 Analysis of variance**

| Source of variance | Sum of squares | Degrees of freedom | Variance |
|---|---|---|---|
| Lack of fit | $\sum_i^p n_i(\bar{y}_i - y_i)^2$ | $\nu_1 = p - k$ | $s_1^2$ |
| Experimental error | $\sum_i^p \sum_j^{n_i} (y_{ij}^* - \bar{y}_i)^2$ | $\nu_2 = \left(\sum_i^p n_i\right) - p$ | $s_2^2$ |
| Total | $\sum_i^p \sum_j^{n_i} (y_{ij}^* - y_i)^2$ | $\nu = \left(\sum_i^p n_i\right) - k$ | $s^2$ |

calculated values of the dependent variable; $y_{ij}^*$ are the experimental values of the dependent variable, and $k$ is the number of parameters being estimated. It should be realized that the total sum of squares shown in Table 7-2

$$\text{Total SS} = \sum_i^p \sum_j^{n_i} (y_{ij}^* - y_i)^2 \tag{7-158}$$

is merely a generalization of (7-105) to apply to both linear and nonlinear models and an extension of that relationship to account for the presence of repeated experimental data.

The ratio of the variances $s_1^2/s_2^2$ has an $F$ distribution with $\nu_1$ and $\nu_2$ degrees of freedom. This ratio must be tested against the $F$ statistic in order to test the hypothesis that the experimental points are adequately represented by the predicted line. For a good fit this ratio should be small, i.e., to accept the hypothesis the following must be true:

$$\frac{s_1^2}{s_2^2} < F_{1-\alpha}(\nu_1, \nu_2) \tag{7-159}$$

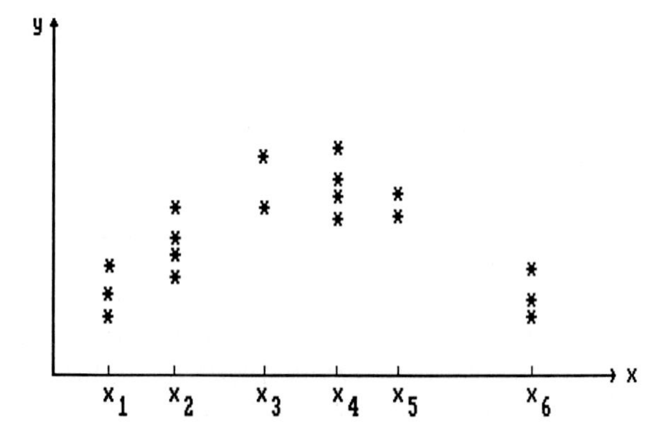


**Figure 7-12** Set of nonlinear data where multiple values of the dependent variable are available from repeated experiments ($p = 6$, $n_1 = 3$, $n_2 = 4$, $n_3 = 2$, $n_4 = 4$, $n_5 = 2$, $n_6 = 3$).

This would mean that the component of the variance due to the lack of fit is small when compared with the variance of the experimental error. In that case, the model adequately represents the data. It is obvious that if the experimental data have a large scatter, then $s_2^2$ is high and the requirements on the model are less stringent. Stating this more simply, almost anything can be fitted through very noisy or sloppy data, but the value of such a model would be marginal.

If more than one model is found to satisfy the above test, the choice of the best one can be facilitated by performing an $F$ test between the values of $s_1^2$ of pairs of models. If the fit of any one of these models is significantly better than that of the others, it will be discovered by this test.

Furthermore, the $F$ test may be used to determine if an experiment whose results deviate from those of other experiments performed under identical conditions should be grouped together with the other ones. To do this, the model is first fitted to each experiment separately. The individual sum of squares from each regression are pooled together as follows:

$$s_{\text{pooled}}^2 = \frac{\sum \text{(individual sum of squares)}}{\sum \text{(degrees of freedom)}} \tag{7-160}$$

Then the model is fitted to the grouped set of experiments to find the variance $s_{\text{grouped}}^2$. Finally, an $F$ test is performed between the pooled and grouped variances. If the inequality

$$\frac{s_{\text{grouped}}^2}{s_{\text{pooled}}^2} < F_{1-\alpha} \tag{7-161}$$

is not satisfied, it means that the model fits the experiments better individually than when grouped together.

A final test can be performed to investigate the lack of fit of the model. In the least squares regression the assumption is made that the model being fitted is the correct one, and that the observations deviate from the model in a random fashion. The residuals between the observations and the model can be either positive or negative, but if these are truly random, the sign of the residuals should change in a random fashion. This randomness (or lack of it) can be detected visually by plotting the residuals $(y^* - y)$ versus the independent variable $x$, and also versus the dependent variable $y$. Figures 7-13 to 7-16 show several different cases of distribution of residuals. Figure 7-13 demonstrates the case where the values of the residuals are randomly distributed around zero. This seems to be a satisfactory fit that would probably pass the randomness test (runs-test) described later in this section.

On the other hand, Fig. 7-14 shows a definite trend in the value of the residuals from positive to negative. The model gives a low prediction of $y$ for low values of $x$ and a high prediction of $y$ for high values of $x$. A correction to the model to remedy this trend seems to be warranted. At first sight, Fig. 7-15

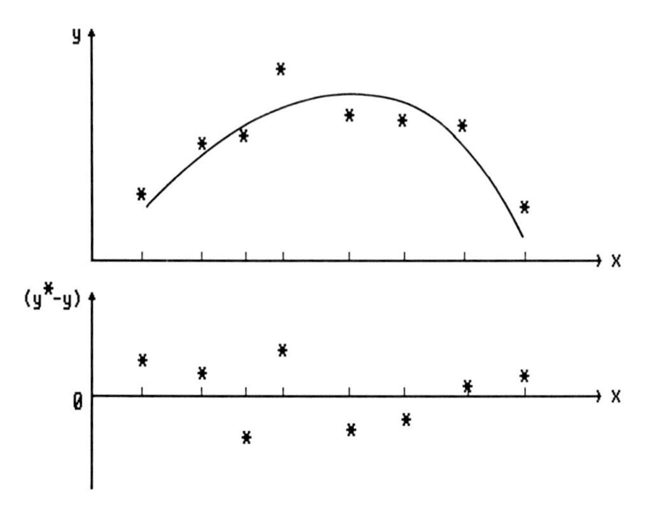


**Figure 7-13** Analysis of residuals showing a random trend.

may seem to give a case of a well-fitting model, but careful examination of the residuals shows that there is an oscillation pattern in the distribution of these residuals around zero. The addition of a term which introduces oscillatory behavior in the model may improve considerably the fit of the model to the data.

Another case is demonstrated in Fig. 7-16, which shows that the value of the residuals grows proportionately to the value of $y$. In such a case, it would

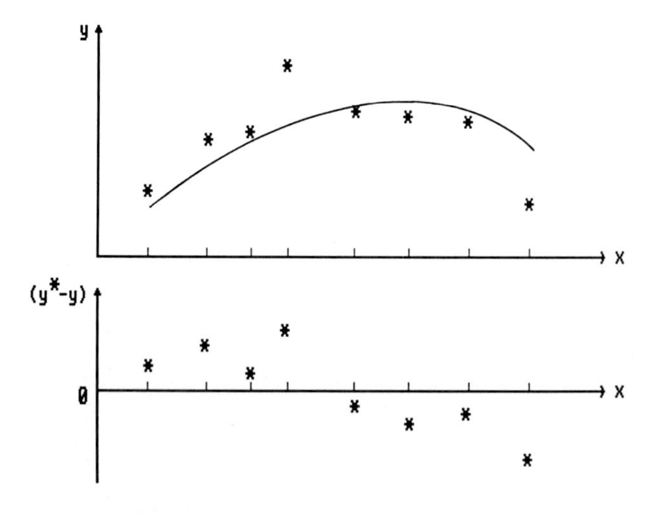


**Figure 7-14** Analysis of residuals showing trend from positive to negative.

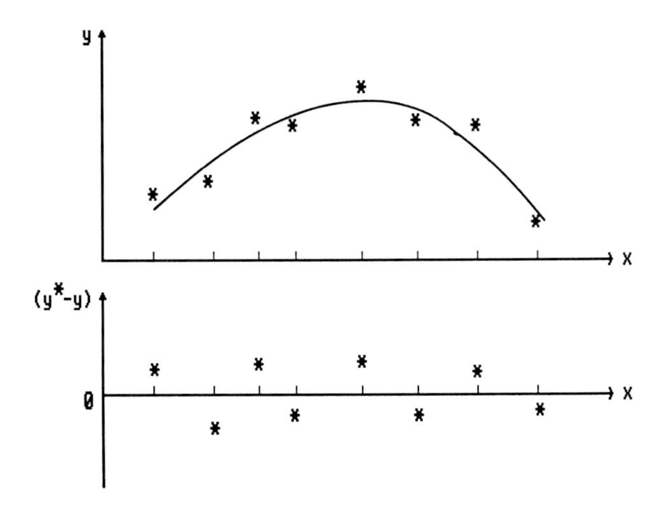


**Figure 7-15** Analysis of residuals showing oscillatory trend.

be more appropriate to normalize the residuals by dividing them by the appropriate value of $y$, that is,

$$\bar{\epsilon} = \frac{\epsilon}{y} \tag{7-162}$$

and then minimize the sum of normalized squared residuals:

$$\bar{\Phi} = \bar{\epsilon}^T \bar{\epsilon} \tag{7-163}$$

The randomness of the distribution of the residuals can be quantified, measured, and tested by the so-called *runs test*. In this test, the total number of positive residuals is represented by $n_1$ and that of negative residuals by $n_2$. The number of times the sequence of residuals changes sign is $r$, which is called the number of runs. The distribution of $r$ is approximated by the normal distribution. Brownlee [3] finds the mean and standard deviation of this variable to be

$$\bar{r} = \frac{2n_1 n_2}{n_1 + n_2} + 1 \tag{7-164}$$

$$\sigma = \sqrt{\frac{2n_1 n_2(2n_1 n_2 - n_1 - n_2)}{(n_1 + n_2)^2(n_1 + n_2 - 1)}} \tag{7-165}$$

The standardized form of the variable is

$$Z = \frac{r - \bar{r}}{\sigma} \tag{7-166}$$

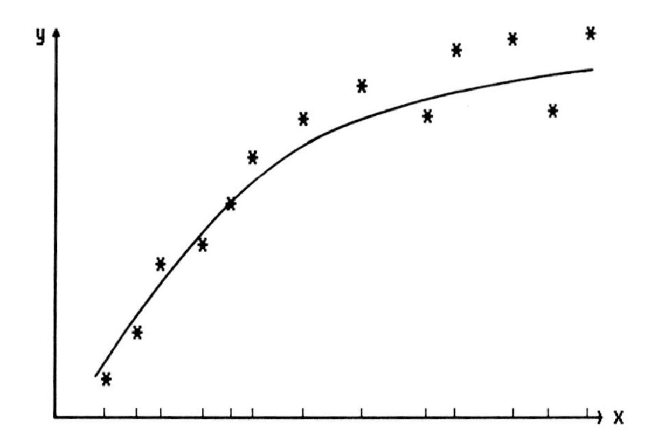


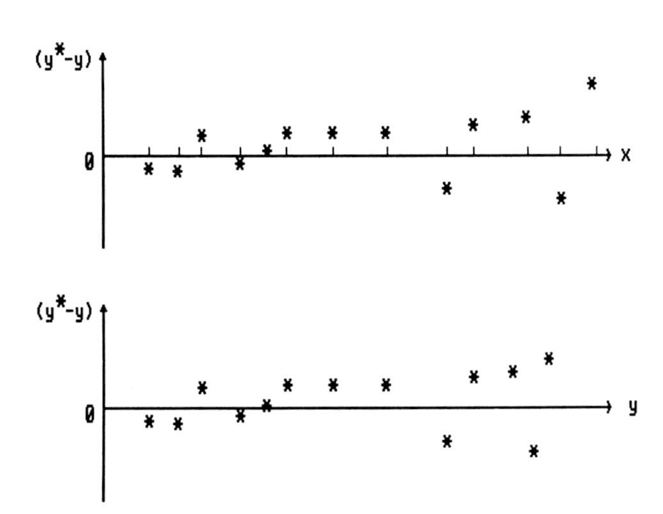


**Figure 7-16** Analysis of residuals showing trend of increasing residuals in proportion to $y$.

which is distributed with zero mean and unit variance. To test the hypothesis that the deviations are random, $Z$ is compared with the standard normal distribution. A two-sided test must be performed because if the value of $Z$ is too low, the model is inadequate; and if $Z$ is too high, then the data contain oscillations which must be accounted for by the model. On the other hand, if the value of $Z$ falls in the region of acceptance for this test, then the hypothesis that the model is the correct one and that the residuals are randomly distributed can be accepted.

## 7-5 THE NONLINEAR REGRESSION PROGRAM

In this section we describe the computer program which applies the Marquardt nonlinear regression method to fit mathematical models to multiresponse data in order to determine a number of unknown parameters. The models can be either ordinary differential equations or algebraic equations. The program can fit any number of dependent variables to one or more sets of experimental data for each such variable. The model may contain more dependent variables than those being fitted to data. In entering the equations into the program, the dependent variables must be numbered Y(1), Y(2), Y(3), etc., according to the following scheme:

First: The dependent variables being fitted to data
Second: The dependent variables that are not being fitted to data (if any)
Last: The variational quantities, that is, $\partial y_i/\partial b_j$, arranged in the order:
1. The first variable with respect to each parameter
2. The second variable with respect to each parameter, etc.

For more details, see Example 7-1.

The program itself instructs the user how to enter the equations, depending on whether the models are algebraic or differential.

The nonlinear regression program is on your diskette labeled "Nonlinear Regression Program for Chap. 7," which was prepared according to instructions given in Sec. 1-6.† This is a self-booting diskette which can be used to boot the personal computer. The AUTOEXEC.BAT file on this diskette contains the following commands:

```
DATE
TIME
BASICA MENU
```

The first two commands are for entering the correct date and time. These steps may be skipped if your computer is equipped with a battery-operated clock-calendar. The command BASICA MENU loads the BASICA interpreter and executes the program MENU.BAS. This menu program displays the directory of the diskette and then executes MAIN.BAS. This main program calls all the other subprograms and performs the Marquardt nonlinear regression method. Control of the program returns to MAIN.BAS at the completion of each subprogram. The use of this program is demonstrated in Example 7-1.

**Example 7-1 Nonlinear regression using the Marquardt method** In Prob. 5-5 we described the kinetics of a fermentation process which manufactures penicillin antibiotics. When the microorganism *Penicillium chrysogenum* is

† If you have copied the programs to a fixed disk, then the nonlinear regression program is in the subdirectory named NLR.

grown in a batch fermentor under carefully controlled conditions, the cells grow at a rate which can be modeled by the logistic law

$$\frac{dy_1}{dt} = b_1 y_1 \left(1 - \frac{y_1}{b_2}\right)$$

where $y_1$ is the concentration of the cell expressed as percent dry weight. In addition, the rate of production of penicillin has been mathematically quantified by the equation

$$\frac{dy_2}{dt} = b_3 y_1 - b_4 y_2$$

where $y_2$ is the concentration of penicillin in units/mL.

The experimental data in Table E7-1*a* were obtained from two penicillin fermentation runs conducted at essentially identical operating conditions. Using the Marquardt method, fit the above two equations to the experimental data and determine the values of the parameters $b_1$, $b_2$, $b_3$, and $b_4$ which minimize the weighted sum of squared residuals.

**Table E7-1*a***

| | Run No. 1 | | Run No. 2 | |
|---|---|---|---|---|
| Time, h | Cell concentration, percent dry weight | Penicillin concentration, units/mL | Cell concentration, percent dry weight | Penicillin concentration, units/mL |
| 0 | 0.40 | 0.0 | 0.18 | 0.0 |
| 10 | | 0.0 | 0.12 | 0.0 |
| 22 | 0.99 | 0.0089 | 0.48 | 0.0089 |
| 34 | | 0.0732 | 1.46 | 0.0642 |
| 46 | 1.95 | 0.1446 | 1.56 | 0.2266 |
| 58 | | 0.5230 | 1.73 | 0.4373 |
| 70 | 2.52 | 0.6854 | 1.99 | 0.6943 |
| 82 | | 1.2566 | 2.62 | 1.2459 |
| 94 | 3.09 | 1.6118 | 2.88 | 1.4315 |
| 106 | | 1.8243 | 3.43 | 2.0402 |
| 118 | 4.06 | 2.2170 | 3.37 | 1.9278 |
| 130 | | 2.2758 | 3.92 | 2.1848 |
| 142 | 4.48 | 2.8096 | 3.96 | 2.4204 |
| 154 | | 2.6846 | 3.58 | 2.4615 |
| 166 | 4.25 | 2.8738 | 3.58 | 2.2830 |
| 178 | | 2.8345 | 3.34 | 2.7078 |
| 190 | 4.36 | 2.8828 | 3.47 | 2.6542 |

METHOD OF SOLUTION The Marquardt method, described in Sec. 7-4.3, and the concept of multiple nonlinear regression, covered in Sec. 7-4.4, have been combined together in this nonlinear regression program. The model being fitted constitutes a set of two simultaneous ordinary differential equations with four unknown parameters. The variational equations, needed for evaluating the matrix **A**, are obtained as shown in Sec. 7-4.1 [Eq. (7-150)]. The statistical tests described in this chapter are implemented by several subprograms (see Program Description for more details).

The complete set of differential equations for this example is shown below:

Model equations:

$$\frac{dy_1}{dt} = b_1 y_1\left(1 - \frac{y_1}{b_2}\right) \tag{1}$$

$$\frac{dy_2}{dt} = b_3 y_1 - b_4 y_2 \tag{2}$$

Variational equations:

$$\frac{d}{dt}\left(\frac{\partial y_1}{\partial b_1}\right) = \left(b_1 - 2\frac{b_1}{b_2}y_1\right)\left(\frac{\partial y_1}{\partial b_1}\right) + y_1 - \frac{y_1^2}{b_2} \tag{3}$$

$$\frac{d}{dt}\left(\frac{\partial y_1}{\partial b_2}\right) = \left(b_1 - 2\frac{b_1}{b_2}y_1\right)\left(\frac{\partial y_1}{\partial b_2}\right) + b_1\left(\frac{y_1}{b_2}\right)^2 \tag{4}$$

$$\frac{d}{dt}\left(\frac{\partial y_1}{\partial b_3}\right) = \left(b_1 - 2\frac{b_1}{b_2}y_1\right)\left(\frac{\partial y_1}{\partial b_3}\right) \tag{5}$$

$$\frac{d}{dt}\left(\frac{\partial y_1}{\partial b_4}\right) = \left(b_1 - 2\frac{b_1}{b_2}y_1\right)\left(\frac{\partial y_1}{\partial b_4}\right) \tag{6}$$

$$\frac{d}{dt}\left(\frac{\partial y_2}{\partial b_1}\right) = b_3\left(\frac{\partial y_1}{\partial b_1}\right) - b_4\left(\frac{\partial y_2}{\partial b_1}\right) \tag{7}$$

$$\frac{d}{dt}\left(\frac{\partial y_2}{\partial b_2}\right) = b_3\left(\frac{\partial y_1}{\partial b_2}\right) - b_4\left(\frac{\partial y_2}{\partial b_2}\right) \tag{8}$$

$$\frac{d}{dt}\left(\frac{\partial y_2}{\partial b_3}\right) = b_3\left(\frac{\partial y_1}{\partial b_3}\right) - b_4\left(\frac{\partial y_2}{\partial b_3}\right) + y_1 \tag{9}$$

$$\frac{d}{dt}\left(\frac{\partial y_2}{\partial b_4}\right) = b_3\left(\frac{\partial y_1}{\partial b_4}\right) - b_4\left(\frac{\partial y_2}{\partial b_4}\right) - y_2 \tag{10}$$

The dependent variables and variational quantities, their correspondence to program variables, and their initial conditions are listed in Table E7-1*b*.

**Table E7-1*b***

| Dependent variable | Program variable | Initial condition |
|---|---|---|
| $y_1$ | Y(1) | 0.29 |
| $y_2$ | Y(2) | 0.0 |
| Variational quantity | | |
| $\frac{\partial y_1}{\partial b_1}$ | Y(3) | 0.0 |
| $\frac{\partial y_1}{\partial b_2}$ | Y(4) | 0.0 |
| $\frac{\partial y_1}{\partial b_3}$ | Y(5) | 0.0 |
| $\frac{\partial y_1}{\partial b_4}$ | Y(6) | 0.0 |
| $\frac{\partial y_2}{\partial b_1}$ | Y(7) | 0.0 |
| $\frac{\partial y_2}{\partial b_2}$ | Y(8) | 0.0 |
| $\frac{\partial y_2}{\partial b_3}$ | Y(9) | 0.0 |
| $\frac{\partial y_2}{\partial b_4}$ | Y(10) | 0.0 |

The initial conditions for $y_1$ and $y_2$ were chosen to be the average values of the corresponding experimental data at $t = 0$.

PROGRAM DESCRIPTION The complete listing of the Nonlinear Regression Program is not given here due to its length (approximately 40 pages). The user of this program can list the main program and the subprograms by utilizing the LOAD and LIST commands of the BASICA interpreter or can print these programs using the LIST, "LPT1:" command. The text file READ.ME contains a brief description of the Nonlinear Regression Program. This file can be read from DOS using the command TYPE READ.ME. The contents of this file are given below.

## THE NONLINEAR REGRESSION PROGRAM

### A Brief Overview

The Nonlinear Regression Program (NLR) has been designed to fit a set of differential or algebraic equations to multiresponse data to determine a number of unknown parameters. The parameter estimation is linked with statistical analysis routines to form a single self-contained package. The input to the program is accomplished through a series of menus.

### Parameter Estimation

The Marquardt algorithm, which utilizes an interpolation technique to combine the Gauss-Newton and steepest descent methods, is used in this program.

### Statistical Analysis

A set of statistical tests accompanies the regression analysis. These are:

1) A linear 95% confidence interval for each parameter is calculated and a t-test is performed to check if the value of the parameter is significantly different than zero. The 95% confidence limits for each variable are calculated.
2) The correlation matrix of the parameters is calculated and the matrix components are tested at the 95% and 99% significance levels to determine the extent of correlation between the parameters.
3) The correlation coefficient matrix defines a hyperellipsoid in parameter space. A sum of squares search is performed in the direction of the major and minor axes of the hyperellipsoid to ascertain whether the minimum sum of squares has been located.
4) A Runs test is performed on the deviations between the experimental and predicted values of the state variables to ensure random distribution, i.e. to test the null hypotheses that the residuals are randomly distributed.
5) Analysis of variance is performed. The F-test is applied on the ratio of the residual variance to the experimental data variance to determine if more scatter exists than can be explained by experimental error.

### Description of the program

The Nonlinear Regression Program consists of the MAIN program, 12 subprograms which are called from the MAIN program through CHAIN commmands, 6 data files, and one equations file. These are described below:

| PROGRAM | DESCRIPTION |
|---|---|
| MAIN.BAS | The main program calls all the other programs and performs the Marquardt method. |
| INPUT.BAS | This program enters all the data, constants and equations for the nonlinear regression. |

| | |
|---|---|
| COMPAR.BAS | Compares the variance due to lack-of-fit with the variance due to experimental error, and performs the F-tests. |
| DATIN.BAS | Determines the variance of the experimental data and the weighting factors. |
| GJLNB.BAS | Uses the Gauss-Jordan reduction method to obtain the parameter increment vector. |
| JSRKB.BAS | This routine evaluates the state and variational equations which are specified by the user through the INPUT.BAS program. If the model contains differential equations, the 4th order Runge-Kutta method is used for the integration. |
| MUS.BAS | Calculates the residuals between the group means of the experimental points and the predicted curve. |
| PHLPA.BAS | A matrix inversion routine to determine the inverse of the ATA matrix. |
| RUNS | This performs a two sided Runs test at the 0.05 significance level. (This routine has been embedded within the main program). |
| SA12V.BAS | A statistical analysis routine that performs a series of tests on the predicted parameters.<br>- Combined residual variance of the data.<br>- Standard errors and covariances of the coefficients.<br>- 95% confidence limits of the coefficients using the Student t distribution, and significance tests. |
| SETPAR.BAS | Searches the sum of squares space at the end of the axes of the hyperellipsoid. |
| SUMS.BAS | This routine calculates the residuals between the experimental and the predicted points, and the weighted sum of squared residuals. |
| SUMS1.BAS | Similar to SUMS, but for unweighted residuals. |
| OUTPUT.BAS | Prints tables and draws graphs (on screen or using an HP 7475A Plotter) of final values of variables and data. |

OTHER FILES ON THE DISKETTE

| | |
|---|---|
| CONS.DAT | Random access file for all nonlinear regression constants. |
| DATA.DAT | Random access file that contains all the experimental data and weights. |

| | |
|---|---|
| DIST.DAT | Read-only file that contains the F-distribution table. |
| GUESS.DAT | Sequential file that contains the initial parameter guesses. |
| INITCOND.DAT | Sequential file for all the initial conditions of the differential equations. |
| OUTPUT.DAT | Sequential output file for the matrix A which contains the profiles of the state and variational variables over the entire integration interval. |
| MODEL.EQU | File containing the state and variational equations. |
| READ.ME | Text file containing this description. Can be read using the command TYPE READ.ME |

Use of the program

The entire program is written in interpretive BASIC and will run under the IBM PC Advanced Basic Interpreter (BASICA). The program MAIN is responsible for the whole package, and it will call upon the appropriate subprogram as needed. NLR is menu driven for data input and adjustment. All the data are stored on the diskette in files with the extension .DAT. The state and variational equations may be entered from the keyboard. These are stored in a file called MODEL.EQU. Whenever new data and/or a new model are used, the previous set of files are overwritten by the new data and model. If the user wants to retain the old data and model, the user must first rename the files, or transfer them to another diskette, before entering the new information.

PROGRAM INPUT The model and variational equations, the regression constants, and the experimental data are entered into the program through a series of menus.

The model and variational equations must strictly adhere to the form shown below:

$$G(i) = f[x, y(1), y(2), \ldots, y(n), b(1), b(2), \ldots]$$

For example,

$$G(1) = b(1)^*x + b(2)^*y(1) - b(3)^*y(1)^*y(2)$$

$$G(2) = b(4)^*y(1)\char`^2 - b(5)^*y(2)/y(1)$$

where G(i) = derivatives $dy(i)/dx$
x = independent variable
y(i) = dependent variables
f[ ] = functions to be integrated
b( ) = parameters to be determined

No other variables should be introduced.

Care must be exercised to enter each equation correctly. Any typing errors must be corrected before each line is terminated by the ENTER key. The equations are automatically saved in a file called MODEL.EQU, which is later merged into the integration program (JSRKB.BAS).†

A complete set of the constants applicable to this example are shown on the computer printout. Modifications to the existing file can be made easily through the options provided with each set of constants. For this example, the total number of dependent variables is 2. The number of variables being fitted to data is also 2. The total number of parameters being estimated is 4. The total number of differential equations (model + variational) is 10. The initial and final values of the independent variable are 0 and 190 h, respectively. This range is dictated by the range of the available experimental data.

The next three constants must be chosen carefully in accordance with the spacing of the experimental data. In order to compute the residuals between calculated and experimental values of the dependent variable, the program must have available to it calculated values of the dependent and variational variables at each point where experimental data can exist. Data are usually unevenly spaced while integrated values are usually evenly spaced. The data in this example were taken at integer values of the independent variable (time); therefore, there are 191 locations of the independent variable where experimental points can exist in the range $0 \leq t \leq 190$. The number of integration steps was chosen at 190, resulting in a $\Delta t = 1.0$ h. This step size provides sufficient accuracy of the integration routine for this particular set of equations. This was verified by executing the program with $\Delta t = 0.5$, i.e., 380 integration steps, not shown here. For other problems, where a finer grid size may be required for integration accuracy, care must be exercised in choosing the number of integration steps so that the grids of calculated and experimental points coincide. With the choice of $\Delta t = 1.0$, the number of integration steps per possible experimental points is 1 for this particular set of data.

In this example, there are two repeated experimental runs which will

† Alternatively, the user may prepare the file MODEL.EQU before executing the regression program. The form of the equations must be the same as above. The statements must be numbered consecutively, starting at 4011 and not exceeding 4995. This file must be saved on the diskette as an ASCII file using the command

```
SAVE "MODEL.EQU", A
```

be used together for fitting the models; therefore, the maximum number of points for any location of the independent variable is 2.

The next set of constants specifies the operation of the nonlinear regression search. The maximum number of iterations is set at 20. When the value of NONEG is set to 1, the search of the parameter space remains in the positive regime, i.e., if any parameters become negative, the program resets them equal to zero and continues the search. The value of NORMAL determines whether a regular sum of squared residuals is calculated (NORMAL = 0) or a sum of *normalized* squared residuals is used (NORMAL = 1).

SIGZR is the initial starting value of $\lambda$ in Eq. (7-154). ALFA is the factor by which $\lambda$ is multiplied, or divided, in order to adjust the direction of the parameter correction vector according to the Marquardt algorithm. BETA is the factor by which the size of the correction vector is multiplied when the value of $\lambda$ becomes greater than 1.

CONV is used by the program to test the convergence of the sum of squared residuals (SSR). If the following inequality is satisfied, the search has converged:

$$\frac{|\text{SSR of previous iteration} - \text{SSR of present iteration}|}{\text{SSR of present iteration}} < \text{CONV}$$

Convergence of the parameter increment vector is also tested using the value of CONV. If the increments of all the parameters satisfy the following inequality, the search has converged:

$$\left|\frac{\Delta b_i}{b_i}\right| < \text{CONV}$$

If $b_i = 0$, then the following inequality is used:

$$|\Delta b_i| < \text{CONV}$$

EPS is an absolute convergence criterion used to eliminate difficulties arising from overflows in the rare case when SSR becomes zero. If the following inequality is satisfied, the search has converged:

$$\text{SSR of present iteration} \leq \text{EPS}$$

The initial conditions of the equations are entered next. These are stored by the program in the file INITCOND.DAT. The initial guesses of the parameters are also entered and stored in file GUESS.DAT. These guesses must be chosen very carefully in order to ensure that the search does not diverge.

After entering the above, we return to previous menus until the DISPLAY/ALTER menu is reached. At this point, the experimental data and weights must be entered. We first enter the number of groups of data points, i.e., the number of positions in the independent variable at which we have experimental points. In this example there are 17 groups of data points for variable 1. (The fact that there is a maximum of two data points for each position of the dependent variable has been specified earlier.)

Next, we are asked to give an estimate of the experimental variance $\sigma_j^2$ of the data. For multiple regression, the variances are used in Eq. (7-157) to calculate the weighting factors $w_j$ which are in turn used in Eq. (7-143) to determine the unbiased weighted sum of squared residuals. In the case where repeated experimental data are available, the program automatically calculates the variance of each dependent variable from these data. Since the set of data in this example contains two repeated experiments, we simply enter a value of unity for the variance, realizing that this will be recalculated by the program. However, in a case where only one experiment is available, we must specify the relative magnitudes of the variances; otherwise a biased sum of squared residuals will be calculated. Of course, knowledge of $\sigma_j^2$ for each curve presupposes the existence of repeated experiments. When these are not available, we can estimate $\sigma_j^2$ from $s^2$ of Eq. (7-123), i.e., the multiple regression can be performed using $\sigma_j^2 = 1$ as a first guess, and then the sum of squared residuals and the number of degrees of freedom, generated by the program for each variable, can be used to estimate $\sigma_j^2$ from Eq. (7-123). The regression must then be repeated using the new values of $\sigma_j^2$. This trial-and-error procedure may have to be repeated several times before a good fit is accomplished.

The experimental data are entered next. Each value must be followed by the ENTER key. A weight of unity should be entered with each existing data point. It should be noted that the set of cell concentration data for run No. 1 is not as complete as that of run No. 2. This can be handled by entering a weight of zero at those positions where data do not exist. The program uses the value of the weight to sense the presence of a data point (when weight $\neq 0$), to count the data points for each curve, and to perform the regression calculations at only those positions where data exist. It is possible to assign a higher (or lower) relative weight for any data point, if it is felt that the particular point is of higher (or lower) importance in the regression analysis.

If any errors are made in entering the experimental data, they can be corrected at the end through the option "ALTER EXISTING EXPERIMENTAL DATA FILE". When everything is entered properly, return to the MAIN MENU and choose option 2 "PERFORM NONLINEAR REGRESSION". This option executes the main program, which reads all the data, prints it out for verification, and proceeds with the application of the nonlinear regression method. Due to the lengthy integration of the differential equations of this example, the program takes about 10 min per iteration on the IBM AT personal computer.

DISCUSSION OF RESULTS The first calculated results given by the nonlinear regression program are the "Statistics of the Experimental Data." The set of data used in this example contains two repeated experiments; therefore the program DATIN.BAS calculates the variance and standard deviation for each variable. The degrees of freedom is the number of points at which there are repeated data (9 for variable 1 and 17 for variable 2). The sum of weights (26 for variable 1 and 34 for variable 2) corresponds to the number of data points (when the weight of unity is entered for each point). The variances, calculated from the data, are used in Eq. (7-157) to readjust the weighting factors from unity to 0.492 and 1.256 (shown later on the printout) for variables 1 and 2, respectively. These weights are actually the square roots of $w_j$ of Eq. (7-157), so they are inversely proportional to the standard deviation of each variable. The weighted sum of squares and weighted variance are calculated using the adjusted weights.

Next, the program prints out the pathway of convergence between the initial set of parameters (iteration 1), which yields a total sum of squares of 69.89, and the converged set of parameters (iteration 6), which results in a total sum of squares of 2.277. The Marquardt method adjusts the value of $\lambda$ in each iteration so that the direction of search in parameter space results in a lower sum of squares.

The experimental data, calculated values of dependent variable, the error, and the adjusted weights are listed for both variables.

The subprogram SA12V.BAS performs a statistical analysis of the converged results. It prints out the variance-covariance matrix $(\mathbf{A}^T\mathbf{A})^{-1}s^2$, calculates the combined (weighted) residual variance $s^2$, the standard deviation $s$, and the 95 percent confidence intervals for the parameters [Eq. (7-127)]. In addition, it performs a significance test ($t$ test) for each parameter according to Eqs. (7-128) to (7-130). Using the residual variances $s_j^2$ the program also calculates the 95 percent confidence limits for each variable, that is, $s_j t_{95\%}$. Finally, this program calculates the correlation coefficient matrix according to Eq. (7-136) and performs the significance tests at the 0.05 and 0.01 levels, as described in Sec. 7-3.2. The correlation between parameters is found to be highly significant. This is not a surprising result for a model of this type.

The subprogram SETPAR.BAS calculates the eigenvalues and eigenvectors of the correlation coefficient matrix and performs a sum-of-squares search at the end of the axes of the parameter hyperellipsoid [see Eq. (7-138)]. The difference between the search sum of squares and the converged sum of squares is calculated for the individual curves and for the total. If any of the differences are negative, this would indicate that the search has found a point in the parameter space where the sum of squares is lower than at the converged point. In this example, no negative differences were found for the total sum of squares, and only two negative differences were found for individual curves; however, these were very small. From this search, we conclude that the converged point is satisfactory.

The subprogram COMPAR.BAS performs the analysis of variance, as described in Sec. 7-4.5. The *F* tests, for both the unweighted and weighted variances, are carried out according to Eq. (7-159). The quantity designated as F-CALC corresponds to the left side of Eq. (7-159), i.e., the calculated ratio of the lack of fit variance to the experimental error variance. The quantity designated as F-95% is the right side of Eq. (7-159), i.e., the value of the *F* statistic at the 95 percent confidence level. All the *F* tests for this example show that

$$\text{F-CALC} < \text{F-95\%}$$

i.e., the inequality in (7-159) is satisfied. From these, we conclude that the model adequately represents the data.

Finally, the program performs the randomness test (runs test) to determine whether the residuals are distributed randomly. Equations (7-164) to (7-166) are used to calculate the variable $Z$, which is then tested against the standard normal distribution at the 95 percent level. In this example, the residuals are randomly distributed for the first variable, but not for the second.

The above series of statistical tests indicates that the program has converged to the minimum sum of squared residuals, the parameters are significantly different from zero but highly correlated to each other, the model adequately represents the data, and the residuals of one variable are randomly distributed. These conclusions are visually verified by examining the plots of model and data for each variable being fitted.

### Program Input

```
*******************************************************************
*                         EXAMPLE 7-1                             *
*                                                                 *
*         NONLINEAR REGRESSION USING THE MARQUARDT METHOD         *
*                                                                 *
*                          (MAIN.BAS)                             *
*******************************************************************
MAIN MENU:
 1            ENTER EQUATIONS, REGRESSION CONSTANTS, AND DATA

 2            PERFORM NONLINEAR REGRESSION

 3            PRINT OR PLOT THE OUTPUT RESULTS

 4            END APPLICATION
CHOICE : 1

INPUT MENU:
 1            DESCRIPTION OF THE DATA AND EQUATION FILES

 2            ENTER MODEL AND VARIATIONAL EQUATIONS

 3            DISPLAY/ALTER DATA FILES

 4            RETURN TO MAIN MENU
CHOICE : 1
FILE                CONTENTS

CONS.DAT      ....  ALL NONLINEAR REGRESSION CONSTANTS

DATA.DAT      ....  EXPERIMENTAL DATA AND WEIGHTS

DIST.DAT      ....  F-DISTRIBUTION TABLE

GUESS.DAT     ....  INITIAL PARAMETER GUESSES

INITCOND.DAT  ....  INITIAL CONDITIONS FOR DIFFERENTIAL EQUATIONS

OUTPUT.DAT    ....  CALCULATED VARIABLES

MODEL.EQU     ....  MODEL AND VARIATIONAL EQUATIONS

PRESS ENTER TO CONTINUE

INPUT MENU:
 1            DESCRIPTION OF THE DATA AND EQUATION FILES

 2            ENTER MODEL AND VARIATIONAL EQUATIONS

 3            DISPLAY/ALTER DATA FILES

 4            RETURN TO MAIN MENU

CHOICE : 2
```

```
                    MODEL AND VARIATIONAL EQUATIONS
                    *******************************
THE MODEL AND VARIATIONAL EQUATIONS MAY BE ENTERED HERE.
EACH LINE-STATEMENT MUST OBEY THE RULES OF BASIC PROGRAMMING.
THE LINES ARE AUTOMATICALLY NUMBERED, STARTING AT 4011.
THESE STATEMENTS ARE STORED IN A FILE CALLED MODEL.EQU. ANY
INFORMATION PREVIOUSLY STORED IN THIS FILE WILL BE ELIMINATED.
THE MODEL.EQU FILE IS MERGED INTO SUBROUTINE JSRKB.BAS.
OPTIONS  :
 1             MODEL CONTAINS ORDINARY DIFFERENTIAL EQUATIONS
 2             MODEL CONTAINS ALGEBRAIC EQUATIONS ONLY
CHOICE : 1
HAVE YOU PREVIOUSLY ENTERED THE EQUATIONS(Y/N)? N


                        *********************
ENTER THE DIFFERENTIAL EQUATIONS IN THE FORM SHOWN BELOW:
     G(i)= f[x, y(1), y(2),....,y(n),b(1), b(2),....]
                   FOR EXAMPLE
     G(1)= b(1)*x + b(2)*y(1) - b(3)*y(1)*y(2)
     G(2)= b(4)*y(1)^2 - b(5)*y(2)/y(1)
WHERE:  G(i)  ARE THE DERIVATIVES: dy(i)/dx
        x     IS THE INDEPENDENT VARIABLE
        y(i)  ARE THE DEPENDENT VARIABLES
        f[ ]  ARE THE FUNCTIONS TO BE INTEGRATED
        b( )  ARE THE PARAMETERS TO BE DETERMINED


                        *********************
GIVE THE EXACT NUMBER OF LINES OF INPUT YOU WILL ENTER(1-980)? 10
SEPARATE EACH STATEMENT WITH THE ENTER(RETURN) KEY.
```

```
4011 G(1)=B(1)*Y(1)*(1-Y(1)/B(2))

4012 G(2)=B(3)*Y(1)-B(4)*Y(2)

4013 G(3)=(B(1)-2*B(1)*Y(1)/B(2))*Y(3)+Y(1)-Y(1)^2/B(2)

4014 G(4)=(B(1)-2*B(1)*Y(1)/B(2))*Y(4)+B(1)*(Y(1)/B(2))^2

4015 G(5)=(B(1)-2*B(1)*Y(1)/B(2))*Y(5)

4016 G(6)=(B(1)-2*B(1)*Y(1)/B(2))*Y(6)

4017 G(7)=B(3)*Y(3)-B(4)*Y(7)

4018 G(8)=B(3)*Y(4)-B(4)*Y(8)

4019 G(9)=B(3)*Y(5)-B(4)*Y(9)+Y(1)

4020 G(10)=B(3)*Y(6)-B(4)*Y(10)-Y(2)
```

```
INPUT MENU:

 1          DESCRIPTION OF THE DATA AND EQUATION FILES

 2          ENTER MODEL AND VARIATIONAL EQUATIONS

 3          DISPLAY/ALTER DATA FILES

 4          RETURN TO MAIN MENU

CHOICE : 3

DISPLAY/ALTER MENU:

 1          ALL NONLINEAR REGRESSION CONSTANTS

 2          EXPERIMENTAL DATA AND WEIGHTS

 3          RETURN TO INPUT MENU

CHOICE : 1

NUMBER                CONSTANT
 1          Total number of dependent variables                      2
 2          Number of variables being fitted to data                 2
 3          Total number of parameters being estimated               4
 4          Number of differential equations (model + variational)  10
 5          Initial value of independent variable                    0
 6          Final value of independent variable                      190
 7          Number of integration steps                              190
 8          Number of locations of indep var where exp pt can exist 191
 9          Number of integration steps per experimental point       1
 10         Max # of points for any location(repeated experiments)   2
```

```
OPTIONS :

 1          MAKE CHANGES TO THE ABOVE CONSTANTS
 2          PROCEED TO NEXT SET OF CONSTANTS
 3          RETURN TO PREVIOUS MENU

CHOICE : 2

NUMBER                CONSTANT
 1          Maximum number of iterations for NLR search            20
 2          NONEG:  0 allows negative parameters
                    1 sets negative parameters to zero             1
 3          NORMAL: 0 for regular sum of squares
                    1 for normalized sum of squares                0
 4          Marquardt constant (SIGZR)                             .1
 5          Marquardt constant (ALFA)                              10
 6          Marquardt constant (BETA)                              .5
 7          NLR Convergence constant (CONV)                        .001
 8          NLR Convergence constant (EPS)                         .00001


OPTIONS  :

 1          MAKE CHANGES TO THE ABOVE CONSTANTS
 2          PROCEED TO NEXT SET OF CONSTANTS
 3          RETURN TO PREVIOUS MENU

CHOICE : 2

HAVE THE INITIAL CONDITIONS FOR THE EQUATIONS BEEN PREVIOUSLY STORED(Y/N)? N
EQUATION #  1 , INITIAL CONDITION? 0.29
EQUATION #  2 , INITIAL CONDITION? 0
EQUATION #  3 , INITIAL CONDITION? 0
EQUATION #  4 , INITIAL CONDITION? 0
EQUATION #  5 , INITIAL CONDITION? 0
EQUATION #  6 , INITIAL CONDITION? 0
EQUATION #  7 , INITIAL CONDITION? 0
EQUATION #  8 , INITIAL CONDITION? 0
EQUATION #  9 , INITIAL CONDITION? 0
EQUATION #  10 , INITIAL CONDITION? 0

EQUATION #   INITIAL CONDITION
 1             .29
 2             0
 3             0
 4             0
 5             0
 6             0
 7             0
 8             0
 9             0
 10            0

OPTIONS :

 1          MAKE CHANGES TO THE ABOVE INITIAL CONDITIONS
 2          PROCEED TO NEXT SET OF CONSTANTS
 3          RETURN TO PREVIOUS MENU

CHOICE : 2
```

```
HAVE THE INITIAL GUESSES OF THE PARAMETERS BEEN PREVIOUSLY STORED(Y/N)? N
PARAMETER #  1 , INITIAL GUESS? 0.1
PARAMETER #  2 , INITIAL GUESS? 4.0
PARAMETER #  3 , INITIAL GUESS? 0.02
PARAMETER #  4 , INITIAL GUESS? 0.02


PARAMETER #   INITIAL GUESS
 1             .1
 2            4
 3             .02
 4             .02


OPTIONS :

 1            MAKE CHANGES TO THE ABOVE PARAMETER GUESSES

 2            RETURN TO PREVIOUS MENU


CHOICE : 2
```

(Return to previous menus until the DISPLAY/ALTER menu is reached)

```
DISPLAY/ALTER MENU:


 1            ALL NONLINEAR REGRESSION CONSTANTS

 2            EXPERIMENTAL DATA AND WEIGHTS

 3            RETURN TO INPUT MENU

CHOICE : 2


OPTIONS :

 1            LIST ALL THE EXPERIMENTAL DATA

 2            ALTER EXISTING EXPERIMENTAL DATA FILE

 3            CREATE A NEW EXPERIMENTAL DATA FILE

 4            RETURN TO PREVIOUS MENU


CHOICE : 3
```

```
ENTER DATA FOR VARIABLE  1


HOW MANY GROUPS OF DATA POINTS ARE THERE IN VARIABLE  1 ? 17


ESTIMATE OF VARIANCE FOR VARIABLE 1 ? 1


VARIABLE  POINT     INDEP. VALUE   DEPEND. VALUE   WEIGHT

  1         1        ? 0            ? 0.4           ? 1
  1         1          0            ? 0.18          ? 1
  1         2        ? 10           ? 0             ? 0
  1         2          10           ? 0.12          ? 1
  1         3        ? 22           ? 0.99          ? 1
  1         3          22           ? 0.48          ? 1
  1         4        ? 34           ? 0             ? 0
  1         4          34           ? 1.46          ? 1
  1         5        ? 46           ? 1.95          ? 1
  1         5          46           ? 1.56          ? 1
  1         6        ? 58           ? 0             ? 0
  1         6          58           ? 1.73          ? 1
  1         7        ? 70           ? 2.52          ? 1
  1         7          70           ? 1.99          ? 1


VARIABLE  POINT     INDEP. VALUE   DEPEND. VALUE   WEIGHT

  1         8        ? 82           ? 0             ? 0
  1         8          82           ? 2.62          ? 1
  1         9        ? 94           ? 3.09          ? 1
  1         9          94           ? 2.88          ? 1
  1         10       ? 106          ? 0             ? 0
  1         10         106          ? 3.43          ? 1
  1         11       ? 118          ? 4.06          ? 1
  1         11         118          ? 3.37          ? 1
  1         12       ? 130          ? 0             ? 0
  1         12         130          ? 3.92          ? 1
  1         13       ? 142          ? 4.48          ? 1
  1         13         142          ? 3.96          ? 1
  1         14       ? 154          ? 0             ? 0
  1         14         154          ? 3.58          ? 1
  1         15       ? 166          ? 4.25          ? 1
  1         15         166          ? 3.58          ? 1
  1         16       ? 178          ? 0             ? 0
  1         16         178          ? 3.34          ? 1
  1         17       ? 190          ? 4.36          ? 1
  1         17         190          ? 3.47          ? 1
```

```
ENTER DATA FOR VARIABLE  2


HOW MANY GROUPS OF DATA POINTS ARE THERE IN VARIABLE  2 ? 17


ESTIMATE OF VARIANCE FOR VARIABLE 2 ? 1


VARIABLE  POINT     INDEP. VALUE    DEPEND. VALUE    WEIGHT
 2          1       ? 0             ? 0              ? 1
 2          1         0             ? 0              ? 1
 2          2       ? 10            ? 0              ? 1
 2          2         10            ? 0              ? 1
 2          3       ? 22            ? 0.0089         ? 1
 2          3         22            ? 0.0089         ? 1
 2          4       ? 34            ? 0.0732         ? 1
 2          4         34            ? 0.0642         ? 1
 2          5       ? 46            ? 0.1446         ? 1
 2          5         46            ? 0.2266         ? 1
 2          6       ? 58            ? 0.5230         ? 1
 2          6         58            ? 0.4373         ? 1
 2          7       ? 70            ? 0.6854         ? 1
 2          7         70            ? 0.6941         ? 1


VARIABLE  POINT     INDEP. VALUE    DEPEND. VALUE    WEIGHT
 2          8       ? 82            ? 1.2566         ? 1
 2          8         82            ? 1.2459         ? 1
 2          9       ? 94            ? 1.6118         ? 1
 2          9         94            ? 1.4315         ? 1
 2          10      ? 106           ? 1.8243         ? 1
 2          10        106           ? 2.0402         ? 1
 2          11      ? 118           ? 2.2170         ? 1
 2          11        118           ? 1.9278         ? 1
 2          12      ? 130           ? 2.2758         ? 1
 2          12        130           ? 2.1848         ? 1
 2          13      ? 142           ? 2.8096         ? 1
 2          13        142           ? 2.4204         ? 1
 2          14      ? 154           ? 2.6846         ? 1
 2          14        154           ? 2.4615         ? 1
 2          15      ? 166           ? 2.8738         ? 1
 2          15        166           ? 2.2830         ? 1
 2          16      ? 178           ? 2.8345         ? 1
 2          16        178           ? 2.7078         ? 1
 2          17      ? 190           ? 2.8828         ? 1
 2          17        190           ? 2.6542         ? 1
```

```
OPTIONS :

 1            LIST ALL THE EXPERIMENTAL DATA

 2            ALTER EXISTING EXPERIMENTAL DATA FILE

 3            CREATE A NEW EXPERIMENTAL DATA FILE

 4            RETURN TO PREVIOUS MENU
CHOICE : 4

DISPLAY/ALTER MENU:

 1            ALL NONLINEAR REGRESSION CONSTANTS

 2            EXPERIMENTAL DATA AND WEIGHTS

 3            RETURN TO INPUT MENU

CHOICE : 3

INPUT MENU:

 1            DESCRIPTION OF THE DATA AND EQUATION FILES

 2            ENTER MODEL AND VARIATIONAL EQUATIONS

 3            DISPLAY/ALTER DATA FILES

 4            RETURN TO MAIN MENU
CHOICE : 4
MAIN MENU:

 1            ENTER EQUATIONS, REGRESSION CONSTANTS, AND DATA

 2            PERFORM NONLINEAR REGRESSION

 3            PRINT OR PLOT THE OUTPUT RESULTS

 4            END APPLICATION

CHOICE : 2

EXPERIMENTAL DATA FOR VARIABLE NUMBER  1

ESTIMATE OF VARIANCE = 1

POSITION    INDEPENDENT    DEPENDENT      WEIGHT
IN MATRIX   VARIABLE       VARIABLE
 1            0              .4             1
 1            0              .18            1
 11           10             .12            1
 23           22             .99            1
 23           22             .48            1
 35           34             1.46           1
 47           46             1.95           1
 47           46             1.56           1
```

| | | | |
|---|---|---|---|
| 59 | 58 | 1.73 | 1 |
| 71 | 70 | 2.52 | 1 |
| 71 | 70 | 1.99 | 1 |
| 83 | 82 | 2.62 | 1 |
| 95 | 94 | 3.09 | 1 |
| 95 | 94 | 2.88 | 1 |
| 107 | 106 | 3.43 | 1 |
| 119 | 118 | 4.06 | 1 |
| 119 | 118 | 3.37 | 1 |
| 131 | 130 | 3.92 | 1 |
| 143 | 142 | 4.48 | 1 |
| 143 | 142 | 3.96 | 1 |
| 155 | 154 | 3.58 | 1 |
| 167 | 166 | 4.25 | 1 |
| 167 | 166 | 3.58 | 1 |
| 179 | 178 | 3.34 | 1 |
| 191 | 190 | 4.36 | 1 |
| 191 | 190 | 3.47 | 1 |

EXPERIMENTAL DATA FOR VARIABLE NUMBER 2

ESTIMATE OF VARIANCE = 1

| POSITION IN MATRIX | INDEPENDENT VARIABLE | DEPENDENT VARIABLE | WEIGHT |
|---|---|---|---|
| 1 | 0 | 0 | 1 |
| 1 | 0 | 0 | 1 |
| 11 | 10 | 0 | 1 |
| 11 | 10 | 0 | 1 |
| 23 | 22 | .0089 | 1 |
| 23 | 22 | .0089 | 1 |
| 35 | 34 | .0732 | 1 |
| 35 | 34 | .0642 | 1 |
| 47 | 46 | .1446 | 1 |
| 47 | 46 | .2266 | 1 |
| 59 | 58 | .523 | 1 |
| 59 | 58 | .4373 | 1 |
| 71 | 70 | .6854 | 1 |
| 71 | 70 | .6943 | 1 |
| 83 | 82 | 1.2566 | 1 |
| 83 | 82 | 1.2459 | 1 |
| 95 | 94 | 1.6118 | 1 |
| 95 | 94 | 1.4315 | 1 |
| 107 | 106 | 1.8243 | 1 |
| 107 | 106 | 2.0402 | 1 |
| 119 | 118 | 2.217 | 1 |
| 119 | 118 | 1.9278 | 1 |
| 131 | 130 | 2.2758 | 1 |
| 131 | 130 | 2.1848 | 1 |
| 143 | 142 | 2.8096 | 1 |
| 143 | 142 | 2.4204 | 1 |
| 155 | 154 | 2.6846 | 1 |
| 155 | 154 | 2.4615 | 1 |
| 167 | 166 | 2.8738 | 1 |
| 167 | 166 | 2.283 | 1 |
| 179 | 178 | 2.8345 | 1 |
| 179 | 178 | 2.7078 | 1 |
| 191 | 190 | 2.8828 | 1 |
| 191 | 190 | 2.6542 | 1 |

## Results

```
STATISTICS OF THE EXPERIMENTAL DATA

UNWEIGHTED

Variable                        1
Sum of squares                  1.38655
Degrees of freedom              9
Variance                        .1540611
Standard deviation              .3925063
Sum of weights                  26

Variable                        2
Sum of squares                  .4019939
Degrees of freedom              17
Variance                        .0236467
Standard deviation              .1537748
Sum of weights                  34

WEIGHTED

Sum of squares                  .9709966
Variance                        3.734603E-02

ITERATION =  1
Sum of squares for variable  1  =  6.552692
Sum of squares for variable  2  =  63.33678
Total sum of squares            =  69.88948

New sum of squares for variable  1  =  2.663645
New sum of squares for variable  2  =  8.718336
New total sum of squares            =  11.38198
Stepsize =  1

Parameter #     Increment        New Value

 1          -6.620906E-02     3.379094E-02
 2          -.5338703     3.46613
 3          -4.182724E-03     1.581728E-02
 4          -1.831215E-03     1.816879E-02

ITERATION =  2
Sum of squares for variable  1  =  2.663645
Sum of squares for variable  2  =  8.718336
Total sum of squares            =  11.38198

New sum of squares for variable  1  =  .7915371
New sum of squares for variable  2  =  1.905436
New total sum of squares            =  2.696973
Stepsize =  1

Parameter #     Increment        New Value

 1           1.184182E-02     4.563276E-02
 2           .2587684     3.724898
 3           7.488785E-04     1.656615E-02
 4           2.725184E-03     2.089397E-02
```

```
ITERATION =  3
Sum of squares for variable  1  =  .7915371
Sum of squares for variable  2  =  1.905436
Total sum of squares            =  2.696973

New sum of squares for variable  1  =  .7077318
New sum of squares for variable  2  =  1.578511
New total sum of squares            =  2.286243
Stepsize =  1

Parameter #    Increment        New Value

 1          -2.998399E-03     4.263436E-02
 2           .2070551     3.931953
 3           9.012144E-04     1.746737E-02
 4           9.823991E-04     2.187637E-02


ITERATION =  4
Sum of squares for variable  1  =  .7077318
Sum of squares for variable  2  =  1.578511
Total sum of squares            =  2.286243

New sum of squares for variable  1  =  .7150831
New sum of squares for variable  2  =  1.562322
New total sum of squares            =  2.277406
Stepsize =  1

Parameter #    Increment        New Value

 1           7.277452E-04     4.336211E-02
 2          -5.509339E-03     3.926444
 3          -1.588923E-04     1.730848E-02
 4          -2.192129E-04     2.165716E-02


ITERATION =  5
Sum of squares for variable  1  =  .7150831
Sum of squares for variable  2  =  1.562322
Total sum of squares            =  2.277406

New sum of squares for variable  1  =  .7138832
New sum of squares for variable  2  =  1.563437
New total sum of squares            =  2.277321
Stepsize =  1

Parameter #    Increment        New Value

 1          -1.332831E-04     4.322883E-02
 2           4.328041E-03     3.930772
 3           9.604745E-05     1.740452E-02
 4           1.551446E-04     .0218123


ITERATION =  6
Sum of squares for variable  1  =  .7138832
Sum of squares for variable  2  =  1.563437
Total sum of squares            =  2.277321
```

```
************************* CONVERGED ******************************

ITERATION =  6              SUM OF SQUARES = 2.277321

Parameter #     Calculated
                Parameters

 1              4.322883E-02
 2              3.930772
 3              1.740452E-02
 4              .0218123
```

VARIABLE 1

| EXPERIMENTAL DATA | | CALCULATED | ERROR | WEIGHT |
|---|---|---|---|---|
| INDEPENDENT | DEPENDENT | | | |
| 0.00000 | 0.40000 | 0.29000 | 0.11000 | 0.49235 |
| 0.00000 | 0.18000 | 0.29000 | -0.11000 | 0.49235 |
| 10.00000 | 0.12000 | 0.42968 | -0.30968 | 0.49235 |
| 22.00000 | 0.99000 | 0.67189 | 0.31811 | 0.49235 |
| 22.00000 | 0.48000 | 0.67189 | -0.19189 | 0.49235 |
| 34.00000 | 1.46000 | 1.01121 | 0.44879 | 0.49235 |
| 46.00000 | 1.95000 | 1.44585 | 0.50415 | 0.49235 |
| 46.00000 | 1.56000 | 1.44585 | 0.11415 | 0.49235 |
| 58.00000 | 1.73000 | 1.94298 | -0.21298 | 0.49235 |
| 70.00000 | 2.52000 | 2.44300 | 0.07700 | 0.49235 |
| 70.00000 | 1.99000 | 2.44300 | -0.45300 | 0.49235 |
| 82.00000 | 2.62000 | 2.88494 | -0.26494 | 0.49235 |
| 94.00000 | 3.09000 | 3.23310 | -0.14310 | 0.49235 |
| 94.00000 | 2.88000 | 3.23310 | -0.35310 | 0.49235 |
| 106.00000 | 3.43000 | 3.48333 | -0.05333 | 0.49235 |
| 118.00000 | 4.06000 | 3.65156 | 0.40844 | 0.49235 |
| 118.00000 | 3.37000 | 3.65156 | -0.28156 | 0.49235 |
| 130.00000 | 3.92000 | 3.75965 | 0.16035 | 0.49235 |
| 142.00000 | 4.48000 | 3.82708 | 0.65292 | 0.49235 |
| 142.00000 | 3.96000 | 3.82708 | 0.13292 | 0.49235 |
| 154.00000 | 3.58000 | 3.86838 | -0.28838 | 0.49235 |
| 166.00000 | 4.25000 | 3.89339 | 0.35661 | 0.49235 |
| 166.00000 | 3.58000 | 3.89339 | -0.31339 | 0.49235 |
| 178.00000 | 3.34000 | 3.90843 | -0.56843 | 0.49235 |
| 190.00000 | 4.36000 | 3.91744 | 0.44256 | 0.49235 |
| 190.00000 | 3.47000 | 3.91744 | -0.44744 | 0.49235 |

VARIABLE 2

| EXPERIMENTAL DATA | | CALCULATED | ERROR | WEIGHT |
|---|---|---|---|---|
| INDEPENDENT | DEPENDENT | | | |
| 0.00000 | 0.00000 | 0.00000 | 0.00000 | 1.25672 |
| 0.00000 | 0.00000 | 0.00000 | 0.00000 | 1.25672 |
| 10.00000 | 0.00000 | 0.05602 | -0.05602 | 1.25672 |
| 10.00000 | 0.00000 | 0.05602 | -0.05602 | 1.25672 |
| 22.00000 | 0.00890 | 0.14392 | -0.13502 | 1.25672 |
| 22.00000 | 0.00890 | 0.14392 | -0.13502 | 1.25672 |
| 34.00000 | 0.07320 | 0.26523 | -0.19203 | 1.25672 |
| 34.00000 | 0.06420 | 0.26523 | -0.20103 | 1.25672 |
| 46.00000 | 0.14460 | 0.43034 | -0.28574 | 1.25672 |
| 46.00000 | 0.22660 | 0.43034 | -0.20374 | 1.25672 |
| 58.00000 | 0.52300 | 0.64404 | -0.12104 | 1.25672 |

| | | | | |
|---|---|---|---|---|
| 58.00000 | 0.43730 | 0.64404 | -0.20674 | 1.25672 |
| 70.00000 | 0.68540 | 0.90117 | -0.21577 | 1.25672 |
| 70.00000 | 0.69430 | 0.90117 | -0.20687 | 1.25672 |
| 82.00000 | 1.25660 | 1.18619 | 0.07041 | 1.25672 |
| 82.00000 | 1.24590 | 1.18619 | 0.05971 | 1.25672 |
| 94.00000 | 1.61180 | 1.47807 | 0.13373 | 1.25672 |
| 94.00000 | 1.43150 | 1.47807 | -0.04657 | 1.25672 |
| 106.00000 | 1.82430 | 1.75718 | 0.06712 | 1.25672 |
| 106.00000 | 2.04020 | 1.75718 | 0.28302 | 1.25672 |
| 118.00000 | 2.21700 | 2.00982 | 0.20718 | 1.25672 |
| 118.00000 | 1.92780 | 2.00982 | -0.08202 | 1.25672 |
| 130.00000 | 2.27580 | 2.22909 | 0.04671 | 1.25672 |
| 130.00000 | 2.18480 | 2.22909 | -0.04429 | 1.25672 |
| 142.00000 | 2.80960 | 2.41356 | 0.39604 | 1.25672 |
| 142.00000 | 2.42040 | 2.41356 | 0.00684 | 1.25672 |
| 154.00000 | 2.68460 | 2.56526 | 0.11934 | 1.25672 |
| 154.00000 | 2.46150 | 2.56526 | -0.10376 | 1.25672 |
| 166.00000 | 2.87380 | 2.68793 | 0.18587 | 1.25672 |
| 166.00000 | 2.28300 | 2.68793 | -0.40493 | 1.25672 |
| 178.00000 | 2.83450 | 2.78591 | 0.04859 | 1.25672 |
| 178.00000 | 2.70780 | 2.78591 | -0.07811 | 1.25672 |
| 190.00000 | 2.88280 | 2.86347 | 0.01933 | 1.25672 |
| 190.00000 | 2.65420 | 2.86347 | -0.20927 | 1.25672 |

STATISTICAL ANALYSIS OF CONVERGED RESULTS

Variance-covariance matrix: Inverse of (A transpose A) * s^2

```
1  7.610023E-06 -2.133171E-04 -6.128263E-06 -1.017917E-05
2 -2.133176E-04  1.985619E-02  1.671893E-04  3.726765E-04
3 -6.12826E-06  1.671886E-04  7.456787E-06  1.271603E-05
4 -1.017917E-05  3.726754E-04  1.271604E-05  2.261971E-05
```

Degrees of freedom = 56

Combined (weighted) residual variance (s^2) = 4.066644E-02

Standard deviation (s) = .2016592

| Parameter | Standard error | 0.95 confidence limits | Confidence-interval lower | upper |
|---|---|---|---|---|
| 4.323E-02 | 2.759E-03 | 5.407E-03 | 3.782E-02 | 4.864E-02 |
| 3.931E+00 | 1.409E-01 | 2.762E-01 | 3.655E+00 | 4.207E+00 |
| 1.740E-02 | 2.731E-03 | 5.352E-03 | 1.205E-02 | 2.276E-02 |
| 2.181E-02 | 4.756E-03 | 9.322E-03 | 1.249E-02 | 3.113E-02 |

SIGNIFICANCE TEST

| Parameter | t-calculated | Is the parameter significantly different from zero? |
|---|---|---|
| 4.323E-02 | 1.567E+01 | yes |
| 3.931E+00 | 2.790E+01 | yes |
| 1.740E-02 | 6.374E+00 | yes |
| 2.181E-02 | 4.586E+00 | yes |

| Measured variable | Degrees of freedom | Residual variance | 95% Confidence limit for each variable |
|---|---|---|---|
| 1 | 24 | 2.974513E-02 | .3559735 |
| 2 | 32 | 4.885742E-02 | .4495895 |

COVARIANCE ANALYSIS

Matrix of Correlation Coefficients

| | | | | |
|---|---|---|---|---|
| 1 | 1 | -.5487628 | -.8135206 | -.7758469 |
| 2 | -.5487642 | 1 | .4344949 | .5560841 |
| 3 | -.8135202 | .4344931 | 1 | .9791119 |
| 4 | -.7758468 | .5560826 | .9791122 | 1 |

Matrix of 0.05 significance test.(Yes means that the correlation is significant)

| | | | | |
|---|---|---|---|---|
| 1 | yes | yes | yes | yes |
| 2 | yes | yes | yes | yes |
| 3 | yes | yes | yes | yes |
| 4 | yes | yes | yes | yes |

Matrix of 0.01 significance test.(Yes means that the correlation is highly significant)

| | | | | |
|---|---|---|---|---|
| 1 | yes | yes | yes | yes |
| 2 | yes | yes | yes | yes |
| 3 | yes | yes | yes | yes |
| 4 | yes | yes | yes | yes |

```
EIGENVALUES for the Correlation Coefficients Matrix

 3.091737
 .6470306
 .2561823
 5.050307E-03


EIGENVECTORS for the Correlation Coefficients Matrix

-.5130412  6.936222E-02  .8452254 -.1325577
 .3907734  .8987579  .141541 -.1396292
 .5358007 -.3821439  .2449261 -.7119653
 .5449791 -.2034426  .4533997  .6753055

SUM OF SQUARES SEARCH
EIGENVALUE      1            3.091737

Parameter       1            3.835127E-02
Parameter       2            4.120543
Parameter       3            2.244691E-02
Parameter       4            3.074496E-02

SUM OF SQUARES
Curve           Converged    Search        Difference
 1              .7138832     .8094425      .0955593
 2              1.563437     1.697457      .1340201
total           2.277321     2.5069        .2295795

EIGENVALUE      1            3.091737

Parameter       1            4.810639E-02
Parameter       2            3.741001
Parameter       3            1.236213E-02
Parameter       4            1.287964E-02

SUM OF SQUARES
Curve           Converged    Search        Difference
 1              .7138832     .8473475      .1334643
 2              1.563437     1.793422      .2299843
total           2.277321     2.640769      .3634486

EIGENVALUE      2            .6470306

Parameter       1            .0435305
Parameter       2            4.13044
Parameter       3            1.575931E-02
Parameter       4            2.028683E-02

SUM OF SQUARES
Curve           Converged    Search        Difference
 1              .7138832     .8841621      .170279
 2              1.563437     1.561002     -2.435684E-03
total           2.277321     2.445164      .1678433

EIGENVALUE      2            .6470306

Parameter       1            4.292716E-02
Parameter       2            3.731104
```

```
Parameter      3             1.904973E-02
Parameter      4             2.333777E-02

SUM OF SQUARES
Curve          Converged     Search        Difference
 1             .7138832      .7968721      8.298898E-02
 2             1.563437      1.643464      8.002651E-02
total          2.277321      2.440336      .1630154

EIGENVALUE     3             .2561823

Parameter      1             4.554194E-02
Parameter      2             3.950558
Parameter      3             1.806802E-02
Parameter      4             2.395152E-02

SUM OF SQUARES
Curve          Converged     Search        Difference
 1             .7138832      .7950975      8.121431E-02
 2             1.563437      1.643615      8.017803E-02
total          2.277321      2.438713      .1613922

EIGENVALUE     3             .2561823

Parameter      1             4.091571E-02
Parameter      2             3.910986
Parameter      3             1.674102E-02
Parameter      4             1.967308E-02

SUM OF SQUARES
Curve          Converged     Search        Difference
 1             .7138832      .7257555      1.187229E-02
 2             1.563437      1.728823      .1653862
total          2.277321      2.454579      .1772585

EIGENVALUE     4             5.050307E-03

Parameter      1             4.317789E-02
Parameter      2             3.928031
Parameter      3             1.713372E-02
Parameter      4             2.225966E-02

SUM OF SQUARES
Curve          Converged     Search        Difference
 1             .7138832      .7126561     -1.227021E-03
 2             1.563437      1.716817      .1533802
total          2.277321      2.429474      .152153

EIGENVALUE     4             5.050307E-03

Parameter      1             4.327976E-02
Parameter      2             3.933513
Parameter      3             1.767532E-02
Parameter      4             2.136494E-02

SUM OF SQUARES
Curve          Converged     Search        Difference
 1             .7138832      .7152378      1.354575E-03
 2             1.563437      1.722703      .1592661
total          2.277321      2.437941      .1606207
```

ANALYSIS OF VARIANCE

UNWEIGHTED STATISTICS

| Source of variance | Variable | Sum of squares | Degrees of freedom | Variance |
|---|---|---|---|---|
| Lack of fit | 1 | 1.558387 | 15 | .1038924 |
| Exp. error | 1 | 1.38655 | 9 | .1540611 |
| Total | 1 | 2.944936 | 24 | .1227057 |
| Lack of fit | 2 | .5879408 | 15 | 3.919606E-02 |
| Exp. error | 2 | .4019939 | 17 | .0236467 |
| Total | 2 | .9899348 | 32 | 3.093546E-02 |

F-Tests

Variable 1 F-CALC ( 15 , 9 ) = .6743585

F-95% ( 15 , 9 ) = 3.01

Variable 2 F-CALC ( 15 , 17 ) = 1.65757

F-95% ( 15 , 17 ) = 2.31

WEIGHTED STATISTICS

| Source of variance | Variable | Sum of squares | Degrees of freedom | Variance |
|---|---|---|---|---|
| Lack of fit | Weighted | 1.306324 | 30 | 4.354413E-02 |
| Exp. error | Weighted | .9709966 | 26 | 3.734603E-02 |
| Total | Weighted | 2.277321 | 56 | 4.066644E-02 |

F-Test

Weighted F-CALC ( 30 , 26 ) = 1.165964

F-95% ( 30 , 26 ) = 1.9

```
RANDOMNESS TEST

Runs test for variable                        1
Number of positive residuals                  8
Number of negative residuals                  9
Number of runs   =                            8
Z  =                                         -.739464

Random at 95% level of confidence


Runs test for variable                        2
Number of positive residuals                  7
Number of negative residuals                  9
Number of runs   =                            3
Z  =                                         -3.092373
Not random at 95% level of confidence

MAIN MENU:

 1            ENTER EQUATIONS, REGRESSION CONSTANTS, AND DATA

 2            PERFORM NONLINEAR REGRESSION

 3            PRINT OR PLOT THE OUTPUT RESULTS

 4            END APPLICATION

CHOICE : 3


            READING LARGE FILES OF VARIABLES AND DATA


                          PLEASE WAIT


            PRINT OR PLOT THE OUTPUT RESULTS

OUTPUT MENU:

 1            PRINT TABLE OF VARIABLES

 2            PRINT TABLE OF DATA

 3            DRAW GRAPHS ON SCREEN

 4            DRAW GRAPHS USING HP PLOTTER

 5            RETURN TO MAIN MENU


CHOICE : 3
```

```
            SCALING OPTIONS

    1. MANUAL CHOICE OF MAXIMA AND MINIMA

    2. AUTOMATIC CHOICE OF MAXIMA AND MINIMA

ENTER YOUR CHOICE(1-2)? 1

          SCALES FOR VARIABLE 1

MINIMUM X=? 0

MAXIMUM X=? 200

SPECIFY X-AXIS TITLE:TIME (Hours)

MINIMUM Y=? 0

MAXIMUM Y=? 5

SPECIFY Y-AXIS TITLE:CELL
```

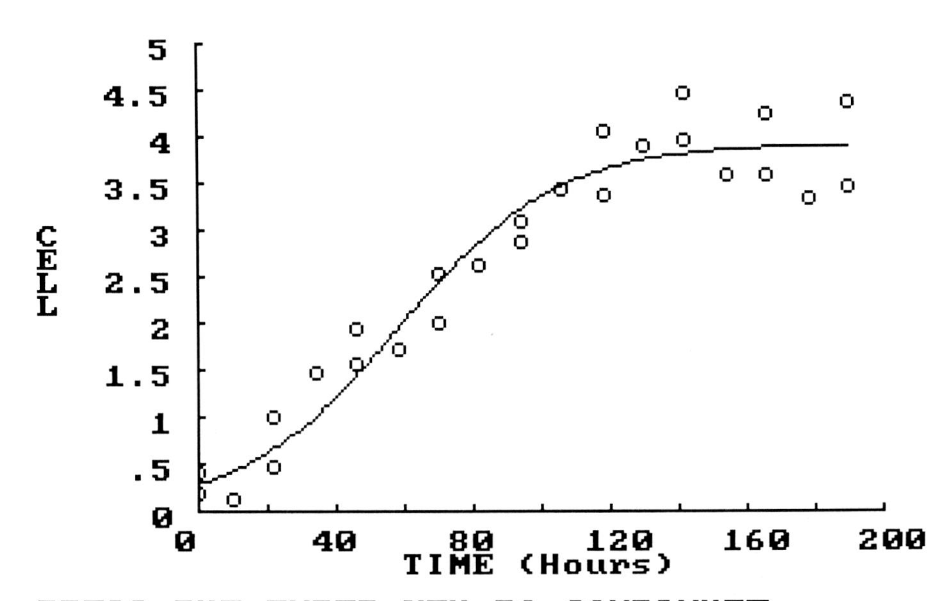

```
          SCALING OPTIONS

    1. MANUAL CHOICE OF MAXIMA AND MINIMA

    2. AUTOMATIC CHOICE OF MAXIMA AND MINIMA

ENTER YOUR CHOICE(1-2)? 1

          SCALES FOR VARIABLE 2

MINIMUM X=? 0

MAXIMUM X=? 200

SPECIFY X-AXIS TITLE:TIME (Hours)

MINIMUM Y=? 0

MAXIMUM Y=? 4

SPECIFY Y-AXIS TITLE:PENICILLIN
```

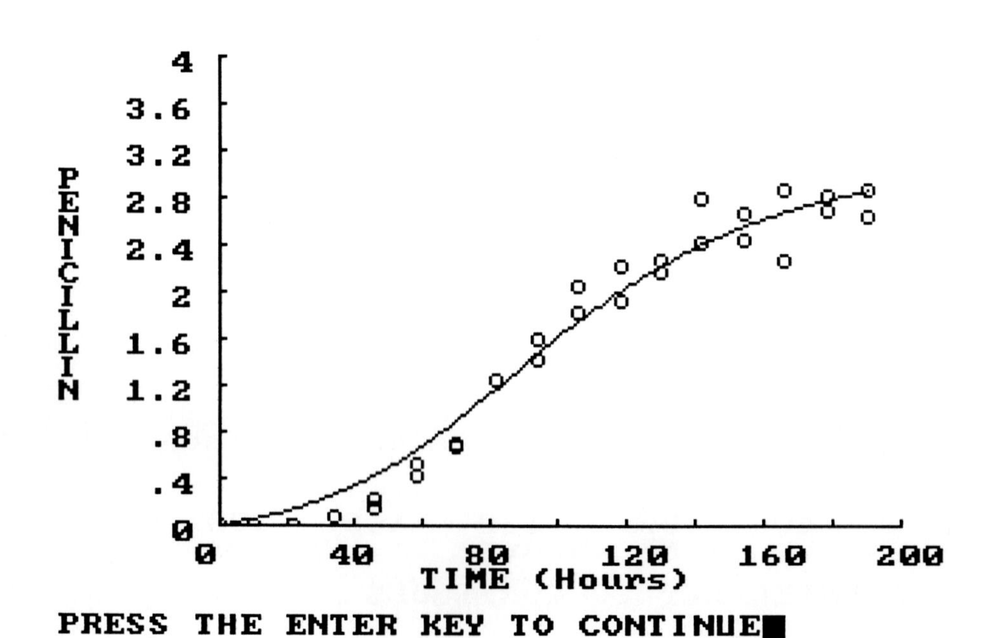

```
            PRINT OR PLOT THE OUTPUT RESULTS

OUTPUT MENU:
 1              PRINT TABLE OF VARIABLES
 2              PRINT TABLE OF DATA
 3              DRAW GRAPHS ON SCREEN
 4              DRAW GRAPHS USING HP PLOTTER
 5              RETURN TO MAIN MENU

CHOICE : 5

MAIN MENU:

 1              ENTER EQUATIONS, REGRESSION CONSTANTS, AND DATA
 2              PERFORM NONLINEAR REGRESSION
 3              PRINT OR PLOT THE OUTPUT RESULTS
 4              END APPLICATION

CHOICE : 4
Ok
```

## PROBLEMS

**7-1** The heat capacity of gases and liquids is a function of temperature. Various forms of polynomial equations have been used to represent this functionality. Two such equations are shown below:

$$c_p = a + bT + cT^2 + dT^3 \tag{1}$$

$$c_p = a + bT + cT^{-2} \tag{2}$$

Using the heat capacity data of Table P7-1, determine the coefficients of the above two equations. Discuss your results and recommend which equation gives the best representation of the data.

**Table P7-1 Simulated heat capacity data**

| Temperature, °C | Heat capacity, J/(g · mol · °C) | | |
|---|---|---|---|
| | Set No. 1 | Set No. 2 | Set No. 3 |
| 100.00 | 29.38 | 30.04 | 28.52 |
| 200.00 | 29.88 | 29.08 | 29.79 |
| 300.00 | 30.42 | 30.18 | 31.41 |
| 400.00 | 30.98 | 30.14 | 31.18 |
| 500.00 | 31.57 | 32.27 | 31.61 |
| 600.00 | 32.15 | 31.79 | 32.81 |
| 700.00 | 32.73 | 32.97 | 32.38 |
| 800.00 | 33.29 | 32.56 | 34.26 |
| 900.00 | 33.82 | 34.24 | 34.72 |
| 1000.00 | 34.31 | 35.27 | 33.69 |

**Table P7-2 Data obtained by varying the temperature at pH 7.0**

| Hours | Cell concentration, U.O.D./mL | Gluconolactone concentration, mg/mL | Gluconic acid concentration, mg/mL | Glucose concentration, mg/mL |
|---|---|---|---|---|
| | | Experiment No. 34 Batch fermentation data at 25.0°C | | |
| 0.0 | 0.56 | 1.28 | 0.16 | 45.00 |
| 1.0 | 0.86 | 2.20 | 1.56 | 43.00 |
| 2.0 | 1.60 | 3.50 | 5.00 | 38.00 |
| 3.0 | 2.60 | 5.60 | 9.50 | 33.00 |
| 4.0 | 3.20 | 7.00 | 16.00 | 25.00 |
| 5.0 | 3.30 | 7.80 | 24.50 | 16.50 |
| 6.0 | 3.50 | 7.20 | 32.00 | 9.00 |
| 7.0 | 3.40 | 6.30 | 45.50 | 4.00 |
| 8.0 | 3.40 | 3.20 | 45.80 | 2.00 |
| | | Experiment No. 33 Batch fermentation data at 28.0°C | | |
| 0.0 | 0.66 | — | — | 48.00 |
| 1.0 | 1.00 | 1.96 | 0.15 | 45.00 |
| 2.0 | 1.60 | 6.67 | 7.00 | 37.50 |
| 3.0 | 2.60 | 10.50 | 15.00 | 28.00 |
| 4.0 | 3.20 | 10.50 | 25.00 | 18.00 |
| 5.0 | 3.30 | 7.58 | 35.00 | 8.00 |
| 6.0 | 3.30 | 2.05 | 42.50 | 3.00 |
| 7.0 | 3.30 | 1.90 | 45.50 | — |
| | | Experiment No. 32 Batch fermentation data at 30.0°C | | |
| 0.0 | 0.80 | 1.34 | 0.95 | 44.50 |
| 1.0 | 1.50 | 4.00 | 4.96 | 37.50 |
| 2.0 | 2.60 | 7.50 | 16.10 | 25.00 |
| 3.0 | 3.50 | 8.00 | 32.10 | 9.00 |
| 4.0 | 3.50 | 5.00 | 43.70 | 3.00 |
| 5.0 | 3.50 | 2.42 | 44.50 | 2.00 |

*Source*: V. R. Rai: "Mathematical Modeling and Optimization of the Gluconic Acid Fermentation," Ph.D. dissertation, Rutgers University, New Brunswick, N.J., 1973.

**7-2** A mathematical model of the fermentation of the bacterium *Pseudomonas ovalis*, which produces gluconic acid, was given in Prob. 5-6. This model, which describes the dynamics of the logarithmic growth phase, can be summarized as follows:

Rate of cell growth:

$$\frac{dy_1}{dt} = b_1 y_1 \left(1.0 - \frac{y_1}{b_2}\right)$$

Rate of gluconolactone formation:

$$\frac{dy_2}{dt} = \frac{b_3 y_1 y_4}{b_4 + y_4} - 0.9082 b_5 y_2$$

Rate of gluconic acid formation:

$$\frac{dy_3}{dt} = b_5 y_2$$

Rate of glucose consumption:

$$\frac{dy_4}{dt} = -1.011\left(\frac{b_3 y_1 y_4}{b_4 + y_4}\right)$$

where $y_1$ = concentration of cell
$y_2$ = concentration of gluconolactone
$y_3$ = concentration of gluconic acid
$y_4$ = concentration of glucose
$b_1$ to $b_5$ = parameters of the system which are functions of temperature and pH.

Using the batch fermentation data given in Table P7-2, determine the values of the parameters, $b_1$ to $b_5$ at the three different temperatures of 25°, 28°, and 30°C.

**7-3** Accurate vapor-liquid equilibrium measurements can be used to compute liquid-phase activity coefficients and excess Gibbs free energies [13]. Consider the data in Table P7-3 for benzene–2,2,4-trimethylpentane (B-TMP) mixtures at constant temperature (55°C).

(*a*) Assume that the gas phase is ideal and neglect any fugacity or Poynting corrections for the liquid phase. Calculate the activity coefficients for B and TMP and the molar excess Gibbs free energy at each experimental point. The vapor pressure of pure B at 55°C is 327.05 mmHg and of pure TMP at 55°C is 178.08 mmHg.

(*b*) If a *three-constant* Redlich-Kister expansion for the excess molar Gibbs free energy is assumed (see Denbigh, Ref. 14, p. 286), evaluate the constants that appear using the data of part *a*; that is, find "best fits" for $A_0$, $A_1$, and $A_2$ (Denbigh notation).

**Table P7-3**

| Liquid phase, mole fraction $x_B$ | Vapor phase, mole fraction $y_B$ | Equilibrium total pressure P mmHg |
|---|---|---|
| 0.0819 | 0.1869 | 201.74 |
| 0.2192 | 0.4065 | 236.86 |
| 0.3584 | 0.5509 | 266.04 |
| 0.3831 | 0.5748 | 270.73 |
| 0.5256 | 0.6786 | 293.36 |
| 0.8478 | 0.8741 | 324.66 |
| 0.9872 | 0.9863 | 327.39 |

*Source*: S. Weissman and S. E. Wood: "Vapor-Liquid Equilibrium of Benzene-2,2,4-trymethylpentane Mixtures," *J. Chem. Phys.*, vol. 32, 1960, p. 1153.

(*c*) Calculate the activity coefficients from your expression in part *b* for B and TMP. (*Hint*: First derive an expression for the activity coefficients assuming a three-constant Redlich-Kister expansion.)

(*d*) Plot the theoretical excess molar Gibbs free energy and theoretical activity coefficients with the experimental data shown as well.

**7-4** Use the data of Prob. 5-9 to fit the Lotka-Volterra predator-prey equation (see Example 5-7) in order to obtain accurate estimates of the parameters of that model. Modify the Lotka-Volterra equations, as recommended in Prob. 5-9, and determine the parameters of your new models. Compare the results of the statistical analysis for each model, and choose the set of equations which gives the best representation of the data.

**7-5** Svirbely and Blaner† modeled the reactions

$$\mathrm{A} + \mathrm{B} \xrightarrow{k_1} \mathrm{C} + \mathrm{F}$$

$$\mathrm{A} + \mathrm{C} \xrightarrow{k_2} \mathrm{D} + \mathrm{F}$$

$$\mathrm{A} + \mathrm{D} \xrightarrow{k_3} \mathrm{E} + \mathrm{F}$$

as follows:

$$\frac{d\mathrm{A}}{dt} = -k_1\mathrm{AB} - k_2\mathrm{AC} - k_3\mathrm{AD}$$

$$\frac{d\mathrm{B}}{dt} = -k_1\mathrm{AB}$$

$$\frac{d\mathrm{C}}{dt} = k_1\mathrm{AB} - k_2\mathrm{AC}$$

$$\frac{d\mathrm{D}}{dt} = k_2\mathrm{AC} - k_3\mathrm{AD}$$

$$\frac{d\mathrm{E}}{dt} = k_3\mathrm{AD}$$

Estimate the coefficients $k_1$, $k_2$, and $k_3$ (all positive) from the data of Table P7-5, and the following initial conditions:

$$\mathrm{C}(0) = \mathrm{D}(0) = 0$$

$$\mathrm{A}(0) = 0.02090 \text{ mol/L}$$

$$\mathrm{B}(0) = (\tfrac{1}{3})\mathrm{A}(0)$$

† W. J. Svirbely and J. A. Blaner, *J. Amer. Chem. Soc.*, vol. 83, 1961, p. 4118. (Problem reproduced from D. M. Himmelblau, Ref. 7, p. 326.)

**Table P7-5**

| Time, min | A × 10³, mol/L | Time, min | A × 10³, mol/L |
|---|---|---|---|
| 4.50 | 15.40 | 76.75 | 8.395 |
| 8.67 | 14.22 | 90.00 | 7.891 |
| 12.67 | 13.35 | 102.00 | 7.510 |
| 17.75 | 12.32 | 108.00 | 7.370 |
| 22.67 | 11.81 | 147.92 | 6.646 |
| 27.08 | 11.39 | 198.00 | 5.883 |
| 32.00 | 10.92 | 241.75 | 5.322 |
| 36.00 | 10.54 | 270.25 | 4.960 |
| 46.33 | 9.780 | 326.25 | 4.518 |
| 57.00 | 9.157 | 418.00 | 4.075 |
| 69.00 | 8.594 | 501.00 | 3.715 |

The estimates reported in the article were

$$k_1 = 14.7$$

$$k_2 = 1.53$$

$$k_3 = 0.294$$

Could estimates be obtained if the initial conditions were unknown?

**7-6** Choose one of the equations given in Prob. 7-1 and perform the following:

(*a*) Fit the equation to the specific heat data given in Prob. 7-1.

(*b*) Add random error (r. e.) to the data, where this error is in the range $-1.0 \leq \text{r.e} \leq 1.0$. and fit the equation to the noisy data.

(*c*) Repeat part *b* with $-5.0 \leq \text{r.e} \leq 5.0$.

(*d*) Repeat part *b* with $-10.0 \leq \text{r.e} \leq 10.0$.

Compare the results of the statistical analysis in parts *a* to *d*. What conclusions do you draw?

# REFERENCES

1. Himmelblau, D. M., and Bischoff, K. B.: *Process Analysis and Simulation: Deterministic Systems*, John Wiley & Sons, Inc., New York, 1968, p. 1.
2. Box, G. E. P., and Hunter, W. G.: "A Useful Method for Model-Building," *Technometrics*, vol. 4, no. 3, 1962, p. 301.
3. Brownlee, K. A.: *Statistical Theory and Methodology in Science and Engineering*, 2d ed., John Wiley & Sons, Inc., New York, 1965.
4. Lipson, C., and Sheth, N. J.: *Statistical Design and Analysis of Engineering Experiments*, McGraw-Hill Book Company, New York, 1973.
5. Draper, N. R., and Smith, H.: *Applied Regression Analysis*, 2d ed., John Wiley & Sons, Inc., New York, 1981.

6. Carnahan, B., Luther, H. A., and Wilkes, J. O.: *Applied Numerical Methods*, John Wiley & Sons, Inc., New York, 1969.
7. Himmelblau, D. M.: *Process Analysis by Statistical Methods*, John Wiley & Sons, Inc., New York, 1970.
8. Johnston, J.: *Econometric Methods*, 3d ed., McGraw-Hill Book Company, New York, 1984.
9. Seinfeld, J. H., and Lapidus, L.: *Mathematical Methods in Chemical Engineeing*, vol. 3, *Process Modeling, Estimation, and Identification*, Prentice-Hall, Inc., Englewood Cliffs, N.J., 1974.
10. Davies, O. L.: *Statistical Methods in Research and Production*, Hafner Publishing Company, Inc., New York, 1957.
11. Box, G. E. P.: "Fitting Empirical Data," *Ann. N. Y. Acad. Sci.*, vol. 86, 1960, p. 792.
12. Marquardt, D. W.: "An Algorithm for Least Squares Estimation of Nonlinear Parameters," *J. Soc. Ind. Appl. Math.*, vol. 11, 1963, p. 431.
13. Pedersen, H.: Private communication, Department of Chemical and Biochemical Engineering, Rutgers University, New Brunswick, N.J., 1986.
14. Denbigh, K.: *The Principles of Chemical Equilibrium*, 3d ed., Cambridge University Press, Cambridge, 1971.

# INDEX